D1748042

Jung-Ki Park

Principles and Applications of Lithium Secondary Batteries

Related Titles

Beguin, F., Frackowiak, E. (eds.)

Supercapacitors

Materials, Systems, and Applications

Series: New Materials for Sustainable Energy and Development

Series edited by Lu, Max

ISBN: 978-3-527-32883-3

Nazri, G.-A., Balaya, P., Manthiram, A., Yamada, A., Yang, Y. (eds.)

Advanced Lithium-Ion Batteries

Recent Trends and Perspectives

Series: New Materials for Sustainable Energy and Development

Series edited by Lu, Max

ISBN: 978-3-527-32889-5

Daniel, C., Besenhard, J. O. (eds.)

Handbook of Battery Materials

2011

ISBN: 978-3-527-32695-2

Aifantis, K. E., Hackney, S. A., Kumar, R. V. (eds.)

High Energy Density Lithium Batteries

Materials, Engineering, Applications

2010

ISBN: 978-3-527-32407-1

Ozawa, K. (ed.)

Lithium Ion Rechargeable Batteries

Materials, Technology, and New Applications

2009

ISBN: 978-3-527-31983-1

Jung-Ki Park

Principles and Applications of Lithium Secondary Batteries

WILEY-VCH

WILEY-VCH Verlag GmbH & Co. KGaA

Prof. Jung-Ki Park
KAIST
Department of Chemical & Biomolecular Eng.
373-1 Guseong-dong, Yuseong-gu
Daejeon 305-701
Republic of Korea

All books published by **Wiley-VCH** are carefully produced. Nevertheless, authors, editors, and publisher do not warrant the information contained in these books, including this book, to be free of errors. Readers are advised to keep in mind that statements, data, illustrations, procedural details or other items may inadvertently be inaccurate.

Library of Congress Card No.: applied for

British Library Cataloguing-in-Publication Data
A catalogue record for this book is available from the British Library.

Bibliographic information published by the Deutsche Nationalbibliothek
The Deutsche Nationalbibliothek lists this publication in the Deutsche Nationalbibliografie; detailed bibliographic data are available on the Internet at http://dnb.d-nb.de.

© 2012 Wiley-VCH Verlag & Co. KGaA, Boschstr. 12, 69469 Weinheim, Germany

All rights reserved (including those of translation into other languages). No part of this book may be reproduced in any form – by photoprinting, microfilm, or any other means – nor transmitted or translated into a machine language without written permission from the publishers. Registered names, trademarks, etc. used in this book, even when not specifically marked as such, are not to be considered unprotected by law.

Cover Design   Adam-Design, Weinheim
Typesetting   Thomson Digital, Noida, India
Printing and Binding   Markono Print Media Pte Ltd, Singapore

Print ISBN: 978-3-527-33151-2
ePDF ISBN: 978-3-527-65043-9
ePub ISBN: 978-3-527-65042-2
mobi ISBN: 978-3-527-65041-5
oBook ISBN: 978-3-527-65040-8

Contents

List of Contributors *XI*
Preface *XIII*

1 Introduction *1*

1.1 History of Batteries *1*
1.2 Development of Cell Technology *3*
1.3 Overview of Lithium Secondary Batteries *3*
1.4 Future of Lithium Secondary Batteries *7*
References *7*

2 The Basic of Battery Chemistry *9*

2.1 Components of Batteries *9*
2.1.1 Electrochemical Cells and Batteries *9*
2.1.2 Battery Components and Electrodes *9*
2.1.3 Full Cells and Half Cells *11*
2.1.4 Electrochemical Reaction and Electric Potential *11*
2.2 Voltage and Current of Batteries *12*
2.2.1 Voltage *12*
2.2.2 Current *14*
2.2.3 Polarization *14*
2.3 Battery Characteristics *15*
2.3.1 Capacity *15*
2.3.2 Energy Density *16*
2.3.3 Power *16*
2.3.4 Cycle Life *17*
2.3.5 Discharge Curves *17*

3 Materials for Lithium Secondary Batteries 21

3.1 Cathode Materials 21

3.1.1 Development History of Cathode Materials 21
3.1.2 Overview of Cathode Materials 23
3.1.2.1 Redox Reaction of Cathode Materials 23
3.1.2.2 Discharge Potential Curves 24
3.1.2.3 Demand Characteristics of Cathode Materials 26
3.1.2.4 Major Cathode Materials 27
3.1.3 Structure and Electrochemical Properties of Cathode Materials 27
3.1.3.1 Layered Structure Compounds 27
3.1.3.2 Spinel Composites 46
3.1.3.3 Olivine Composites 52
3.1.3.4 Vanadium Composites 57
3.1.4 Performance Improvement by Surface Modification 58
3.1.4.1 Layered Structure Compounds 60
3.1.4.2 Spinel Compound 61
3.1.4.3 Olivine Compounds 64
3.1.5 Thermal Stability of Cathode Materials 65
3.1.5.1 Basics of Battery Safety 65
3.1.5.2 Battery Safety and Cathode Materials 68
3.1.5.3 Thermal Stability of Cathodes 69
3.1.6 Prediction of Cathode Physical Properties and Cathode Design 75
3.1.6.1 Understanding of First-Principles Calculation 77
3.1.6.2 Prediction and Investigation of Electrode Physical Properties Using First-Principles Calculation 79
References 84

3.2 Anode Materials 89

3.2.1 Development History of Anode Materials 89
3.2.2 Overview of Anode Materials 90
3.2.3 Types and Electrochemical Characteristics of Anode Materials 91
3.2.3.1 Lithium Metal 91
3.2.3.2 Carbon Materials 92
3.2.3.3 Noncarbon Materials 118
3.2.4 Conclusions 137
References 137

3.3 Electrolytes 141

3.3.1 Liquid Electrolytes 142
3.3.1.1 Requirements of Liquid Electrolytes 142
3.3.1.2 Components of Liquid Electrolytes 143

3.3.1.3	Characteristics of Liquid Electrolytes	147
3.3.1.4	Ionic Liquids	149
3.3.1.5	Electrolyte Additives	153
3.3.1.6	Enhancement of Thermal Stability for Electrolytes	157
3.3.1.7	Development Trends of Liquid Electrolytes	161
3.3.2	Polymer Electrolytes	162
3.3.2.1	Types of Polymer Electrolytes	162
3.3.2.2	Preparation of Polymer Electrolytes	169
3.3.2.3	Characteristics of Polymer Electrolytes	171
3.3.2.4	Development Trends of Polymer Electrolytes	173
3.3.3	Separators	173
3.3.3.1	Separator Functions	173
3.3.3.2	Basic Characteristics of Separators	174
3.3.3.3	Effects of Separators on Battery Assembly	176
3.3.3.4	Oxidative Stability of Separators	176
3.3.3.5	Thermal Stability of Separators	178
3.3.3.6	Development of Separator Materials	179
3.3.3.7	Separator Manufacturing Process	180
3.3.3.8	Prospects for Separators	181
3.3.4	Binders, Conducting Agents, and Current Collectors	181
3.3.4.1	Binders	181
3.3.4.2	Conducting Agents	189
3.3.4.3	Current Collectors	191
	References	192
3.4	**Interfacial Reactions and Characteristics**	**195**
3.4.1	Electrochemical Decomposition of Nonaqueous Electrolytes	195
3.4.2	SEI Formation at the Electrode Surface	200
3.4.3	Anode–Electrolyte Interfacial Reactions	203
3.4.3.1	Lithium Metal–Electrolyte Interfacial Reactions	204
3.4.3.2	Interfacial Reactions at Graphite (Carbon)	209
3.4.3.3	SEI Layer Thickness	211
3.4.3.4	Effect of Additives	212
3.4.3.5	Interfacial Reactions between a Noncarbonaceous Anode and Electrolytes	214
3.4.4	Cathode–Electrolyte Interfacial Reactions	216
3.4.4.1	Native Surface Layers of Oxide Cathode Materials	217
3.4.4.2	SEI Layers of Oxide Cathodes	218
3.4.4.3	Interfacial Reactions at Oxide Cathodes	218
3.4.4.4	Interfacial Reactions of Phosphate Cathode Materials	223
3.4.5	Current Collector–Electrolyte Interfacial Reactions	225
3.4.5.1	Native Layer of Aluminum	225
3.4.5.2	Corrosion of Aluminum	226
3.4.5.3	Formation of Passive Layers on Aluminum Surface	228
	References	229

4 Electrochemical and Material Property Analysis 231

4.1 Electrochemical Analysis 231

- 4.1.1 Open-Circuit Voltage 231
- 4.1.2 Linear Sweep Voltammetry 232
- 4.1.3 Cyclic Voltammetry 232
- 4.1.4 Constant Current (Galvanostatic) Method 234
- 4.1.4.1 Cutoff Voltage Control 234
- 4.1.4.2 Constant Capacity Cutoff Control 236
- 4.1.5 Constant Voltage (Potentiostatic) Method 236
- 4.1.5.1 Constant Voltage Charging 236
- 4.1.5.2 Potential Stepping Test 236
- 4.1.6 GITT and PITT 238
- 4.1.6.1 GITT 238
- 4.1.6.2 PITT 239
- 4.1.7 AC Impedance Analysis 239
- 4.1.7.1 Principle 239
- 4.1.7.2 Equivalent Circuit Model 241
- 4.1.7.3 Applications in Electrode Characteristic Analysis 247
- 4.1.7.4 Applications in Al/LiCoO$_2$/Electrolyte/Carbon/Cu Battery Analysis 249
- 4.1.7.5 Applications in Al/LiCoO$_2$/Electrolyte/MCMB/Cu Cell Analysis 253
- 4.1.7.6 Relative Permittivity 254
- 4.1.7.7 Ionic Conductivity 256
- 4.1.7.8 Diffusion Coefficient 257
- 4.1.8 EQCM Analysis 257
- References 260

4.2 Material Property Analysis 263

- 4.2.1 X-ray Diffraction Analysis 263
- 4.2.1.1 Principle of X-ray Diffraction Analysis 263
- 4.2.1.2 Rietveld Refinement 265
- 4.2.1.3 In Situ XRD 267
- 4.2.2 FTIR and Raman Spectroscopy 269
- 4.2.2.1 FTIR Spectroscopy 270
- 4.2.2.2 Raman Spectroscopy 275
- 4.2.3 Solid-State Nuclear Magnetic Resonance Spectroscopy 280
- 4.2.4 X-ray Photoelectron Spectroscopy (XPS) 282
- 4.2.5 X-ray Absorption Spectroscopy (XAS) 285
- 4.2.5.1 X-ray Absorption Near-Edge Structure (XANES) 287
- 4.2.5.2 Extended X-ray Absorption Fine Structure (EXAFS) 288
- 4.2.6 Transmission Electron Microscopy (TEM) 292

4.2.7	Scanning Electron Microscopy (SEM)	*296*
4.2.8	Atomic Force Microscopy (AFM)	*300*
4.2.9	Thermal Analysis	*301*
4.2.10	Gas Chromatography-Mass spectrometry (GC–MS)	*306*
4.2.11	Inductively Coupled Plasma Mass Spectroscopy (ICP-MS)	*311*
4.2.12	Brunauer–Emmett–Teller (BET) Surface Analysis	*311*
	References	*315*
5	**Battery Design and Manufacturing**	*319*
5.1	Battery Design	*319*
5.1.1	Battery Capacity	*320*
5.1.2	Electrode Potential and Battery Voltage Design	*321*
5.1.3	Design of Cathode/Anode Capacity Ratio	*323*
5.1.4	Practical Aspects of Battery Design	*325*
5.2	Battery Manufacturing Process	*327*
5.2.1	Electrode Manufacturing Process	*328*
5.2.1.1	Preparation of Electrode Slurry	*328*
5.2.1.2	Electrode Coating	*329*
5.2.1.3	Roll Pressing Process	*330*
5.2.1.4	Slitting Process	*330*
5.2.1.5	Vacuum Drying Process	*331*
5.2.2	Assembly Process	*331*
5.2.2.1	Winding Process	*331*
5.2.2.2	Jelly Roll Insertion/Cathode Tab Welding/Beading Process	*332*
5.2.2.3	Electrolyte Injection Process	*334*
5.2.2.4	Cathode Tab Welding/Crimping/X-Ray Inspection/Washing Process	*334*
5.2.3	Formation Process	*334*
5.2.3.1	Purpose of the Formation Process	*334*
5.2.3.2	Procedures and Functions	*334*
	References	*335*
6	**Battery Performance Evaluation**	*337*
6.1	Charge and Discharge Curves of Cells	*337*
6.1.1	Significance of Charge and Discharge Curves	*337*
6.1.2	Adjustment of Charge/Discharge Curves	*339*
6.1.3	Overcharging and Charge/Discharge Curves	*340*
6.2	Cycle Life of Batteries	*342*
6.2.1	Significance of Cycle Life	*342*
6.2.2	Factors Affecting Battery Cycle Life	*342*
6.3	Battery Capacity	*344*
6.3.1	Introduction	*344*
6.3.2	Battery Capacity	*345*

6.3.3	Measurement of Battery Capacity	*346*
6.4	Discharge Characteristics by Discharge Rate	*347*
6.5	Temperature Characteristics	*349*
6.5.1	Low-Temperature Characteristics	*349*
6.5.2	High-Temperature Characteristics	*350*
6.6	Energy and Power Density (Gravimetric/Volumetric)	*351*
6.6.1	Energy Density	*351*
6.6.2	Power Density	*351*
6.7	Applications	*351*
6.7.1	Mobile Device Applications	*352*
6.7.2	Transportation	*352*
6.7.3	Others	*353*

Index *355*

List of Contributors

Chil-Hoon Doh
Korea Electrotechnology
Research Institute (KERI)
Battery Piezoelectric Research Center 12
Bulmosan-ro 10beon-gil, Seongsan-gu
Changwon-si, Gyeongsangnam-do,
642-120 Republic of Korea

Kyoo-Seung Han
Chungnam National Univ.
Department of Fine Chemical
Engineering and Applied Chemistry
99 Daehak-ro, Yuseong-gu
Daejeon, 305-764
Republic of Korea

Young-Sik Hong
Seoul National Univ. of Education
Department of Science Education
96 Seochojungang-ro, Seocho-gu
Seoul, 137-742
Republic of Korea

Kisuk Kang
Seoul National Univ.
Department of Materials
Science and Eng.
1 Gwanak-ro, Gwanak-gu
Seoul, 151-742
Republic of Korea

Dong-Won Kim
Hanyang Univ.
Department of Chemical Engineering
222 Wangsimni-ro, Seongdong-gu
Seoul, 133-791
Republic of Korea

Jae Kook Kim
Chonnam National Univ.
Department of Materials
Science and Eng.
77 Yongbong-ro, Buk-gu
Gwangju, 500-757
Republic of Korea

Sung-Soo Kim
Chungnam National Univ.
Graduate School of Green Energy Tech.
99 Deahak-ro, Yuseong-gu
Daejeon, 305-764
Republic of Korea

Chang Woo Lee
Kyunghee Univ.
Department of Chemical Engineering
26 Kyunghee-daero, Dongdaemun-gu
Seoul, 130-701
Republic of Korea

Sung-Man Lee
Kangwon National Univ.
Department of Advanced Materials
Science and Eng.
1 Kangwondaehak-gil, Chuncheon-si
Gangwon-do, 200-701
Republic of Korea

Sang-Young Lee
Kangwon National Univ.
Department of Chemical Engineering
1 Kangwondaehak-gil, Chuncheon-si
Gangwon-do, 200-701
Republic of Korea

Young-Gi Lee
Electronics and Telecommunications
Research Institute (ETRI)
Power Control Device Research Team
218 Gajeong-ro, Yuseong-gu
Daejeon, 305-700
Republic of Korea

Yong Min Lee
Hanbat National Univ.
Department of Chemical and Biological
Eng. 125 Dongseo-daero, Yuseong-gu
Daejeon, 305-719
Republic of Korea

Hong-Kyu Park
LG Chem
Battery Research Institute
104-1 Munji-dong, Yuseong-gu
Daejeon, 305-380
Republic of Korea

Jung-Ki Park
Korea Advanced Institute of Science
and Technology (KAIST)
Department of Chemical and
Biomolecular Eng.
291 Deahak-ro, Yuseong-gu
Daejeon, 305-701
Republic of Korea

Seung-Wan Song
Chungnam National Univ.
Department of Fine Chemical
Engineering and Applied Chemistry
99 Deahak-ro, Yuseong-gu
Daejeon, 305-764
Republic of Korea

Preface

With lithium secondary batteries being considered a key energy storage system, there has been growing expectations for revolutionary technological developments in lithium secondary batteries. However, it is not easy to come across a systematic and logical textbook on the principles and applications of lithium secondary batteries and I felt the pressing need for a comprehensive textbook on lithium secondary batteries for a wide range of readers. I partnered with other experts who shared the same vision and began working on this book. Our writers are researchers of lithium secondary batteries from universities, research centers, and industries.

Before writing, we agreed on a writing philosophy that extends beyond facts to provide straightforward fundamental explanations throughout all the chapters in a single, uniform pattern. In this respect, I myself, as a representing author, reviewed all the content and remodeled all the materials of the writers on the same philosophy. I am afraid our goal has not been fully attained but it is suffice to say that we have worked to the best of our abilities.

Many discussions were held concerning the structure and organization of this book. The introduction is followed by the basics of electrochemical reactions occurring within the battery. Next, we present the structure and properties of key components such as the anode, cathode, and electrolyte, as well as their interfacial reactions. In addition, we cover techniques used in the analysis of electrochemical and physical properties related to batteries. The book also includes sections on battery design, manufacturing, and performance evaluation.

This book was first conceptualized in early 2006 and took almost 4 years to complete. It has been a long journey with tens of meetings between all of our writers. We managed to get through this difficult process through mutual encouragement and in the hope of making significant contributions in relevant fields. We usually ordered packed lunches for these meetings, and I think many of our participating writers will look back with fond memories.

Particularly, I would like to express my appreciation to all who have made this book possible. I would like to thank Professor Yong-Mook Kang who spared no effort in reviewing our sections on cathodes, Professor Hochun Lee for his faithful review of electrochemical analysis techniques, Professor Doo-Kyung Yang for preparing the draft on NMR analysis, and Professor Nam-Soon Choi for her dedication and review of electrolytes and interfacial reactions. I appreciate the support of Miss Myung-su Lee, who was willing to facilitate our countless weekend meetings. Last but not least, it is the readers we are most thankful to.

From my laboratory at KAIST,
Jung-Ki Park, Representing Author

1
Introduction

With the proliferation of mobile telecommunication devices arising from remarkable developments in information technology (IT), the twenty-first century is moving toward a ubiquitous society, where high-quality information services are available regardless of time and place. The establishment of a ubiquitous society can be traced back to lithium secondary batteries, which were first commercialized in the early 1990s. Compared to other secondary batteries, lithium secondary batteries not only have higher working voltage and energy density, but also have long service life. Such superior characteristics enable secondary lithium batteries to fulfill complex requirements for diversified growth in devices. Global efforts are underway to further develop the existing technology of lithium secondary batteries and expand their use from eco-friendly transportation to various fields, such as power storage, health care, and defense.

A fundamental and systematic understanding of lithium secondary batteries is essential for the continuous development of related technologies along with technological innovation.

1.1
History of Batteries

A battery can be defined as a system that uses electrochemical reaction to directly convert the chemical energy of an electrode material into electric energy. The battery was first described in an 1800 study by Volta, an Italian professor at the University of Pavia, and published by the Royal Society of London. In 1786, Galvani of Italy discovered that touching a frog's leg with a metal object caused muscular convulsions. He claimed that "animal electricity" was generated from within the frog and transported through its muscles. Volta, who doubted the credibility of animal electricity, confirmed that the animal's body fluid merely served as an electrolyte between two different metals. In 1800, Volta invented the voltaic pile, in which an electric current is produced by connecting the two ends of a stack of two metal disks separated by cloth soaked in an alkaline solution. This was the first form of the battery as we know it today [1].

Figure 1.1 The voltaic pile (a) and the Baghdad battery (b).

A 2000-year-old clay jar, believed to be the earliest specimen of a battery, was discovered at a historic site near Baghdad in 1932 (Figure 1.1b). This clay jar had a height of 15 cm and contained a copper cylinder that was held in place by copper and iron rods. The rods had been corroded by acid. Although the artifact is accepted by some scholars as a primitive cell, it is uncertain whether it was indeed used for such a purpose.

Batteries can be classified into primary batteries, which are used once and disposed, and secondary or rechargeable batteries, which can be recharged and used multiple times. Since the invention of the voltaic pile, various batteries have been developed and commercialized.

The first widely used primary battery was the Leclanché (or manganese) cell invented in 1865 by Leclanché, a French engineer. The Leclanché cell, containing a zinc anode, a manganese dioxide (MnO_2) cathode, and an acidic aqueous electrolyte of ammonium chloride (NH_4Cl) and zinc chloride ($ZnCl_2$), had a wide range of applications with an electromotive force of 1.5 V. Later, the aqueous electrolyte in the Leclanché cell was replaced with an alkaline electrolyte of potassium hydroxide (KOH). This became the alkaline battery, which enhanced capacity and discharge, with the same voltage. New types of primary batteries later emerged, such as zinc–air batteries (1.4 V) and silver oxide batteries (1.5 V). The performance of primary batteries was greatly improved in the 1970s when 3 V lithium primary batteries with lithium as an anode became commercialized.

The oldest type of secondary batteries is lead–acid battery, invented by French physicist Planté in 1859. Lead–acid batteries have a lead peroxide anode, a lead cathode, and weak sulfuric acid as an electrolyte. With an electromotive force of 2 V per cell, they are commonly used as storage batteries in motor vehicles. When NiCd (1.2 V) batteries became widespread in 1984, they began to replace primary batteries in small electric appliances [2]. However, owing to the harmful environmental effects of cadmium, NiCd batteries are not as widely used today.

In the early 1990s, NiMH (1.2 V) cells were favored over NiCd batteries for their eco-friendliness and enhanced performance. This was followed by the emergence of 3 V lithium secondary batteries with greatly improved energy density. Compact and lightweight lithium secondary batteries soon dominated the market for portable devices, including cell phones, laptops, and camcorders [3].

1.2
Development of Cell Technology

After the invention of the voltaic pile in 1800, two significant milestones were reached in the 200-year history of cell technology. One was the development of primary batteries into secondary batteries and the other was the advancement to a working voltage of 3 V. Lithium secondary batteries, which use lithium ions as the main charge carrier, can maintain a high average discharge voltage of 3.7 V despite being lightweight. With the highest energy density among all currently available batteries, lithium secondary batteries have led the revolution of cell technology.

Looking at the changes in energy density with developments in secondary cell technology, lead–acid batteries have a specific energy of 30 Wh/kg and energy density of 100 Wh/l, whereas the energy density of lithium secondary batteries has shown an annual increase of 10%. At present, cylindrical lithium secondary batteries have a specific energy of 200 Wh/kg and energy density of 600 Wh/l (Figure 1.2). The specific energy of lithium secondary batteries is five times that of lead–acid batteries and three times higher than that of NiMH cells [4].

NiMH cells, a type of secondary batteries, have limited working voltage and energy density, but are attractive in terms of application to hybrid electric vehicles (HEVs) owing to their high stability. Recently, the advent of plug-in hybrid electric vehicles (PHEVs) and electric vehicles (EVs) has brought greater attention to lithium secondary batteries, which have higher energy and output compared to NiMH cells. Because secondary batteries in electric vehicles should offer fast charging time, lightweight, and high performance, future technology developments surrounding lithium secondary batteries are likely to be highly competitive. We can expect revolutionary and continuous technological progress that will overcome current limitations.

1.3
Overview of Lithium Secondary Batteries

For a cell to be characterized as a secondary battery, the anode and cathode have to repeat charging and discharging. The electrode structure should be kept stable during the insertion and extraction of ions within electrodes, while an electrolyte acts as an ion transfer medium. The charging of a lithium secondary battery is illustrated in Figure 1.3.

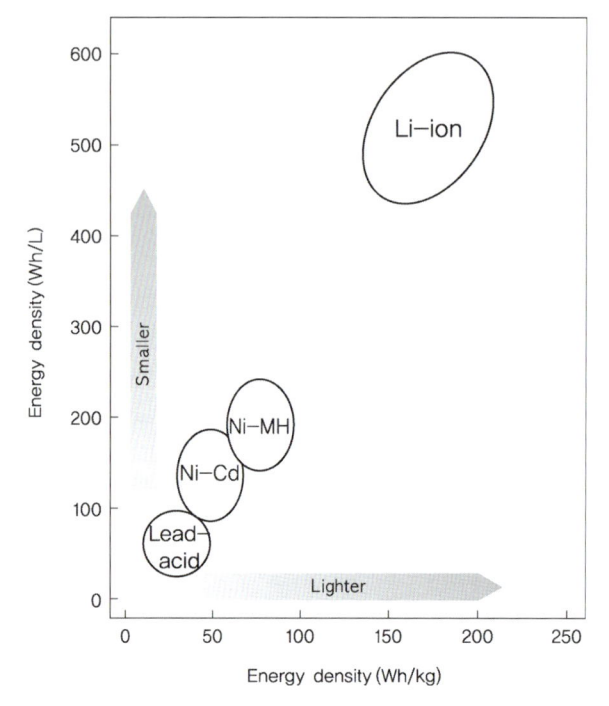

Figure 1.2 Changes in energy density with developments in cell technology.

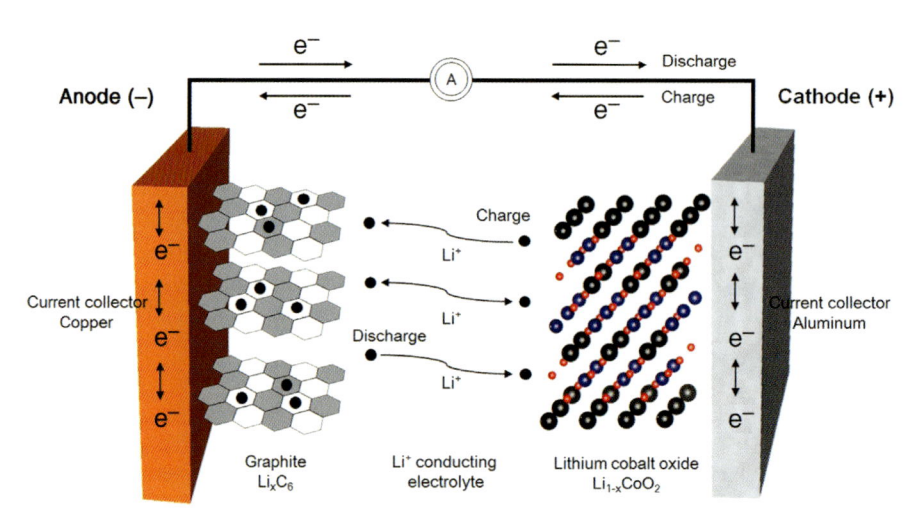

Figure 1.3 Movement of Li$^+$ in an electrolyte and insertion/extraction of Li$^+$ within electrodes in a lithium secondary battery.

1.3 Overview of Lithium Secondary Batteries

Charge neutrality occurs when ions flowing into electrodes collide with electrons entering through a conductor, thus forming a medium to store electric energy in the electrodes. Furthermore, the rate of reactions is increased as ions from the electrolyte are drawn to the electrodes. In other words, the overall reaction time of a cell heavily depends on the movement of ions between electrolyte and electrodes. The amount of ions inserted into electrodes for charge neutrality determines the electrical storage capacity. Ultimately, the types of ions and materials are main factors that influence the amount of electric energy to be stored. Cells based on lithium ions (Li^+) are known as lithium secondary batteries.

Lithium, which is the lightest of all metals and has the lowest standard reduction potential, is able to generate a working voltage greater than 3 V. With a high specific energy and energy density, it is suitable for use as an anode material. Since the working voltage of lithium secondary batteries is greater than the decomposition voltage for water, organic electrolytes should be used instead of aqueous solutions. Materials that facilitate the insertion and extraction of Li^+ ions are appropriate as electrodes.

Lithium secondary batteries use a transition metal oxide as an cathode and carbon as a anode. The electrolyte of lithium ion batteries (LIBs) is held in an organic solvent, while that of lithium ion polymer batteries (LIPBs) is a solid polymer composite.

As shown in Figure 1.4, commercialized lithium secondary batteries can be classified according to cell shape and component materials. The various forms of

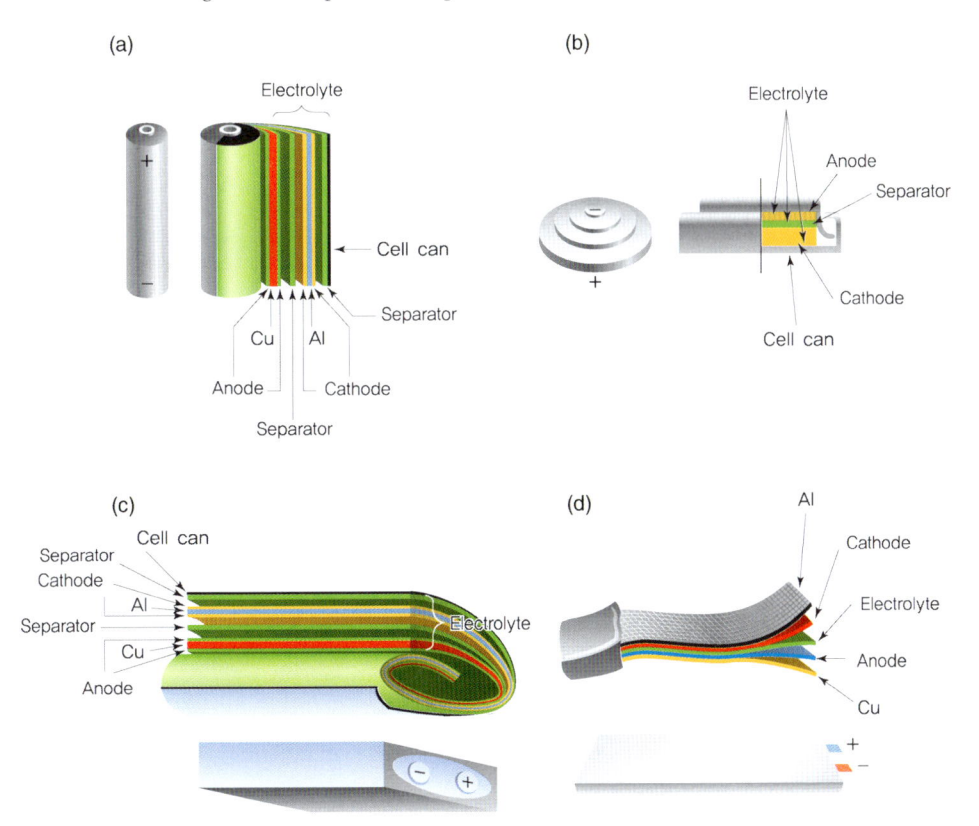

Figure 1.4 Different shapes of lithium secondary batteries: (a) cylindrical, (b) coin, (c) prismatic, (d) pouch [4]. Reprinted by Permission from Macmillan Publishers Ltd: [4], copyright 2001.

Table 1.1 Characteristics and examples of key components in a lithium secondary battery.

	Component	Material/characteristics	Example
Electrode	Cathode active material	Transition metal oxide/cell capacity	$LiCoO_2$, $LiMn_2O_4$, $LiNiO_2$, $LiFePO_4$
	Anode active material	Carbon/noncarbon alloy/reversion reaction with electrodes	Graphite, hard (soft) carbon, Li, Si, Sn, lithium alloy
	Conductive agent	Carbon/electron conductivity	Acetylene black
	Binder	Polymer/binding property	Polyvinylidene fluoride (PVdF), SBR/CMC
	Current collector	Metal film/formation of pole plates	Cu (−), Al (+)
Electrolyte	Separator	Polymer/separation of cathode and anode, prevention of short circuits	Polyethylene (PE), polypropylene (PP), PVdF
	Lithium salt	Organic and inorganic lithium compound/ion conduction	$LiPF_6$, $LiBF_4$, $LiAsF_6$, $LiClO_4$, $LiCF_3SO_3$, $Li(CF_3SO_2)_2N$
	Electrolyte solvent	Nonaqueous organic solvent/dissolution of lithium salt	Ethylene carbonate (EC), propylene carbonate (PC), dimethyl carbonate (DMC), diethyl carbonate (DEC), ethyl methyl carbonate (EMC)
	Additive	Organic compounds/SEI formation and overcharging protection	Vinylene carbonate (VC), biphenyl (BP)
Others	Tab	Metal/pole socket	Al (+), Ni (−)
	Outer casing	Cell protection and casing	Mo-rich stainless steel, Al pouch
	Safety component	Overcharging and overdischarge protection, safety device	Safety vent, positive temperature coefficient (PTC) device, protective circuit module (PCM)

batteries include cylindrical laptop batteries, prismatic cells for portable devices, single-cell coin-shaped batteries, and pouch-shaped cells cased in aluminum plastic composites [4].

Table 1.1 shows the key components of a lithium secondary battery. Its materials can be described as follows. Because lithium is removed from the lattice structure and released as ions, stable transition metal oxides are used as cathodes. Anode materials should have a standard reduction potential similar to that of lithium so as to stabilize the released ions and provide a large electromotive force. The electrolyte consists of a lithium salt in an organic solvent, thus maintaining electrochemical and thermal stability within the range of the working voltage. In addition, separators made of polymers or ceramics have a high-temperature melt integrity, which prevents short circuits caused by electrical contact between the cathode and the anode.

1.4
Future of Lithium Secondary Batteries

To date, the development of lithium secondary batteries has focused on small electric appliances and portable IT devices. Lithium secondary batteries are expected to build upon these achievements and create new applications described by buzzwords, such as "green energy," "wireless charging," "self-development," "recycle," "from portable to wearable," and "flexible." It is an important task to make advance preparations for future batteries by considering the functions of these applications.

Among the future applications of lithium secondary batteries, medium- and large-sized cells show great promise. Energy storage systems, in the form of batteries for electric vehicles and robots, or high-performance lithium secondary batteries capable of storing alternative energy such as solar, wind, and marine energy, are viewed as a key component of next-generation smart grid technology.

Other types of future lithium secondary batteries are microcells and flexible batteries. Microcells can be applied to RFID/USN, MEMS/NEMS, and embedded medical devices, while flexible batteries are mainly used in wearable computers and flexible displays. The structural control and manufacturing process of these batteries will be very different from today's methods.

The development of all-solid-state lithium secondary batteries is highly anticipated as well. With massive recalls caused by frequent battery explosions, an important challenge that lies ahead is to resolve the instability problem of existing liquid electrolytes through application of electrolytes consisting of polymers or organic/inorganic composites, along with the development of suitable electrode materials and processes.

References

1 Vincent, C.A. and Scrosati, B. (1997) *Modern Batteries: An Introduction to Electrochemical Power Sources*, 2nd edn, Arnold, London.

2 Besenhard, J. (1998) *Handbook of Battery Materials*, Wiley-VCH Verlag GmbH.

3 Mizushima, K., Jones, P.C., Wiseman, P.J., and Goodenough, J.B. (1980) *Mater. Res. Bull.*, **15**, 783.

4 Tarascon, J.M. and Armand, M. (2001) *Nature*, **414** (15), 359.

2
The Basic of Battery Chemistry

Electrochemistry is the study of electron transfer caused by redox reactions at the interface of an electron conductor, such as a metal or a semiconductor, and an ionic conductor, such as an electrolyte. Technologies based on electrochemistry include batteries, semiconductors, etching, electrolysis, and plating. In this book, electrochemistry refers to the conversion of chemical energy into electric energy in various systems such as primary batteries, secondary batteries, and fuel cells. In particular, this chapter describes the electrochemical aspects of secondary batteries.

2.1
Components of Batteries

2.1.1
Electrochemical Cells and Batteries

An electrochemical cell is the smallest unit of a device that converts chemical energy to electric energy, or vice versa. In general, a battery has multiple electrochemical cells, but it may be used to refer to a single cell. An electrochemical cell consists of two different electrodes and an electrolyte. The two electrodes of different electric potential create a potential difference when immersed in the electrolyte. This potential difference is also known as electromotive force.

Electric potential, denoted by V, is the potential energy of a unit charge within an electric field, and electromotive force drives current in an electric circuit. Redox reactions occur at each electrode due to this force and the generated electrons pass through the external circuit. To maintain charge neutrality of the electrolyte, redox reactions continue at the electrodes until the cell reaches electrochemical equilibrium.

2.1.2
Battery Components and Electrodes

As explained above, a battery (or an electrochemical cell) is a device that converts chemical energy to electric energy, or vice versa, using redox reactions. Figure 2.1

Principles and Applications of Lithium Secondary Batteries, First Edition. Jung-Ki Park.
© 2012 Wiley-VCH Verlag GmbH & Co. KGaA. Published 2012 by Wiley-VCH Verlag GmbH & Co. KGaA.

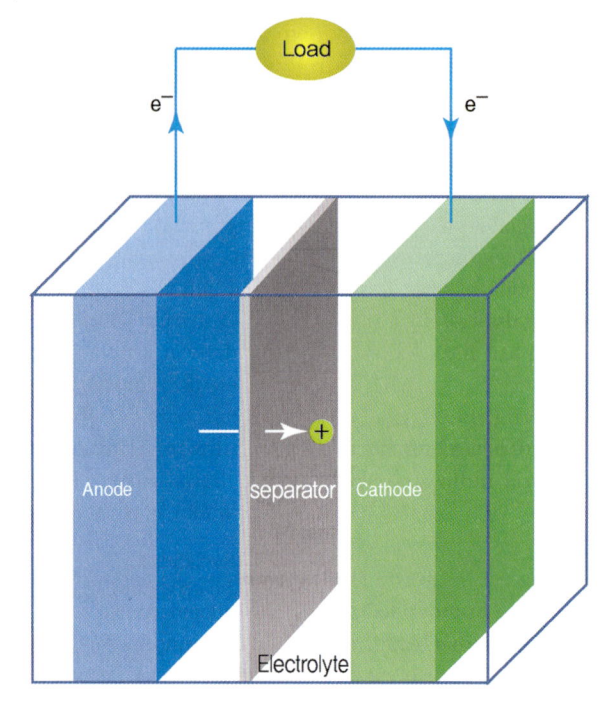

Figure 2.1 A battery or electrochemical cell (discharge).

shows the components of a cell, including the cathode, anode, electrolyte, and a separator to prevent short circuits between the electrodes.

When electrochemical redox reactions occur at the electrodes, ions are shuttled between the anode and the cathode through the electrolyte. At the same time, electron transfer takes place between the two electrodes. These electrons travel through the external wire connecting the two electrodes, thus forming a closed circuit.

In a discharging battery, electrochemical oxidation (oxidation, $A \rightarrow A^+ + e^-$) of the electrode proceeds at the negative terminal, which is termed an anode. Discharging is the process of converting chemical energy carried by the battery into electric energy. Electrons transferred from the negative terminal through the external circuit engage in reduction (reduction, $B^+ + e^- \rightarrow B$) at the positive terminal, which is known as a cathode. The electrolyte serves as an ionic conductor between the two electrodes and should be distinguished from an electron conductor.

While redox reactions at the electrodes are irreversible for primary batteries, they are reversible and repeatable in secondary batteries. Here, "reversible" means that redox reactions are repeated within the same electrode. One advantage of secondary batteries over primary batteries is that they can be recharged repeatedly. In the case of secondary batteries, both oxidation and reduction reactions can occur at the same electrode. It means that a cathode during discharging can be an anode during charging. However, in the conventional point of view, the terms remain the same for both charging and discharging, with the oxidative electrode as the anode and the reductive electrode as the cathode during discharging, spontaneous electrochemical reaction.

2.1.3
Full Cells and Half Cells

Electrochemical cells, in the form of a full cell or a half cell, are often used to analyze the electrochemical properties of batteries. A full cell takes on the complete form of a battery, where electrochemical reactions occur at both the cathode and the anode, and allows direct measurement of battery characteristics and performance. With additional use of a reference electrode, the full cell is able to obtain the electric potential difference between the two electrodes through individual measurements at the cathode and anode. On the other hand, the counter electrode of a half cell is used as the reference electrode to facilitate measurement and analysis of activities at the working voltage. This is useful in understanding the basic properties of each electrode material. Experiments may be performed using a full cell or a half cell, depending on one's purpose.

2.1.4
Electrochemical Reaction and Electric Potential

The electrochemical reaction occurring during discharge is related to the amount of electric energy that the battery can deliver. Consider the following electrochemical reaction at a given electrode:

$$pA + qB = rC + sD \tag{2.1}$$

Here, p, q, r, and s are stoichiometric coefficients of A, B, C, and D, which are different chemical species. Gibbs free energy for the above equation is given by Eq. (2.2), where a is the activity.

$$\Delta G = G^\circ + RT \ln(a_C^r a_D^s / a_A^p a_B^q) \tag{2.2}$$

The electric work (W_{rev}) at the equilibrium state is the maximum possible electric energy (W_{max}). When the battery is undergoing a chemical reaction, this can be expressed using ΔG, the change in Gibbs free energy.

$$W_{rev} = W_{max} \tag{2.3}$$

$$-W_{max} = \Delta G \tag{2.4}$$

Meanwhile, electric energy is associated with the charge Q (unit coulomb, C) and electric potential (E) as follows:

$$-W_{max} = QE \tag{2.5}$$

Q can be represented as a product of the number of electrons within the cell and the elementary charge. The number of electrons, n_e, is the number of moles multiplied by the Avogadro constant (N_A, 6.023×10^{23}). Q, in terms of moles and elementary charge, is written as follows:

$$Q = n_e e \tag{2.6}$$

$$Q = nN_A e \tag{2.7}$$

Q can also be described by the following equation.

$$Q = nF \tag{2.8}$$

Here, F is the Faraday constant, which is the elementary charge per mole of electrons (96 485 C/mol). The movement of n moles of electrons due to the potential difference between the two electrodes results in the following expression:

$$W_{max} = nFE \tag{2.9}$$

$$\Delta G = -nFE \tag{2.10}$$

This shows the relationship between the change in Gibbs free energy during equilibrium and electromotive force of the cell.

When all reactants and products are at a standard state, the standard potential is denoted by $E°$.

$$\Delta G° = -nFE° \tag{2.11}$$

Equations (2.2) and (2.11) lead to the following Nernst equation, where the difference in electric potential is affected by the concentration of components involved in the electrochemical reaction:

$$E = E° - RT \ln(a_C^r a_D^s / a_A^p a_B^q) \tag{2.12}$$

2.2
Voltage and Current of Batteries

2.2.1
Voltage

Voltage is the electrical driving force and is equal to the electric potential difference between two points in an electric circuit. It is also known as electromotive force and measured in volts (V). Since the actual voltage of a cell is subject to various conditions such as temperature and pressure, a reference point is needed. This corresponds to the standard state (1 bar, 25 °C, and 1 mol/dm^3) of an electrode. The standard electric potential, which is the measure of electric potential under equilibrium conditions, sets the basis for the electric potential at each electrode. The actual difference in electric potential between two electrodes can be expressed as follows:

$$E_{rxn} = E_{right} - E_{left} \tag{2.13}$$

E_{rxn} is the potential difference arising from chemical reactions, while E_{right} and E_{left} correspond to the electric potential at each electrode. For a galvanic cell, where redox reactions occur spontaneously, E_{rxn} takes a positive value. For nonspontaneous redox reactions in an electrolytic cell, the value of E_{rxn} is negative.

When the battery is in an equilibrium state with little or no current flow, it can provide electric energy equal to the amount of ΔG. Since current continues to flow in the battery during discharge, it is considered to be in a nonequilibrium state according to thermodynamics. The maximum possible energy cannot be used, because the voltage at this time is always less than the open circuit voltage (OCV). The open circuit voltage is the difference in the electric potential between two terminals of a device without an external load. The lower value of operating voltage compared to the open circuit voltage can be explained by ohmic polarization and by similar polarization effects from the movement of electric charge at the interface of the electrode/electrolyte. On the other hand, the voltage for the reverse charging reaction is higher than the open circuit voltage. This is due to internal resistance, overcharging caused by activation polarization, ionic conductivity lower than electron conductivity, impurities in electrode materials, and concentration polarization from varying diffusion speeds of lithium ions on the surface and inside the electrodes.

Figure 2.2 shows the voltage induced by periodic current pulses during the actual charging/discharging process. The voltage of the charging/discharging curve was measured by applying current slowly at the standard state. The open circuit voltage profile was recorded in the equilibrium state without permitting any additional flow of current. As demonstrated above, the difference in voltage between actual measurements during charging/discharging and the open circuit voltage can be interpreted as a result of polarization.

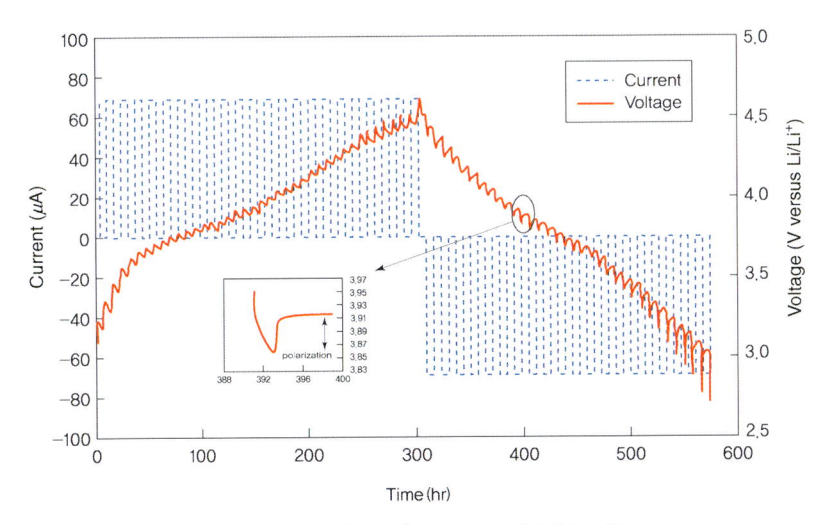

Figure 2.2 Charging/discharging voltage of a battery and OCV profile.

2.2.2
Current

Current, which is the rate of flow of electric charge, is closely related to the rate of electrochemical reactions at the electrodes. The rate of electrode reactions is determined by the transfer of electrons from the electrolyte to the electrodes and at the surface of the electrode active material.

At the electrodes, the reactants O and the product R undergo a reversible electrochemical reaction, as shown in Eq. (2.14). The relationship between current and the rate of reactions is given by the Nernst equations in Eqs. (2.15) and (2.16).

$$O + ne \leftrightarrow R \tag{2.14}$$

$$v_f = k_f C_o(0, t) = i_c / nFA \tag{2.15}$$

$$v_b = k_b C_R(0, t) = i_a / nFA \tag{2.16}$$

Here, v_f and v_b represent the speed of forward and backward reactions, respectively, while k_f and k_b are the respective rate constants. C_o and C_R are the concentration of oxidation and reduction substances, respectively, and $C_o(x, t)$ is the concentration as a function of time t and distance x from the electrode surface. Cathodic and anodic currents are denoted by i_c and i_a, respectively. The number of moles, Faraday constant, and electrode surface area are given as n, F, and A, respectively.

The net reaction rate is the difference in forward and backward reaction rates.

$$v_{net} = v_f - v_b = k_f C_o(0, t) - k_b C_R(0, t) = i / nFA \tag{2.17}$$

In other words, the current generated at the electrodes is highly dependent on the net reaction rate. When the forward and backward reactions proceed at the same rate during equilibrium, the net reaction rate v_{net} and the net current flow become 0.

2.2.3
Polarization

Polarization is a lack or excess of electrode potential at equilibrium. Since each battery component undergoes charge transfer at different rates, the slowest transfer becomes the rate-limiting process. When current flows between the two terminals of the battery, the actual potential E is always larger (charging) or smaller (discharging) than the equilibrium potential E_{eq}. Overpotential refers to this potential difference between the actual potential and the equilibrium potential. It is used as a measure of the extent of polarization. The relationship between actual potential (E), equilibrium potential (E_{eq}), and overpotential (η) can be expressed as follows:

$$\eta = E - E_{eq} \tag{2.18}$$

As shown in Figure 2.3, polarization is classified into ohmic polarization (iR drop), activation polarization, and concentration polarization.

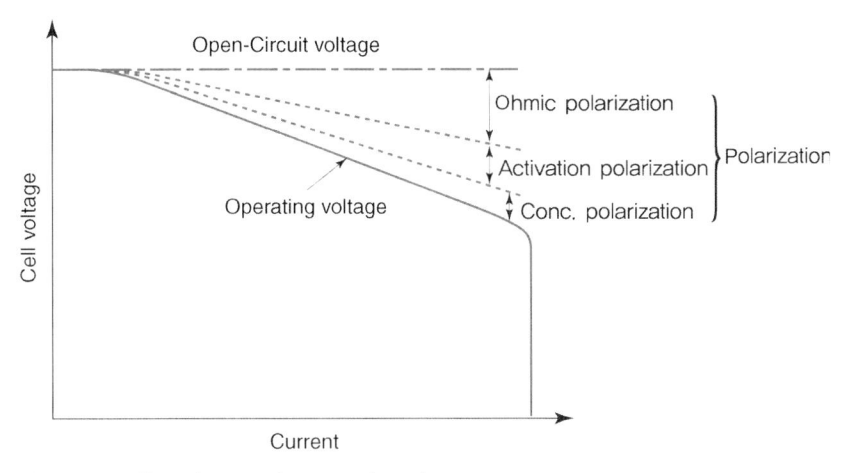

Figure 2.3 Effect of current density on the polarization.

Here, iR drop is associated with the electrolyte instead of resistance from electrode reactions. Considering that the iR drop increases proportionally to current density, a drastic decrease in working voltage at high current density conditions can be prevented by minimizing internal resistance.

On the other hand, activation polarization is closely related to electrode characteristics. As an inherent property of active materials, it is strongly influenced by temperature. Concentration polarization results from the concentration gradient of reactants at the surface of active materials. However, these different types of polarization are difficult to distinguish in an actual battery.

2.3
Battery Characteristics

2.3.1
Capacity

The capacity of a battery is the product of the total amount of charge, when completely discharged under given conditions, and time. The theoretical capacity C_T is determined by the amount of active materials and is calculated as follows:

$$C_T = xF \tag{2.19}$$

Here, F is the Faraday constant and x is the number of moles of electrons produced from the discharging process. The practical capacity, C_p, is smaller than the theoretical capacity because reactants are not 100% utilized in discharge. As the rate of charging/discharging increases, the practical capacity is further reduced due to the iR drop.

In general, the rate of charging/discharging is denoted by C_{rate}. The battery capacity and current drawn from charging/discharging are related by the following equation:

$$h = C_p/i \tag{2.20}$$

Here, h is the time (in hours) taken to completely discharge (or charge) a battery, i is the current drawn (A), and C_p is the battery capacity (Ah). The reciprocal of h is given by C_{rate}. In other words, as C_{rate} increases, the battery requires less time to be charged or discharged. Battery capacity can be measured using gravimetric specific capacity (Ah/kg, mAh/g) or volumetric specific capacity (Ah/l, mAh/cm³).

2.3.2
Energy Density

Energy density, an important factor in determining battery performance, is the amount of energy stored per unit mass or volume. The maximum energy that can be obtained from 1 mol of reactant is given as follows:

$$\Delta G = -FE = \varepsilon_T \tag{2.21}$$

Here, E is the electromotive force of the battery and ε_T is the theoretical energy (unit Wh, 1 Wh = 3600 J) for the cell reaction of 1 mol.

The actual energy ε_p, which varies according to the discharging method, from 1 mol of reactant is derived as follows:

$$\varepsilon_p = \int E \, dq = \int (E_i) dt = -F(E_{eq} - \eta) \tag{2.22}$$

As the rate of discharge or discharge current per unit time increases, the electric potential of the battery departs further from the equilibrium potential. Similarly to battery capacity, energy density is measured in Wh/kg, mWh/g or Wh/l, and mWh/cm³.

2.3.3
Power

The power of a battery refers to the energy that can be derived per unit time. The power P is the product of current i and electric potential E.

$$P = i \cdot E \tag{2.23}$$

Electric power is a measure of the amount of current flowing at a given electric potential. When the current increases, the power rises to a peak and declines. The battery voltage drops when the current goes beyond a certain limit, thus leading to a decrease in power. This phenomenon of polarization is related to the diffusion of lithium ions and the internal resistance of the battery. To improve power, it is necessary to enhance the diffusion rate of lithium ions and the electrical conductivity.

Similarly to battery capacity and energy, power per unit mass or volume is described as power density.

2.3.4
Cycle Life

Cycle life is the number of charge and discharge cycles that a battery can achieve before its capacity is depleted. A high-performance battery should be able to maintain its capacity even after numerous charge and discharge cycles. The cycle life of lithium secondary batteries strongly depends on the structural stability of electrode active materials during the charging/discharging process. Irreversible capacity, which is the amount of charge lost, is usually observed after the first charge/discharge cycle and results from the formation of a new layer at the interface of electrodes and electrolyte.

After N charge/discharge cycles, the capacity retention is given by C_N/C_1 (%) and the relative capacity decrease is $(C_1 - C_N)/C_1$. The remaining capacity after N charge/discharge cycles and 1 cycle is C_N and C_1, respectively. The cycle life is generally affected by the depth of discharge, which varies according to battery type. Lithium secondary batteries tend to display a longer cycle life when charging is repeated at a low depth of discharge, such that capacity is not fully depleted.

2.3.5
Discharge Curves

Repeated cycles of charging and discharging affect the discharge characteristics of a battery, allowing the discharge curve to take various forms depending on discharge conditions, electrical properties, and other measurement variables. Constant current, constant power, and constant external resistance are normal discharge conditions. Electrical properties to be measured include the battery voltage, current, and power, while measurement variables are discharge time, capacity, and lithium ion occupancy. With the same battery consisting the same materials and cell design, various discharge curves can be produced according to measurement conditions. It is essential to compare these discharge curves to obtain a more accurate understanding of battery properties. An actual battery gives a wide range of discharge curves according to battery components. The typical discharge curves of a battery are shown in Figure 2.4.

Figure 2.4 is a plot of changes in battery voltage as a function of capacity when the battery is discharged under constant current. Since capacity is directly proportional to the time during which current is applied, Figure 2.4 implies the change in voltage with elapsed time. Furthermore, the battery voltage represents the open circuit voltage when the battery is not connected to an external load, or the working voltage when the circuit is closed. The battery voltage at the point at which discharge is complete is known as the cutoff voltage.

In the case of curve I in Figure 2.4, the battery voltage is hardly influenced by reactions occurring within the battery during discharge. Curve II shows two flat

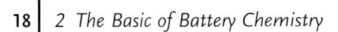

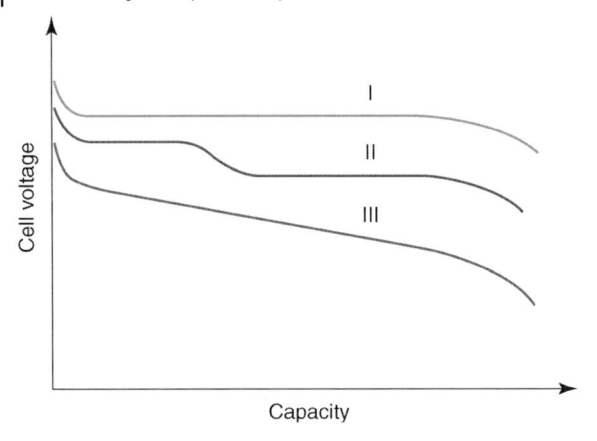

Figure 2.4 Discharge curves obtained from electrochemical reactions.

regions due to a change in the reaction mechanism. In curve III, the reactants, products, and internal resistance of the battery are continuously changing throughout the discharge process.

For lithium secondary batteries, the change in battery voltage after charging/discharging is given by the Armand equation.

$$E_{cell} = E^{\circ}_{cell} - (nRT/F)\ln(y/1-y) + ky \tag{2.24}$$

In Eq. (2.24), y is the lithium ion occupancy and ky is the effect of interactions between intercalated lithium ions on the battery voltage. The change in the gradient of the battery voltage based on capacity is determined by direct factors such as the rate

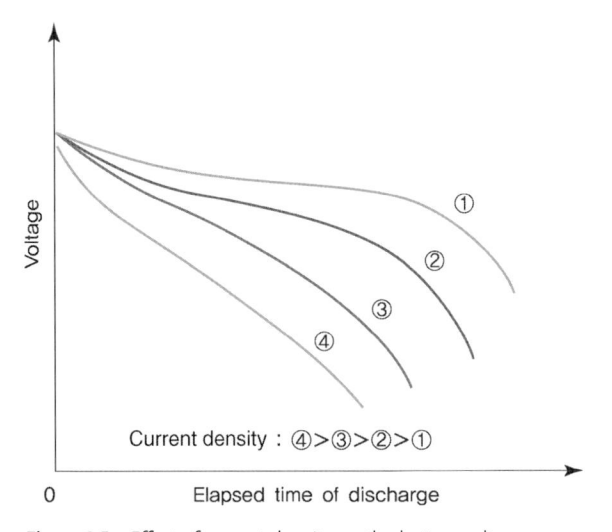

Figure 2.5 Effect of current density on the battery voltage.

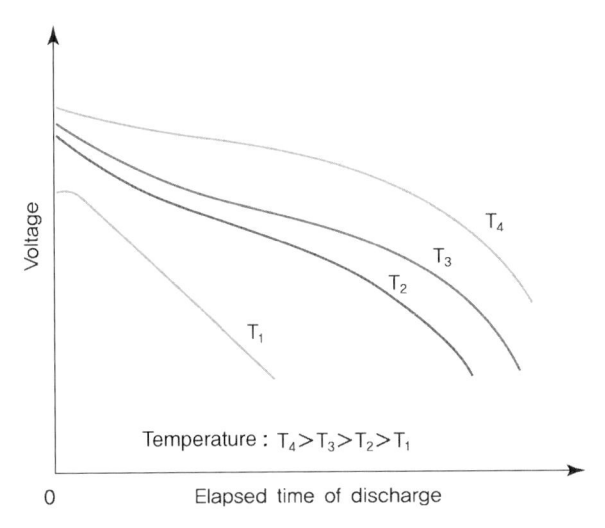

Figure 2.6 Effect of temperature on the battery capacity.

of lithium ion diffusion, phase transition, change in lattice structure, and dissolution, as well as additional factors such as the particle size of electrode active materials, temperature, electrolyte characteristics, and porosity of the separator. These factors may change the values of y and k in the Armand reaction.

Under low current density conditions, both voltage and discharge capacity of the working battery come close to the theoretical equilibrium. However, from the trends of ① to ④ in Figure 2.5, we can see that the battery voltage decreases during discharge since the iR drop and overvoltage from polarization increase with discharge current. The battery capacity also decreases when the battery is discharged beyond its cutoff voltage, owing to the rising gradient of discharge curves. These characteristics of the discharge voltage vary greatly with temperature.

As shown in Figure 2.6, when the battery is discharged at a low temperature, the reduced chemical activity of the reactants leads to less internal resistance. This gives rise to a sharp decline in the battery voltage along with a decrease in capacity. At higher temperatures, there is a capacity increase caused by lower internal resistance and higher discharge voltage. However, if the temperature is too high, increased chemical activity may result in self-discharge and other unwanted chemical reactions.

3
Materials for Lithium Secondary Batteries

3.1
Cathode Materials

3.1.1
Development History of Cathode Materials

Pioneering work on lithium batteries began in the 1910s under G. N. Lewis, and the first Li/(CF)$_n$ primary batteries were sold in the 1970s. The cathode material (CF)$_n$ is an intercalation compound of fluorine, which has the highest electrical conductivity of all elements, and carbon. Attempts were made in the United States to develop Li/MnO$_2$ batteries, but failed due to problems with moisture control, cell structure, and assembly technology. In 1973, Japan became the first nation to commercialize Li/MnO$_2$ primary batteries, which gained recognition for their high working voltage of 3 V, obtained through the use of an organic solvent instead of an aqueous solution. These developments laid the foundation for the commercialization of lithium secondary batteries in Japan.

Depending on the cathode material, lithium primary batteries are available in various forms such as Li/(CF)$_n$, Li/MnO$_2$, Li/SO$_2$, and Li/SOCl$_2$ batteries. Among these, Li/MnO$_2$ batteries are the most widespread. Despite the excellent low-temperature performance and durability of Li/SO$_2$ and Li/SOCl$_2$ batteries, they are limited to military use due to the presence of harmful substances.

Following the discovery in 1970s of the cathode material (CF)$_n$, which forms the intercalation layers in primary batteries, various intercalation experiments have been performed to combine high electrical conductivity with high electrochemical reactivity. Studies confirmed that chalcogen compounds such as TiS$_2$ are able to engage in intercalation reactions between layers. This served as the basis for the acquisition of commercial technology for lithium secondary batteries. As shown in Figure 3.1, TiS$_2$ is a light semimetal with a layered structure and can be used as an electrode material without any conducting agent. Since its structure remains unchanged during battery charging/discharging, the lithium intercalation–deintercalation cycle is reversible. However, its difficult synthesis and high cost restricted commercialization. In 1989,

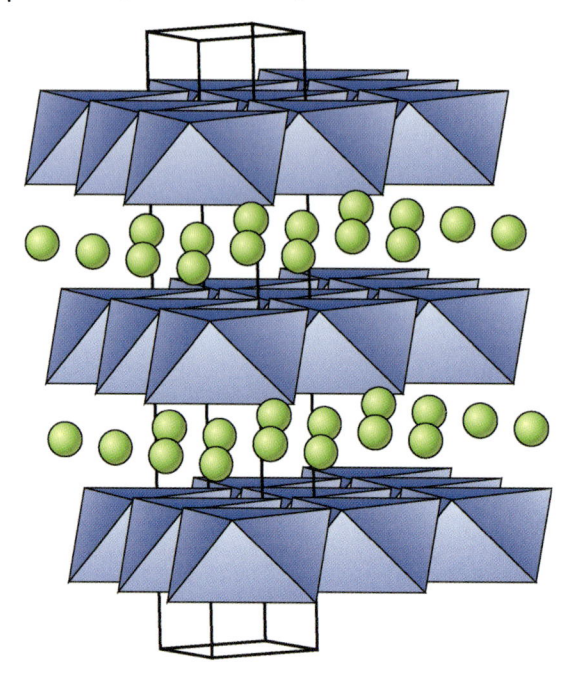

Figure 3.1 Structure of cathode material Li_xTiS_2. Adapted with permission from American Chemical Society Copyright 2004.

Canada's Moli Energy developed lithium metal secondary batteries using MoS_2 as the cathode material, but dendrite growth on the lithium anode led to safety problems such as internal short circuits and combustion.

Besides chalcogen compounds with low electrode potential, other oxides were considered as cathode materials but failed to be commercialized. With the 1991 commercialization of lithium secondary batteries using $LiCoO_2$ as the cathode and carbon as the anode, various studies were conducted on cathode materials [1, 2]. When $LiCoO_2$ is used as the cathode, Li_xC_6 compounds are formed from the intercalation of lithium ions into carbon, and the lithium metal anode is eliminated due to internal short circuits caused by dendrite growth. The reduction potential of carbon, which is 0.1–0.3 V higher than that of lithium metal, is balanced by the relatively higher potential of LiCoO2, and the average voltage of carbon/$LiCoO_2$ becomes 3.7 V. $LiCoO_2$ has a lithium ion diffusion coefficient of 5×10^{-9} cm^2/s, which is similar to that of $LiTiS_2$ (10^{-8} cm^2/s). Its electrical conductivity, depending on the amount of intercalated ions, lies between the ranges of semiconductors and those of metals [2].

Following the commercialization of $LiCoO_2$, research on other types of cathode materials was pursued. Notable high-capacity materials include spinel $LiMn_2O_4$, which is structurally stable, and $LiNiO_2$, where more than 70% of lithium can be charged and discharged reversibly. However, spinel $LiMn_2O_4$ has a small capacity and lowered performance resulting from the migration of manganese ions at high temperatures, while $LiNiO_2$ was found to have safety issues. To resolve these problems, the advantages of $LiCoO_2$, $LiNiO_2$, and $LiMn_2O_4$ have been integrated

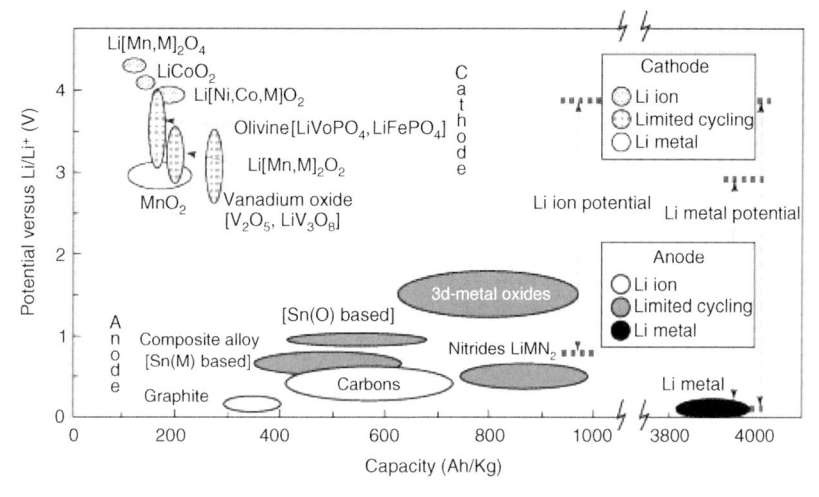

Figure 3.2 Electrode potential versus capacity of anode/cathode materials [3]. Reprinted by Permission from Macmillan Publishers Ltd: [3], copyright 2001.

into the three-component Li[Ni, Mn, Co]O$_2$. Olivine LiFe-PO$_4$, including iron, is also being actively studied. Figure 3.2 shows the electrode potential and capacity of electrode active materials in the range of 3–4 V [3].

3.1.2
Overview of Cathode Materials

3.1.2.1 Redox Reaction of Cathode Materials

In the case of the cathode material LiCoO$_2$, the lithium ions of LiCoO$_2$ are deintercalated during the charging process, and electrons are transported through the external circuit as Co^{+3} is oxidized into Co^{+4}. When the battery is discharging, Co^{+4} of Li$_{1-x}$CoO$_2$ is reduced to Co^{+3} by the inward flow of electrons from the external circuit, and lithium ions are intercalated. The reducible Li$_{1-x}$CoO$_2$ is therefore included as part of the cathode during spontaneous discharge.

Figure 1.3 shows the redox reactions of a lithium secondary battery, which consists of a carbon anode and a LiCoO$_2$ cathode. During the charging process, the deintercalated lithium ions from LiCoO$_2$ and emitted electrons are transported to the anode through the electrolyte and external circuit, respectively, before recombining inside carbon. These nonspontaneous reactions are based on electric energy from an external power supply and correspond to nonspontaneous oxidation reactions of the cathode material. Electrons and lithium ions produced from oxidation reactions are each delivered to the anode through the external circuit and electrolyte. They are stored as chemical energy as a result of the nonspontaneous oxidation of the anode material.

On the other hand, as spontaneous reactions occur during discharging, anode materials engage in reduction reactions due to the potential difference (or electromotive force) between the two charged cathode materials. Electrons produced from

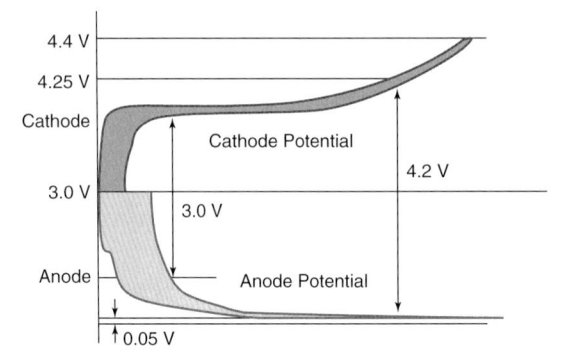

Figure 3.3 Effect of charge/discharge on the battery voltage.

the spontaneous oxidation of anode materials activate the device by moving through the external circuit and contribute to the reduction of the cathode material. At the same time, deintercalated lithium ions from the anode material pass through the electrolyte and intercalate into the cathode material. As indicated by the arrows in Figure 3.3, the electric potential of the cathode material decreases while that of the anode material increases, thus causing the battery voltage to drop from a fully charged level of 4.2 to 3.0 V.

3.1.2.2 Discharge Potential Curves

Since the voltage of a battery changes with the state of charge/discharge, the midpoint of the range of discharge is referred to as the nominal voltage. The battery voltage is affected by changes in electron arrangement and orbital energy levels during the intercalation/deintercalation of lithium into crystals of anode and cathode materials. In other words, the battery voltage during discharge is determined by lithium ion occupancy, which involves changes in the Fermi levels of electrode active materials and interactions between intercalated lithium ions. This is represented in the Armand equation (2.24).

The battery voltage gradually decreases (Figure 2.5) during discharge. The change in the gradient of the discharge curves is influenced by the rate of diffusion of lithium ions on the surface of the cathode material, phase transition of the active material, disintegration of the crystal structure, and the migration of transition metal ions into the electrolyte. When subject to the same rate-limiting step, discharge curves may vary according to the particle size and distribution of the cathode material, temperature, mixing conditions for cathode material/conductive agent/binder, electrolyte characteristics, and pore structure of the separator.

The voltage can also be interpreted as the chemical potential difference of lithium ions at the cathode and anode. The chemical potential (μ) of lithium ions is defined as the partial derivative of free energy of electrode materials with respect to lithium ion concentration. The equilibrium potential or open-circuit voltage (OCV) can be obtained from the following equation, where Li_xMX is the lithium composition, z is the oxidation number of lithium ions in the electrolyte, and e is the charge:

$$V(x) = -\left(\mu_{Li}^{cathode}(x) - \mu_{Li}^{anode}\right)/ze \qquad (3.1)$$

In the above equation, $V(x)$ is the equilibrium potential, with x being the amount of lithium, and μ is the lithium chemical potential at each electrode. In general, the equilibrium potential of a battery is derived as shown below. Assuming that the anode consists of lithium metal, the change in cathode from x_1 to x_2 can be described by the following:

$$Li_{x1} MX(cathode) + (x_2 - x_1)Li(anode) \rightarrow Li_{x2} MX \qquad (3.2)$$

With ΔG as the Gibbs free energy, the equilibrium potential (V) of a battery is

$$V = -\Delta G/(x_2 - x_1)ze$$

The discharge voltage is directly affected by the various factors that determine the chemical potential of lithium ions. For example, a change in lithium chemical potential from the displacement of lithium ions within the lattice changes the discharge voltage. In other words, the relative redox energy varies with site energy, which is determined by the placement of lithium in the crystals of the active material. As such, the two lithium ions at different positions in the $Li_2Mn_2O_4/LiMn_2O_4$ spinel structure have different potentials. The same Co ion has a potential of 3.6–3.7 V in the layered $LiCoO_2$ but a higher potential of 4.5 V in the spinel-type structure $LiMnCoO_4$.

The distance between lithium ions within the crystal structure also changes the discharge potential by affecting chemical potential. For instance, lithium ions in the layered Li_2NiO_2 and $Immm$ structure of Li_2NiO_2 exert different repulsions and thus result in different discharge potentials. Another factor that influences discharge potential is the redox potential of the transition metal, arising from electron interactions between the p-orbital of the oxygen and the d-orbital of the transition metal in the cathode material. The p-orbital electrons of the oxygen are -2 with almost uniform energy, but the d-orbital electrons of the transition metal show huge variations in energy. From Figure 3.4, we can see that the electric potential increases with the number of d-orbital electrons belonging to the same period. If the period increases in the same family, the binding energy of electrons decreases, and the electric potential of the cathode material having a 3d outermost shell becomes higher than that having a 4d shell.

In the case of $LiCoO_2$ and $LiNiO_2$, $LiCoO_2$ has a higher voltage despite the greater number of d-orbital electrons in Ni. This is because the energy levels of Co^{3+} and Ni^{3+} are partially reversed, as shown in Figure 3.5. All 6 Co^{3+} in $LiCoO_2$ exist in their low spin states of the t_{2g} orbital, whereas the 7 d-orbital electrons of Ni^{3+} in $LiNiO_2$ are split into $6t_{2g}$ and $1e_g$. Since electrons are more easily released at higher energies, the cell potential of Ni^{3+} is lowered.

The energy difference caused by the inductive effect also affects the discharge potential. For example, in the Fe—O—P bond of $LiFePO_4$, the strong P—O covalent

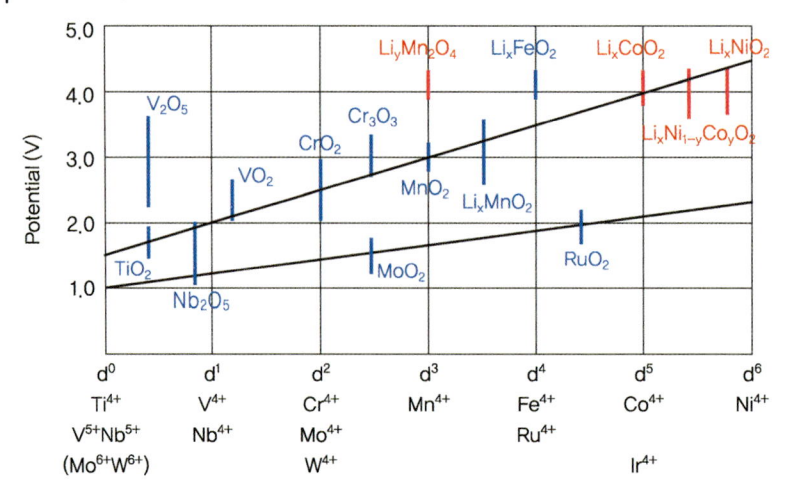

Figure 3.4 Correlation between working voltage of the cathode material and the number of d electrons.

bond of PO_4^{2-} attracts the d-orbital electrons of iron ions, hindering oxidization of iron. Thus, $LiFePO_4$ has a higher voltage than $LiFeO_2$.

Nonequilibrium polarization is another influential factor determining battery voltage. In particular, when there is high resistance to the movement of lithium ions or electrons at the cathode, electrochemical equilibrium is not established, and the voltage decreases with iR drop. To overcome the iR drop, it is necessary to minimize resistance to the transfer of electrons or lithium ions at the cathode material.

3.1.2.3 Demand Characteristics of Cathode Materials

1) With the intercalation–deintercalation of large amounts of lithium ions, cathode materials should display reversible behavior and a flat potential so as to enhance energy efficiency during charge/discharge.
2) Cathode materials should be light and densely packed to allow high capacity per weight or volume, and have high electrical and ionic conductivities for high power.
3) High cycle efficiency must be maintained. Side reactions unrelated to lithium ion circulation at the anode or cathode decrease the efficiency.

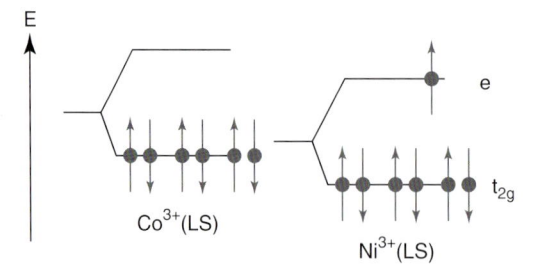

Figure 3.5 d-orbital energy levels for Co^{3+} and Ni^{3+} ions occupying octahedral sites.

4) Since cycle life is shortened by irreversible phase transitions of the crystal structure, such phase transitions should not occur in cathode materials during charge/discharge. If the crystal lattice undergoes a large change in volume, the active material is desorbed from the current collector and the battery capacity is reduced.
5) To prevent reactions with the electrolyte, cathode materials should have electrochemical and thermal stability.
6) The particles of cathode materials must be globular with a narrow grain size distribution, so that the aluminum film remains intact when making electrodes. This also improves both particle-to-particle contact and electrical conductivity.

3.1.2.4 Principle Cathode Materials

3d transition metals are often used as cations of cathode materials. Compared to 4d and 5d transition metals, 3d transition metals have a higher electrode potential and higher capacity per weight/volume due to their light weight and small size. Chalcogens, especially oxygen, are more suitable as cations than halogens in terms of structural stability.

To maintain the charge neutrality of transition metal oxides when charging or discharging, the intercalation/deintercalation of cations into the cathode material and transport through the electrolyte should take place rapidly. For fast charging/discharging based on the wide redox potential range of cations or to minimize changes in the crystal structure of cathode materials, low-charge cations that are small and form weak bonds with active materials should be selected. With a coordination number of 6, beryllium (Be^{2+}) has an ionic radius of 0.45 Å, which is smaller than that of lithium at 0.76 Å, but bonds strongly with oxygen due to its high charge. The high resistance to other cations causes beryllium to travel slowly within the crystal structure, making it inadequate as a cathode material. On the other hand, lithium is a suitable cathode material since it is smaller than other candidate atoms, except beryllium, and has a low charge. Lithium has a standard reduction potential of -3.040 V, which is much lower compared to that of beryllium at -1.847 V, and thus its electric potential can be increased with greater ease.

Common cathode materials are lithium transition metal oxides such as layered $LiMO_2$ and spinel LiM_2O_4, and lithium transition metal phosphates such as olivine $LiMPO_4$. Research is being actively pursued to utilize these different structures by mixing, surface coating, or compound formation. For example, when an inexpensive, stable cathode material with low electrical conductivity and high capacity is surface coated using an expensive material with high electrical conductivity and small capacity, a material that has both high capacity and high power is obtained. Table 3.1 shows various cathode materials and their characteristics including discharge capacity and potential.

3.1.3 Structure and Electrochemical Properties of Cathode Materials

3.1.3.1 Layered Structure Compounds

Lithium transition metal oxides in the form of layered $LiMO_2$ have strong ionic character and the most densely packed crystal structure. A dense layer is first

3 Materials for Lithium Secondary Batteries

Table 3.1 Battery characteristics of various cathode materials.

Cathodes	Theoretical capacity (mAh/g)	[a)]Practical capacity (mAh/g)	Avg. potential (V versus Li/Li+)	True density (g/cc)
$LiCoO_2$	274	~150	3.9	5.1
$LiNiO_2$	275	215	3.7	4.7
$LiNi_{1-x}Co_xO_2$ (0.2 ≤ x ≤ 0.5)	~280	~180	3.8	4.8
$LiNi_{1/3}Mn_{1/3}Co_{1/3}O_2$	278	~154	3.7	4.8
$LiNi_{0.5}Mn_{0.5}O_2$	280	130–140	3.8	4.6
$LiMn_2O_4$	148	~130	4.0	4.2
$LiMn_{2-x}M_xO_4$	148	~100	4.0	4.2
$LiFePO_4$	170	~160	3.4	3.6

a) Commercially available capacity.

formed by oxygen ions, which have the largest ionic radius among ions of lithium, transition metals, and oxygen. The spaces between oxygen ions are filled with lithium ions and transition metal ions, so as to increase the bulk density. The most densely packed layer of oxygen ions, as shown in Figure 3.6, can be achieved through hexagonal close packed (HCP) and cubic close packed (CCP) arrangements. Both structures have a high packing density of 0.7405. In these structures, the empty tetrahedral and octahedral sites between oxygen ions are filled by 3d

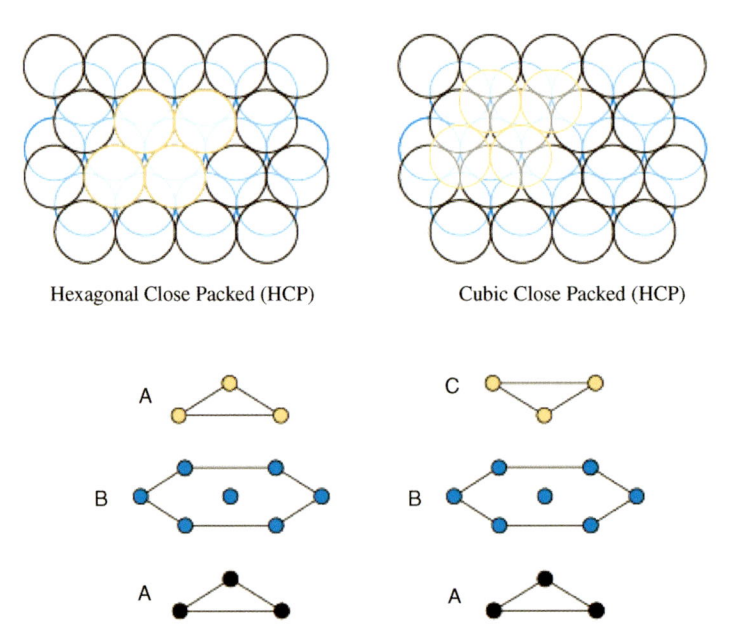

Figure 3.6 Structure of the densest oxygen layer (each color represents the distribution of oxygen layers in c-axis).

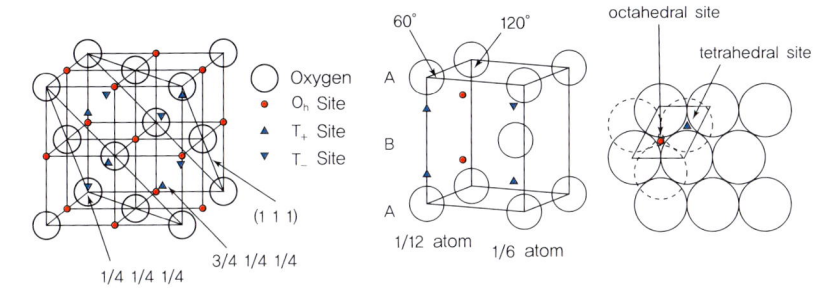

Figure 3.7 Octahedral and tetrahedral sites in cubic close packed and hexagonal close packed arrangements (T_1 and T_2 vertices are in opposite directions when the tetrahedron is viewed as a triangular pyramid).

transition metals (when CN = 6, Co^{3+}: 0.685 Å, Ni^{3+}: 0.700 Å, Mn^{3+}: 0.720 Å, Fe^{3+}: 0.690 Å) and lithium ions (0.900 Å when CN = 6, 0.730 Å when CN = 4) with an ionic radius of 0.680–0.885 Å.

For n number of oxygens in a unit cell, there are $2n$ tetrahedral and n octahedral sites (Figure 3.7). It follows that the layered $LiMO_2$ has four tetrahedral and two octahedral sites. In consideration of the geometrical arrangement of cations, tetrahedral sites are occupied by ions with an ionic radius ratio of $0.225 \leq r/R < 0.414$, while octahedral sites are filled with $0.414 \leq r/R < 0.732$. Since the ionic radius ratio of 3d transition metal ions is $0.5397 \leq r(M^{3+})/R(O^{2-}) \leq 0.7024$, and that of lithium ions is $r(\text{lithium ion})/R(O^{2-}) = 0.7143$, these ions reside in the two octahedral sites of $LiMO_2$. Because the octahedral and tetrahedral sites are close to one another, it is difficult to further intercalate lithium ions.

In addition, the two-dimensional corner sharing not only contributes to a layered structure but also enhances electrical conductivity through direct M–M mutual interaction, thus reducing volume change caused by charging/discharging. As such, compounds denoted by $LiMO_2$ take on a layered structure with the regular arrangement of O–Li–O–M–O–Li–O–M–O, with lithium, transition metals, and oxygen along the [111] side of rock salt. In other words, lithium ions and transition metal ions each constitute 50% of octahedral sites in the face-centered cubic structure of ABCABC. For three repeats of the MO_2 layer within the unit cell, lithium occupies an octahedral site. Since an octahedral site occupied by lithium is represented as O, it is known as the O3 structure in reference to the three repeats.

The structure of the layered $LiMO_2$ cathode material is shown in Figure 3.8. The metal oxide slab, consisting of transition metals and oxygen, and oxide octahedral slab surrounding lithium are arranged one after another. Strong ionic bonds are formed within MO_2, and the Coulomb repulsion between MO_2 layers allows the intercalation/deintercalation of lithium ions. The ions diffuse along the two-dimensional plane, leading to a higher ionic conductivity.

Lithium ions on the particle surface are deintercalated during charging, and ions close to empty octahedral sites gradually diffuse and undergo deintercalation. During discharge, lithium ions are intercalated into empty octahedral sites on the particle surface. When lithium moves between layers, it cannot pass through empty

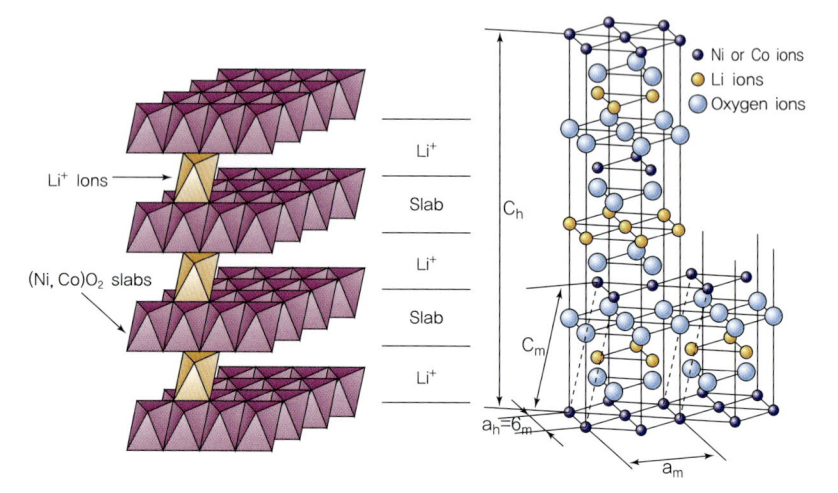

Figure 3.8 Structure of layered LiMO$_2$ cathode material [4]. Reproduced by permission of ECS – The Electrochemical Society.

tetrahedral sites to reach empty octahedral sites. Since tetrahedral sites share faces with transition metal octahedra in other layers, activation energy is required for lithium ions to diffuse through the region of electrostatic repulsion. Because the ionic conductivity is lower than the electrical conductivity at the initial stage of discharge, lithium ions are densely populated on the surface of the active material and diffused until equilibrium. This leads to a potential difference between the open and the closed circuits in the charge/discharge process.

When lithium moves out during the charging process, the lattice expands due to repulsion between oxygen atoms in the MO$_2$ layer. When lithium is completely removed, various changes can be observed in the layered LiMO$_2$ active material, which contracts rapidly in the c-axis. Depending on lithium content during charging, the cathode material reconfigures into a stable lattice structure. At this point, the region of the active material should be single phase. Figure 3.9 shows the phase transition of lithium transition metal oxides with charge/discharge.

LiCoO$_2$ changes from an O3- to P3-type structure when lithium content falls below 0.5. From Figure 3.10, which shows various possible forms of LiMO$_2$ layered oxides, we can see that O3 and P3 are comprehensively different structures. Reduced lithium content in LiCoO$_2$ leads to a change in structure. As shown in Figure 3.9, lithium content has little effect on the O3-type structure of Li[Ni,Mn]O$_2$. It is expected to maintain such characteristics even if a higher charging voltage is applied.

LiCoO$_2$ LiCoO$_2$, with a *R3m* rhombohedral structure, is widely used as a cathode material in commercial lithium secondary batteries due to its suitability for mass production, despite its high cost. Depending on the heat treatment temperature, LiCoO$_2$ adopts two different structures. Layered LiCoO$_2$ is formed when synthesized at a temperature greater than 800 °C through solid-state reactions, while spinel Li$_2$Co$_2$O$_4$ is obtained at around 400 °C [5]. When synthesis occurs at low temperatures, electrochemical characteristics are degraded by defects within the crystal and

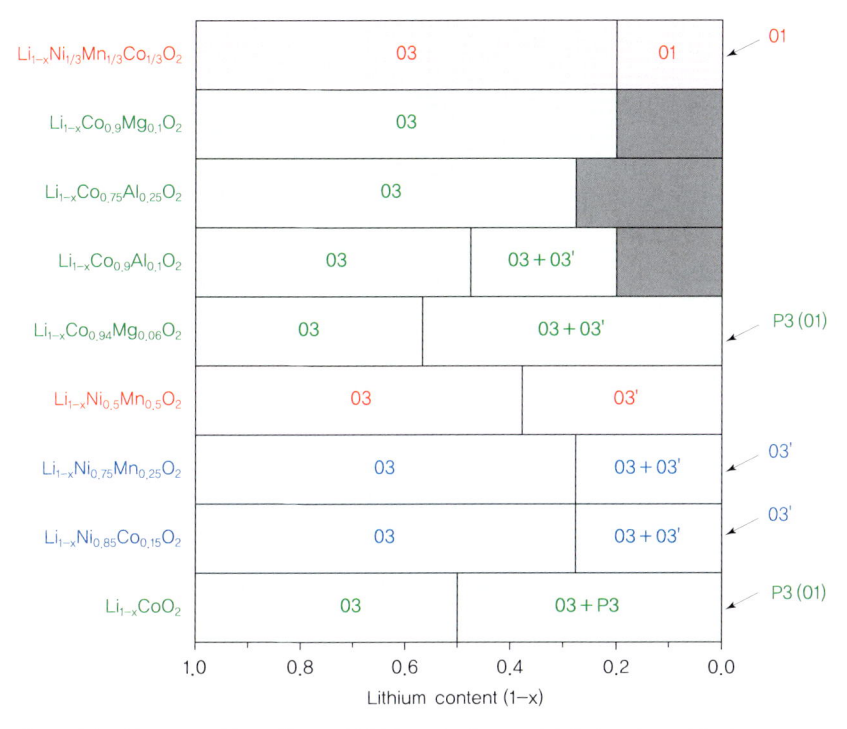

Figure 3.9 Phase transition of layered LiMO$_2$ cathode material during charge/discharge.

low crystallinity. As such, layered LiCoO$_2$ prepared at high temperatures is used as active cathode materials [6].

Taking into account the crystal field stabilization energy (CFSE) of Co ions, the relative energy levels of Co^{2+} ($t_{2g}^6 e_g^1$), Co^{3+} (t_{2g}^6), and Co^{4+} (t_{2g}^5) are $-18Dq + P$, $-24Dq + 2P$, and $-20Dq + 2P$, respectively. If the pairing energy p is relatively small, Co^{2+} ($t_{2g}^6 e_g^1$) can be the most stable. However, for Co^{2+} to exist, nonstoichiometric Li$_{1-x}$Co$_{1+x}$O$_2$ must be formed to obtain charge neutrality. Considering the ionic radius, $r(Co^{2+})/r(O^{2-}) = 0.627$ is within the octahedral range of 0.414–0.732, but $r(Co^{2+})/r(Li^+) = 0.878$ gives 12.2%, which is close to the limit of 15% according to the Hume–Rothery rules. The synthesis of stoichiometric LiCoO$_2$ is thus more stable.

Figure 3.11 illustrates the charge/discharge curves of the LiCoO$_2$ cathode material with lithium as the anode. A high voltage of approximately 4 V is shown, and the discharge voltage does not vary greatly over time. There is high electrical conductivity due to the direct Co–Co interaction of partially filled t_{2g}^{6-x} orbitals in Co$^{3+}/^{4+}$. Irreversible phase transitions are unlikely to occur during charge/discharge because of low spin and stabilized Co^{4+} in octahedral sites.

However, for $x > 0.5$ in Li$_{1-x}$CoO$_2$, phase transitions may be irreversible if the mixed layers of O3 and P3 assume a monoclinic structure. Less than 50% of lithium ions in LiCoO$_2$ undergo reversible intercalation–deintercalation. In CoO$_2$, where

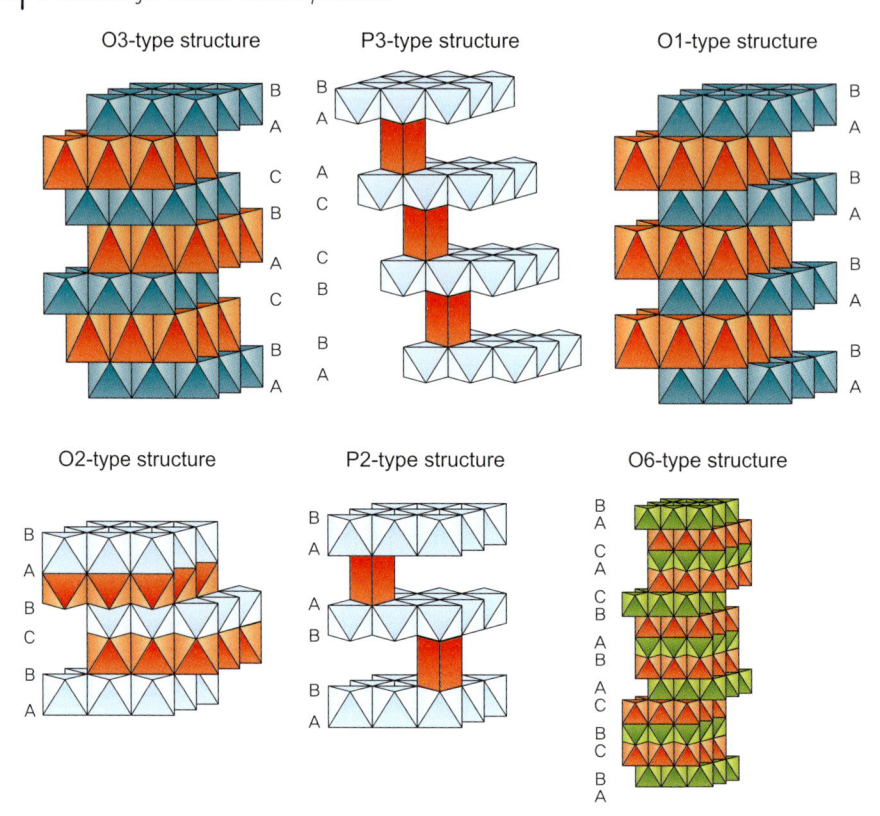

Figure 3.10 Possible structures of layered LiMO$_2$ during phase transition.

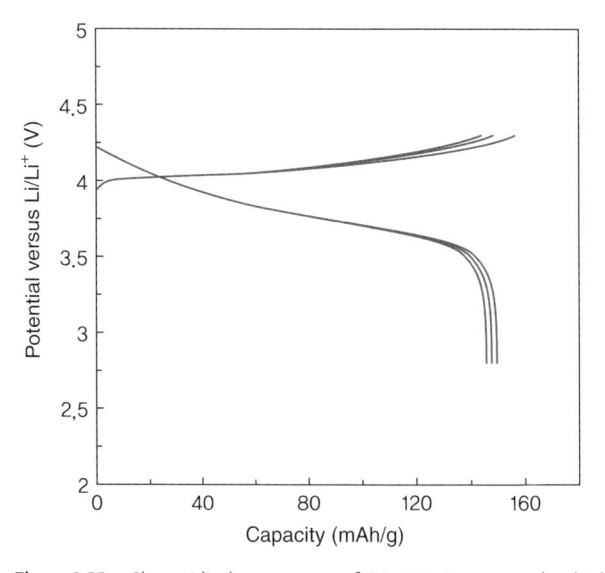

Figure 3.11 Charge/discharge curves of O3–LiCoO$_2$ prepared at high temperatures.

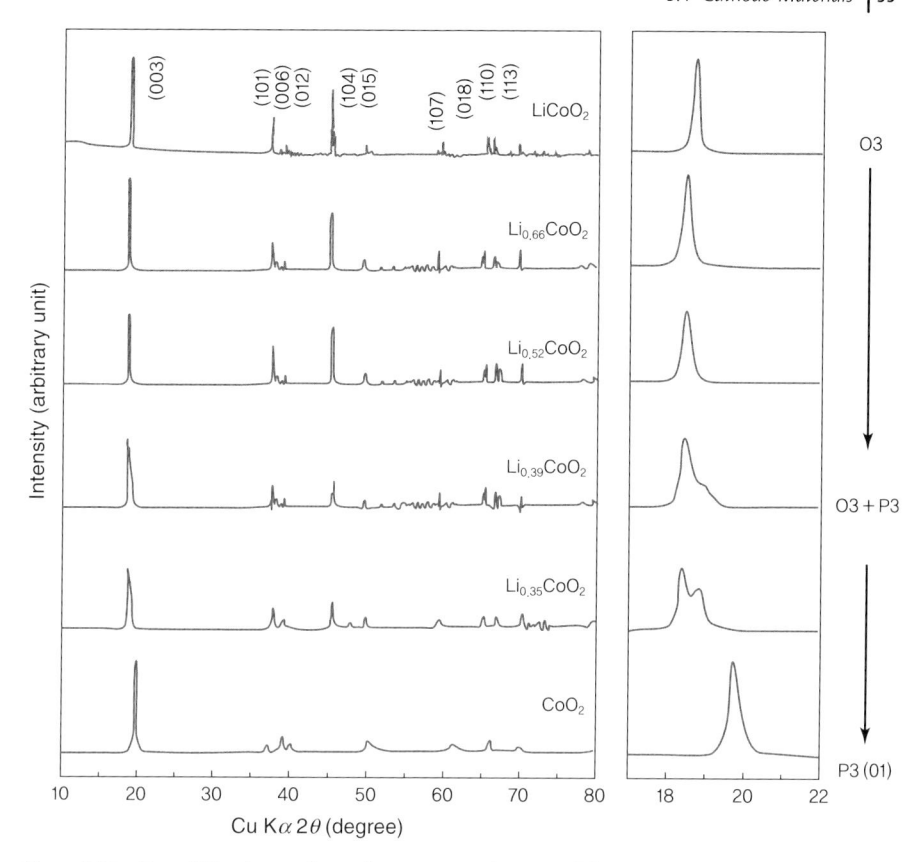

Figure 3.12 X-ray diffraction analysis of $Li_{1-x}CoO_2$ with varying lithium content [7].

lithium ions are completely deintercalated, a hexagonal close packed O1 layered structure is obtained with irreversible phase transitions. The theoretical capacity of $LiCoO_2$ is 274 mAh/g, but its actual capacity of 145 mAh/g is slightly higher than 50% of the theoretical capacity at 137 mAh/g. Figure 3.12 presents the results of X-ray diffraction, indicating the structural change in $Li_{1-x}CoO_2$ during the charge/discharge process. Upon charging, CoO_2 transforms into a completely different structure as the small peak to the right of peak (003) becomes larger.

During charging, phase transition of the $Li_{1-x}CoO_2$ structure takes place as O3 → [O3 + P3] → P3(O1). Since the P3(O1) structure consists of AABBCC and ABABAB oxygen arrangements along the c-axis, Co—O bonds are maintained, as shown in Figure 3.13, while the CoO_2 layer slides into a new structure.

With overcharging to about 4.5 V, the increased oxidation number of Co causes a decrease in the O—Co—O bond length and a 0.3 % reduction in the lattice constant a of the x–y plane. However, the overall volume expands because the lattice constant c increases by more than 2% due to the repulsion between CoO_2 layers. Note that as x becomes closer to 1, the lattice constant c drops rapidly.

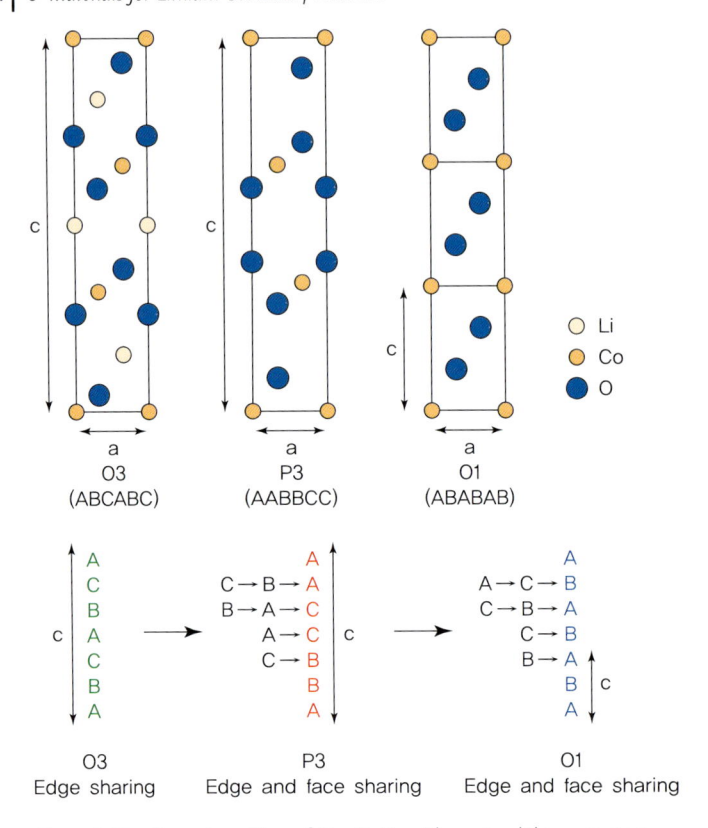

Figure 3.13 Phase transition of $Li_{1-x}CoO_2$ with varying lithium content.

For $x > 0.72$, reversible capacity is reduced as charge is compensated by the release of oxygen within the lattice instead of oxidized Co. At temperatures above 50 °C, the structure collapses as oxygen is released in greater amounts. Also, the disintegrated electrolyte from side reactions with the cathode material produces gases, expanding the battery and making it unsafe.

$LiNi_{1-x}Co_xO_2$ O3-$LiNiO_2$ has a R-$3m$ rhombohedral structure with $a = 2.887$ Å, $c = 14.227$ Å, $c/a = 4.928$, and a unit cell volume of ($V = \sqrt{3/2}a^2c$) 102.69 Å^3. The layered $LiNiO_2$ has been studied as an alternative cathode material to $LiCoO_2$, as it is 20% larger in actual capacity. Due to the stable Ni^{2+} in $LiNiO_2$, nonstoichiometric $Li_{1-y}Ni_{1+y}O_2$ is often synthesized. This is because some of the Ni^{3+} octahedral sites as well as lithium ion sites are replaced with Ni^{2+} to preserve charge neutrality. Ni^2 is easily formed as low-spin Ni^{3+} and becomes relatively unstable with unpaired spin in octahedral sites. As such, the synthesis of $LiNiO_2$ usually involves oxidation.

Looking at the crystal field stabilization energy of Ni ions, the relative energy levels of Ni^{4+} (t_{2g}^6), Ni^{3+} ($t_{2g}^6 e_{g1}$), and Ni^{2+} ($t_{2g}^6 e_g^2$) are $-24Dq + 2P$, $-18Dq + P$, and $-12Dq$, respectively. Assuming that the pairing energy P is relatively small, the electron arrangement of Ni^{2+} can be the most stable. Similar to Co, when considering the

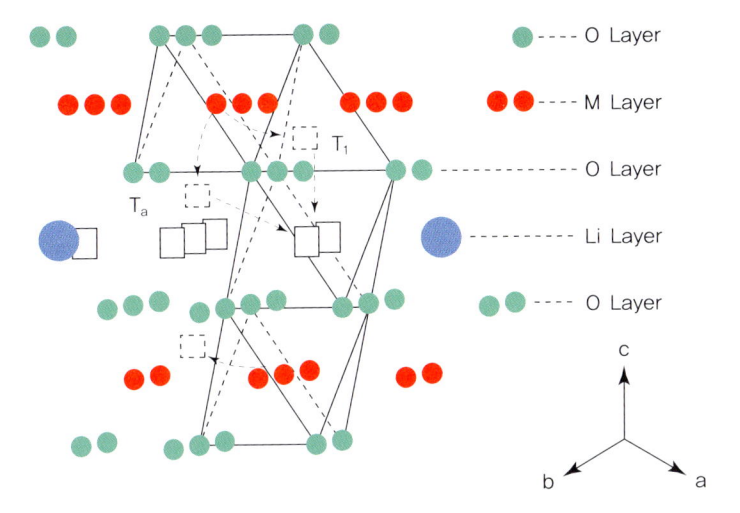

Figure 3.14 Mimetic diagram of cation migration path.

ionic radius, $r(Ni^{2+})/r(O^{2-}) = 0.659$ is within the octahedral range of 0.414–0.732, but $r(Ni^{2+})/r(Li^{+}) = 0.922$ gives a radii difference of 7–8% according to the Hume–Rothery rules. Compared to Co^{2+}, the radius of Ni^{2+} is closer to that of lithium ions. Ni^{2+} is thus readily found in lithium ion layers and enables the synthesis of non-stoichiometric $Li_{1-x}Ni_{1+x}O_2$.

Figure 3.14 shows the ion migration path between transition metals and lithium when Ni^{2+} substitutes lithium ions. Between the cubic close packed oxygen atoms, transition metals and lithium each occupy $3b$ and $3a$ octahedral sites. All tetrahedral sites are left empty. Transition metals in $3b$ octahedral sites move to the closest T_1 tetrahedral sites, and then migrate to $3a$ octahedral sites located farther away. Another path is to first move to T_2 tetrahedral sites and subsequently to $3a$ tetrahedral sites. The two paths have the same total distance.

In nonstoichiometric $Li_{1-x}Ni_{1+x}O_2$, NiO_2 layers form a local three-dimensional structure since lithium is replaced by Ni^{2+}. This hinders the diffusion of lithium ions and leads to reduced efficiency. The battery capacity also decreases due to irreversible phase transitions during the charge/discharge process. Stoichiometric $LiNiO_2$ can be synthesized through hydrothermal methods or redox ion exchange, but has low reproducibility. In addition, bonding increases in the z-axis with the Jahn–Teller effect because of the low-spin Ni^{3+} (d^7) electron arrangement. The Jahn–Teller effect is demonstrated when d-orbitals are split into t_{2g} and e_g due to CFSE in octahedral MO_6, increasing or decreasing the bond distance with metals or oxygen in the z-axis. The crystal gains additional stabilization energy as t_{2g} gives d_{xy}, d_{yz} and d_{xy}, while e_g splits into d_{z2} and d_{x2-y2}. In this case, electrical conductivity is lowered from repeated expansion and contraction in the z-axis, causing electrode characteristics to deteriorate.

Figure 3.15 is the cyclic voltammetry of $LiNiO_2$. For $0 \leq x \leq 0.75$ of $Li_{1-x}NiO_2$, battery characteristics are negligibly affected by the three irreversible phase transitions. When charged up to 4.2 V, the actual capacity of 160 mAh/g is greater than that

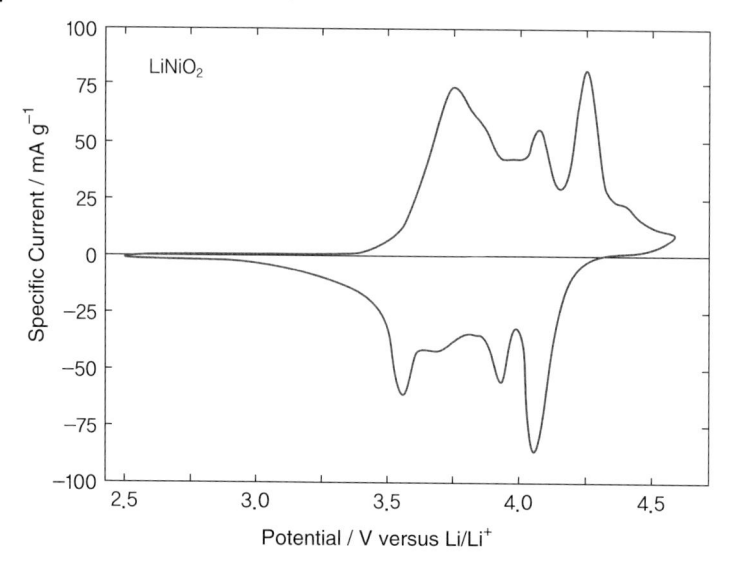

Figure 3.15 Cyclic voltammetry of $LiNiO_2$.

of $LiCoO_2$. Phase transitions occur into rhombohedral for $0 < x < 0.25$, monoclinic for $0.25 < x < 0.55$, and rhombohedral for $0.55 < x < 0.75$. Upon charging above 4.2 V with $x > 0.75$, battery characteristics are reduced, as lithium ions cannot be intercalated in the resulting NiO_2.

Despite $LiNiO_2$ having a larger actual capacity than $LiCoO_2$, it is not used as a cathode material. First, in nonstoichiometric $Li_{1-y}Ni_{1+y}O_2$, the diffusion of lithium ions is hindered by Ni ions at the lithium ion layer, leading to a decrease in capacity. Second, when the amount of lithium falls with charging, structural instability causes oxidized substances to break up and increases oxygen partial pressure. The battery becomes unsafe from chemical reactions with the organic electrolyte. Instead of $LiNiO_2$, $LiNi_{1-x}Co_xO_2$ or $LiNi_{1-x}Co_xO_2$ are used since nickel ions are partially replaced with transition metal ions [4]. When another M^{3+} ion with a stable oxidation number is added, it is difficult for Ni^{2+} to substitute lithium ions to maintain charge neutrality. The crystal structure is stabilized as Co ions prevent Ni^{2+} from taking the place of lithium, and Al, a nonactive material, prevents the intercalation/deintercalation of lithium ions. The production of Ni^{2+} can be inhibited by compensating for lithium volatilization through reactions with massive amounts of lithium salt during synthesis at high temperatures.

Because transition metal ions have different redox potentials, the electrode material may deteriorate during charge/discharge if compounding is nonuniform. Instead of solid-state reactions, uniformity can be achieved through other methods of synthesis such as coprecipitation sintering, coprecipitation molten salt, sol–gel, and spray pyrolysis.

In $LiNi_{1-x}Co_xO_2$, the lattice constant decreases with an increase in the amount of substituted Co, and thus the reduced volume raises the volumetric specific capacity.

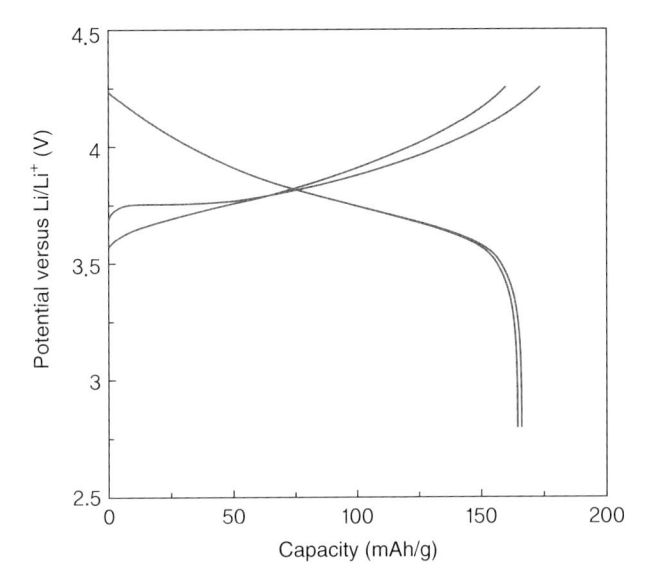

Figure 3.16 Charge/discharge curves of LiNi$_{0.85}$Co$_{0.15}$O$_2$.

Figure 3.16 shows the charge/discharge curves of LiNi$_{0.85}$Co$_{0.15}$O$_2$. LiNi$_{0.85}$Co$_{0.15}$O$_2$ has a theoretical capacity of 274 mAh/g, which is 174 mAh/g higher than its actual capacity. Due to Co substitution, the Jahn–Teller effect is weakened and fewer Ni^{2+} ions (or formation of NiO) are exchanged. This results in superior electrochemical properties, as phase transitions are suppressed during charge/discharge of LiNiO$_2$.

During charging in Li$_{1-x}$Ni$_{1-y}$Co$_y$O$_2$, the O3 layered structure is maintained with $0 \leq x \leq 0.70$, and phase transition to O3 and P3 layered structures occurs with $0.70 < x < 1$. In Ni$_{1-y}$Co$_y$O$_2$, where lithium ions are completely deintercalated, a new O3 layered structure is formed. Figure 3.17 presents the X-ray diffraction results, which can be seen as the effect of varying lithium content on the structure of Li$_{1-x}$Ni$_{0.85}$Co$_{0.15}$O$_2$.

Li$_{1-x}$CoO$_2$ transforms into O3- and P3-type structures with $0.50 < x < 1$, but the same occurs for Li$_{1-x}$Ni$_{1-y}$Co$_y$O$_2$ when $x = 0.70$. When charged, the theoretical capacity of Li$_{1-x}$Ni$_{0.85}$Co$_{0.15}$O$_2$ is ~233 mAh/g since Ni^{3+} is oxidized but not Co^{3+}. Its actual capacity is 163 mAh/g if phase transitions are taken into account. However, the capacity becomes 174 mAh/g when O3 and P3 layered structures are involved in irreversible charging/discharging. Low-spin Ni^{3+} (d^7) is oxidized into Ni^{4+} as charging proceeds, and the Jahn–Teller distorted NiO$_6$ octahedron turns into a symmetric form.

LiMO$_2$ (M = Mn, Fe) Synthesis of layered LiMnO$_2$ is known to be difficult as it is thermodynamically metastable, but it is possible through the exchange of lithium ions and sodium ions in the α-NaMnO$_2$ precursor [8]. Instead of a well-developed stoichiometric LiMnO$_2$ layered structure, nonstoichiometric oxides with 3–10% of manganese at lithium sites are obtained. Li$_{1-x}$Mn$_{1+x}$O$_2$, in which lithium is

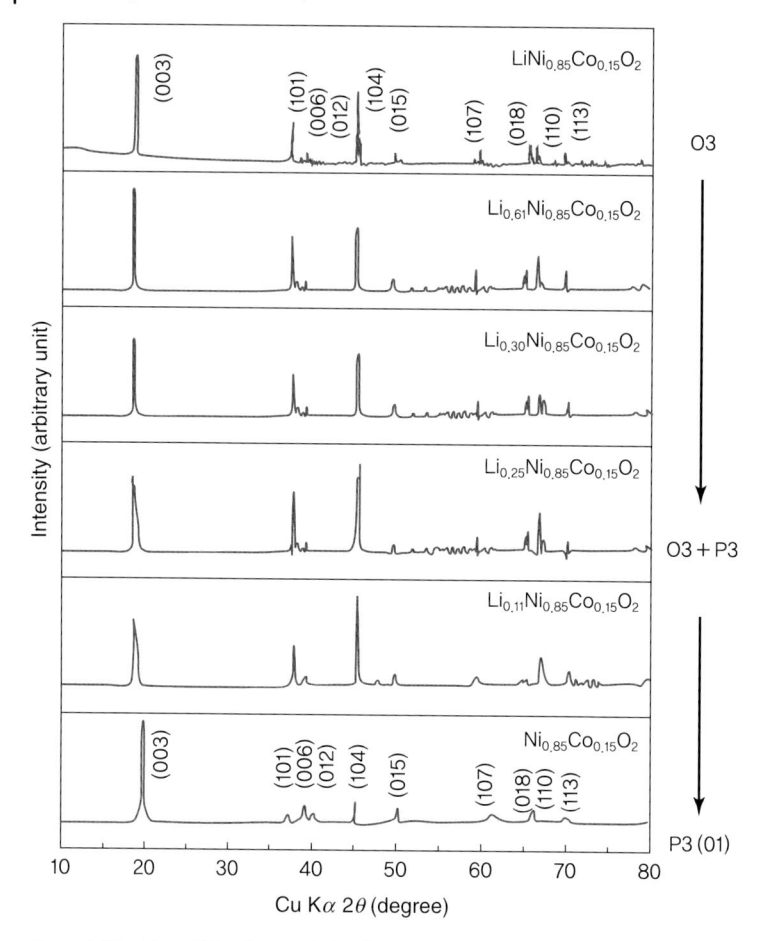

Figure 3.17 X-ray diffraction analysis of $Li_{1-x}Ni_{0.85}Co_{0.15}O_2$ with varying lithium content.

substituted with 3% manganese, transforms from a layered to spinel-like structure, as shown in Figure 3.18, and exhibits a flat voltage at 4 V and 3 V. For the layered $LiMnO_2$, although lithium ions can be 100% deintercalated, an irreversible phase transition into a spinel structure occurs during charge/discharge. Despite this transformation, an initial increase in capacity is observed, followed by a decrease over the cycle [9].

The orthorhombic $LiMnO_2$, with a rock salt structure, has a high capacity of 200 mAh/g when the initial charge capacity is in the range of 2.0–4.5 V. Since the structure of $Li_{1-x}MnO_2$ formed from lithium ion deintercalation is very unstable, the capacity stabilizes with the phase transition into spinel $Li_{1-x}Mn_2O_4$ (Figure 3.18). Surface modification and improved synthesis methods have been attempted in efforts to enhance the electrochemical performance of $LiMnO_2$ [10, 11].

While numerous studies have been carried out on layered $LiFeO_2$ prepared from ion exchange of α-$NaFeO_2$, it transforms into a spinel during the initial charging

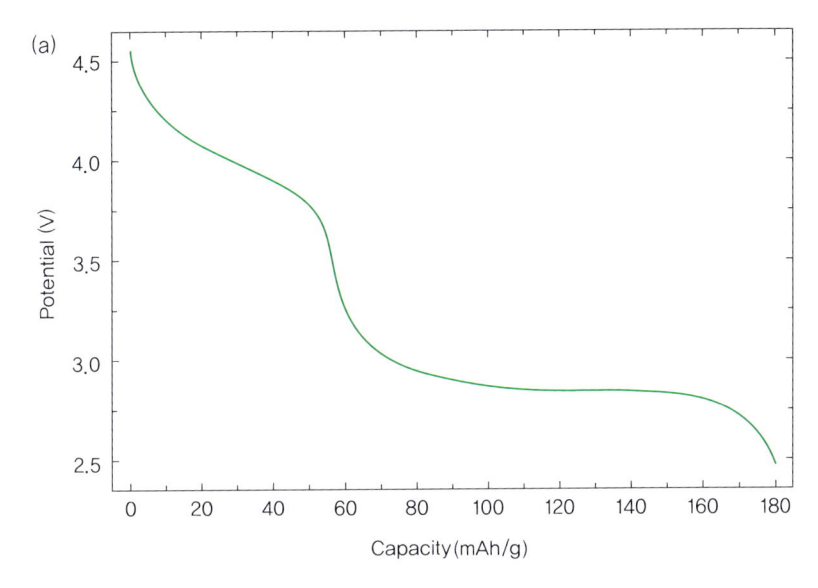

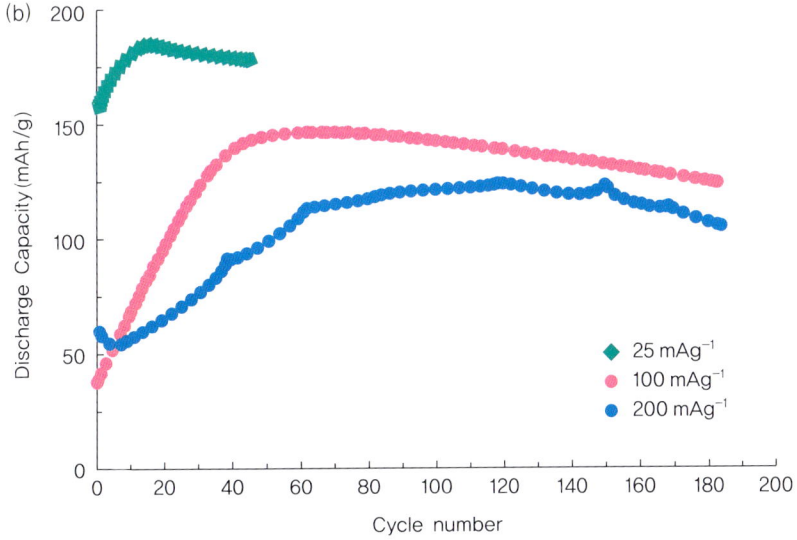

Figure 3.18 (a) 50th discharge curve (25 mA/g) Adapted with permission from American Chemical Society Copyright 2002 and (b) cycle characteristics of $Na_{0.069}Li_{0.59}Mn_{0.94}O_2$ produced from ion exchange carried out at 80 °C of $Na_{0.51}Mn_{0.90}O_2$ synthesized at 600 °C [9]. Adapted with permission from [9] Copyright 2002 American Chemical Society.

stage due to its metastable structure [12]. First, from the general relationship between the structure and the relative size of cations, $LiCoO_2$ and $LiNiO_2$ satisfy the stabilization condition of $r_B/r_A < 0.8$ with a radius ratio of 0.76 and 0.78, respectively, whereas $LiFeO_2$ exceeds the limit with $rFe^{3+}/rLi^+ = 0.87$. Second, the high-spin Fe^{3+} moves easily to tetrahedral sites, as it has zero crystal field stabilization energy. Variants in the form of corrugulated-, goethite-, and tunnel-, and iron

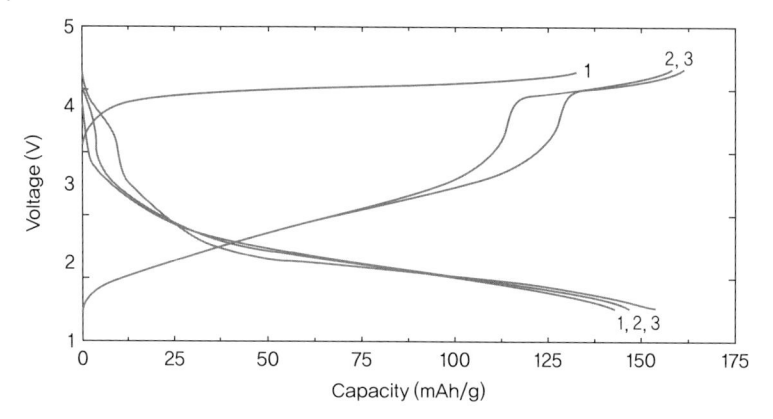

Figure 3.19 Charge/discharge curves of Li/LiFeO$_2$.

compounds such as FePS$_3$, FeOCl, and FeOOH show poor electrochemical performance [13]. Compared to LiCoO$_2$, the average voltage of iron-containing materials is either too high or low (Figure 3.19) depending on the redox pair of Fe^{3+}/Fe^{2+} or Fe^{4+}/Fe^{3+}. This is due to the relatively large mutual interaction between iron and electrons in iron compounds containing oxygen anions. Also, the Fermi potential of lithium differs greatly from Fe^{4+}/Fe^{3+} (e$_g$: 3d^{5}σ*), but not as much with Fe^{3+}/Fe^{2+} (e$_g$:3d^{5}π*).

Ni-Co-Mn Three-Component System The three-component system Li[Ni$_x$Co$_{1-2x}$Mn$_x$]O$_2$ exhibits outstanding electrochemical properties by combining the high capacity of LiNiO$_2$ and the thermal stability and low cost of manganese in LiMnO$_2$, and stable electrochemical characteristics of LiCoO$_2$. According to the first-principles calculation, LiNi$_{1/3}$Co$_{1/3}$Mn$_{1/3}$O$_2$ consisting of low-spin Co^{3+}, Ni^{2+}, and Mn^{4+} is more stable than a 1:1:1 mixture of LiNi^{3+}O$_2$, LiCo^{3+}O$_2$, and LiMn^{3+}O$_2$, and is characterized by a space group of symmetry $P3_112$. This implies that LiNi$_{1/3}$Co$_{1/3}$Mn$_{1/3}$O$_2$ can be synthesized using appropriate methods. In the case of low-spin mixtures, the lattice constant a at 4.904 Å is $\sqrt{3}$ times larger than that of LiMO$_2$ due to the regularization of transition metals, as shown in Figure 3.20, and manganese raises the constant c to 13.884 Å [14].

In general, a solid solution is formed between LiCoO$_2$ and LiNiO$_2$, but not with LiMnO$_2$. This is also observed between LiNiO$_2$ and LiMnO$_2$. When preparing LiNi$_{1/3}$Co$_{1/3}$Mn$_{1/3}$O$_2$, a regular mixture of transition metals is more important than in the synthesis of LiNi$_{1-x}$Co$_x$O$_2$. As such, the coprecipitation method is widely used. When a precursor is obtained from hydroxide coprecipitation, MnOOH or MnO$_2$ is oxidized even with Mn(OH)$_2$ precipitation. The Mn^{2+} state of the precipitate can be preserved by sintering via carbonate coprecipitation. However, battery characteristics may be limited, as impurities from NiO and Li$_2$MnO$_3$ cannot be completely removed.

If charge neutrality of the compound is preserved, nickel and manganese ions favor the electron arrangement of Ni^{2+} and Mn^{4+} over Ni^{3+} and Mn^{3+}. In LiNi$_{1/3}$Co$_{1/3}$Mn$_{1/3}$O$_2$, the oxidation number is 2+ for nickel, 3+ for cobalt, and 4+ for

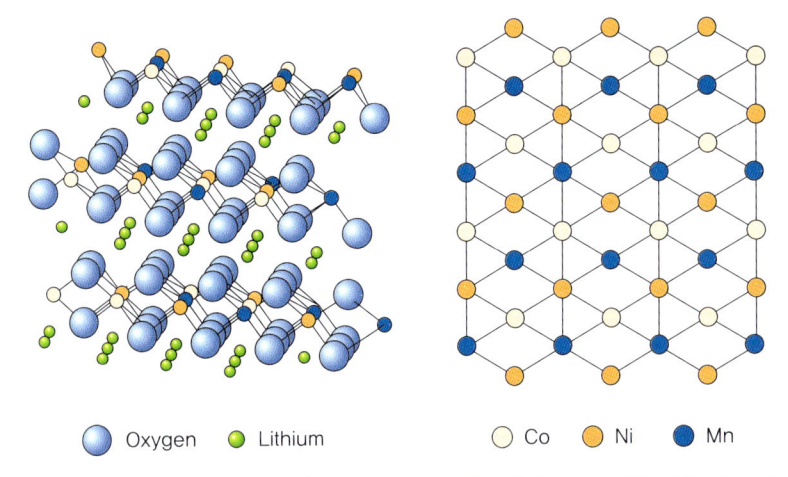

Figure 3.20 Mimetic diagram of LiNi$_{1/3}$Co$_{1/3}$Mn$_{1/3}$O$_2$ with the superlattice [Ni$_{1/3}$Co$_{1/3}$Mn$_{1/3}$O$_2$] layer in $[\sqrt{3} \times \sqrt{3}]$ R30° form [14]. Reproduced by permission of ECS – The Electrochemical Society.

manganese. Ni^{2+} is usually involved in charge/discharge while Co^{3+} becomes active toward the end of the charging process. On the other hand, Mn^{4+} does not participate in charging/discharging but contributes to overall structural stability with crystal field stabilization energy at octahedral sites.

Li$_{1-x}$Ni$_{1/3}$Co$_{1/3}$Mn$_{1/3}$O$_2$ undergoes structural change with charge/discharge such that c increases with reduced lithium content, but decreases when $(1-x) < 0.35$ due to the presence of oxygen. However, the opposite is observed for changes in the lattice constant a [15]. Ultimately, it undergoes minimal volume change and offers considerable advantages as an active material. Li$_{1-x}$CoO$_2$ has a P3 structure for $x=1$, whereas Li$_{1-x}$Ni$_{1/3}$Co$_{1/3}$Mn$_{1/3}$O$_2$ assumes an O1 structure for $x=1$, and the same O3 structure in the range of $0 \leq x \leq 0.8$, indicating that a stable structure is maintained throughout charging. Its superior charge/discharge characteristics produce batteries with long life and enhanced safety. Figure 3.21 illustrates the discharge curves of the LiNi$_{1/3}$Co$_{1/3}$Mn$_{1/3}$O$_2$ cathode material with current density.

While this material has a similar capacity to existing LiCoO$_2$, it is a promising alternative in terms of performance, safety, and cost. Recently, various material combinations are being developed to expand capacity by increasing the amount of Ni while retaining its advantages.

Ni-Mn System LiNi$_{1/2}$Mn$_{1/2}$O$_2$ has lattice constants of $a = 2.889$ Å, $c = 14.208$ Å, and $c/a = 4.918$, and a unit cell volume of 102.697 Å^3. Similar to LiNi$_{1/3}$Co$_{1/3}$Mn$_{1/3}$O$_2$, LiNi$_{1/2}$Mn$_{1/2}$O$_2$ contains nickel and manganese in a ratio of 1:1, and thus battery characteristics are lost with impurities such as NiO and Li$_2$MnO$_3$. The theoretical capacity of LiNi$_{1/2}$Mn$_{1/2}$O$_2$ is 280 mAh/g. Figure 3.22 shows the charge/discharge curves of LiNi$_{1/2}$Mn$_{1/2}$O$_2$, while Figure 3.23 displays the X-ray diffraction analysis results of Li$_{1-x}$Ni$_{1/2}$Mn$_{1/2}$O$_2$ with varying lithium content. Li$_{1-x}$Ni$_{1/2}$Mn$_{1/2}$O$_2$ maintains a uniform O3 structure in the range of $0 \leq x \leq 1$ and has an actual capacity of 260 mAh/g (60 °C).

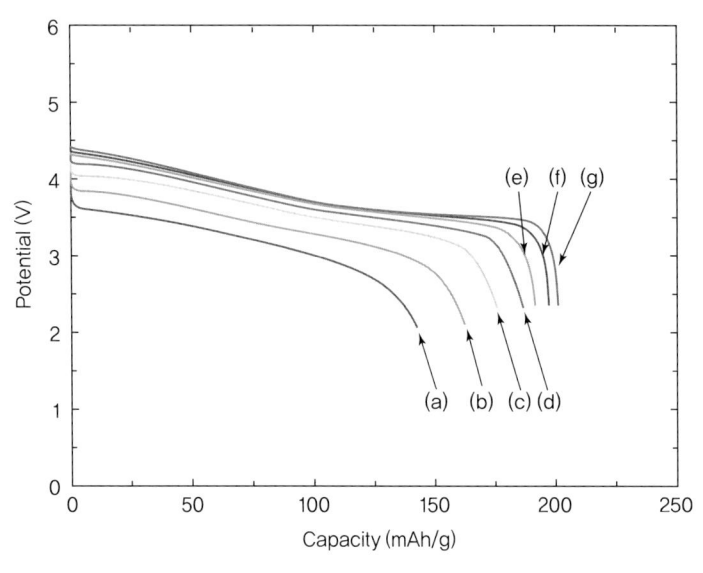

Figure 3.21 Rate-limiting characteristics of $LiNi_{1/3}Co_{1/3}Mn_{1/3}O_2$ charged until 4.6 V. (a) 2400 mA/g, (b) 1600, (c) 800, (d) 400, (e) 200, (f) 100, and (g) 50. Active material 88%; acetylene black 6%; PVdF 6%.

The $LiNiO_2$–$LiMnO_2$ solid solution is fundamentally different from $LiCoO_2$–$LiMnO_2$ and $LiCrO_2$–$LiMnO_2$ solid solutions. While some Ni^{2+} and Mn^{4+} are formed in the $LiNiO_2$–$LiMnO_2$ solid solution, Co^{2+}/Mn^{4+} is not readily found in $LiCoO_2$–$LiMnO_2$ [16]. In addition, $LiCrO_2$–$LiMnO_2$ assumes a monoclinic crystal

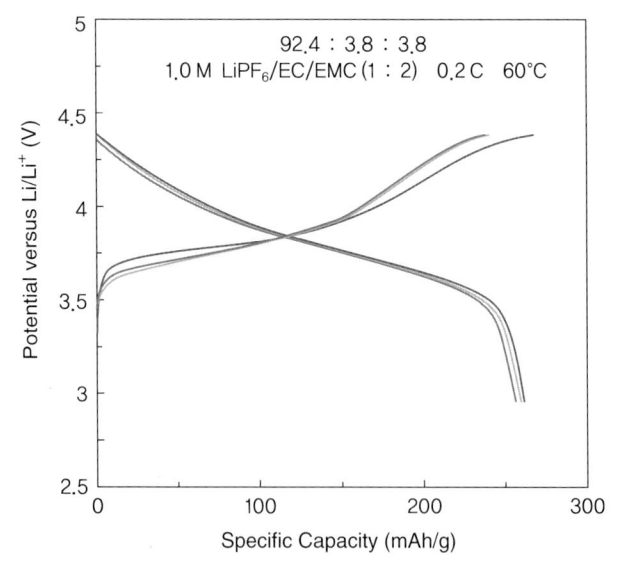

Figure 3.22 Charge/discharge curves of $LiNi_{1/2}Mn_{1/2}O_2$.

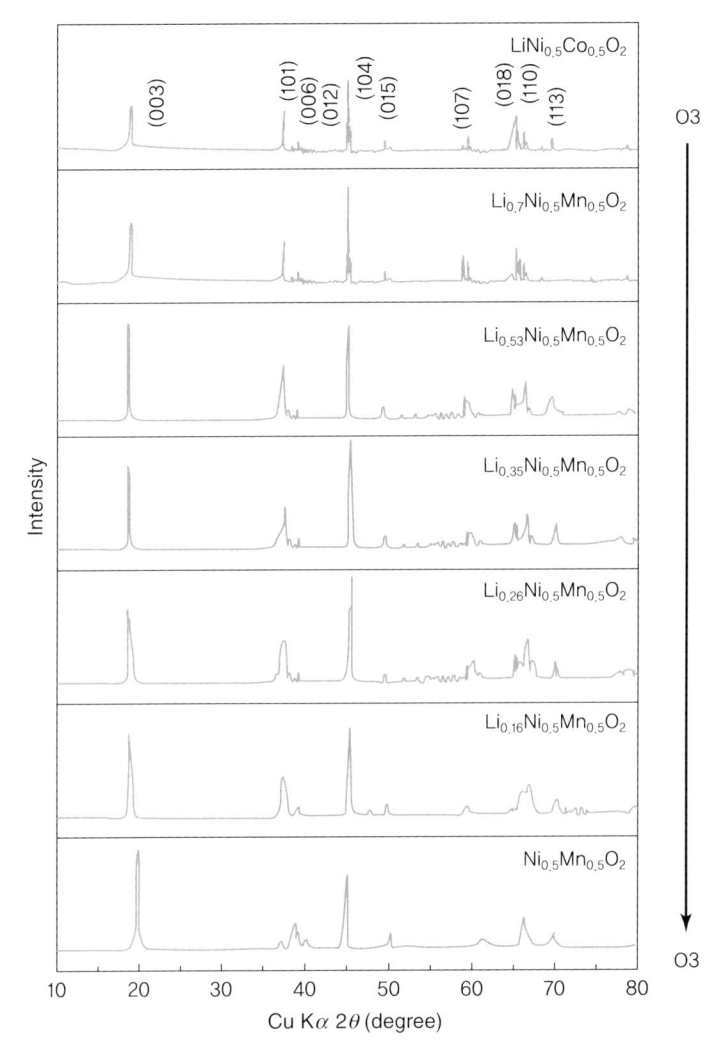

Figure 3.23 X-ray diffraction analysis results of $Li_{1-x}Ni_{1/2}Mn_{1/2}O_2$ with varying lithium content.

structure with oxidized Mn^{3+} when the amount of $LiMnO_2$ exceeds 30% due to the stability of Cr^{3+}.

In the $LiNiO_2$–$LiMnO_2$ solid solution, there is no layered-to-spinel phase transition because Mn^{4+} is not involved in redox reactions during charge/discharge. While charge/discharge progresses with redox reactions of $Ni^{2+/4+}$, the reversible capacity of $Li[Ni_{1/2}Mn_{1/2}]O_2$ is similar to the working voltage of $Ni^{3+/4+}$ in $LiNiO_2$. Based on theoretical calculations, changes in Mn–Ni interaction when $2+/4+$ becomes $4+/4+$ can be traced to the high working voltage of $Li[Ni_{1/2}Mn_{1/2}]O_2$. However, the synthesis of $Li[Ni_{1/2}Mn_{1/2}]O_2$ is complicated by impurities of the spinel structure and the mixture of Ni^{2+} ions in lithium layers. To overcome these weaknesses, an excess amount of lithium is incorporated into reactions.

Lithium-Rich Phases Since the development of $Li_xCr_yMn_{2-y}O_{4+z}$, there has been active research on the $(1-x)Li_2MnO_3-xLiMO_2$(M = Ni, Co, Cr) solid solution (or nanocomposite) [17]. Li_2MnO_3, which is electrochemically nonactive in the Mn^{4+} oxidation state, and $Li[Ni_xLi_{1/3-2x/3}Mn_{2/3-x/3}]O_2$, the solid solution of $Li[Ni_{1/2}Mn_{1/2}]O_2$, have a high capacity of 250 mAh/g. The $LiMO_2$–$LiMnO_2$ solid solution is transformed into a lithium stoichiometric phase, $LiMO_2$–Li_2MnO_3 into a lithium-saturated phase, and $LiMO_2$–$LiMnO_2$–Li_2MnO_3 into a lithium-rich phase [18]. The occupancy of cations at 3b sites is 1, and the sum of oxidation numbers is $+3$. In the existing manganese oxide, redox reactions do not occur for $Mn^{3+/4+}$. Because these cathode materials maintain the oxidation state of Mn^{4+}, the Jahn–Teller effect of Mn^{3+} is not as significant and electrode performance is unaffected.

Although the above oxides are known as solid solutions, TEM analysis on $Li[Cr_xLi_{(1-x)/3}Mn_{2(1-x)/3}]O_2$, while varying their composition, revealed characteristics of both solid solutions and composites (Figure 3.24). Once a critical composition is reached, they change from a solid solution to a composite. Upon charging up to 4.6–4.8 V, high charge and discharge capacities of 352 and 287 mAh/g are obtained, with a flat potential observed at around 4.5 V (Figure 3.25). This nonreversible capacity

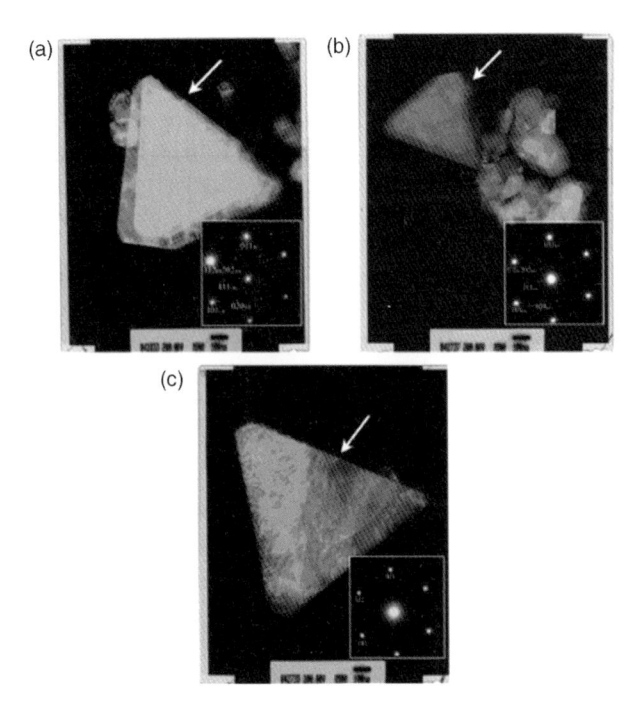

Figure 3.24 Diffraction patterns in an electron microscope: (a) $Li[Cr_{0.211}Li_{0.268}Mn_{0.520}]O_2$, (b) $Li[Cr_{0.290}Li_{0.240}Mn_{0.470}]O_2$, and (c) $Li[Cr_{0.338}Li_{0.225}Mn_{0.436}]O_2$ [21]. Reprinted from [21] Copyright 2007, with permission from Elsevier.

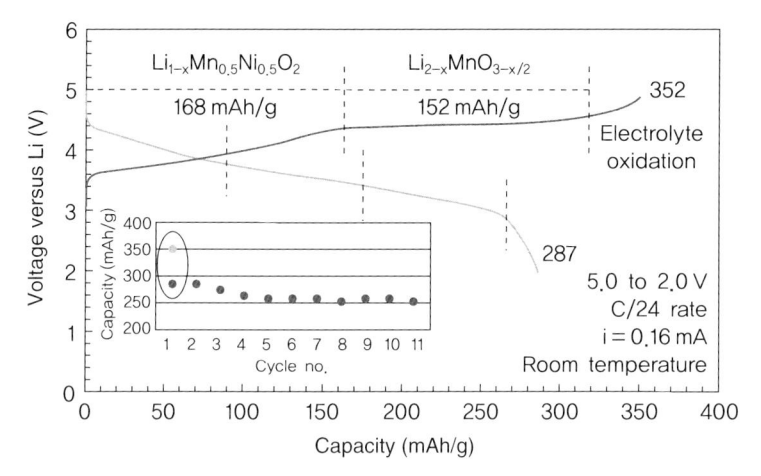

Figure 3.25 Initial charge/discharge curves of Li/0.3Li$_2$MnO$_3$·0.7LiMn$_{0.5}$Ni$_{0.5}$O$_2$ (5.0–2.0 V) [19]. Reprinted from [19] Copyright 2004, with permission from Elsevier.

loss is caused by the production of Li$_2$O from Li$_2$MnO$_3$ during charging. In the case of LiNi$_{0.20}$Li$_{0.20}$Mn$_{0.6}$O$_2$, Ni^{2+} is oxidized into Ni^{4+} and continued charging produces Li$_2$O, which leads to Ni$^{4+}{}_{0.20}$Li$_{0.20}$Mn$_{0.60}$O$_{1.7}$. During discharge, Mn$^{4+/3+}$ is usually reduced, but the cause remains unknown [19, 20].

According to recent studies [22], xLi$_2$MnO$_3$·$(1-x)$LiMO$_2$ takes on the structure of xLi$_2$MnO$_3$·$(1-x)$MO$_2$ with lithium deintercalation from LiMO$_2$ when charged until 4.4 V, and forms $(x-\delta)$Li$_2$MnO$_3$·δMnO$_2$·$(1-x)$MO$_2$ with Li$_2$O at above 4.4 V (Figure 3.26). At a voltage higher than 4.4 V, Li$_2$O and MnO$_2$ are both produced

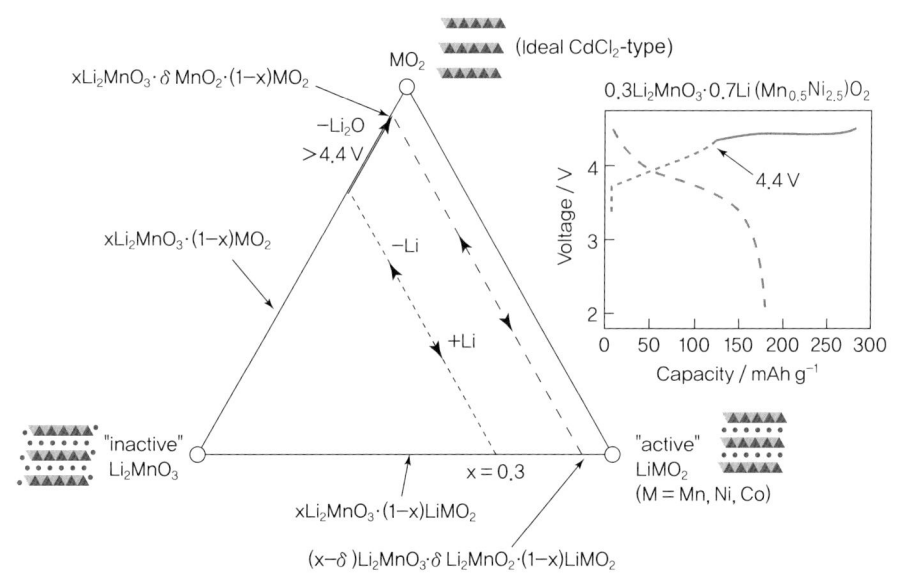

Figure 3.26 Compositional change and electrochemical reaction paths based on the three-component phase diagram of xLi$_2$MnO$_3$·$(1-x)$LiMO$_2$ [22]. Reprinted with permission from The Royal Chemical Society, Copyright 2007.

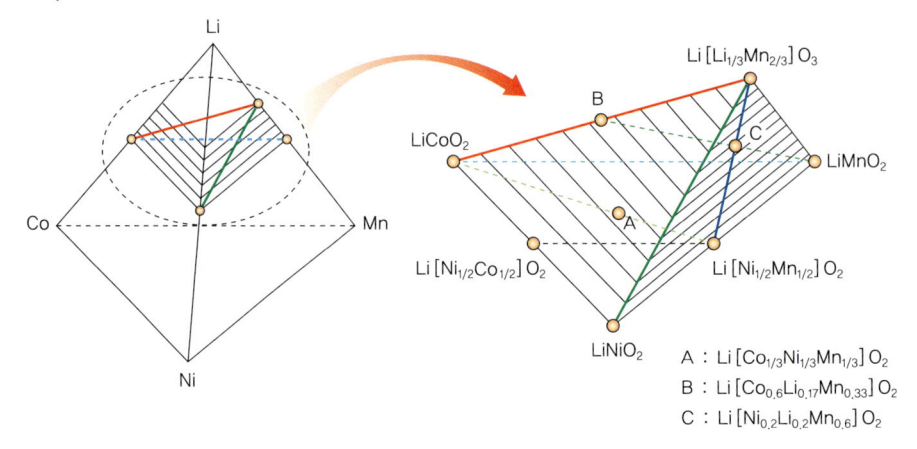

Figure 3.27 Phase diagram of the tetrahedral Li_2MnO_3–$Li[Ni_{1/2}Mn_{1/2}]O_2$–$LiNiO_2$–$LiCoO_2$ [24].

A : $Li[Co_{1/3}Ni_{1/3}Mn_{1/3}]O_2$
B : $Li[Co_{0.6}Li_{0.17}Mn_{0.33}]O_2$
C : $Li[Ni_{0.2}Li_{0.2}Mn_{0.6}]O_2$

due to lithium deintercalation in Li_2MnO_3 and increased oxygen. This process can be represented as $Li_2MnO_3 \rightarrow xLi_2O + yMnO_2 + (1 - xy)Li_2MnO_3$. Since the amount of Li_2MnO_3 is determined by the cutoff voltage, a higher cutoff voltage reduces the remaining portion. It is necessary to maintain structural stability through control of the electrochemical reactions of Li_2MnO_3 by setting an appropriate cutoff voltage. In addition, $(x - \delta)Li_2MnO_3 \cdot (x - \delta)Li_2MnO_3 \cdot \delta LiMnO_2 \cdot (1 - x)LiMnO_2$ is produced when discharged after initial charging.

Some recent studies have derived new cathode materials such as Li_2MnO_3–$Li[Ni_{1/2}Mn_{1/2}]O_2$–$LiCoO_2$ by adding $LiCoO_2$ to the Li_2MnO_3–$Li[Ni_{1/2}Mn_{1/2}]O_2$ solid solution for improved electrical conductivity [23].

Li_2MnO_3 is added to the $LiCoO_2$–$LiNiO_2$–$LiMnO_2$ oxide, which has an oxidation state of M^{3+}. This results in a solid solution oxide (Li_2MnO_3–$LiCoO_2$–$LiNiO_2$–$LiMnO_2$) as the cathode material assumes various oxidation states such as Ni^{2+} and Mn^{4+}. As shown in Figure 3.27, it is possible to synthesize several forms of active cathode materials with the addition of Li_2MnO_3 to $Li[Ni_{1/2}Mn_{1/2}]O_2$–$LiNiO_2$–$LiCoO_2$.

While the above oxides are more stable compared to $LiCoO_2$, they require a stable electrolyte above 4.6 V in order to be used as cathode materials. As shown in Figure 3.28, $LiNi_{0.20}Li_{0.20}Mn_{0.60}O_2$, with an excess of lithium ions, has a high capacity of 220 mAh/g, but cannot be utilized as it has a lower tap density than $LiCoO_2$. However, these oxides are highly practical materials for batteries that demand high safety and capacity. Recent studies have focused on substitution with Li_2TiO_3, which has an oxidation number of 4+ for manganese, based on consideration of various factors such as manufacturing cost, ease of synthesis, and toxicity [26, 27].

3.1.3.2 Spinel Composites

LiMn₂O₄ In the cubic spinel structure of LiM_2O_4 (M = Ti, V, Mn), oxygen is cubic close packed in an ABCABC sequence, as shown in Figure 3.29, and situated in the

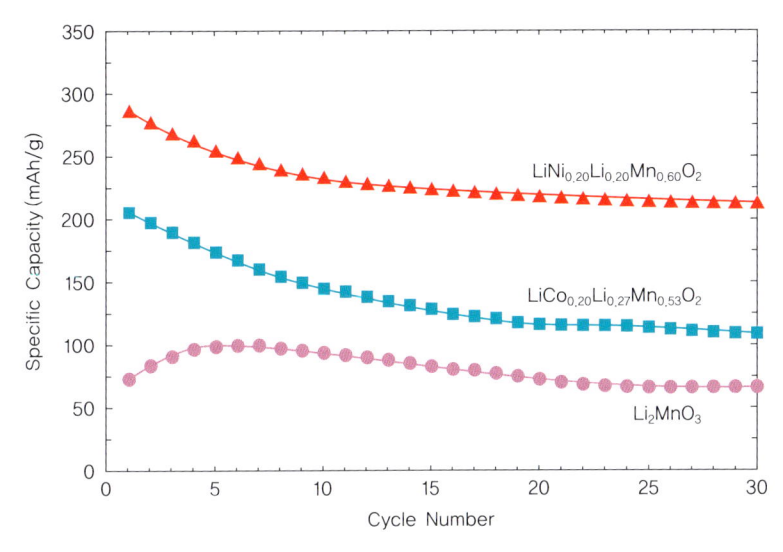

Figure 3.28 Discharge capacity (@20 mA/g, 2.0–4.8 V) of LiNi$_{0.20}$Li$_{0.20}$Mn$_{0.60}$O$_2$ and LiCo$_{0.20}$Li$_{0.27}$Mn$_{0.53}$O$_2$ [25]. Reprinted from [25] Copyright 2005, with permission from Elsevier.

32e site. LiMn$_2$O$_4$, with a lattice constant of $a = 8.245$ Å, is a representative spinel active material. A high capacity can be achieved with Mn as it is easily available, inexpensive, and environmentally benign.

However, the oxidation number of Mn may vary from 2+ to 4+, depending on the composition of reactants, heat treatment temperature, and conditions for synthesis. Because Mn can be located at tetrahedral or octahedral sites in the spinel, the synthesized oxides have various compositions and electrochemical properties are also affected. As such, the synthesis of LiMn$_2$O$_4$ is more complicated than in the cases

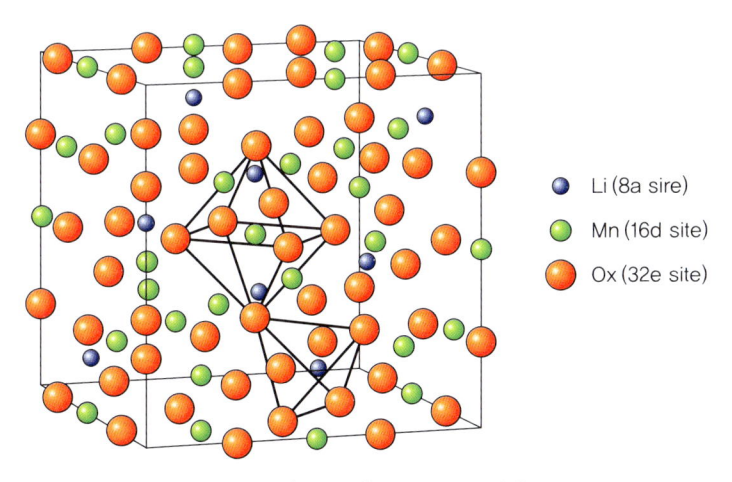

Figure 3.29 Lattice points in the spinel (Li)$_{8a}$[Mn$_2$]$_{16d}$[O$_4$]$_{32e}$.

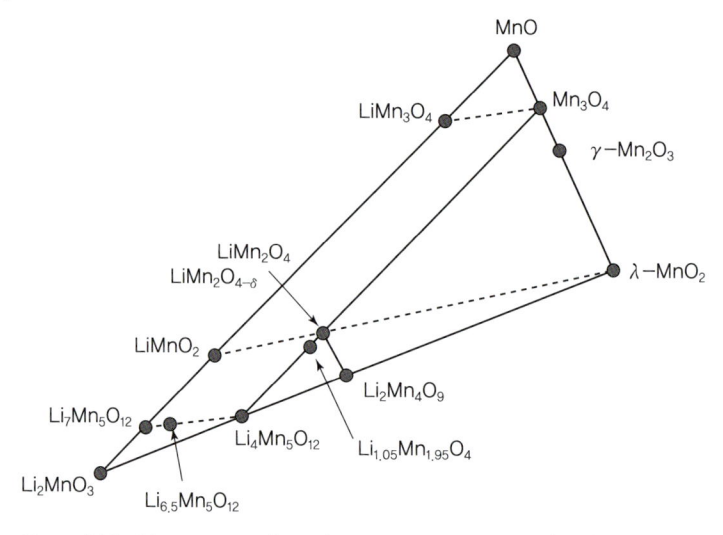

Figure 3.30 Manganese oxides with varying composition and oxidation number.

of other cathode materials and involves various phase transitions. Figure 3.30 shows manganese oxides with different compositions and oxidation numbers.

The arrangement of transition metals and lithium ions, which occupy the empty sites between oxygen in the spinel LiM_2O_4, is determined by electrostatic attraction, repulsion, and ionic radius. 3d transition metals and oxygen ions with an oxidation number of 3+ or 4+ have an ionic radius ratio of $0.476 \leq r(M^{3+})/R(O^{2-}) \leq 0.702$ and $0.492 \leq r(M^{4+})/R(O^{2-}) \leq 0.591$, and are thus located at $16d$ octahedral sites. Similarly, all transition metal ions of layered and spinel cathode materials occupy octahedral sites. The six neighboring MO_6 around a single MO_6 are arranged two-dimensionally in a layered structure, whereas the same arrangement is three-dimensional for a spinel (Figure 3.31). This leads to a difference in the oxidation number of transition metal ions. Because of the difference in charge distribution and M−O bond

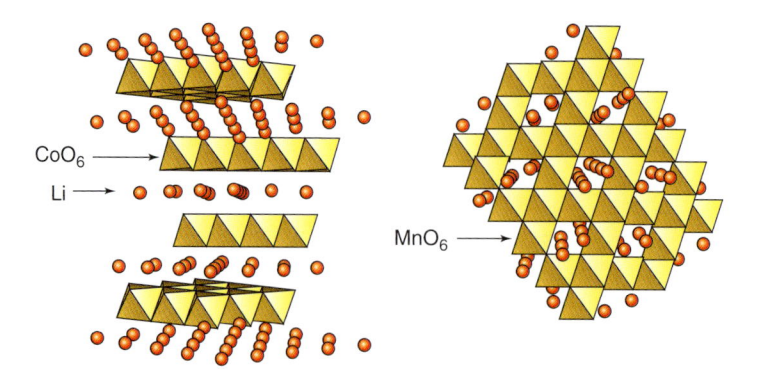

Figure 3.31 Arrangement of MO_6 octahedron in a layered/spinel structure.

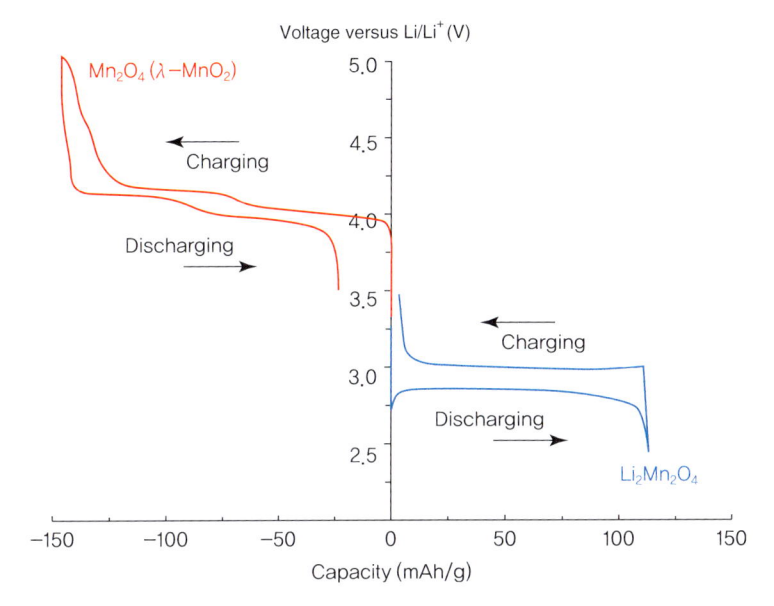

Figure 3.32 Charge/discharge curves of LiMn$_2$O$_4$.

length between M^{3+}O$_6$ and M^{4+}O$_6$, a three-dimensional octahedral arrangement allows greater uniformity than two-dimensional arrangements.

In the three-dimensional structure of (M$_2$O$_4$)$^{1-}$, lithium ions are farthest from M^{3+} and M^{4+} for charge neutrality, and placed at 8a tetrahedral sites, where electrostatic repulsion is minimal. With 8a tetrahedral sites and 16d octahedral sites occupied by lithium ions and M^{3+} and M^{4+}, the (Li)$_{8a}$[M$_2$]$_{16d}$[O$_4$]$_{32e}$ crystallographic distribution is assumed. With the three-dimensional connection of face-sharing octahedra, the spinel composite provides a path for lithium migration during charge/discharge.

In general, changes to the electrochemical characteristics of the Mn spinel proceed in two stages [28]. For $0 \leq 1x \leq 1$ in Li$_{1-x}$Mn$_2$O$_4$, lithium ions undergo intercalation/deintercalation around 4 V to maintain a cubic structure. For $1 \leq 1 + x \leq 2$ in Li$_{1+x}$Mn$_2$O$_4$, lithium ions at 16c octahedral sites undergo intercalation/deintercalation at 3 V with phase transitions between the cubic LiMn$_2$O$_4$ and tetragonal Li$_2$Mn$_2$O$_4$ (Figures 3.32 and 3.33). In other words, redox reactions for the same Mn$^{3+/4+}$ cause a potential difference of 1 V due to the difference in potential energy of lithium at 8a tetrahedral sites in the cubic LiMn$_2$O$_4$ and 16c tetrahedral sites in the tetragonal Li$_2$Mn$_2$O$_4$.

For $0 \leq x \leq 0.73$ in Li$_{1-x}$Mn$_2$O$_4$, reversible intercalation/deintercalation of lithium ions occurs. The electrode potential of Li$_{1-x}$Mn$_2$O$_4$ is influenced by the average oxidation number of Mn up to 50% deintercalation of lithium ions at 8a tetrahedral sites. However, for $x > 0.5$, it is also affected by the change in deintercalation energy of lithium ions caused by the rearrangement of remaining lithium ions. These changes in Li$_{0.5}$Mn$_2$O$_4$ appear as two flat potentials at 4.0–4.2 V.

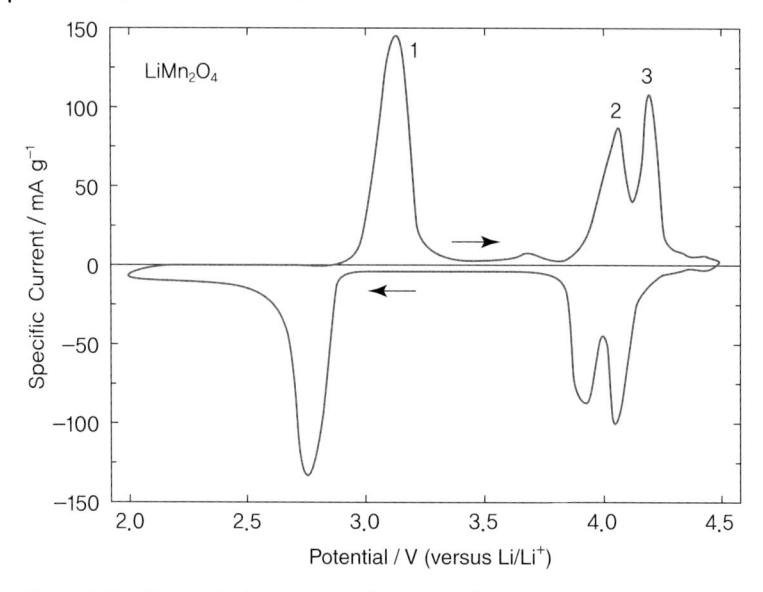

Figure 3.33 Charge/discharge cyclic voltammetry of LiMn$_2$O$_4$.

The lattice constant of LiMn$_2$O$_4$ and Li$_{0.5}$Mn$_2$O$_4$ is 8.245 and 8.029 Å, respectively. When the average oxidation number of Mn ions is above 3.5, charge/discharge cycle characteristics can be improved by suppressing Mn dissolution or the Jahn–Teller effect caused by Mn^{3+}. With an excess amount of lithium ions inserted in Li$_{1+x}$Mn$_{2-x}$O$_4$, lithium moves to 16d sites and increases the Mn oxidation number, thus showing a stable reversible capacity during charge/discharge for $0.03 \leq x \leq 0.05$. Here, the lattice constant is calculated by $a = 8.4560 - 0.2176x$ and the average oxidation number is greater than 3.58. This implies that the lattice constant is an important variable in charge/discharge as it directly affects the average Mn oxidation number. However, since the amount of Mn^{3+} involved in redox reactions decreases, the theoretical capacity of Li$_{1.05}$Mn$_{1.95}$O$_4$ at 4 V is 128 mAh/g. Similar to Li$_{1.06}$Mn$_{1.95}$Al$_{0.05}$O$_4$, the oxidation number of Mn can be increased for greater stability by substituting Mn with Al [29].

Initially, the average oxidation number of Mn in [Li$^+$]$_{8a}$[Mn^{3+}Mn^{4+}]$_{16d}$O$_4$ is 3.5. Because 8a tetrahedral sites are located close to 16c octahedral sites that are 50% occupied, the spinel structure of [M$_2$]O$_4$ is maintained and lithium ions follow the reversible path of 8a → 16c → 8a → 16c → 8a. The three-dimensional spinel provides a short diffusion path for lithium ions and high ionic conductivity, and is thermally stable during charging. When Li$_{1+x}$Mn$_2$O$_4$ is discharged and additionally intercalated lithium ions occupy the empty 16c octahedral sites, there is strong electrostatic repulsion between 8a tetrahedral sites and lithium ions, as 8a tetrahedral sites and 16c octahedral sites are in proximity close enough for face sharing. As such, lithium ions in 8a tetrahedral sites move to 16c octahedral sites to form the rock salt structure of (Li$_2$)$_{16c}$[M$_2$]$_{16d}$[O$_4$]$_{32e}$. As the proportion of x in Li$_{1+x}$Mn$_2$O$_4$ increases with additional lithium, most of the Mn at 16d octahedral sites changes to the

Mn^{3+} (d^4) state. With the structural transition from cubic to tetragonal by the Jahn–Teller effect, the c/a ratio increases by 16%, and unit cell volume by 6.5%, along with a rapid decrease in capacity.

In the early stage, the surface of cathode materials is in an overdischarge state due to excess lithium. The thermal equilibrium at the surface is disturbed and results in a nonreversible phase transition from cubic to tetragonal. $LiMn_2O_4$ is used only at 4 V and has an actual capacity of 120 mAh/g. When $Li_{1-x}Mn_2O_4$ is charged at 4 V, the crystal structure is transformed with nonreversible phase transitions due to the lack of lithium ions at the surface.

During discharge, the disproportionation ($2Mn^{3+} = Mn^{2+} + Mn^{4+}$) of Mn ions at the electrode surface produces Mn^{2+}. The amount of $LiMn_2O_4$ active material is reduced as Mn^{2+} dissolves in the acidic electrolyte. The dissolved manganese disrupts the flow of lithium ions with electrodeposition at the anode or decreases capacity by working as a catalyst of electrolyte decomposition. Capacity is greatly reduced at high temperatures where the catalytic reaction is enhanced.

$LiM_xMn_{2-x}O_4$ (M = Transition Metal) To effectively reduce the Jahn–Teller distortion by Mn^{3+} ions and attain structural stability, Mn^{3+} is substituted with transition metal ions (M = Co^{2+}, Ni^{2+}, Mg^{2+}, Cu^{2+}, Zn^{2+}, Al^{3+}, Cr^{3+}) with an oxidation number below $3+$ or lithium ions. Another method is to replace oxygen with fluorine (F). When a transition metal with an oxidation number of $2+$ or $3+$ substitutes Mn, the average valence of Mn increases, leading to greater stability and enhanced life. Figure 3.34 shows the change in capacity of $LiMn_{2-y}M_yO_4$ with charge/discharge cycles. By comparing the two graphs, it can be seen that Mn substitution provides superior life characteristics. Life can be further improved by applying transition metal oxides such as Al_2O_3, $LiCoO_2$, MgO, and ZrO_2 to the surface of $LiMn_2O_4$ to suppress interfacial reaction with the electrolyte. However, Mn substitution comes with a decrease in capacity as the average valence is increased. It is more effective to modify the surface in contact with the electrolyte, so as to minimize the loss of capacity.

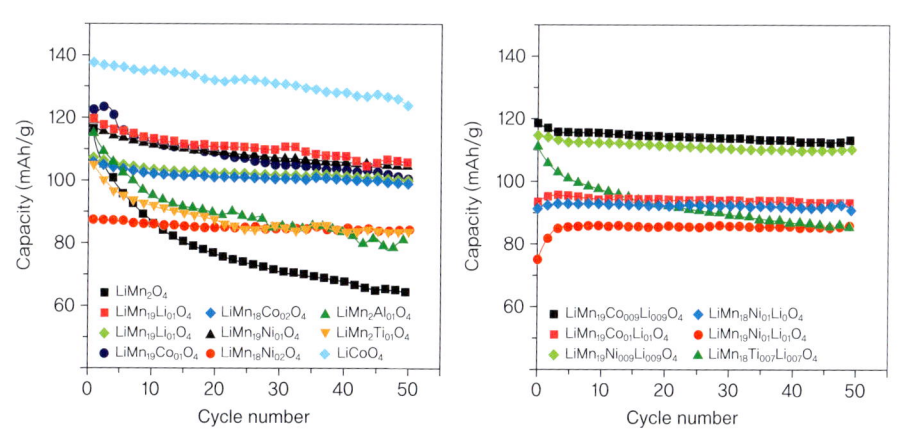

Figure 3.34 Stability of the reversible capacity of substituted $LiMn_{2-y}M_yO_4$. Adapted with permission from American Chemical Society Copyright 2003.

Spinel oxides stabilized by the above substitution or modification methods are useful as high-power lithium secondary batteries for hybrid electric vehicles. Battery capacity has increased along with expanded applications in the transportation industry, and safety has become a priority. The Mn spinel is more stable compared to existing layered structures, but it is accompanied by various problems such as deterioration at high temperatures and self-discharge from Mn elution. However, it is considered the most appropriate cathode material for batteries used in the transportation industry.

$Li_4Mn_5O_{12}$ is the case where $x = 0.33$ in $Li_{1+x}Mn_{2-x}O_4$ and results when 1/6 of lithium ions in $LiMn_2O_4$ at 16d octahedral sites are substituted to yield $Li[Li_{1/3}Mn_{5/3}]O_4$. Since the oxidation number of Mn in $Li_4Mn_5O_{12}$ is 4 +, lithium ions are electrochemically intercalated. The increase in Mn^{3+} induces the Jahn–Teller effect but nonreversible phase transitions from cubic to tetragonal do not affect $Li_{6.5}Mn_5O_{12}$, wherein Mn has an average oxidation number of 3.5 +. As such, $Li_4Mn_5O_{12}$ can be used as a cathode material at 3 V with a theoretical capacity of 163 mAh/g and actual capacity of 130–140 mAh/g. However, its small capacity compared to $LiMn_2O_4$ makes it inadequate as a replacement for existing materials.

3.1.3.3 Olivine Composites

LiFePO$_4$ Iron (Fe) is an abundant metal that is cheaper and more environmentally benign than cobalt (Co). Research on the development of iron-containing cathode materials has found olivine $LiFePO_4$ to be the most promising. $LiFeO_2$, from which $LiFePO_4$ is derived, is not being used owing to its poor electrochemical performance. $LiFeO_2$ has a low working voltage of 3.2 V due to the similar Fermi potential between Fe^{3+}/Fe^{2+} and lithium, but a higher voltage of 3.4 V may be obtained by substituting oxygen in $LiFeO_2$ with the polyanion XO_4^{y-} (X = S, P, As; $y = 2, 3$). The strong X–O bond in XO_4^{y-} reduces the Fe–O bond in Fe–O–X, increasing the ionization tendency of Fe^{3+}/Fe^{2+} and improving voltage [30–32].

$LiFePO_4$ is structurally and chemically stable. The disadvantages of LiFe–PO$_4$ are a low electron conductivity and slow diffusion of lithium ions. The electron conductivity of $LiFePO_4$ can be improved by coating it with conducting agents such as carbon or silver nanoparticles.

In a general M_2XO_4 olivine structure, 50% of octahedral sites are filled with M while 1/8 of tetrahedral sites in the hexagonal close packed oxygen are occupied by X. This corresponds to the hexagonal structure in the cubic spinel $Li[Mn_2]O_4$. If X has a small ionic radius, as in Be^{2+}, B^{3+}, Si^{4+}, or P^{5+}, an olivine structure is formed before a spinel. Table 3.2 provides a comparison of spinel and olivine structures. In the inverse spinel structure, the flow of lithium ions is disrupted by the irregular arrangement of lithium ions and nickel ions at octahedral sites. In the spinel structure, lithium ions follow the path of 16c-8a-16c as lithium ions occupy tetrahedral sites, while manganese and cobalt ions are located at octahedral sites. In the olivine structure, the two octahedral sites form a uniform structure despite having different crystallography and size, thus allowing lithium ions to diffuse in a one-dimensional path.

Table 3.2 Comparison of spinel, inverse spinel, and olivine structures.

Structure	Spinel	Inverse spinel	Olivine
Crystal structure	Cubic	Cubic	Orthorhombic
Space group	$Fd3m$	$Fd3m$	$Pmnb$
Li ion Site	Tetrahedral	Octahedral	Octahedral
Oxygen	ccp	ccp	hcp
Materials	$LiCoMnO_4$	$LiNiVO_4$	$LiFePO_4$

In the *Pmnb* structure of olivine, Fe occupies M_2 tetrahedral sites while lithium ions take up M_1 octahedral sites.

Figure 3.35 schematically illustrates the hexagonal close packed structure of olivine $LiFePO_4$. Lithium ions form a linear chain of the edge-sharing octahedron in the *c*-axis, while the FeO_6 octahedron consists of zigzag rows. Each lithium ion shares an edge with two iron ions and two XO_4 tetrahedra.

The distortion of oxygen in the hexagonal close packed structure arises from electrostatic repulsion between cations that cut across shared edges. In the edge-sharing octahedral chain, the intercalation/deintercalation of lithium ions is similar to two-dimensional intercalation/deintercalation in the layered $LiMO_2$ (M = Co, Ni). The free volume of lithium ions between MO_2 layers is restricted by the Li–O bond in $LiMO_2$, and also by the XO_4 tetrahedron that connects Fe edges in $LiMPO_4$.

$LiFePO_4$ has a theoretical density of 3.6 g/cm³, which is smaller than that of $LiCoO_2$ (5.1 g/cm³), $LiNiO_2$ (4.8 g/cm³), and $LiMn_2O_4$ (4.2 g/cm³), but greater than $LiMnPO_4$ (3.4 g/cm³) and NaSICON (Na superionic conductor). Its theoretical capacity of 170 mAh/g (2.0–4.2 V) and average working voltage of 3.4 V are not high enough for electrolyte decomposition, yet it maintains energy density, making it an excellent cathode material among iron compounds. From the charge/discharge curves of

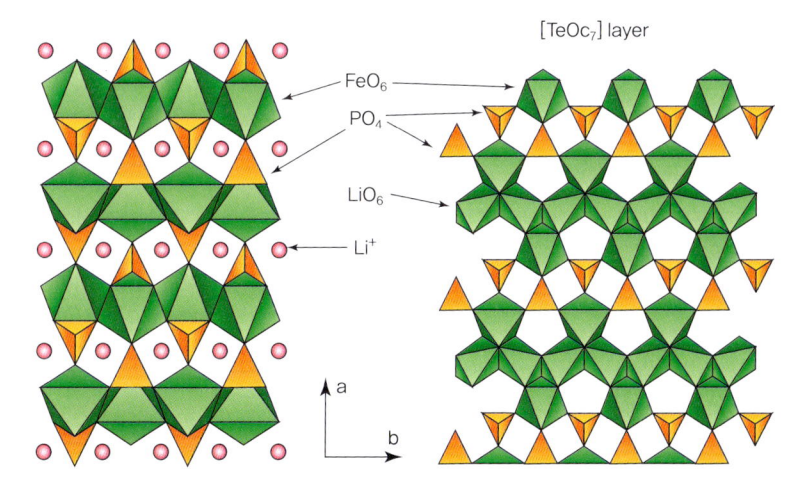

Figure 3.35 Structure of olivine $LiFePO_4$.

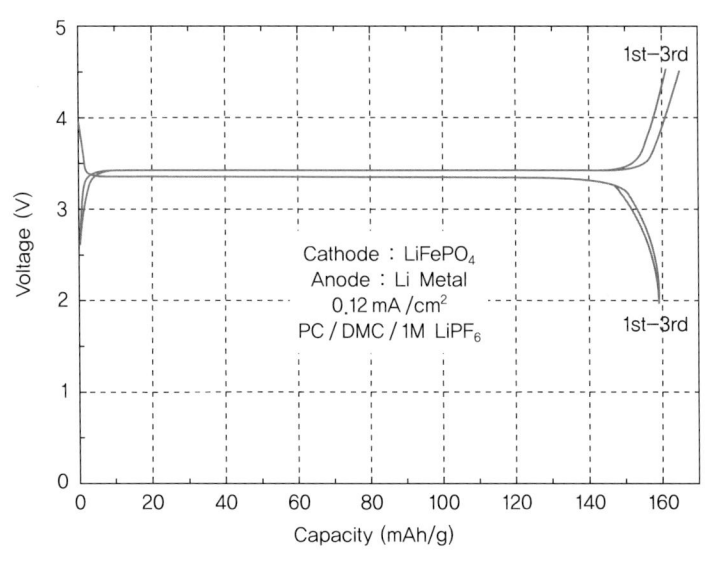

Figure 3.36 Charge/discharge potential of LiFePO$_4$.

LiFePO$_4$ in Figure 3.36, it can be seen that Li$_{1-x}$FePO$_4$ gives a flat voltage curve for a wide range of x. Also observed in Li$_4$Ti$_5$O$_{12}$ and Li$_4$Mn$_5$O$_{12}$, this phenomenon is related to the diffusion-controlled or phase boundary-controlled movement of lithium ions during charge/discharge. According to the phase rule, the intercalation/deintercalation of LiFePO$_4$ and FePO$_4$ entail two-phase redox reactions. Since the chemical potential of lithium remains unchanged in the Gibbs free energy curve for two-phase reactions, the corresponding voltage is uniform.

The intercalation/deintercalation reactions of lithium in LiFePO$_4$ can be described as follows. The crystal structures of fully discharged LiFePO$_4$ and fully charged FePO$_4$ are the same. The speed of charge/discharge reactions is influenced by the rate of phase boundary movement. In general, the movement of lithium within the crystal structure during charging and discharging can be divided into two types. The first is diffusion reactions. In this case, the crystal structure is unchanged and the diffusion of lithium appears as a gradient in the charge/discharge curve. The lattice constant may change with mutual interactions between particles in the structure during charging but the crystal structure is preserved. The second is the phase boundary movement of two-phase reactions. Figure 3.37 is a schematic representation of the phase boundary movement.

When lithium at the particle surface undergoes deintercalation during charging, the A phase (FePO$_4$) without lithium and the stoichiometric B phase (LiFePO$_4$) coexist. As charging progresses, the A phase increases and B decreases until only the A phase remains. With lithium deintercalation, the phase boundary shifts continuously. The capacity of Li$_{1+x}$Mn$_2$O$_4$ at 3 V is sharply reduced but LiFePO$_4$ exhibits superior life characteristics. This phenomenon can be explained by the structural stability of the olivine LiFePO$_4$ and by the similar crystal structures between discharged LiFePO$_4$ and charged FePO$_4$.

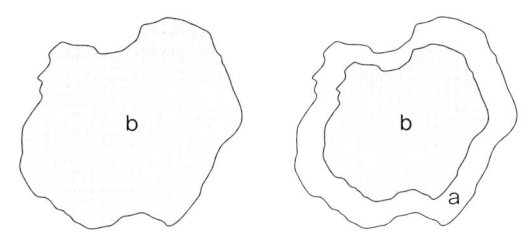

Figure 3.37 Schematic diagram of phase boundary movement.

The weight of FePO$_4$ after lithium deintercalation is unchanged up to 350 °C in a nitrogen atmosphere. The results of X-ray diffraction analysis show that there is hardly any structural change after heat treatment. In addition, FePO$_4$ contains no impurities and is stable at high temperatures for most lithium salts and electrolytes. Unlike other materials, charging reduces the volume of LiFePO$_4$ by approximately 6.8%. This is advantageous in the design of lithium ion batteries, as it compensates for carbon anode expansion during charging. Due to its structural and thermal stability, LiFePO$_4$ is expected to see wide use in various fields requiring batteries with a high capacity, large volume, and long life, namely, hybrid electric vehicles, power tools, and power storage.

In LiFePO$_4$, lithium diffuses through a one-dimensional channel and thus is affected by material defects. When a lithium diffusion channel is occupied by other cations such as iron, the channel is blocked, restricting lithium movement. Because of the low mobility of iron, the lithium channel remains inactive and cannot engage in electrochemical reactions. In the synthesis of LiFePO$_4$, it is important to create a flawless crystal structure.

The greatest disadvantage of LiFePO$_4$ is its low electrical conductivity. This is a general characteristic of materials containing polyanions such as PO$_4^{2-}$ and may lead to extreme polarization during charge/discharge. In addition, its capacity will be markedly reduced if the conducting agent is not uniformly reduced.

Various methods are being developed to enhance the rate-controlling properties of LiFePO$_4$, such as adjusting the particle size for improved ionic conductivity and applying a carbon coating on particles for higher electrical conductivity [33]. Another method is to use Nb as a doping element. This leads to phases with high electrical conductivity such as Fe$_2$P, which is difficult to repeat and has unknown effects on electrochemical properties [34, 35].

To prevent the production of Fe^{3+} ions in high-temperature solid phase synthesis, an excess of nitrogen should be supplied. Source materials must be homogeneous in order to minimize the formation of impurities such as Fe$_2$O$_3$ and Li$_3$Fe$_2$(PO$_4$)$_3$. When samples are prepared at high temperatures, there is an increase in particle size, a decrease in surface area, a less diffusion of lithium ions, and a deteriorated battery performance. Recent studies have investigated methods of low-temperature synthesis to suppress particle growth [36].

LiMPO$_4$ (M = Mn, Co, Ni) LiFe$_{1-x}$Mn$_x$PO$_4$, which is a solid solution of LiFePO$_4$ and LiMnPO$_4$, can also be charged and discharged. The lattice constant of LiFe$_{1-x}$Mn$_x$O$_4$

satisfies Vegard's law while mutual interaction within $Fe^{3+}-O-Mn^{2+}$ allows a high voltage of 4.1 V for Mn^{3+}/Mn^{2+} reduction, as shown in Figure 3.38. When iron is found in the vicinity of Mn in $LiFe_{1-x}Mn_xPO_4$, Mn^{3+}/Mn^{2+} is electrochemically activated. In the fully charged state of $(Mn_y^{3+} Fe_{1-y}^{3+})PO4$, electrochemical activation is influenced by a severe lattice distortion with $y > 0.75$ from the Mn^{3+} Jahn–Teller effect.

Comparing the VO_4 tetrahedron and PO_4 tetrahedron, an olivine structure is favored over a spinel structure due to PO_4 sharing. This stabilizes redox energy at octahedral sites and increases the Mn^{3+}/Mn^2 potential from 3.7 V in $V[LiMn]O_4$ to 4.1 V in $LiFe_{0.5}Mn_{0.5}PO_4$. Region I at 4.1 V has a flat two-phase region, while region II

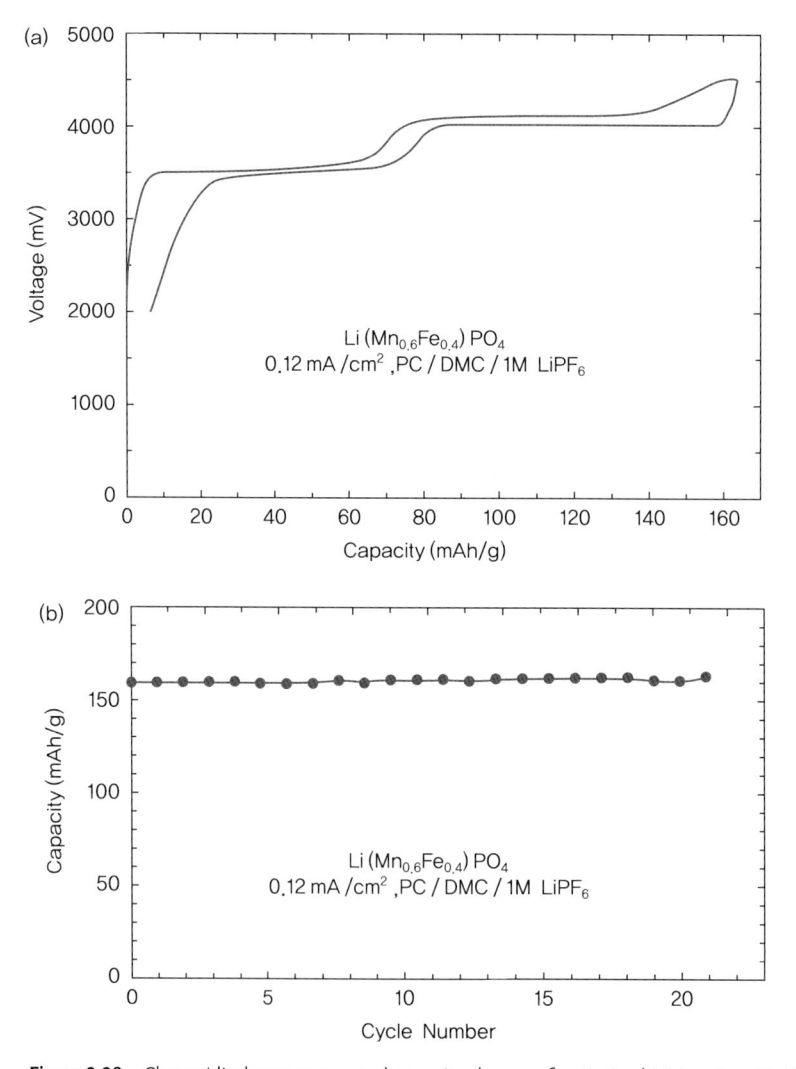

Figure 3.38 Charge/discharge curves and capacity change of optimized $LiMn_{0.6}Fe_{0.4}O_4$ [32].

consists of a S-shaped single phase. There is more overvoltage at region I than region II. This is due to resistance against lithium ions passing through the mixed phase and the larger effective mass of 3d electrons in Mn^{3+} from the Jahn–Teller effect. Region II has a slightly higher voltage of 3.5 V compared to 3.4 V in Li_xFePO_4 because the Fe^{3+}–O–Mn^{2+} superexchange lowers the potential of Fe^{3+}/Fe^{2+} and increases that of Mn^{3+}/Mn^{2+}. The lattice constant in region I is the same for the two phases but varies in region II.

By optimizing the electrical conductivity and particle size of $LiCoPO_4$, which is charged/discharged at 5.1/4.8 V, it may be used as a high-voltage cathode material along with the spinel $LiNiVO_4$ and $LiCr_xMn_{2-x}O_4$. When charged up to 5.1 V with a current density of 0.2 mA/cm², it has a charge capacity of 100 mAh/g and a 4.6% change in volume. It has an energy density similar to $LiCoO_2$ but better power than $LiFePO_4$. For charged $Li_{0.4}CoPO_4$, the first phase transition occurs at 290 °C, which is similar to the phase transition temperature of $FePO_4$ (315 °C). According to magnetic moment measurements, changes in electrochemical characteristics contribute to the ratio of high spin $Co^{2+}(t_{2g})4(e_g)^2$ and low spin $Co^{3+}(t_{2g})^6$. In the case of $LiMnPO_4$, lithium deintercalation was not possible due to the thermal instability of $MnPO_4$ but it was recently found that capacity of 140 mAh/g may be obtained [32, 37]. In addition, a capacity of 70 mAh/g was achieved by direct precipitation and ball milling with carbon, and $MnPO_4$ was presented in a thermally stable phase [38]. Based on first-principles calculations, the equilibrium voltage of $LiMnPO_4$ is 4.1 V. Low electrical conductivity was expected from the large bandgap and slow polaron movement [39], and actual experiments revealed a low power density.

While there has not been extensive research on $LiNiPO_4$, the predicted equilibrium voltage of 5.1 V from first-principles calculations was recently proven [40].

3.1.3.4 Vanadium Composites

Vanadium oxide and its derivatives can be found in various phases and crystal structures, including V_2O_5, V_2O_3, VO_2 (B), V_6O_{13}, V_4O_9, V_3O_7, $Ag_2V_4O_{11}$, $AgVO_3$, $Li_3V_3O_5$, $\delta\text{-}Mn_yV_2O_5$, $\delta\text{-}NH_4V_4O_{10}$, $Mn_{0.8}V_7O_{16}$, LiV_3O_8, $Cu_xV_2O_5$, and $Cr_xV_6O_{13}$ [41–46].

The V_2O_5 oxide obtained from polyol-mediated synthesis exhibits several morphologies such as globular, nanorod, or nanowire, and has an initial capacity of 250–400 mAh/g. The discharge curves of these oxides tend to be widely spread across 1.5–4.0 V, as shown in Figure 3.39 [47, 48].

VO_2(B) is an unstable phase that easily changes to the VO_2 rutile at high temperatures above 300 °C [51]. When synthesized at low temperatures based on redox reactions, the resulting nanosized oxides display gradual discharge curves with a high capacity of 300 mAh/g (Figure 3.40) [51]. The aerosol form of VO_2(B), obtained from freeze drying and low-temperature heating, possesses a high capacity of 300–520 mAh/g. The temperature of heat treatment is known to have a significant influence in creating the flat region of discharge curves [52].

In addition, vanadium oxides synthesized into silver-containing nanowires or nanorods ($Ag_2V_4O_{11}$, $AgVO_3$) have a capacity greater than 300 mAh/g [54].

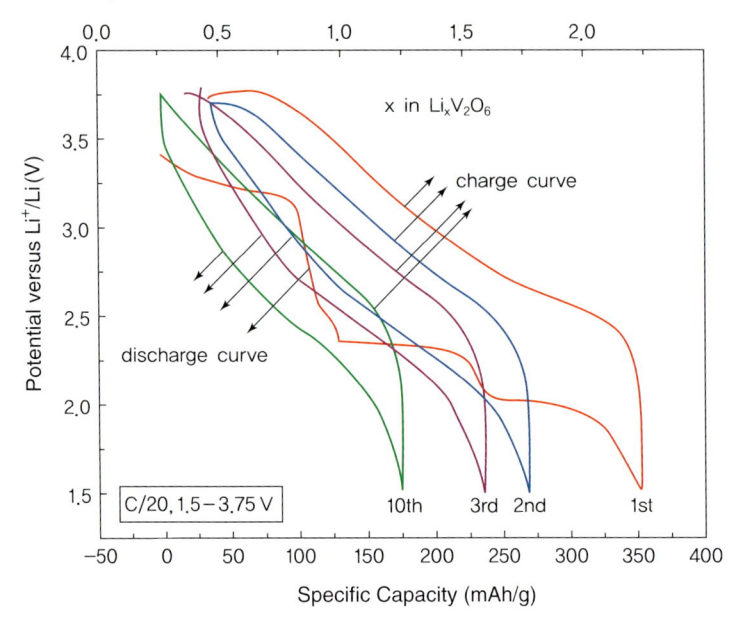

Figure 3.39 Charge/discharge curves of nanowire V_2O_5 at a current density of C/20 [47–50].

Figure 3.41 shows the discharge curves when various mixed phases are synthesized as nanowires or nanorods.

LiV_3O_8 active material, synthesized using the sol–gel method and 100 nm in size, showed a capacity greater than 300 mAh/g and a gradual discharge curve [55]. Chromium-added $Cr_{0.36}V_6O_{13}$ has a high capacity of 380 mAh/g and a relatively flat discharge curve at 3 V. However, its production involves the complicated process of sol–gel synthesis and Cr doping through ion exchange followed by acid washing.

While vanadium can serve as a high-capacity electrode material, vanadium oxides that are microsized and high crystalline show a distinct flat potential during redox reactions for each oxidation number. Recent studies have attempted to synthesize nanosized vanadium oxides with gradual discharge curves by achieving a metastable phase that lies between amorphous and high crystalline phases [47–54].

3.1.4
Performance Improvement by Surface Modification

In layered oxides such as $LiMO_2$ (M = Co, Ni, Mn), there is a significant decrease in reversible capacity when charged higher than 4.3 V due to the elution of transition metals and ion exchange between lithium and transition metal ions. Li_xMO_2, from which lithium has been deintercalated, is subject to surface deterioration and structural damage, and safety issues may arise. The reversible capacity of the cubic spinel $LiMn_2O_4$ is also reduced during charge/discharge with manganese elution and the Jahn–Teller effect. The olivine $LiFePO_4$ is structurally and chemically stable but has very low electrical and ionic conductivity.

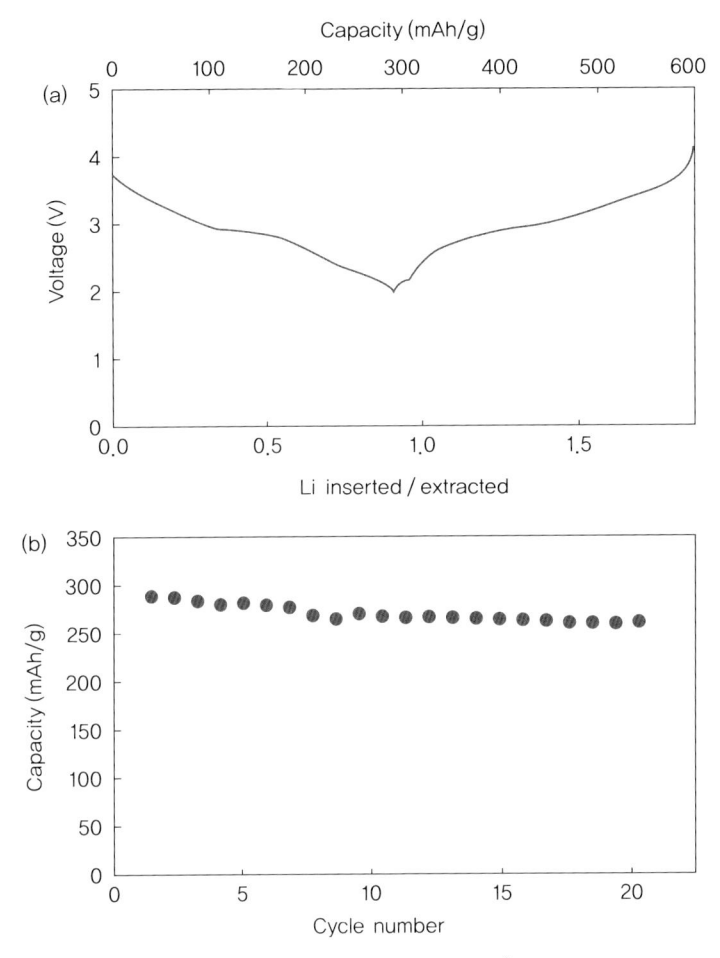

Figure 3.40 (a) Charge/discharge curves at 0.5 mA/cm² current density and (b) cycle characteristics of nanocrystalline VO$_2$(B) [53].

In an effort to resolve this problem, various surface modification methods have been introduced. Several methods and materials are available for different purposes. For example, the reversible capacity of LiCoO$_2$ can be enhanced by coating it with metal oxides, metal flourides, or mixed metal oxides for greater stability at high voltage [56–60]. This method suppresses the elution of transition metals or improves surface stability and enables charging and discharging at high voltage to obtain a larger capacity. In the case of spinel LiMn$_2$O$_4$, structural stability is achieved by controlling the elution of metal ions through the insertion of two different elements or surface modification. The surface modification method also provides enhanced cycle efficiency at high rate discharge, thermal stability, high capacity, high power, and extended life. While LiFePO$_4$ has fewer eluting metal ions and better high-voltage stability compared to other oxides, surface modification aimed at improving

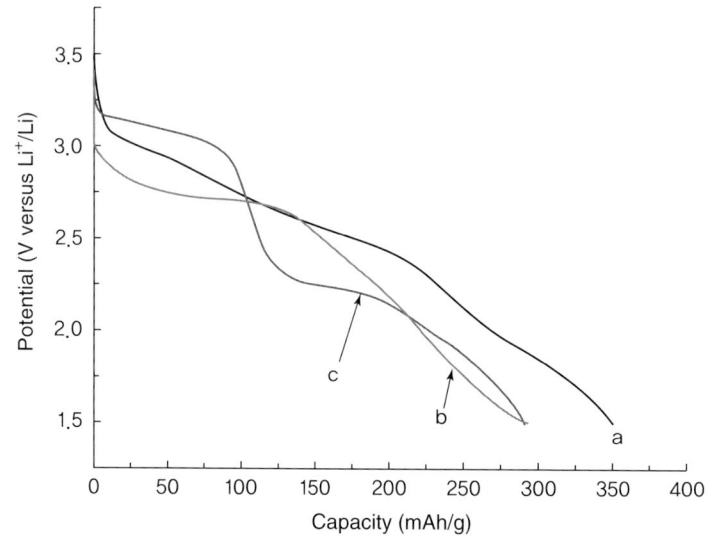

Figure 3.41 Discharge curves (current density: 0.1 mA) of AgVO$_3$ electrodes in different phases: as-prepared (a) Ag$_2$V$_4$O$_{11}$ nanowire, (b) α-AgVO$_3$ nanorod, and (c) β-AgVO$_3$ nanowire [55]. Adapted with permission from [55] Copyright 2006 American Chemical Society.

conductivity is the most effective approach, given its low electrical and ionic conductivity [61]. As such, conducting materials such as carbon [62, 63] or nanosized metal particles [64] are used as surface modification materials for LiFePO$_4$.

However, the problems accompanying surface modification must be resolved as well. For example, adding a new material may reduce specific capacity, while using a material with low ionic conductivity disrupts the movement of lithium ions during charge/discharge. High-rate characteristics can be reduced with a decrease in the reaction area of lithium intercalation/deintercalation at the material surface. In terms of manufacturing cost, it is more expensive to add a precursor for the surface modification process.

3.1.4.1 Layered Structure Compounds

When the surface of LiCoO$_2$ is coated with a chemically safe material, significant improvements are shown in the charge/discharge cycle. When lithium is deintercalated from the layered LiCoO$_2$ during charge/discharge, the internal energy of Li$_x$CoO$_2$ rises along with a change in the crystal structure. In the overcharged and unstable Li$_x$CoO2, cobalt is eluted from reactions between the cobalt ions and the electrolyte. In contrast, cobalt elution is suppressed for surface-modified LiCoO$_2$ even at a high voltage above 4.3 V and temperatures exceeding 60 °C [65].

Figure 3.42 shows the charge/discharge characteristics of LiCoO$_2$ at 2.75–4.4 V when coated with various oxides. There is very little change in the reversible capacity when ZrO$_2$ is used for surface modification. Depending on the material, cycle characteristics follow the ascending order of B$_2$O$_3$ < TiO$_2$ < Al$_2$O$_3$ < ZrO$_2$. This is known to be closely related to the toughness of the selected materials [66]. In general, the coating layer does not cover the whole surface but only a small portion of the

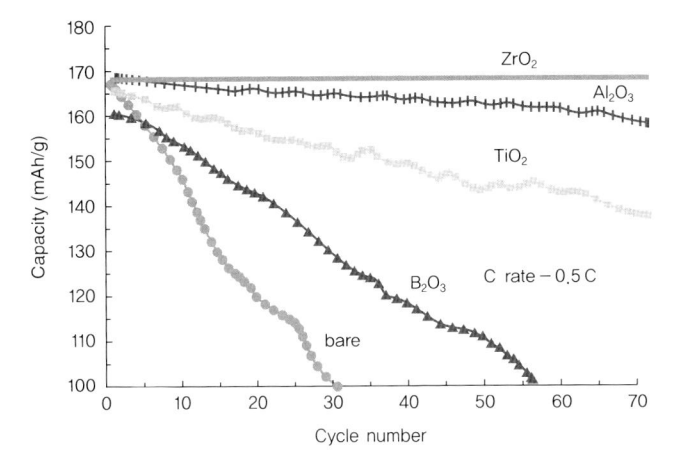

Figure 3.42 Charge/discharge capacity maintenance characteristics of LiCoO$_2$ cathode material surface modified with ZrO$_2$, Al$_2$O$_3$, TiO$_2$, B$_2$O$_3$, and so on [57]. Reproduced from [57], with permission from Wiley-VCH.

cathode material. This prevents the eluting of transition metals by modifying the surface energy below or around the coating layer.

For Li$_x$NiO$_2$ ($x < 0.5$) in the lithium-deintercalated state, structural instability at the surface upon overcharging may heat up the battery and cause safety problems. Similar to LiCoO$_2$, electrochemical characteristics can be improved with surface modification using Li$_2$O·2B$_2$O$_3$, MgO, AlPO$_4$, SiO$_2$, TiO$_2$, ZrO$_2$, and so on. By achieving structural stability through surface modification, activation energy is increased to suppress phase transitions. From the electrochemical characteristics of LiNiO$_2$ coated with ZrO$_2$ oxide illustrated in Figure 3.43, we can see a stable charge/discharge curve in the initial stage with enhanced cycle characteristics.

Similar surface modification results can be obtained for LiNi$_{0.8}$Co$_{0.2}$O$_2$, which mainly consists of nickel. By protecting its surface with a chemically and thermally safe coating material, the thermal stability of the electrode is enhanced with a decrease in reaction area and increase in interfacial stability. This can be easily confirmed using thermal analysis methods such as DSC (differential scanning calorimetry). Figure 3.44 shows the results of a thermal analysis for CeO$_2$-coated LiNi$_{0.8}$Co$_{0.2}$O$_2$ when charged at a high voltage. With a significant drop in the exothermic peak, the exothermic onset temperature is raised.

For three-component cathode materials, research is underway to achieve both high capacity and stability with greater lithium deintercalation at high voltages through surface modification. With application of an AlF$_3$ coating, it is possible to attain superior cycle characteristics and stable battery performance under high voltage. In the photograph shown in Figure 3.45, a substantial uniform coating of 2–3 nm has been applied [59].

3.1.4.2 Spinel Compound

The spinel LiMn$_2$O$_4$ is environmentally benign, is inexpensive, is structurally stable, and has excellent rate-controlling characteristics. However, research on surface

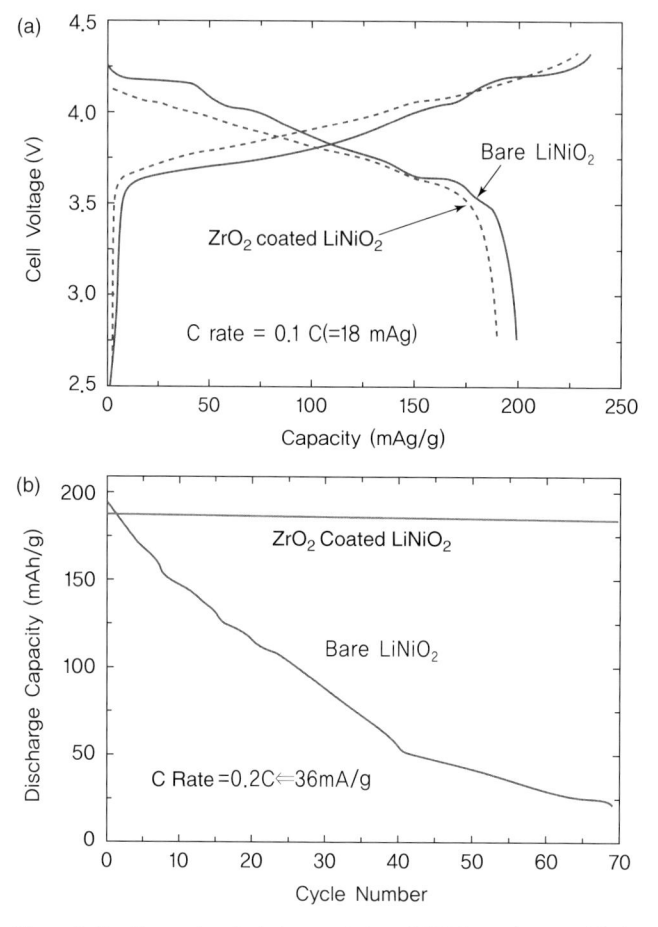

Figure 3.43 Electrochemical characteristics of LiNiO$_2$ surface-modified with ZrO$_2$ [58].

modification is being conducted to overcome its poor electrochemical properties at high temperatures. Despite its outstanding charge/discharge and cycle characteristics at room temperature, the performance of LiMn$_2$O$_4$ is reduced at high temperatures due to the elution of Mn ions from the electrode, the Jahn–Teller effect of Mn^{3+}, and oxidation reactions with the electrolyte at the surface of particles. Since these problems are related to electrochemical reactions at the surface and Mn valence, attempts are being made to suppress reactions with the electrolyte through surface modification.

To reduce the concentration of Mn^{3+} at the surface of the LiMn$_2$O$_4$ spinel compound, elements of different valances including Al, Mg, and lithium are used together. Surface modification is particularly effective in preventing performance deterioration arising from electrochemical reactions between the electrolyte and the LiMn$_2$O$_4$ particles at high temperatures. For example, when Mn valence in LiMn$_2$O$_4$ increases during charging, Mn^{4+} becomes unstable and the electrolyte is

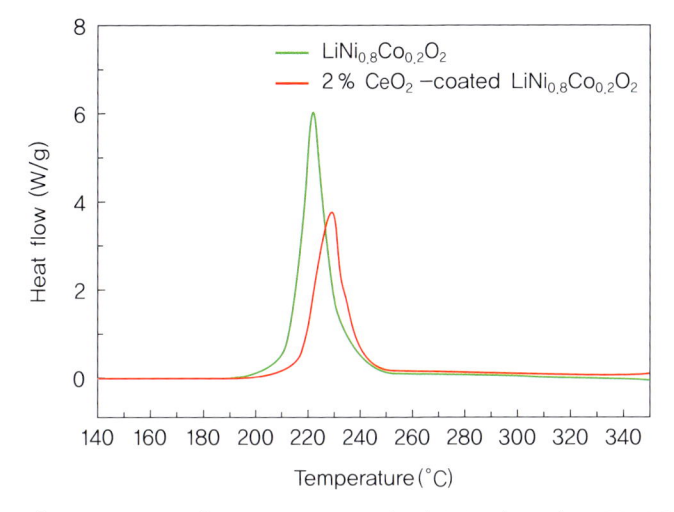

Figure 3.44 DSC of LiNi$_{0.8}$Co$_{0.2}$O$_2$ coated with CeO$_2$ charged at 4.5 V [59]. Reprinted from [59] Copyright 2005, with permission from Elsevier.

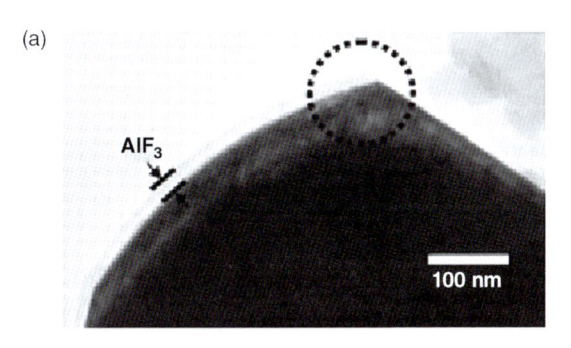

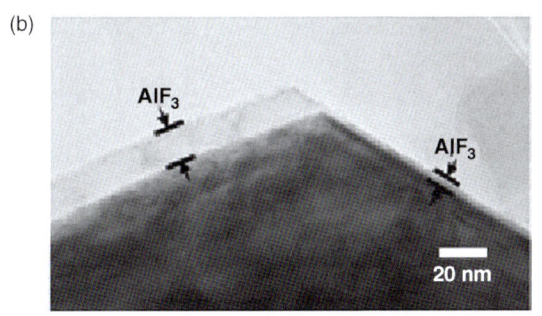

Figure 3.45 (a) Surface of AlF$_3$-coated LiNi$_{1/3}$Co$_{1/3}$Mn$_{1/3}$O$_2$ observed by TEM after charge/discharge. (b) A partial magnification of (a) [60].

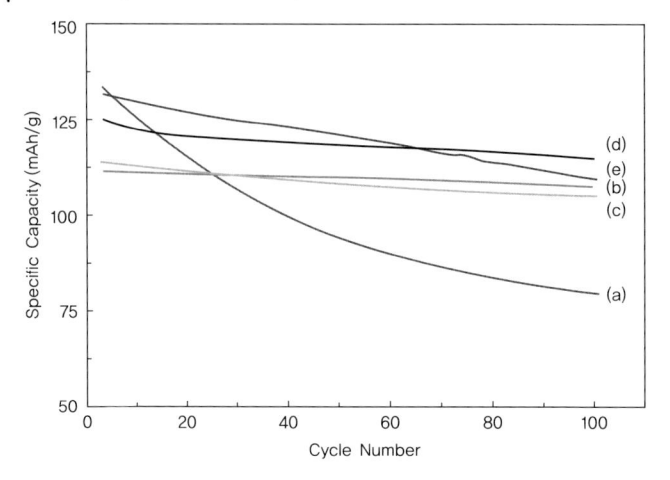

Figure 3.46 Life characteristics of surface-modified LiMn$_2$O$_4$ at high temperatures: (a) uncoated LiMn$_2$O$_4$, (b) LiCo$_{1/2}$Ni$_{1/2}$O$_2$ coating, (c) LiCoO$_2$ coating, (d) Li$_{0.75}$CoO$_2$ coating, and (e) Al$_2$O$_3$ coating [67].

decomposed, which rapidly impairs battery performance even as the active material remains unaffected. Figure 3.46 shows the cycle characteristics of LiMn$_2$O$_4$ that has been coated with various oxides. For LiCo$_{1/2}$Ni$_{1/2}$O$_2$ coated with LiMn$_2$O$_4$, a high reversible capacity of 97.2% (110 mAh/g) is maintained for 100 cycles at 60 °C.

Other methods are being developed to not only suppress surface reactions but also enhance rate-controlling performance through the application of materials with high conductivity such as ITO (indium tin oxide). The use of LiMn$_2$O$_4$ also reduces capacity per unit weight mass as the average valence of Mn increases with doping or surface coating.

3.1.4.3 Olivine Compounds

LiFePO$_4$ surface modification is used to improve the electrical conductivity of particles. Combining a carbon precursor with active material particles followed by heat treatment leads to the formation of a thin carbon layer of tens to hundreds of nanometers at the surface of particles. This carbon layer endows LiFePO$_4$ with higher electrical conductivity and thus enhances its rate-controlling characteristics. The increase in electrical conductivity at the particle surface does not improve the electrical conductivity within particles or the ionic conductivity of lithium. The size of particles should be minimized in order to maintain short diffusion paths for lithium ions. Smaller particle size reduces the diffusion paths of lithium within the active material and increases surface area for reactions. As shown in Figure 3.47, a theoretical capacity higher than 95% can be achieved by coating the small particles with carbon [67]. Similar effects may be achieved by using silver instead of carbon, but this entails a higher manufacturing cost [68]. Obtaining nanosized particles and surface modification using carbon are practical methods for the commercialization of olivine compounds.

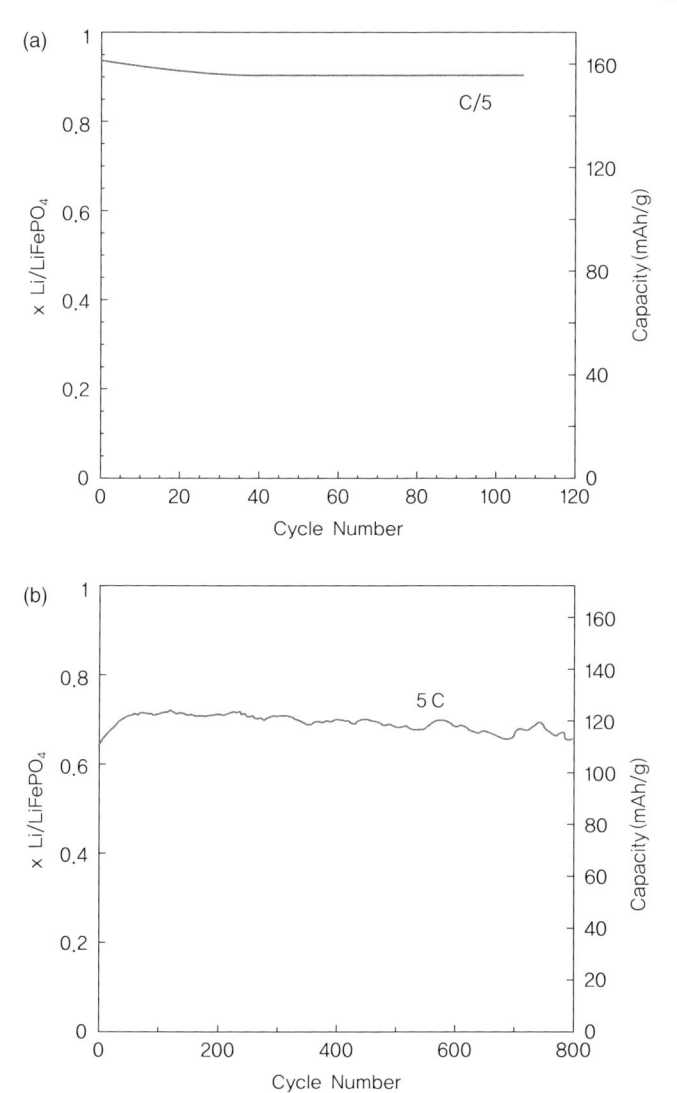

Figure 3.47 Cycle characteristics of LiFePO$_4$ coated with carbon gel: (a) C/5 and (b) 5C [69].

3.1.5
Thermal Stability of Cathode Materials

3.1.5.1 Basics of Battery Safety

Battery safety issues concern fires and explosions that may arise from various chemical reactions due to abnormal energy conversion and increased temperatures within the cell. Figure 3.48 is a mimetic diagram of battery safety. Charging and discharging of a battery is kept under control in the normal operating range, but under conditions of abuse, rapid energy conversion produces reactant by-products.

3 Materials for Lithium Secondary Batteries

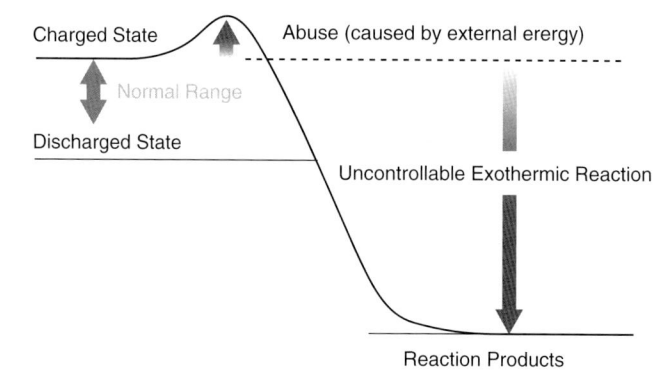

Figure 3.48 Mimetic diagram of battery safety.

To secure battery safety, temperature increases in the battery should be suppressed, as most dangerous situations such as heat generation, heat dissipation, fires, and explosions are triggered by heat. When there is a temperature increase, lithium ion batteries engage in spontaneous heat reactions that further elevate the temperature. Incorrect use of lithium ion batteries or manufacturing defect presents a risk of fire because of an imbalance in the amount of heat dissipation and heat generation.

Figure 3.49 illustrates the relationship between heat dissipation/generation and safety in a battery. If the rate of heat dissipation is faster than heat generation, thermal runaway is avoided and the battery is stabilized. If the rate of heat generation exceeds heat dissipation, energy is accumulated and the battery becomes unsafe with an increase in temperature over time.

With W_e (J/s) as the energy entering from outside the battery, W_i (J/s) as spontaneous heat reactions within the battery, W_d (J/s) as heat dissipation, C_b [(J/s)/T] as the thermal capacity of 1 battery, t (s) as time, T_0 as external temperature,

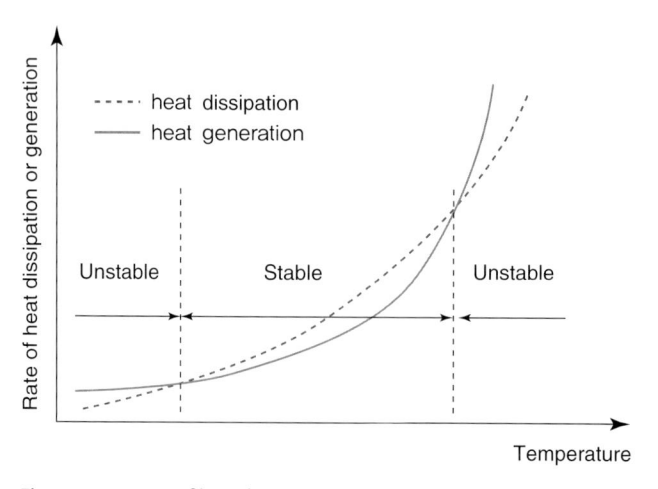

Figure 3.49 Rate of heat dissipation/generation and safety within a battery.

and k (1/s) as the heat dissipation constant, battery temperature (T) can be obtained from the following equation:

$$dT/dt = (W_e + W_i)/C_b - W_d/C_b = (W_e + W_i)/C_b - k(T - T_o)$$

Since minimizing the value of $(W_e + W_i)/C_b$ is important for safety enhancement, we should first look at spontaneous reactions (W_i) in the above equation. The battery temperature is largely determined by heat generation in the battery and heat dissipation to its surroundings. Spontaneous heat generation occurs with exothermic pyrolysis reactions above a certain temperature. Such reactions become more frequent as temperature increases and can result in thermal runaway.

In general, W_i is proportionate to body capacity, while W_d and W_e are proportionate to the surface area of the battery. Thermal runaway problems are more likely to arise in larger batteries where the amount of heat generation ($W_e + W_i$) is greater than heat dissipation (W_d). Some examples of exothermic reactions are reduction reactions of the electrolyte at the anode, pyrolysis reactions of the electrolyte, oxidation reactions with the electrolyte at the cathode, thermal decomposition of the cathode, and deoxygenation pyrolysis reactions of high-voltage metal oxides. Short-circuiting at the melting of the separator also contributes to heat generation. The PE separator melts at 135 °C and the PP at 165 °C. Figure 3.50 shows the general process of thermal reactions within a battery.

The films at the anode and cathode grow thicker with electrolyte decomposition at 60 °C and thus influences battery performance. When the temperature exceeds 100 °C, the SEI (solid electrolyte interphase) film is decomposed followed by heat generation. The melting of the separator may induce short circuits in the battery. During short circuits, electrons diffuse rapidly from the anode to the cathode,

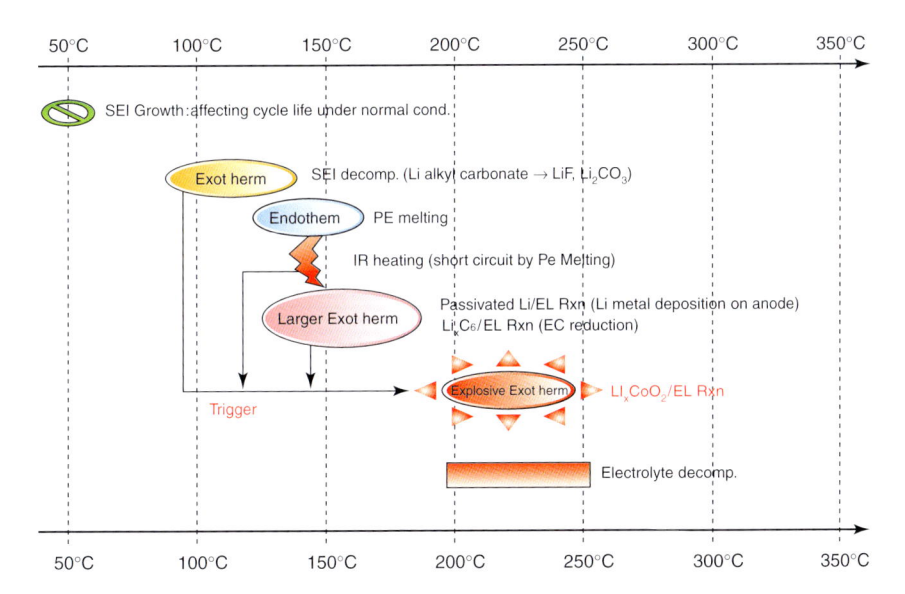

Figure 3.50 Thermal reactions by temperature within a battery.

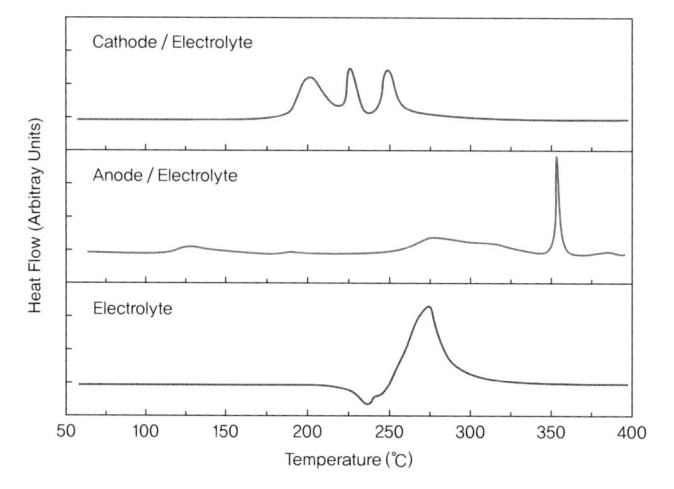

Figure 3.51 DSC analysis of cathode, anode, and electrolyte.

producing iR heat with electrical resistance. More heat is generated from reactions of the anode with the electrolyte. This serves as a catalyst that speeds up explosive heat reactions at the cathode. At this stage, it is most important to secure thermal stability of the cathode. Figure 3.51 presents the results of a DSC analysis of charged $LiCoO_2$ cathode, charged graphite anode, and electrolyte. The cathode (Li_xCoO_2) exhibits exothermic reactions at relatively low temperatures (180–260 °C) while exothermic reactions of the anode are found at 100–150 due to SEI decomposition. At 360 °C, a large exothermic peak is observed in the anode from the decomposition of LiC_6. The electrolyte also shows exothermic reactions at 250–300 °C. An accurate analysis of heat generation is difficult to achieve in actual batteries due to these complex reactions.

3.1.5.2 Battery Safety and Cathode Materials

Factors related to battery safety can be organized into three categories, as shown in Figure 3.52.

Overcharging refers to the charging of a battery beyond its working voltage and is usually caused by malfunctioning of the charger. Heat generation occurs with abnormal electrochemical reactions within the battery, while short-circuiting arises from manufacturing defects or battery abuse. These reactions may have different causes but are closely related to heat production and contribute to explosive heat reactions in the anode. Some related processes are the surface reaction between the electrolyte and the cathode, pyrolysis reaction of the cathode (oxygen production), oxidation reaction of the electrolyte, and pyrolysis reaction of the electrolyte. The respective chemical equation and energy are as follows.

1) Cathode/electrolyte surface reaction
2) Pyrolysis reaction of the cathode (oxygen production)
3) Oxidation reaction of the electrolyte
4) $\Delta H = -0.139$ kJ/g (electrolyte) of pyrolysis reaction of the electrolyte

```
                          Safety
         ┌─────────────────┼─────────────────┐
     Overcharge      Safety at high temp    Short Circuit
```

- Li metal deposition: anode
- Fast iR heating
- Electrolyte decomposition
  →T&P↑
  →Separator shrinkage
    (2nd internal short circuit)
  →Rxn b/w decomposed
    Products and Li
- Structural Destruction of cathodes
- Al Deposition

- Exothermic Rxn on anode
  by Heat
  →Rxn b/w electrolyte and LiC_6
- Internal short circuit
  →Separator shrinkage
  →Separator meltdown
- Thermal congestion (cathode)
  →Rxn b/w $LiCoO_2$ and electrolyte
  →Electrolyte decomposition itself

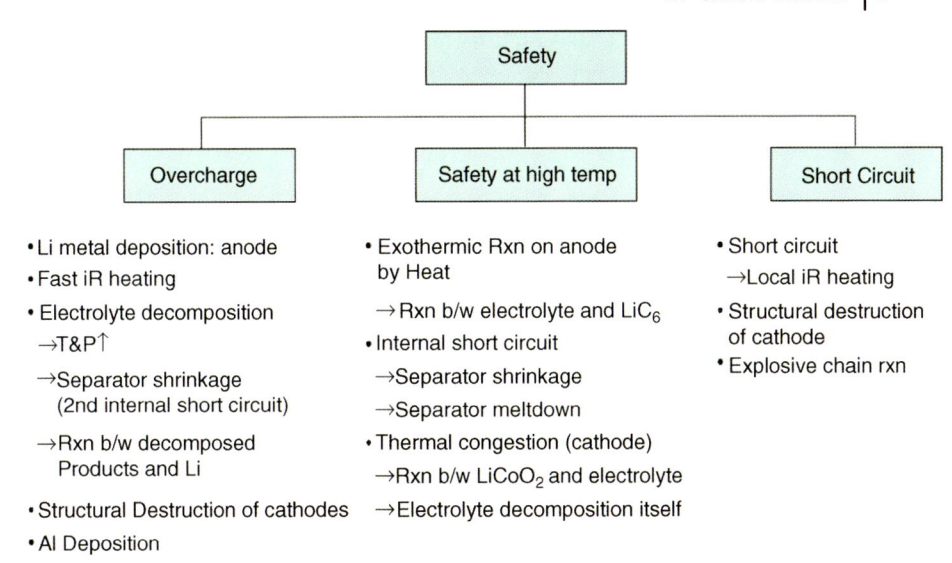

- Short circuit
  →Local iR heating
- Structural destruction
  of cathode
- Explosive chain rxn

Figure 3.52 Reactions involving battery safety.

Electrolyte decomposition does not occur under normal conditions since the electrolyte has a higher decomposition potential than the cathode. The overcharged state, when the potential of the cathode rises beyond the electrolyte decomposition potential, results in an oxidation reaction of the electrolyte accompanied by heat generation. Thermal decomposition of the cathode releases a great amount of heat and oxygen during the structural change in transition oxides used as cathode materials at high temperatures. Depending on the active material, the initial decomposition temperature is in the ascending order of $LiNiO_2 < LiCoO_2 < LiMn_2O_4$.

3.1.5.3 Thermal Stability of Cathodes

As explained earlier, the thermal stability of cathodes plays an important role in battery safety. In particular, the thermal stability of a charged cathode is used as a scale in measuring battery safety. It is also related to the structural stability of cathode materials. The active material exists in a stable state immediately after synthesis and when discharged. However, as lithium diffuses out of the cathode during the charging process, it becomes thermodynamically unstable and enters a metastable state. When an energy greater than the activation energy is applied, it changes into a stable phase and releases a great amount of heat through exothermic reactions. To secure battery safety, transitions to an unstable state can be avoided by reducing the amount of heat generation or by increasing the activation energy. While there are various methods to attain battery safety, the most important is to reduce exothermic energy arising from structural change in cathode materials. In other words, structural stability of cathode materials should be secured in the charged state. Recently, there have been demands for higher charging voltage of the cathode in order to increase battery capacity. Since a greater amount of energy is concentrated in the cathode at high charging voltage, more heat is likely to be produced from a change in the structure of active materials.

Methods of measuring thermal safety of cathodes are as follows. After two–three charge/discharge cycles, the cathode is separated from the charged cell. Extra care must be taken during this step to prevent short-circuiting, and the separation should proceed inside a glove box so as to minimize exposure to air. The separated cathode is then washed with an adequate electrolyte, and a thermal analysis is conducted to observe any change in temperature from exothermic reactions. It should be noted that the reactivity of the cathode is affected by the type of washing solution and washing methods. During the cathode analysis, a separate review of electrolyte types, amount, and binder influence may lead to a better understanding of thermal stability. The next section describes the thermal stability of various types of cathode materials.

$LiCoO_2$ Complex Oxide Figure 3.53 shows the thermal stability of $LiCoO_2$ according to charge voltage with (a) in the discharged state. Thermal stability rises from (b) until (f) as charge increases. In the discharged state, no exothermic peak is observed and it is found to be thermally stable. The peak increases at higher voltages while the exothermic onset point becomes lower. As charging progresses, the structural stability of the material is reduced. When more than 55% of lithium is deintercalated, the exothermic onset point and exothermic peak are similar in strength. In other words, if the amount of lithium falls below a certain point, the structural stability of $LiCoO_2$ is maintained. Thermal stability should be carefully analyzed as the strength or shape of exothermic peaks may vary according to the electrolyte type or content.

Figure 3.54 presents the thermal analysis results of the $LiCoO_2$ active material in the charged state before and after washing with an organic solvent. Measurements were taken after washing the electrode for 36 h using DEC solvent followed by 12–14 h of drying at 65 °C in a vacuum oven. As shown in the figure, the exothermic peak before washing differs in shape and size from that after washing. This implies that the lithium salt remaining in the electrolyte contributes to heat generation of the cathode. In the washed sample, two exothermic peaks are observed with the exothermic onset point at 178 °C, while Li_xCoO_2 is decomposed at 178–250 and 250–400 °C. The decomposition of Li_xCoO_2 proceeds according to the following chemical equation.

$$Li_xCoO_2 \rightarrow x\, LiCoO_2 + (1-x)/3\, Co_3O_4 + (1-x)/3\, O_2(g)\uparrow$$

As outlined above, the structural change in Li_xCoO_2 is exothermic and releases oxygen. The exothermic onset point for the unwashed sample is lowered to 160 °C, with a large peak appearing at 167–250 °C. This is not only entirely due to the decomposition of Li_xCoO_2 but also a result of synergy reactions with the electrolyte. Exothermic reactions cannot be found between the electrolyte and the oxygen, indicating that heat production is not caused by such reactions. Since thermal stability is affected by structural stability and reactions with the electrolyte at the surface, differences may arise from the specific surface area of the material.

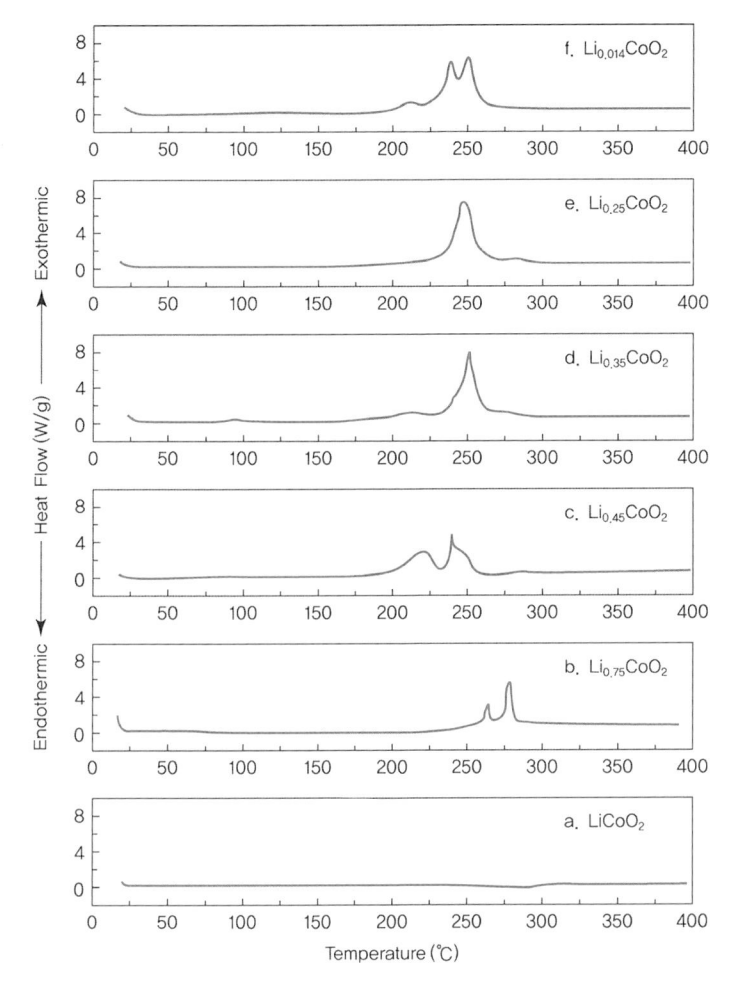

Figure 3.53 DSC analysis of Li_xCoO_2 containing an electrolyte [70]. Reprinted from [70] Copyright 1998, with permission from Elsevier.

Ni-Co-Mn Three-Component Oxide The three-component system of Ni, Mn, and Co exists in various compositions, but we shall focus on $Li[Ni_{1/3}Mn_{1/3}Co_{1/3}]O_2$. This complex oxide has the same structure as Li_xCoO_2 but is far more stable with a superlattice from the different valences between the three elements. Figure 3.55 compares the results of a thermal analysis for $Li_x[Ni_{1/3}Mn_{1/3}Co_{1/3}]O_2$ and Li_xCoO_2.

As shown in the figure above, $Li_x[Ni_{1/3}Mn_{1/3}Co_{1/3}]O_2$ has a higher exothermic onset point and a smaller peak compared to Li_xCoO_2. This can be explained by the structural stability of its superlattice. The exothermic onset point is similar or higher in the Ni–Co–Mn three-component oxide as Ni content increases, while the exothermic peak becomes larger and more frequent. As shown in Figure 3.56 (as-prepared), the exothermic onset point of the Li_xCoO_2 active material has an extremely narrow range.

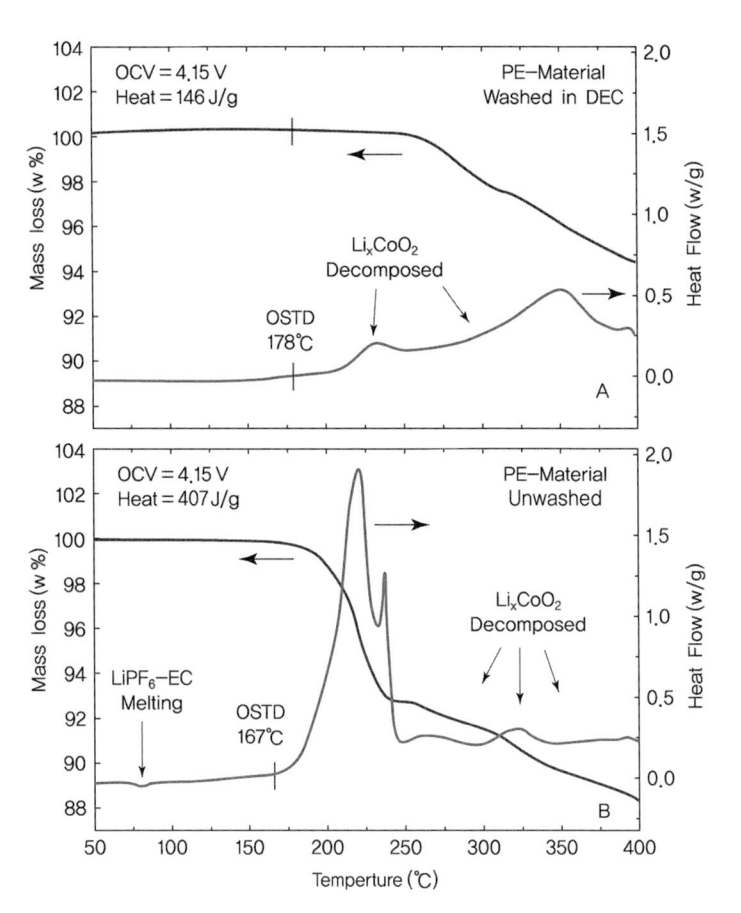

Figure 3.54 DSC curve of Li$_x$CoO$_2$ before and after washing [71].

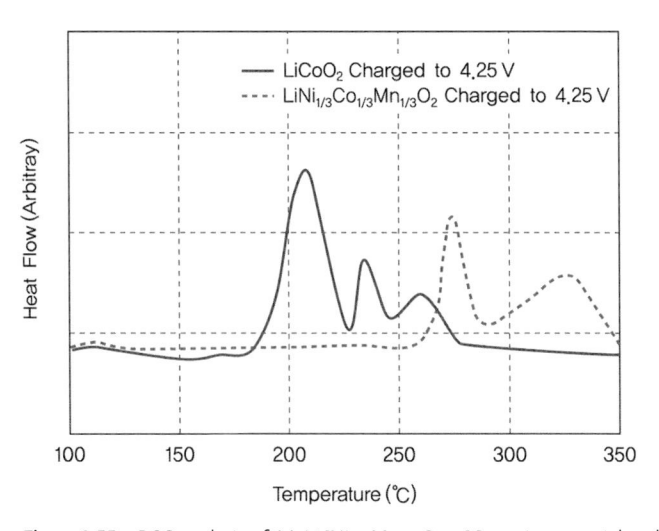

Figure 3.55 DSC analysis of (a) Li$_x$[Ni$_{1/3}$Mn$_{1/3}$Co$_{1/3}$]O$_2$ active material and (b) Li$_x$CoO$_2$ [72].

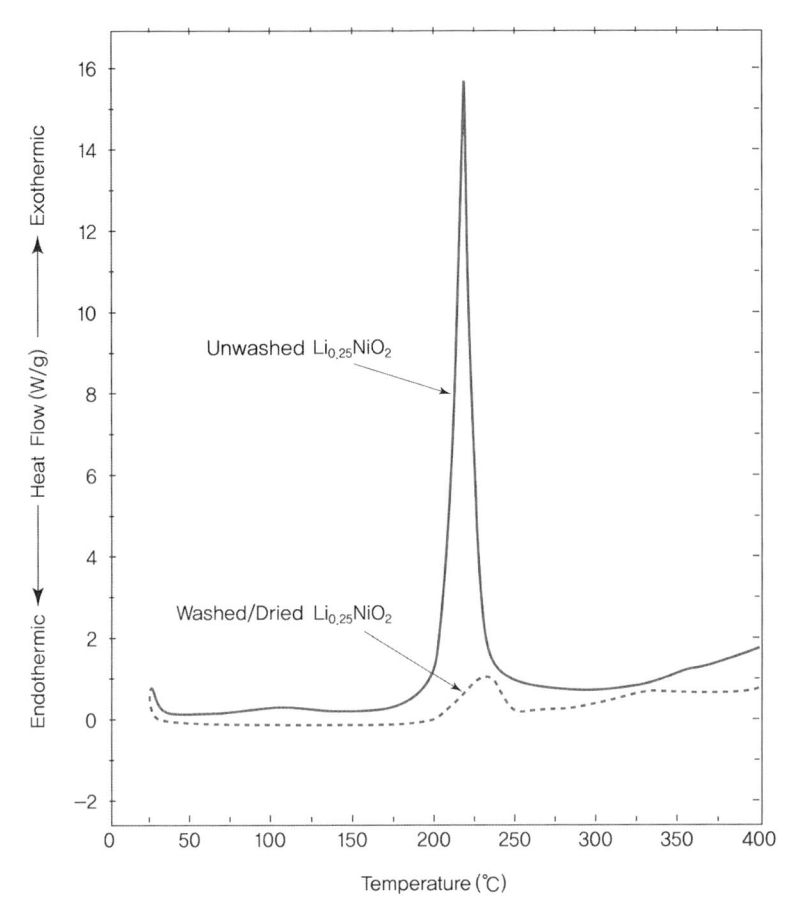

Figure 3.56 DSC analysis of washed/unwashed Li_xNiO_2 [70]. Reprinted from [70] Copyright 1998, with permission from Elsevier.

Even under the same charge voltage, a greater amount of lithium is deintercalated from Li_xNiO_2 due to its larger specific capacity, thus making it more unstable than the three-component system. As such, its exothermic reactions are more significant than those of Li_xCoO_2. In the Li_xNiO_2 active material, these reactions may be reduced by washing its surface with organic solvent. This is because lithium salts or other ingredients in the electrolyte participate in exothermic reactions at the cathode. Figure 3.57 shows the thermal analysis results of Li_xCoO_2 and Li_xNiO_2 obtained under the same charging and electrolyte conditions. In the figure, a larger peak can be seen for the $Li_xNi_{0.86}Co_{0.1}Al_{0.05}O_2$ active material.

Spinel $LiMn_2O_4$ The spinel $LiMn_2O_4$ is thermodynamically stable in both the charged and the discharged state with no heat production from structural change. Figure 3.58 compares the results of thermal analyses of $LiCoO_2$ and spinel $LiMn_2O_4$.

The spinel $LiMn_2O_4$ has a higher exothermic onset point than $LiCoO_2$, but its exothermic peak is significantly large. Considering that there is no structural change in the spinel $LiMn_2O_4$, most heat is produced from reactions between surface

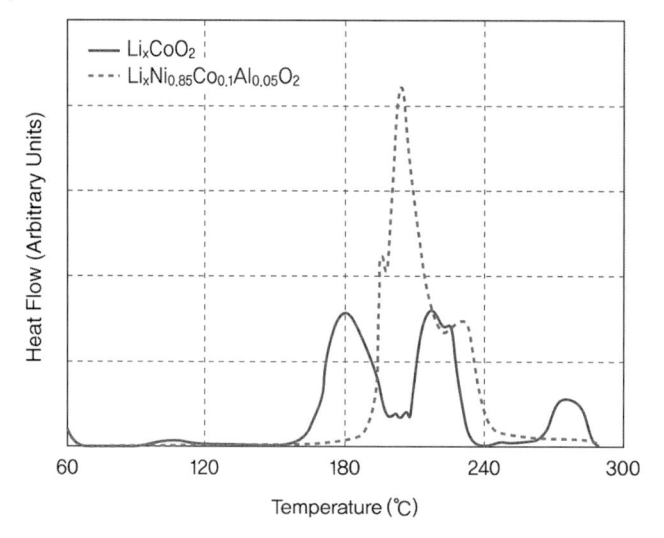

Figure 3.57 DSC analysis of Li_xCoO_2 and Li_xNiO_2 active materials.

particles of the active material and the electrolyte. The thermal stability of the spinel $LiMn_2O_4$ may be enhanced by making adjustments to the electrolyte composition or by reducing the specific surface area of the active material.

$LiFePO_4$ Active Material Since the structure of $LiFePO_4$ is unaffected by charging or heating, it provides outstanding battery stability. Figure 3.59 presents a comparison of the thermal analysis results of $LiCoO_2$ and $LiFePO_4$. The $LiFePO_4$ active material remains unchanged up to 230 °C but exhibits some endothermic reactions with electrolyte vaporization or binder decomposition at around 250 °C. With no

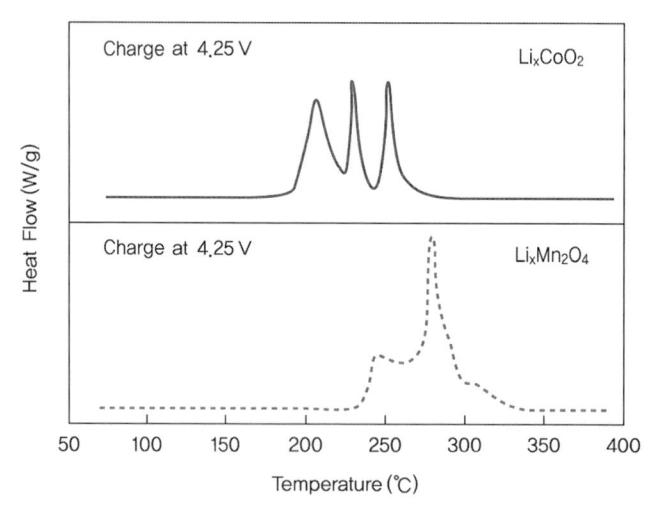

Figure 3.58 DSC analysis of Li_xCoO_2 and $Li_xMn_2O_4$ charged to 4.25 V.

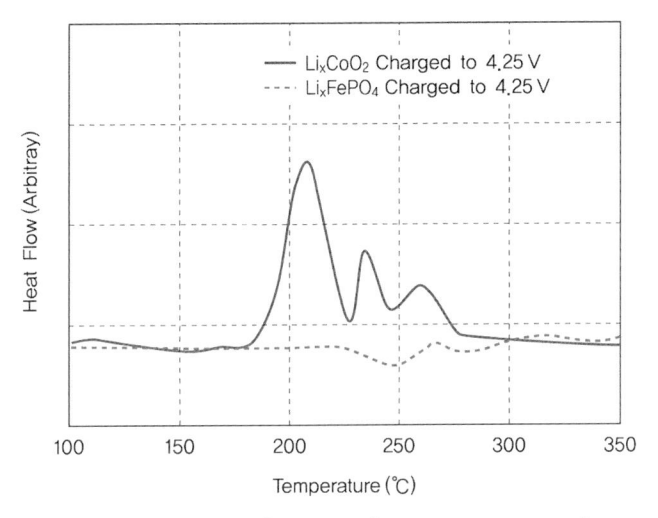

Figure 3.59 DSC analysis of Li_xCoO_2 and Li_xFePO_4 active materials.

exothermic reactions observed between the electrolyte and the cathode, the surface structure of $LiFePO_4$ appears to be extremely stable and shows a great promise for future applications requiring stability.

3.1.6
Prediction of Cathode Physical Properties and Cathode Design

In the development of cathode materials, the charge/discharge working voltage must be considered. The cathode design varies greatly with the working and cutoff voltage of the battery. By examining the design of the cathode, the possibility of new materials may be explored. From the energy levels shown in Figure 3.60, the potential difference of the battery can be inferred on the basis of the difference in Fermi energy between lithium and transition metal ions. In addition, the electrical conductivity of a specific cathode material can be obtained from the energy gap between the redox couple of transition metals and the 2p orbitals of oxygen. This energy band provides more information on the possibility of synthesizing new materials.

Figure 3.61 shows the potential difference of lithium metals in representative cathode materials and the energy levels for redox couples. For instance, since the $3+/4+$ redox couple of nickel ions overlaps with the 2p band of oxygen, O^{2-} decomposes into O_2 gas during charging with the oxidation of nickel and release of electrons from the 2p band. As such, the synthesis of NiO_2 is not possible. When $LiNiO_2$ is charged, nickel ions are oxidized to $4+$ while gas is released from reactions with the 2p band of oxygen.

To develop new cathode materials, it is necessary to predict the possibility of synthesis and the electric potential of the material. It is necessary to carry out calculations on the energy state of electrons. The energy levels of surrounding

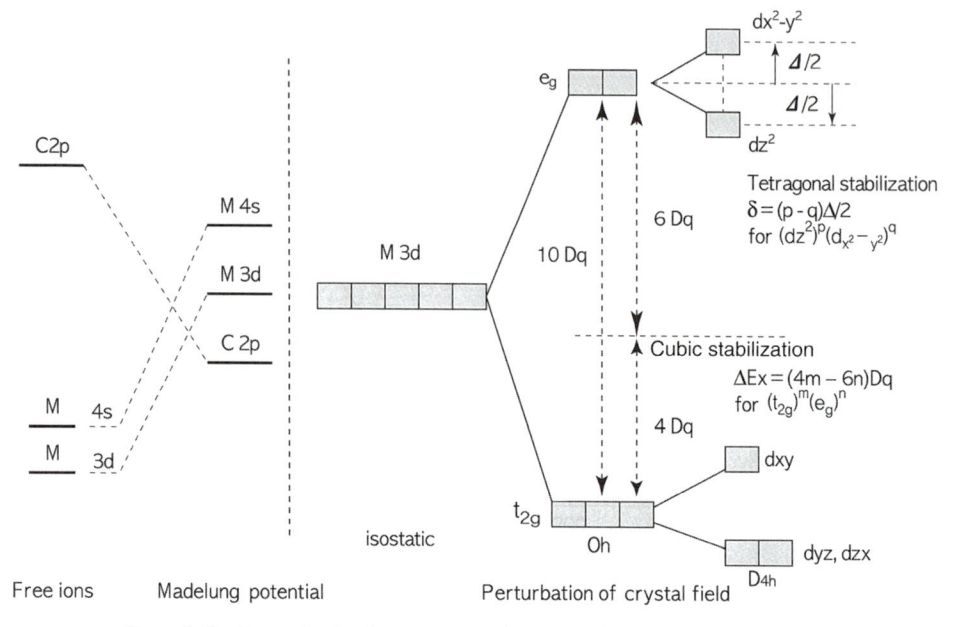

Figure 3.60 Energy levels of transition metal compounds.

electrons are calculated from the wavefunction and current density. For example, the wavefunction of lithium is expressed as $(1s)^2(2s)^1$ and that of each electron can be derived from the Schrödinger wave equation based on the spherical symmetric field of isolated atoms.

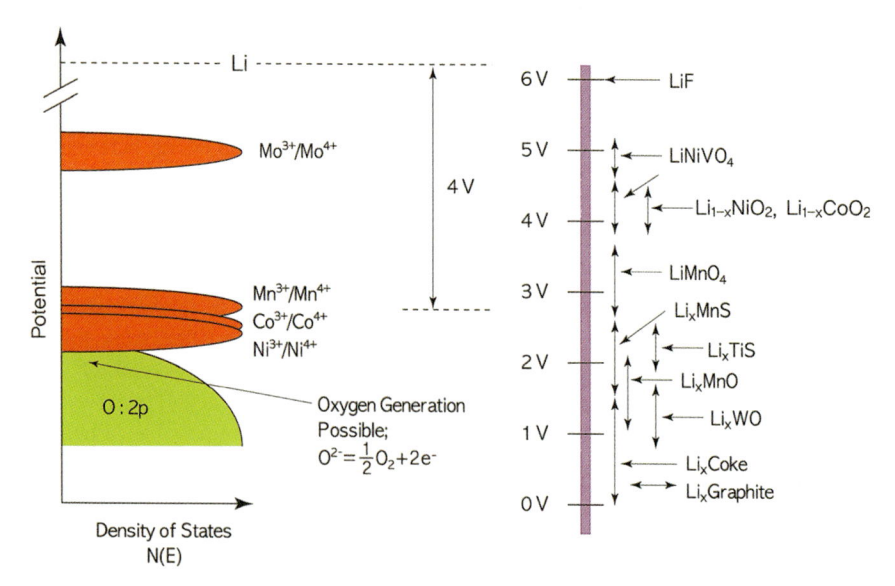

Figure 3.61 Potential difference and redox couple of cathode materials [73]. Reprinted from [73] Copyright 1994, with permission from Elsevier.

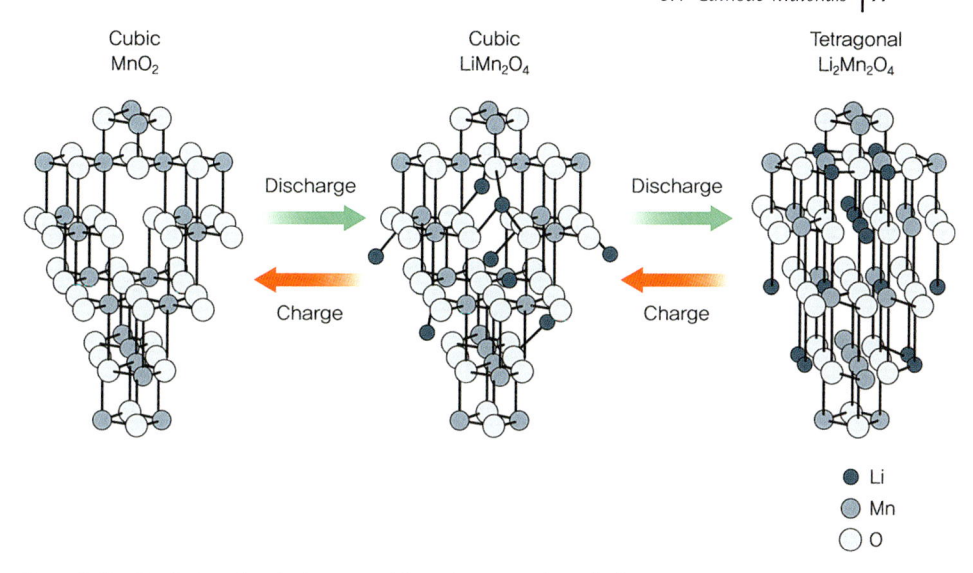

Figure 3.62 Prediction of cathode material from first-principles calculation.

The wavefunction of molecules is found from the linear combination of each atomic orbital function. If we assume that mutual interactions between the electrons of hydrogen atoms A and B are negligible in a hydrogen molecule, variables such as angle θ and φ for the atomic function $X(r)$ need not be considered when examining the 1s spherical symmetric field. Here, the molecular orbital is expressed as $p(r) = C_A X_A(r) + C_B X_B(r)$. However, atomic orbitals are complicated for period 2 elements containing more electrons. For example, the atomic orbital of carbon monoxide includes 10 variables, and its eigenfunction and energy have to be calculated from the determinant of a 10 by 10 matrix. Complex calculations are required even for a simple molecule. To simplify this process, binding mechanisms and chemical reactivity are being studied through experimental methods. With recent advancements in technology, nonexperimental methods such as the first-principles calculation or *ab initio* calculation have been proposed to calculate material characteristics using atomic number and composition. Figure 3.62 shows the prediction results for cathode materials.

3.1.6.1 Understanding of First-Principles Calculation

The first-principles calculation derives physical properties of substances using quantum dynamics based on the most basic information. Unlike other experiential methods, physical properties can be calculated from atomic number and material composition even without experimental data. By applying the principles of quantum dynamics, various properties such as energy and structure can be obtained.

The purpose of the first-principles calculation is to obtain the wavefunction containing all information on the given system by solving a time-dependent Schrödinger equation. However, a time-independent Schrödinger equation is sufficient in the development of cathode materials for secondary batteries. This time-independent Schrödinger equation is written as follows.

$$H\psi = E\psi$$

In the above equation, H is the Hamiltonian operator, ψ is the wavefunction, and E is the total energy of the system. In a system with various electrons and atomic nuclei, it is almost impossible to solve the equation for all particles. Thus, a number of assumptions must be made. According to the Born–Oppenheimer approximation, the state of electrons surrounding the nuclei is immediately determined by nuclear coordinates since electrons move at a faster rate. The position of the nuclei ($\rightarrow Rn$) in the Hamiltonian can be seen as a parameter instead of as a variable. We can rewrite the Hamiltonian in more detail as follows:

$$H = T + V_{ee} + \sum_j v(\vec{r}_j) + \sum_n \sum_{m<n} \frac{Z_m Z_n}{|\vec{R}_m - \vec{R}_n|}$$

Here, T is the kinetic energy, V_{ee} is the electrostatic interactions between electrons, and the third term is the Coulomb interactions between the electrons and the nuclei. The fourth term is known as the Ewald energy or Madelung energy, which arises from Coulomb interactions between atomic nuclei of charge Z. This term can be regarded as a constant and is negligible in the calculation of the wavefunction.

To solve the Schrödinger equation with the above assumptions, the Rayleigh–Ritz variational theorem is used [71]. According to this theorem, the ground-state energy of the Hamiltonian can be found from the following Eq. (3.6):

$$\frac{\langle \varphi | H | \varphi \rangle}{\langle \varphi | \varphi \rangle} = E[\varphi] \geq E_0$$

In other words, $E[\phi]$ becomes the ground-state energy when the square integrable function ϕ is varied to minimize the energy. The single-electron function of the Slater determinant is used to vary the value of ϕ, and this is known as the Hartree–Fock method [72, 74–76]. This allows us to obtain the ground-state energy and the corresponding wavefunction, which contains information on the system's physical properties.

Another method of solving the Schrödinger equation in Eq. (3.4) is the density functional theory with current density as a variable [77, 78]. Dealing with a three-dimensional problem is much simpler than the Hartree–Fock method, in which the wavefunction is of 4 N (N: total number of electrons in the system) dimension. The Hohenberg–Kohn theorem provides a theoretical basis for the reduced dimension. According to this theorem, energy and physical properties during the ground state are determined by current density [77]. Hence, the ground-state energy is given as follows:

$$E[\varrho] = F[\varrho] + \int \varrho(\vec{r}) v(\vec{r}) d\vec{r}$$

$$F[\varrho] = T[\varrho] + V_{ee}[\varrho]$$

Here, $F[\varrho]$ is an independent function of the system, and $v(\rightarrow r)$ is the Coulomb potential energy between atomic nuclei. Energy is no longer a functional of the wavefunction ($E[\phi]$) but of the current density ($E[(\varrho)]$). According to the density functional theory, the ground-state energy of the system is determined by individual potential energy if the value of the independent $F[\varrho]$ is known. Since it is difficult to accurately calculate $F[\varrho]$, an approximation is used. Kohn and Sham used the following method to obtain an approximate value [79]:

$$F[\varrho] = T_s[\varrho] + J[\varrho] + E_{xc}[\varrho]$$

$F[\varrho]$ is calculated by summing three functions, where $T_s[\varrho]$ is the kinetic energy of noninterfering electrons, $J[\varrho]$ is the traditional Coulomb energy known as the Hartree potential, and $E_{xc}[\varrho]$ is the exchange–correlation energy from the quantum effect. This exchange–correlation energy is the difference in kinetic energy of electrons in Eq. (3.8) and that of noninterfering electrons. It accounts for the different values of $J[\varrho]$ arising from mutual interactions between electrons. The kinetic energy $T_s[\varrho]$ is derived using the Slater determinant of an independent electron orbital function, while the Hartree potential $J[\varrho]$ can be found by traditional methods of calculating Coulomb energy.

By minimizing the system energy $E(\varphi)$ through the functional theory, we obtain the following single-electron equations known as the Kohn–Sham equations. Because v_{eff} is adjusted according to $\varrho(r)$, the equation is solved by self-iteration.

$$H\varphi(\vec{r}) = \left[-\frac{1}{2}\nabla^2 + v_{\text{eff}}(\vec{r})\right]\varphi_j(\vec{r}) = \varepsilon_j\varphi_j(\vec{r})$$

$$v_{\text{eff}}(\vec{r}) = v(\vec{r}) + \frac{\delta J[\varrho]}{\delta \varrho} + \frac{\delta E_{xc}[\varrho]}{\delta \varrho}$$

In solving these equations, approximations are needed for the exchange–correlation energy. A widely used method is the LDA (local density approximation), which was first proposed by Kohn and Sham [79]. Under this assumption, the exchange–correlation energy is given as follows:

$$E_{xc}[\varrho] = \int \varrho(\vec{r})\varepsilon_{xc}(\varrho)d\vec{r}$$

Here, ε_{xc} is the exchange–correlation energy for each electron in a uniformly distributed electron cloud. The GGA (generalized gradient approximation) method, which accounts for current density and gradient, was suggested to solve the problems of LDA for nonuniform electron clouds in transition metal oxides [80, 81]. However, since GGA is not always more accurate than LDA, appropriate methods should be applied according to system characteristics.

By applying the wavefunction obtained from the Hartree–Fock method or density functional theory, we can obtain the ground-state energy, which allows calculations of various physical properties. In the next section, we will look at calculations of physical properties based on ground-state energy.

3.1.6.2 Prediction and Investigation of Electrode Physical Properties Using First-Principles Calculation

We can accurately predict the important properties of materials that are electrochemical in nature by solving the Kohn–Sham equations [82–85]. Among the various physical properties of secondary batteries, we focus on intercalation voltage between layers, structural stability of electrode active materials, and lithium diffusion.

Battery Voltage When the cathode changes from $x = x_1$ to x_2, the voltage of the lithium metal anode is given by Eq. (3.13). Here, ΔG is the change in Gibbs free energy for the reaction in Eq. (3.15).

$$V = \frac{-\Delta G}{(x_2 - x_1)ze}$$

$$\text{Li}_{x1}\text{MX(cathode)} + (x_2 - x_1)\text{Li(anode)} \rightarrow \text{Li}_{x2}\text{MX}$$

As can be seen from ΔG in the above reaction, there are three components. The change in internal energy denoted as ΔE can be easily obtained from the difference in the ground-state energy of electrode materials through the first-principles calculation. In Eq. (3.14), after finding the ground-state energy of Li_{x1}MX, Li (metal), and Li_{x2}MX, the change in internal energy is calculated by subtracting the ground-state energy of Li_{x2}MX from the sum of the ground-state energy of Li_{x1}MX and Li (metal). $P\Delta V$ is left out of the equation since it is negligible in solid-state reactions. In this type of solid-state reactions, the value of ΔE is 3–4 eV per molecule, whereas that of $P\Delta V$ is 5–10 eV. $T\Delta S$ is caused by heat energy and has a negligible value of 0.025 eV. The change in Gibbs free energy ΔG is approximated from the change in internal energy ΔE, which can be derived from the first-principles calculation [84, 86, 87].

Structural Stability of Electrode Materials The phase transition that occurs with a change in lithium content in electrode active materials can be obtained using the first-principles calculation. By comparing the thermodynamic energy of substances to be formed from phase transitions, we are able to predict the possible reactions and the products from such reactions. The thermodynamic energy is approximated with the ground-state energy derived from the first-principles calculation. However, much effort is required in predicting phase changes if the structure of by-products is unknown. This is because a comparison must be made between all possible ground-state configurations with varying lithium content. A new approach is needed to simplify this process. For instance, experimental information on possible configurations for a specific amount of lithium may be useful in studying phase changes. For $x = 0.5$ in Li_xMO_2, the possible configurations are spinel, layered, and rock salt. By finding their ground-state energy for the transition metal M, it is easy to predict the respective phase changes of lithium transition metal oxides. We can also study the thermodynamic reactive energy and phase change mechanisms [88].

For cases other than $x = 0$, 0.5, or 1, cluster expansion is recommended, as the arrangement of lithium atoms within the structure becomes irregular [89–91]. New phases may be formed depending on the arrangement of lithium sites and non-lithium sites during charge/discharge, thus affecting battery voltage and stability. With cluster expansion, the ground-state energy of various configurations is first obtained and that of the new structure is predicted based on a geometric relationship. This method allows a more efficient study of the relationship between lithium content and ground-state energy [92].

Lithium Diffusion Lithium diffusion within the electrode active material can be predicted from the first-principles calculation. While diffusion refers to the movement of atoms according to the distribution gradient of the chemical potential during nonequilibrium, it also helps to maintain equilibrium if the motion is not far from the equilibrium state [93]. Here, motion variables such as the lithium diffusion coefficient can be obtained by considering the amplitude of the fluctuation when the system is in equilibrium [73, 94–102]. Most of the lithium is found in crystallographic sites within the electrode active material and is placed on nonequilibrium sites for a very short time. The movement of lithium is assumed to be a continuous process of jumping from one crystallographic site to another, and can be treated statistically. The transition state theory is used to find the frequency of lithium ions jumping between crystallographic sites during equilibrium [103]. This theory represents the various trajectories of lithium as statistical values. The lithium jump frequency is calculated by averaging the movement of lithium and can be written as follows:

$$\Gamma = v^* exp\left(\frac{-\Delta E_b}{k_B T}\right)$$

Here, v^* is the vibration prefactor, and ΔE_b is the activation energy of lithium diffusion. Using the energy obtained from the first-principles calculation, there are

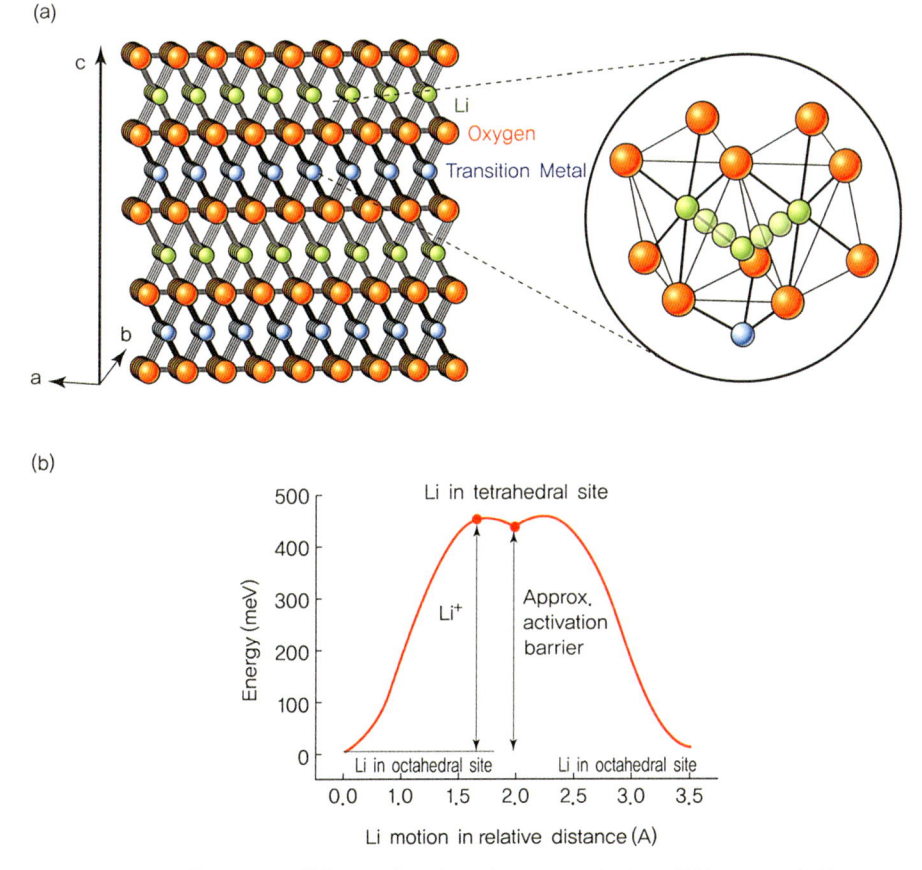

Figure 3.63 (a) Diffusion path of lithium within a layered structure [105]. From [105] Reprinted with permission from AAAS. (b) Change in energy (activation energy) with relative distance.

```
SYSTEM = Rhodium surface calculation

Start parameter for this Run:
ISTART  =   0       Jab   : 0-new 1-cont 2-samecut
ICHARG  =   2       charge: 1-file 2-atom 10-const
INIWAV  =   1       electr: 0-liwe 1-rand

Electronic Relaxation 1
ENCUT   = 200.00 eV
IALGO   =   18      algorithm NEIM = 60; NELMIN = 0; NELMDL = 3 # of
                    ELM steps m
EDIFF   =   1E-04 stopping-criterion for ELM
BMIX = 2.0
TIME = 0.05

Ionic Relaxation
EDIFFG  =  .1E-02  stopping-curterion for IOM
NSW     =   9      number of steps for IOM
IBRION  = 2

POTTM   =   10.0   tine-step for ion-motion

POMASS  = 102.91
ZVAL    = 11.0

DDS related values:
SIGMA = 0.4; ISMEAR = 1 broad. in eV, -4-tet-1-fermi 0-gaus
```

Figure 3.64 INCAR calculations for rhodium surface.

several methods of finding the lithium diffusion path and activation energy. The first is the elastic band method, which predicts the intermediate state of lithium movement by extrapolating the initial and final state [104]. The transition states are metastable, as they are connected by energy values acting as an elastic band. As such, the accurate diffusion path and change in energy can be calculated. Figure 3.63 shows the diffusion of lithium in a layered structure and the corresponding activation energy [105].

```
Cubic BN
   3.57
0.0 0.5 0.5
0.5 0.0 0.5
0.5 0.5 0.0
   1 1
Direct
0.00 0.00 0.00
0.25 0.25 0.25
```

Figure 3.65 POSCAR containing the initial structural information of BN.

Application Programs VASP (Vienna *Ab-Initio* Simulation Package) is a commercial program that is widely used for first-principles calculations based on the functional theory. More details on VASP are available at http://cms.mpi.univie.ac.at/vasp. Some examples of VASP calculations are listed below. The actual calculations begin from four input files (INCAR, POSCAR, POTCAR, and KPOINTS).

INCAR File INCAR is the central input file of VASP. Figure 3.64 shows the general format of an INCAR file. Each line consists of a tag, =, and a corresponding value. Under Ionix Relaxation, NSW = 9 sets the number of ionic steps to 9. Tags and appropriate values are described in http://cms.mpi.univie.ac.at/vasp/vasp/node81.html. In most cases, the default values are used.

POSCAR File The POSCAR file contains the lattice geometry and ionic positions. While it is possible to obtain accurate information on the position of atoms based on the first-principles calculation, it is more efficient to make use of known atomic arrangements. Figure 3.65 shows a general POSCAR file. The first line indicates that the calculation represents a cubic BN structure. The second line provides a scaling factor for lattice constants, while the following three lines give the Bravais lattice vector.

Figure 3.65 shows the three face-centered vectors, each being 3.57 Å. In the sixth line, "1 1" is the number of atoms per atomic species. If it is "2 1 1," there is one species containing two atoms, and two species with one atom each. The lines 7–9 represent the arrangement of these atoms. "Direct" means that the atomic positions are provided in direct coordinates. The last two lines give the three coordinates for each atom.

```
VASP. 4.4.3 10Jun99
POSCAR found: 1 types and 2 ions
LDA part: xc-table for CA standard interpolation
file 10 ok, starting setup
WARNING: warp around errors must be expected
entering main loop
      N       E            dE         d eps    ncg    rms         rms(c)
CG: 1   0.1209934E+02   0.120E+02  -0.175E+03  165  0.475E+02
CG: 2  -0.1644093E+02  -0.285E+02  -0.661E+01  181  0.741E+01
CG: 3  -0.2047323E+02  -0.403E+01  -0.192E+00  173  0.992E+00   0.416E+00
CG: 4  -0.2002923E+02   0.444E+00  -0.915E+01  175  0.854E+00   0.601E-01
CG: 5  -0.2002815E+02   0.107E-02  -0.266E-03  178  0.475E-01   0.955E-02
CG: 6  -0.2002815E+02   0.116E-05  -0.307E-05  119  0.728E-02
    1 F= -.20028156E+02 EO=-.20028156E+02  d E=0.000000E+00
  wrting wavefunctions
```

Figure 3.66 OSZICAR obtained from energy calculations for a diamond structure.

KPOINTS File Energy calculations in VASP are carried out in a reciprocal lattice. KPOINTS determines the density of energy information to be collected from the reciprocal lattice. A small value of KPOINTS is used for a large unit lattice vector, and vice versa. A large KPOINTS involves complex calculations due to the dense amount of information but gives more accurate results. KPOINTS should be set according to the purpose of calculations and material characteristics.

POTCAR File The POTCAR file decides the type of atoms presented in the POSCAR file, and contains the pseudopotential for each atomic species used in the Hamiltonian. VASP supplies the pseudopotential for all atoms in the periodic table, and the user organizes the necessary information for calculations into a single file. This is easily obtained using the Unix command of cat.

After saving the four initial files into a single folder, the >vasp command is entered to begin calculations. Among the various output files generated in the folder, the important ones are OSZICAR, CHG, CHGCAR, and DOSCAR. The OSZICAR file provides important information for the calculation of electrode voltage or the activation energy of lithium diffusion. Figure 3.66 is an example of an OSZICAR file.

The process of finding an optimal electron arrangement is represented by lines containing CG, which indicates the change in energy according to the position of electrons. The last line gives the optimized energy for the ionic arrangement, while E0 is the energy that determines the physical properties of the material. This energy is used to predict the characteristics of electrode active materials. Files such as DOCAR, CHG, and CHGCAR contain information on the electron/spin structure and can predict the electrical conductivity of electrode active materials. This method of predicting the physical properties of active materials through the first-principles calculation can be applied to both the anode and the cathode.

References

1 Mizushima, K., Jones, P.C., Wiseman, P.J., and Goodenough, J.B. (1993) *Mater. Res. Bull.*, **15**, 1159.
2 Winter, M., Besenhard, J.O., Spahr, M.E., and Novak, P. (1998) *Adv. Mater.*, **10**, 725.
3 Tarascon, J.M. and Armand, M. (2001) *Nature*, **414** (15), 359.
4 Ohzuku, T., Ueda, A., and Nagayama, M. (1993) *J. Electrochem. Soc.*, **140**, 1862.
5 Gummow, R.J., Liles, D.C., and Thackeray, M.M. (1993) *Mater. Res. Bull.*, **28**, 235.
6 Gummow, R.J., Thackeray, M.M., David, W.I.F., and Hull, S. (1992) *Mater. Res. Bull.*, **27**, 327.
7 Reimers, J.N. and Dahn, J.R. (1992) *J. Electrochem. Soc.*, **139**, 2091.
8 Armstrong, A.R. and Bruce, P.G. (1996) *Nature*, **381**, 499.
9 Armstrong, A.R., Paterson, A.J., Robertson, A.D., and Bruce, P.G. (2002) *Chem. Mater.*, **14**, 710.
10 Cho, J.P., Kim, Y.J., Kim, T.J., and Park, B. (2002) *J. Elelctrochem. Soc.*, **149** (2), A127.
11 Komaba, S., Myung, S.T., Kumagai, N., Kanouchi, T., Oikawa, K., and Kamiyama, T. (2002) *Solid State Ionics*, **152–153**, 311.
12 Lee, Y.S., Sato, S., Tabuchi, M., Yoon, C.S., Sun, Y.K., Kobayakawa, K., and Sato, Y. (2003) *Electrochem. Commun.*, **5**, 549.
13 Matsumura, T., Kanno, R., Inaba, Y., Kawamoto, Y., and Takano, M. (2002) *J. Electrochem. Soc.*, **149** (12), A1509.

14 Koyama, Y., Yabuuchi, N., Tanaka, I., Adachi, H., and Ohzuku, T. (2004) *J. Electrochem. Soc.*, **151** (10), A1545.
15 Choi, J. and Manthiram, A. (2005) *J. Electrochem. Soc.*, **152** (9), A1714.
16 Yoon, W.S., Paik, Y., Yang, X.Q., Balasubramanial, M., McBreen, J., and Grey, C.P. (2002) *Electrochem. Solid State Lett.*, **5**, A263.
17 (1) Lu, Z., MacNeil, D.D., and Dahn, J.R. (2001) *Electrochem. Solid State Lett.*, **4**, A191;(2) MacNeil, D.D. and Dahn, J.R. (2002) *J. Electrochem. Soc.*, **149**, A912;(3) Ohzuku, T., and Makimura, Y. (2001) *Chem. Lett.*, **7**, 642.
18 Lu, Z., MacNeil, D.D., and Dahn, J.R. (2001) *Electrochem. Solid State Lett.*, **4**, A200.
19 Lu, Z. and Dahn, J.R. (2002) *J. Electrochem. Soc.*, **149**, A815.
20 (1) Robertson, A.D. and Bruce, P.G. (2003) *Chem. Mater.*, **15**, 1984;(2) Hong, Young-Sik, Park, Yong Joon, Ryu, Kwang Sun, Chang, Soon Ho, and Kim, Min Kyu (2004) *J. Mater. Chem.*, **14** (9), 1424–1429.
21 Park, C.W. et al. (2007) *Mater. Res. Bull.*, **42**, 1374.
22 Thackeray, M. et al. (2007) *J. Mater. Chem.*, **17**, 3112.
23 (1) Jiang, J., Eberman, K.W., Krause, L.J., and Dahn, J.R. (2005) *J. Electrochem. Soc.*, **152**, A1879;(2) Tran, N., Groguennec, L., Labrugere, C., Jordy, C., Biensan, Ph., and Delmas, C. (2006) *J. Electrochem. Soc.*, **153**, A261.
24 Hong, Young-Sik, (2006) Chemworld, **46** (8), 45.
25 Hong, Young-Sik, Park, Yong Joon, Ryu, Kwang Sun, and Chang, Soon Ho (2005) *Solid State Ionics*, **176** (11–12), 1035.
26 Kim, J.S., Johnson, C.S., Vaughey, J.T., Thackeray, M.M., and Hackney, S.A. (2004) *Chem. Mater.*, **16** (10), 1996.
27 Johnson, Christopher S., Kim, J.S., Kropf, A. Jeremy, Kahaian, A.J., Vaughey, J.T., and Thackeray, M.M. (2002) *Electrochem. Commun.*, **4** (6), 492.
28 Tarascon, J.M., Wang, E., Shokoohi, F.K., McKinnon, W.R., and Colson, S. (1991) *J. Electrochem. Soc.*, **138**, 2859.
29 Myung, S.T., Komaba, S., and Kumagai, N. (2001) *J. Electrochem. Soc.*, **148** (5), A482.
30 Goodenough, J.B., Padhi, A.K., Nanjundaswamy, K.S., and Masquelier, C. (1999) U.S. Patent No. 5910382.
31 Yamada, A., Chung, S.C., and Hinouma, K. (2001) *J. Electrochem. Soc.*, **148** (3), A224.
32 Yamada, A., Kudo, Y., and Liu, K.Y. (2001) *J. Electrochem. Soc.*, **148** (7), A747.
33 Huang, Y.H., Park, K.S., and Goodenough, J.B. (2006) *J. Electrochem. Soc.*, **153** (12), A2282.
34 Chung, S.Y., Bloking, J.T., and Chiang, Y.M. (2002) *Nat. Mater.*, **1** (2), 123.
35 Herle, P.S., Ellis, B., Coombs, N., and Nazar, L.F. (2004) *Nat. Mater.*, **3** (3), 147.
36 Franger, S., Le Cras, F., Bourbon, C., and Rouault, H. (2003) *J. Power Sources*, **119**, 252.
37 Yamada, A. et al. (2001) *J. Electrochem. Soc.*, **148**, A960.
38 Li, G. et al. (2002) *Electrochem. Solid State Lett.*, **5**, A135.
39 Delacourt, C. (2004) *Chem. Mater.*, **16**, 93.
40 Zhou, F. et al. (2004) *Solid State Commun.*, **132** (3–4), 181.
41 Zhou, F. et al. (2004) *Electrochem. Commun.*, **6** (11), 1144; Wolfenstine, J. (2005) *J. Power Sources*, **142**, 389–390.
42 Manthiram, A. and Kim, J. (1998) *Chem. Mater.*, **10**, 2895.
43 Zhang, S., Li, W., Li, C., and Chen, J. (2006) *J. Phys. Chem. B*, **110**, 24855.
44 Whittingham, M.S. (2004) *Chem. Rev.*, **104**, 4271.
45 Liu, G.Q., Xu, N., Zeng, C.L., and Yang, K. (2002) *Mater. Res. Bull.*, **37**, 727.
46 Xia, H., Jiao, L.F., Yuan, H.T., Zhao, M., Zhang, M., and Wang, Y.M. (2007) *Mater. Lett.*, **61**, 101.
47 Leger, C., Bach, S., and Ramos, J. (2007) *J. Solid State Electrochem.*, **11**, 71.
48 Xiao, K., Wu, G., Shen, J., Xie, D., and Zhou, B. (2006) *Mater. Chem. Phys.*, **100**, 26.
49 Cao, A.M., Hu, J.S., Liang, H.P., and Wan, L.J. (2005) *Angew. Chem. Int. Ed.*, **44**, 4391.

50 Sudant, G., Baudrin, E., Dunn, B., and Tarascon, J.M. (2004) *J. Electrochem. Soc.*, **151**, A666.
51 Li, X., Li, W., Ma, H., and Chen, J. (2007) *J. Electrochem. Soc.*, **154**, A39.
52 Kannan, A.M. and Manthiram, A. (2003) *Solid State Ionics*, **159**, 265.
53 Tsang, C. and Manthiram, A. (1997) *J. Electrochem. Soc.*, **144**, 520.
54 Baudrin, E., Sudant, G., Larcher, D., Dunn, B., and Tarascon, J.M. (2006) *Chem. Mater.*, **18**, 4374.
55 Zhang, S., Li, W., Li, C., and Chen, J. (2006) *J. Phys. Chem. B*, **110**, 24855.
56 Callister, W.D., Jr. (1997) *Materials Science and Engineering: An Introduction*, John-Wiley & Sons, Inc., New York, p. 787.
57 Cho, J. et al. (2001) *Angew. Chem. Int. Ed.*, **40**, 3367.
58 Cho, J. et al. (2001) *Electrochem. Solid State Lett.*, **4**, A159.
59 Kim, K. et al. (2005) *Electrochim. Acta*, **50**, 3764.
60 Sun, Y.K. et al. (2007) *J. Electrochem. Soc.*, **154** (3), A168.
61 Gu, Y., Chen, D., Jiao, X., and Liu, F. (2006) *J. Mater. Chem.*, **16**, 4361.
62 Kim, D.H. and Kim, J. (2006) *Electrochem. Solid State Lett.*, **9** (9), A439.
63 Ravet, N., Goodenough, J.B., Besner, S., Simoneau, M., Hovington, P., and Armand, M. (1999) The Electrochemical Society Meeting Abstract, Honolulu, HI, Oct. 17–22, Abstract 127.
64 Huang, H., Yin, S.C., and Nazar, L.F. (2001) *Electrochem. Solid State Lett.*, **4**, A170.
65 Park, K.S., Son, J.T., Chung, H.T., Kim, S.J., Lee, C.H., Kang, K.T., and Kim, H.G. (2004) *Solid State Commun.*, **129**, 311.
66 Amatucci, G.G., Tarascon, J.M., and Klein, L.C. (1996) *Solid State Ionics*, **83**, 168.
67 Kannan, A.M. and Manthiram, A. (2002) *Electrochem. Solid State Lett.*, **5** (7), A167.
68 Park, K.S., Son, J.T., Chung, H.T., Kim, S.J., Lee, C.H., Kang, K.T., and Kim, H.G. (2004) *Solid State Commun.*, **129** (5), 311.
69 Huang, H. et al. (2001) *Electrochem. Solid State Lett.*, **4**, A170.
70 Zhang, Z. et al. (1998) *J. Power Sources*, **70**, 16.
71 Maleki, H., Deng, G., Anani, A., and Howard, J. (1999) *J. Electrochem. Soc.*, **146** (9), 3224.
72 Bransden, B.H. and Joachin, C.J. (1989) *Introduction to Quantum Mechanics*, Longman Group UK Limited, p. 379.
73 Goodenough, J.B. (1994) *Solid State Ionics*, **69** (3–4), 184.
74 Lowe, J.P. (1993) *Quantum Chemistry*, Academic Press, Inc., p. 627.
75 Bethe, H.A. and Jackiw, R. (1997) *Intermediate Quantum Mechanics*, Addison–Wesley Longman, p. 55.
76 Gross, E.K.U., Runge, E., and Heinonen, O. (1991) *Many-Particle Theory*, IOP Publishing, p. 51.
77 Parr, R.G. and Yang, W. (1989) *Density-Functional Theory of Atoms and Molecules*, Oxford University Press, p. 7.
78 Hohenberg, P. and Kohn, W. (1964) *Phys. Rev.*, **136** (3B), 864.
79 Jones, R.O. and Gunnarsson, O. (1989) *Rev. Mod. Phys.*, **61**, 689.
80 Kohn, W. and Sham, L.J. (1965) *Phys. Rev.*, **140** (4A), 1133.
81 Perdew, J. and Yue, W. (1986) *Phys. Rev. B*, **33**, 8800.
82 Perdew, J., Burke, K., and Ernzerhof, M. (1996) *Phys. Rev. Lett.*, **77**, 3865.
83 Van der Ven, A. and Ceder, G. (2000) *Electrochem. Solid State Lett.*, **3** (7), 301.
84 Van der Ven, A. and Ceder, G. (2001) *Phys. Rev. B*, **64** (18), 184307.
85 Aydinol, M.K. et al. (1997) *Phys. Rev. B*, **56** (3), 1354.
86 Kang, K. et al. (2004) *Chem. Mater.*, **16**, 2685.
87 Aydinol, M.K. and Ceder, G. (1997) *J. Electrochem. Soc.*, **144** (11), 3832.
88 Aydinol, M.K., Kohan, A.F., and Ceder, G. (1997) *J. Power Sources*, **68** (2), 664.
89 Reed, J., Ceder, G., and Van Der Ven, A. (2001) *Electrochem. Solid State Lett.*, **4** (6), A78.
90 Tepesch, P.D., Garburlsky, G.D., and Ceder, G. (1995) *Phys. Rev. Lett.*, **74**, 2272.
91 de Fontaine, D. (1994) *Solid State Physics: Advances in Research and Applications* (eds H. Ehrenreich and D. Turnbull), vol. **47**, Academic, New York, pp. 33–176.

92 Sanchez, J.M., Ducastelle, F., and Gratias, D. (1984) *Physica A*, **128**, 334.
93 Van der Ven, A. *et al.* (1998) *Phys. Rev. B*, **58** (6), 2975.
94 de Groot, S.R. and Mazur, P. (1984) *Non-Equilibrium Thermodynamics*, Dover Publications, New York.
95 Onsager, L. (1931) *Phys. Rev.*, **37**, 405.
96 Onsager, L. (1931) *Phys. Rev.*, **38**, 2265.
97 Callen, H.B. and Welton, T.A. (1951) *Phys. Rev.*, **83**, 34.
98 Callen, H.B. and Greene, R.F. (1952) *Phys. Rev.*, **86**, 702.
99 Green, M.S. (1952) *J. Chem. Phys.*, **20**, 1281.
100 Green, M.S. (1954) *J. Chem. Phys.*, **22**, 398.
101 Kubo, R. (1957) *J. Phys. Soc. Jpn.*, **12**, 570.
102 Kubo, R., Yokota, M., and Nakajima, S. (1957) *J. Phys. Soc. Jpn.*, **12**, 1957.
103 Zwanzig, R. (1964) *J. Chem. Phys.*, **40**, 2527.
104 Zwanzig, R. (1965) *Annu. Rev. Phys. Chem.*, **16**, 67.
105 Kang, K., Meng, Y.S., Breger, J., Grey, C.P., and Ceder, G. (2006) *Science*, **311**, 977.

3.2
Anode Materials

3.2.1
Development History of Anode Materials

In the early stage of development, lithium metal was used as an anode material in lithium secondary batteries. While lithium metal is able to achieve a high specific capacity during repeating charge and discharge, the formation of dendrites on the surface of the metal leads to safety issues of Li batteries such as short circuits [1–3]. For this reason, there are many constraints to the production of lithium metal batteries, necessitating extra care. Also, it is necessary to avoid the occurrence of excessive exothermic reactions with moisture.

In efforts to overcome these problems, many studies were conducted in the 1970–1980s to replace lithium metal as an anode material. Such research was focused on the reaction of lithium ions with carbon materials such as graphite, metals, and metal compounds.

The use of carbon-based anode materials helps to resolve safety issues at the lithium metal electrode, since lithium ions can be inserted within this type of anode, thereby maintaining a stable state. In this case, the electrochemical reaction potential of carbon-based materials with lithium ions is close to that of lithium metal. It is also possible to obtain continuous, repeated redox reactions since there is no significant effect on the change in the crystal structure of carbon-based anode materials during the intercalation and deintercalation of lithium ions. These key factors allow the realization of high-energy density and long cycle life in lithium secondary batteries and finally led to their commercialization in 1991.

Modification of carbon-based anode materials, where various structures of carbons change the storage mechanism of lithium, has been carried out in order to enhance their storage capacity and thereby realize high-performance lithium secondary batteries. The market for large-scale secondary batteries including electric vehicles requires lithium secondary batteries offering high electric power characteristics. Also, in order to realize high specific capacity in anodes, studies of noncarbon-based materials such as silicon and tin are being carried out. Furthermore, to achieve high output power anodes, it is necessary to develop carbon-based materials with outstanding electronic and ionic conductivities.

3.2.2
Overview of Anode Materials

In lithium secondary batteries, oxidation reactions occur spontaneously in anode materials during discharge, while reduction reactions spontaneously take place in cathode materials. For instance, in a lithium secondary battery consisting of $Li_xC/Li_{1-x}CoO_2$, the anode material Li_xC donates electrons and lithium ions. In this case, Li_xC itself oxidizes. Similarly, the cathode material $Li_{1-x}CoO_2$ accepts electrons and lithium ions. In this case, $Li_{1-x}CoO_2$ itself reduces. At the anode, lithium ions are stored and released during charge and discharge, respectively.

In the case of the graphite anode, one lithium reacts theoretically with six carbons, as shown in the equation below. The voltage of the graphite anode in comparison with the lithium electrode (Li^+/Li) ranges from 0.0 to 0.25 V and its theoretical specific capacity is 372 mAh/g. The potential of pristine graphite is 3.0 V but rapidly declines when lithium is intercalated. For both the cathode and the anode, the potential of the lithium electrode decreases by increasing lithium within the electrode active material, finally reaching 0 V.

$$Li_xC_6 \rightarrow C_6 + xLi^+ + xe^- \qquad 0.00-0.25 \text{ V (versus } Li^+/Li)$$

Decomposition of the electrolyte occurs at the surface of the anode during charging since its reduction potential is relatively higher than that of lithium. Such decomposition not only causes the formation of solid electrolyte interphase at the surface of the electrode but also suppresses the electron transfer reactions between the anode and the electrolyte, thus preventing further electrolyte decomposition. The performance of Li batteries is strongly influenced by the characteristics of the SEI film deposited at the electrode surface. Many attempts are being made to produce a more compact SEI layer with outstanding electrochemical properties by introducing additives that induce decomposition prior to the electrolyte decomposition.

Anode materials influence the performance of lithium secondary batteries, including energy density, power density, and cycle life. To maximize the performance of lithium secondary batteries, anode materials should fulfill the following conditions:

1) Anode materials should have a low potential corresponding to a standard electrode and provide a high cell voltage with the cathode. The potential relating to electrochemical reactions must be a close approximation of the electrochemical potential of lithium metal.
2) No significant change in the crystal structure should occur during reactions with lithium ions. Change in structure leads to the accumulation of crystal strain and hinders the reversibility of electrochemical reactions, thus resulting in poor cycle life characteristics.
3) Anode materials should engage in highly reversible reactions with lithium ions. Ideal reversible reactions have 100% charge/discharge efficiency, indicating no change in reaction efficiency with advanced cycles.

Table 3.3 Characteristics of anode materials.

Anode	Theoretical capacity (mAh/g)	a)Practical capacity (mAh/g)	Avg. potential (V)	True density (g/cc)
Li metal	3800	—	0.0	0.535
Graphite	372	~360	~0.1	2.2
Cokes	—	~170	~0.15	<2.2
Silicon	4200	~1000	~0.16	2.36
Tin	790	~700	~0.4	7.30

a) Practical capacity: Commercially available capacity.

4) Fast diffusivity of lithium ions is required within the active electrode material at the anode since this is particularly important to realize cell performance.
5) High electronic conductivity is necessary to facilitate the movement of electrons during electrochemical reactions.
6) Active anode materials should be sufficiently dense so as to obtain a high electrode density. This is an important design factor that is considered to enhance battery energy. For example, graphite material shows a theoretical density of 2.2 g/ml and a theoretical capacity density of 818 mAh/ml.
7) Materials should store a large amount of charge (Coulomb) per unit mass.

Other important factors that determine energy density and power are specific surface area, tap density, particle size, and distribution. Because the anode has a large specific capacity per unit mass, it is more difficult to intercalate or deintercalate lithium ions in comparison with the cathode. As such, the design of the anode should take into account the fast movement of lithium ions in order to enhance the performance of Li batteries. Table 3.3 shows the main characteristics of common anode material.

3.2.3
Types and Electrochemical Characteristics of Anode Materials

3.2.3.1 Lithium Metal
Lithium metal has a body-centered cubic (bcc) with a high tendency of ionization and an atomic radius of 0.76 Å. It has a small atomic weight (6.941), low density (0.534 g/cc), and a very low standard electrode potential (-3.04 V SHE) that leads to a high specific capacity of 3860 mAh/g. However, batteries using lithium metal electrode have not been commercialized due to the metal's low melting point of 180.54 °C and safety issues caused by dendritic growth. Recent studies have attempted to stabilize lithium metal by coating the surface with polymeric or inorganic substances. Despite these efforts, lithium metal still accompanies many difficulties such as the risk of explosive reactions, when exposed to water (moisture), and the complex process of electrode manufacturing.

By overcoming the problems of the lithium metal electrode, it could potentially be used as an anode of lithium secondary batteries. Some companies have recently attempted to use lithium metal as a supplementary material. By adding lithium metal, the initial irreversibility of the anode is compensated by the oxidation of lithium metal. Therefore, overconsumption of the lithium resource of cathode should be prevented in order to increase the energy density of the battery.

3.2.3.2 Carbon Materials

There are several types of carbon materials, for example, graphite with an sp^{2-} hybrid orbital, diamond with a sp^{3-} hybrid orbital, and carbine with a sp^- hybrid orbital. Such carbon allotropes exhibit various structures and physical/chemical characteristics, with some capable of reversible intercalation/deintercalation of lithium. They can be used as anode materials in lithium ion batteries and are classified into graphite and nongraphite. This section describes graphite and nongraphite carbon materials.

Graphite

Structure of Graphite In graphite, graphene layers are conductive with carbon atoms of the sp^{2-} hybrid orbital layered along a hexagonal plane. In addition, delocalized π electrons have van de Waals bondings between graphene layers. Since π electrons have freedom to move between graphene layers, graphite has a good electronic conductivity. The π electrons are comprised of weak van der Waals interactions, whereas the graphene layers are anistropic with strong covalent bonds. Lithium ions are intercalated and deintercalated between these graphene layers.

As shown in Figure 3.67, graphite usually has a hexagonal structure arranged in ABAB along the *c*-axis and also takes the form of a rhombohedral structure in ABCABC stacking.

The graphite crystal is anistropic due to the perpendicular basal plane and the parallel edge plane with respect to the *c*-axis. This anistropic behavior of graphite

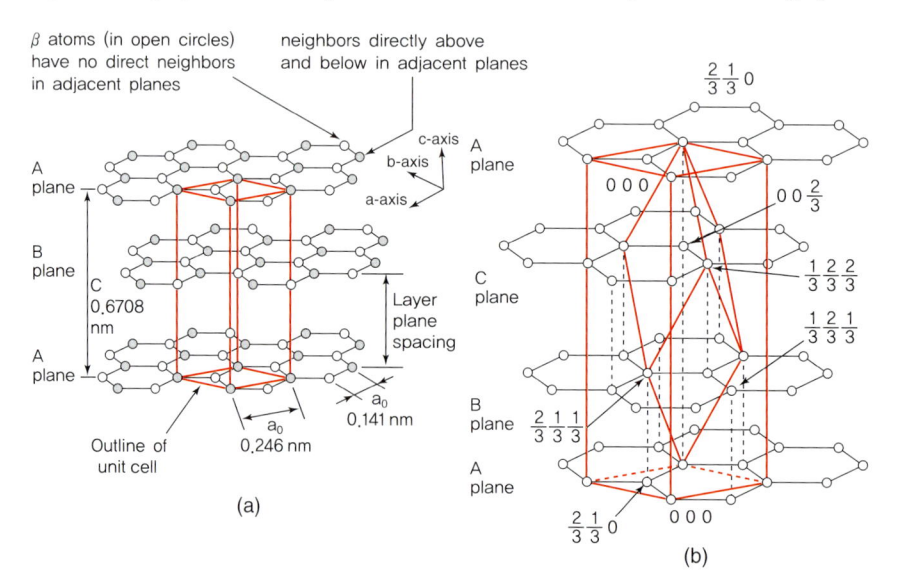

Figure 3.67 (a) Hexagonal unit cell [1] and (b) rhombohedral unit cell [2] of graphite.

affects electrochemical reactions at the anode of lithium secondary batteries. While the basal plane is relatively nonactive for electrochemical reactions, the edge plane shows a great amount of activity. As such, electrochemical characteristics are determined by the ratio of the basal plane and edge plane. Furthermore, the high reactivity of the graphite edge plane expedites the formation of surface functional groups including oxygen atoms. For this reason, artificial (synthetic or pyrolytic) graphite having many edge planes is produced by heating pitch cokes at temperature higher than 2500 °C.

Carbon that can be graphitized under heat treatment is called graphitizable carbon. Graphite is more easily formed at high temperature as the arrangement of atoms facilitates layer structuring. This material is also known as soft carbon. Carbon that does not transform into graphite structure at temperature above 2500 °C is called nongraphitizable carbon or hard carbon. At relatively low temperature (below 1000 °C), small graphite planes in graphitizable carbon are stacked parallel but show turbostratic disorder along the c-axis. As the temperature increases, the graphite plane becomes larger and the planes are stacked in order. For soft carbon, turbostratic disorder is less significant at temperatures higher than 2000 °C. Heat treatment of soft carbon at around 3000 °C produces well-structured graphite, as shown in Figure 3.68.

To facilitate carbonization, the carbon precursor should include high-density aromatic hydrocarbons that are suitable for transformation into graphite planes. Also, graphite planes that are close to one another should be properly aligned. Such weak links between carbon layers in graphitizable carbon allow rearrangement into a graphite structure.

In general, the graphitization process includes the expansion and stacking of graphitic planes, in a three-dimensional manner. During graphitization, graphite undergoes an increase in density and crystallite sizes (L_a and L_c), respectively and a decrease in (002) interplanar spacing. Here, L_a is the crystallite size parallel to the basal plane and L_c is the crystallite size perpendicular to the basal plane. L_c and L_a are the crystallite sizes in the a- and c-axis, respectively. The crystal structure is

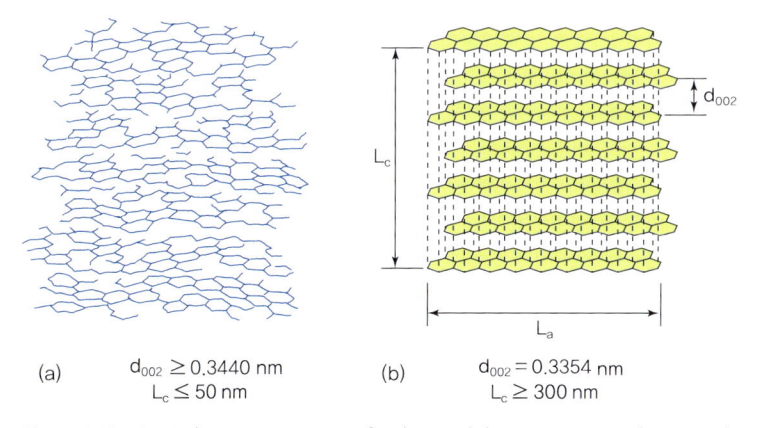

(a) $d_{002} \geq 0.3440$ nm
$L_c \leq 50$ nm

(b) $d_{002} = 0.3354$ nm
$L_c \geq 300$ nm

Figure 3.68 (a) Turbostratic structure of carbon and (b) comparison with 3D graphite lattice [3].

determined by various factors such as crystallite size, d_{002} interplanar spacing, and the graphitization temperature. By studying these values, we are able to predict the usability of carbon materials as anodes in lithium secondary batteries. L_a and L_c can be obtained through an XRD analysis and the equation $t = k\lambda/\beta \cos(\theta)$, where t is the height of stacked layers of L_a or L_c, k is the shape factor of crystallites, and L_c and L_a are 0.9 and 1.84, respectively [3]. Here, λ is the X-ray wavelength, β is the width of the diffraction peak, and θ is the angle of incidence.

Electrochemical Reaction of Graphite During charging, carbon materials engage in reduction reactions and lithium ions are intercalated to form Li_xC compounds. During discharge, oxidation reactions occur and lithium ions are removed from carbon materials. In these charge/discharge reactions, the electrochemical characteristics of carbon materials such as reaction potential and lithium storage capacity differ depending on crystallinity, microstructure, and the shape of particles.

In graphite, lithium ions are intercalated through the edge plane or structural defects in the basal plane. Most intercalation occurs at potentials below 0.25 V. When there is a low concentration of lithium ions during the initial intercalation stage, a single lithium ion layer is formed, and lithium ions are not intercalated into adjacent graphene layers. Graphene layers without lithium ions are periodically arranged, and the concentration of lithium ions in graphite increases. With more intercalation of lithium ions, the number of unfilled graphene layers between lithium ion layers is reduced. In the composition of LiC_6, which has the highest amount of lithium intercalation, the layers of lithium ions and graphite are arranged one after another. This gradual lithium intercalation to graphene layers is known as the phenomenon of staging, which is schematically represented in Figure 3.69 [4].

As shown in Figure 3.69, staging from the intercalation of lithium ions is observed in the form of a continuous flat potential in the charging potential curve measured under a constant charge. The flat potential implies that two phases coexist. The high stage changes into low stage with an increased concentration of lithium, and inverse reactions occur during deintercalation (discharge). When lithium ions are intercalated, graphene layers shift from ABAB to AAAA. In the fully charged stage 1 of LiC_6, the two adjacent graphene layers exist in a high sequence order [5], and the interplanar spacing is 0.370 nm, as shown in Figure 3.70a [6, 7]. In LiC_6, lithium ions intercalated between graphene layers are not arranged adjacently, and the distance between lithium ions is about 0.430 nm, as shown in Figure 3.70b. The specific capacity per unit mass is 372 mAh/g and corresponds to the theoretical capacity of graphite.

The design of lithium secondary batteries should consider the expansion in volume arising from the structural change of graphite during charging. If this factor is not taken into account, the expanded anode may cause distortion of electrodes, thus hindering cycle life and other battery performance metrics.

Design of Graphite Particles One example of artificial graphite is MCMB (mesophase carbon microbeads). An anisotropic small sphere is formed when raw materials such as coal tar, coal tar pitch, and petroleum are heated to 400 °C.

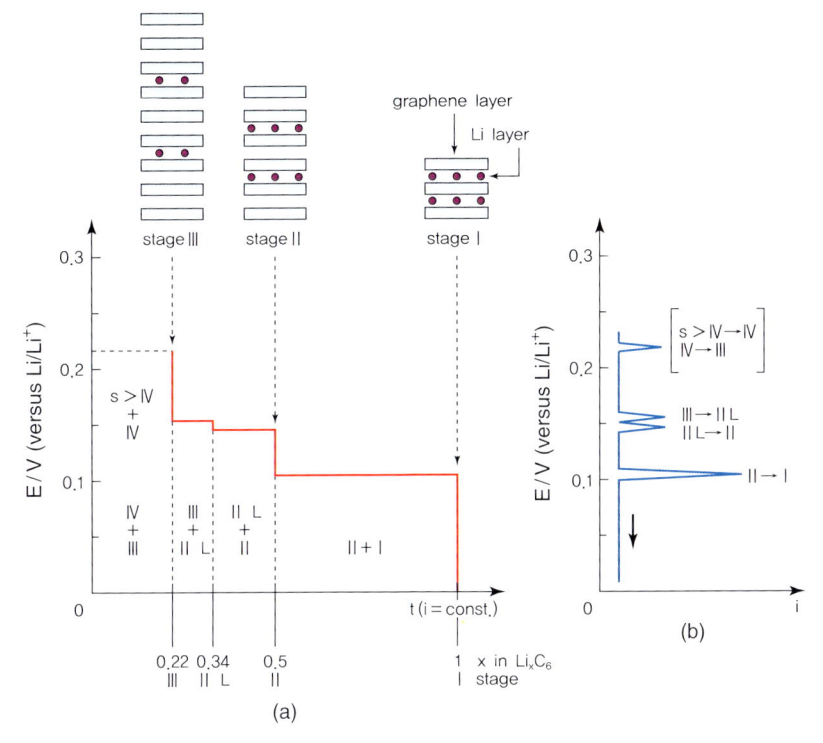

Figure 3.69 Staging effects during lithium intercalation between graphene layers: (a) schematic galvanostatic curve; (b) schematic voltammetric curve [4]. Reproduced from [4], with permission from Wiley-VCH.

With thermal decomposition and condensation reactions, planar molecules of the polycyclic aromatic compound are stacked in the same direction, resulting in the lamellar structure illustrated in Figure 3.71. Due to the difference in viscosity with its surroundings, the sphere takes the form of a liquid-type crystal. When the pitch becomes semicoke, the small sphere is transformed into a mesophase with optical anisotropy and liquidity. This mesophase sphere is known as MCMB, which is a type of artificial graphite.

Figure 3.72 shows SEM images of a commercialized graphite anode of MCMB-25-28 (average grain size 25 μm, graphitization temperature 2800 °C). From the magnified view of Figure 3.72a in Figure 3.72b, we can see that spherical particles have formed from the liquid-type crystal.

For MCMB grown in the liquid medium, graphene layers are not directly exposed to the electrolyte due to amorphous phases on the surfaces of particles. The capacity of MCMB is typically around 320 mAh/g but may be increased to 340 mAh/g by reducing the amount of amorphous phases. Figure 3.73 shows the charge and discharge curves for MCMB artificial graphite. When charging begins, there is a sharp drop in voltage followed by a flat potential, which indicates that the crystal is well developed. Note that its capacity is about 325 mAh/g. Before natural graphite became available, MCMB was used as an anode material in lithium secondary

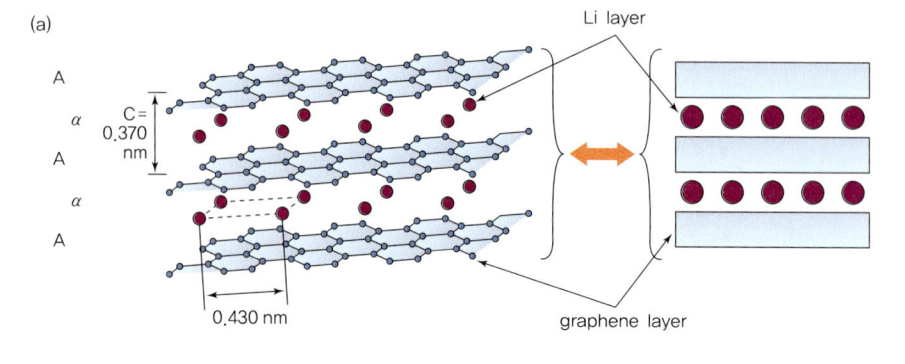

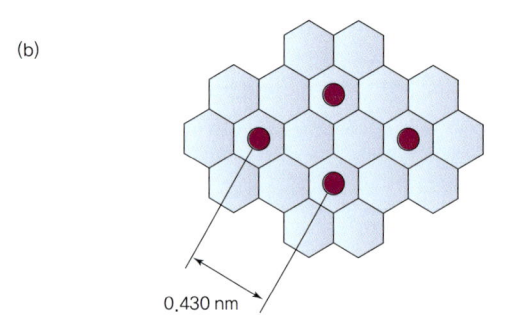

Figure 3.70 In-plane structure in stage 1 [8]. (a) Side view and (b) top view.

batteries due to its high capacity. The shape of particles in MCMB is also suitable for the manufacture of electrodes.

Figure 3.74 shows a schematic diagram of particle shape and arrangement of graphene layers in mesophase pitch-based carbon fiber (MPCF), a type of fibrous

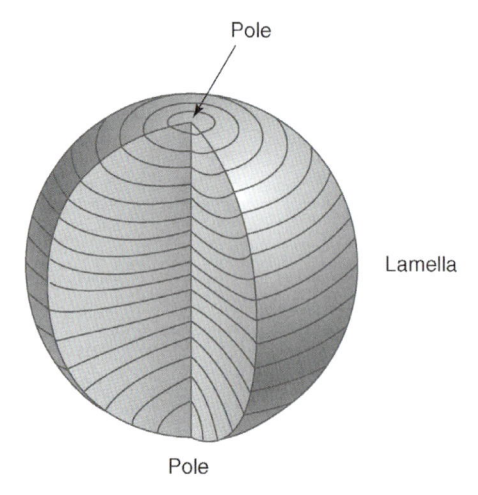

Figure 3.71 Schematic diagram of particle structure in MCMB artificial graphite.

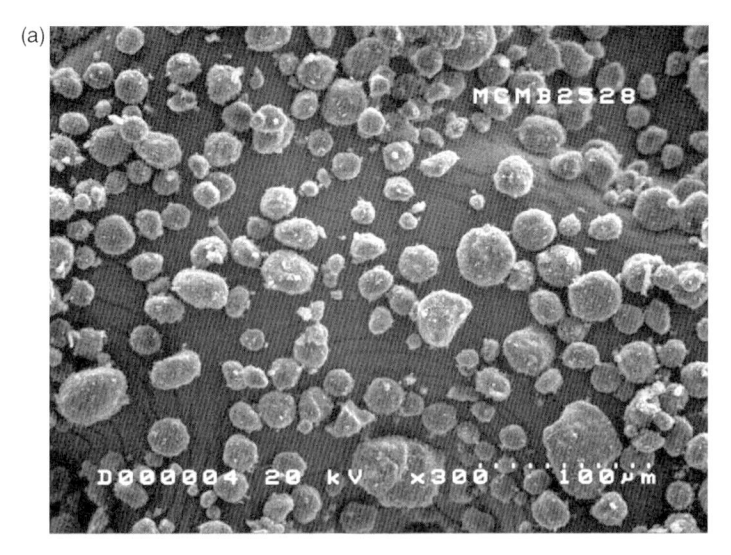

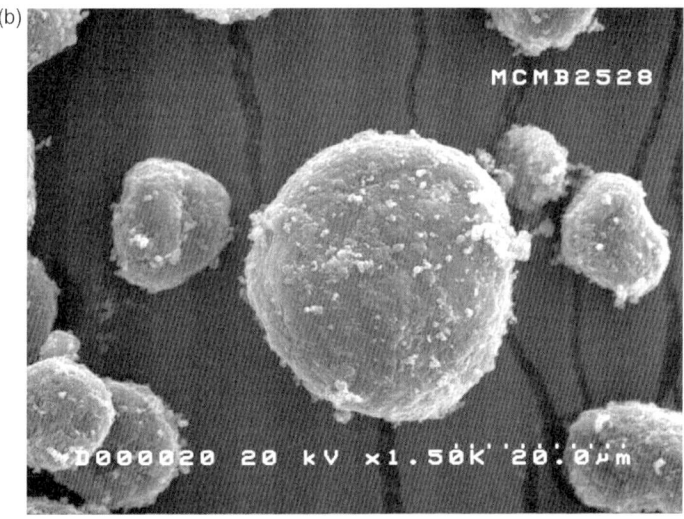

Figure 3.72 SEM images of MCMB-25-28 artificial graphite.

artificial graphite. We can see that various types of MPCF are synthesized depending on the arrangement of carbon domains. A radial or line-origin shape is preferred for better rate capability characteristics, but the exposure of the edge leads to an increase in irreversible capacity. On the other hand, an onion-skin morphology greatly reduces irreversible capacity by minimizing edge exposure but restricts lithium diffusion to the zone among graphene layers. Fibrous anode materials have a low rolling density and tend to spring back after rolling, which causes electrodes to expand. This low energy density leads to many constraints in terms of application as anode materials in

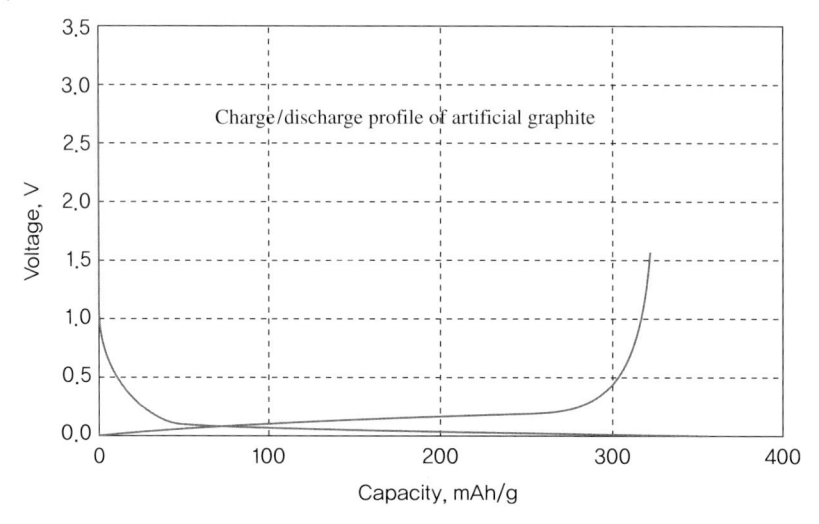

Figure 3.73 Charge/discharge curves of artificial graphite.

mobile phones or laptops. Recently, these materials have been noted for obtaining outstanding characteristics in HEVs.

MPCF-3000, prepared via thermal treatment at 3000 °C, is an example of commercialized fibrous artificial graphite. Figure 3.75a shows the shape of synthesized MPCF-3000, while Figure 3.75b shows the material after zirconia ball milling. The milling density can be increased by adjusting the size of fibers. Excessively large fibers impede the homogeneity of slurry viscosity during slurry manufacturing or electrode coating. Other examples of graphitized materials include graphitized black, graphite nanofibers, and multiwalled carbon nanotubes.

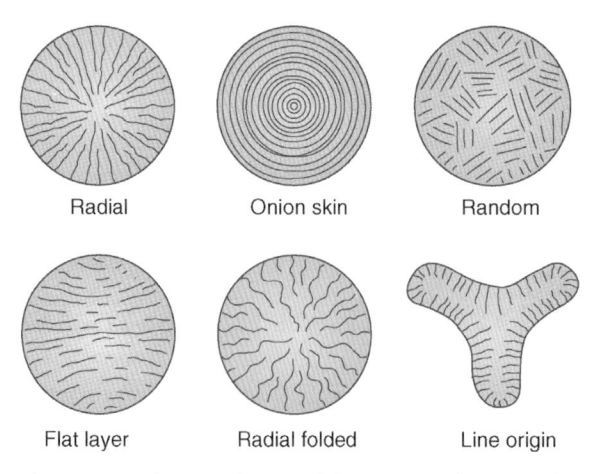

Figure 3.74 Schematic diagram of the structure of MPCF artificial graphite.

(a) MPCF–3000   (b) Ball–milled MPCF–3000

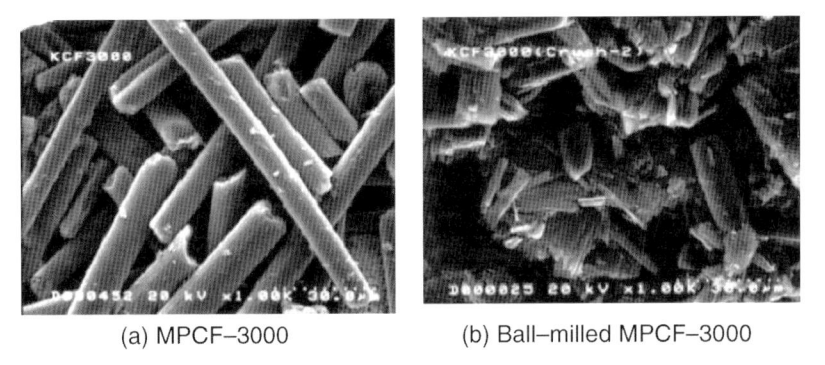

Figure 3.75 SEM images of MPCF-3000 artificial graphite.

Figure 3.76 shows SEM images of natural graphite. From Figure 3.76a, we can see that unprocessed natural graphite exists as flat, plate-like particles. While the capacity of natural graphite is close to its theoretical capacity, irreversible reactions may occur with electrolyte decomposition caused by the large surface area of irregular plates and direct exposure of the edges. The plate-like particles hinder the coating process of the active material slurry during electrode production. Furthermore, milling difficulties encountered after the coating step may result in undesirable electrode densities. As such, to reduce irreversible reactions and enhance processability, natural graphite is subjected to particle grinding and reassembly in order to obtain a smooth surface, as shown in Figure 3.76b. Irreversible reactions that destroy the edge can be reduced by coating the pitch to prevent direct exposure. The electrochemical properties of natural graphite are thus enhanced. The scaly particles of natural graphite can also be grinded and reassembled into the spherical form shown in Figure 3.76c. This process minimizes specific surface area, reduces electrolyte decomposition at the surface of active materials, increases electrode packing density, and improves the uniformity of the electrode coating.

Natural graphite extracted applying the above methods may be used for commercial purposes. Its capacity of 365 mAh/g contributes to improved battery capacity. However, rate capability is reduced with high capacity due to the greater amount of lithium ions per unit surface area.

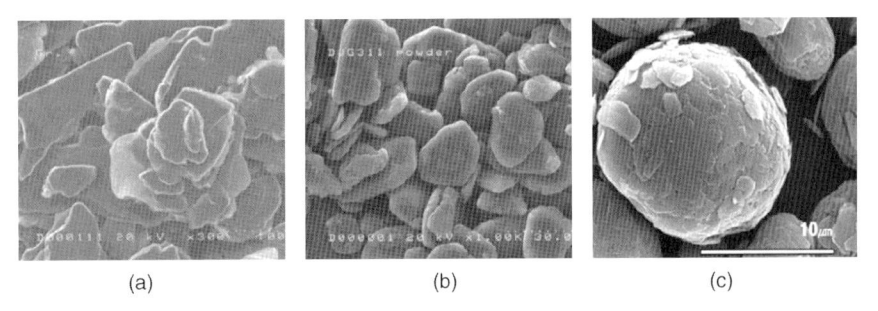

(a)   (b)   (c)

Figure 3.76 Particle shapes of natural graphite: (a) unprocessed natural graphite, (b) processed natural graphite, and (c) natural graphite in spherical form. Reproduced by permission of ECS – The Electrochemical Society.

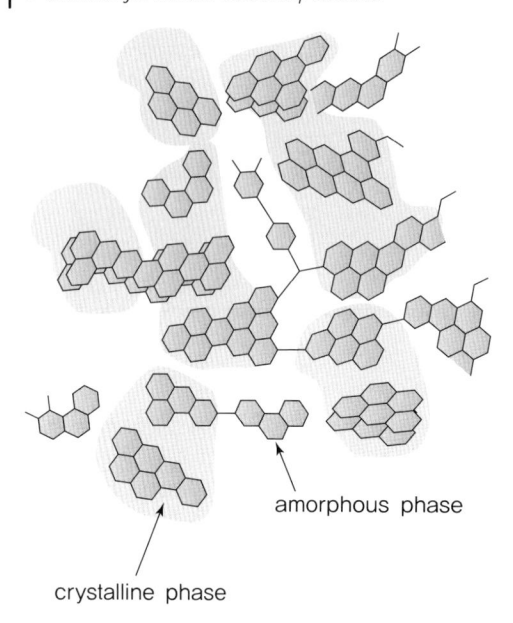

Figure 3.77 Schematic diagram of nongraphitic carbon [9]. Reproduced from [9], with permission from Wiley-VCH.

Amorphous Carbon

Structure of Amorphous Carbon As shown in Figure 3.77, nongraphitic carbon consists of small hexagonal networks and exhibits a disorderly structure that is poorly developed in the c-axis. Crosslinking is established between crystallites, and these crystallites exist together with amorphous phases.

Nongraphitic carbon can be grouped into graphitizable and nongraphitizable carbon depending on graphitizability. Figure 3.78 shows the structure of graphitizable and nongraphitizable carbon [10]. This structural difference contributes to the rearrangement of crystallites during the carbonization process of carbon precursors.

In nongraphitizable carbon, the carbonization process suppresses the stacking of graphene layers and causes crosslinking between crystallites. Due to the small crystallites and a disorderly structure, it is difficult to rearrange the crystals for graphitization even at high temperatures above 2500 °C. On the other hand, graphene layers in graphitizable carbon are arranged in a parallel manner, thus facilitating graphitization. While different carbon materials are produced depending on raw materials or the carbonization process, the size of carbon crystallites (L_a and L_c) tends to increase with heat treatment temperature.

Figure 3.79 presents the changes in L_a and L_c of graphitizable and nongraphitizable carbon with varying heat treatment temperatures. From the graphs, we can see that L_a and L_c values are higher in graphitizable carbon than in nongraphitizable carbon, and this difference becomes more pronounced at higher temperatures [11]. While graphitizable carbon graphitizes rapidly at a certain temperature range, the increase in temperature does not help graphitization of nongraphitizable carbon.

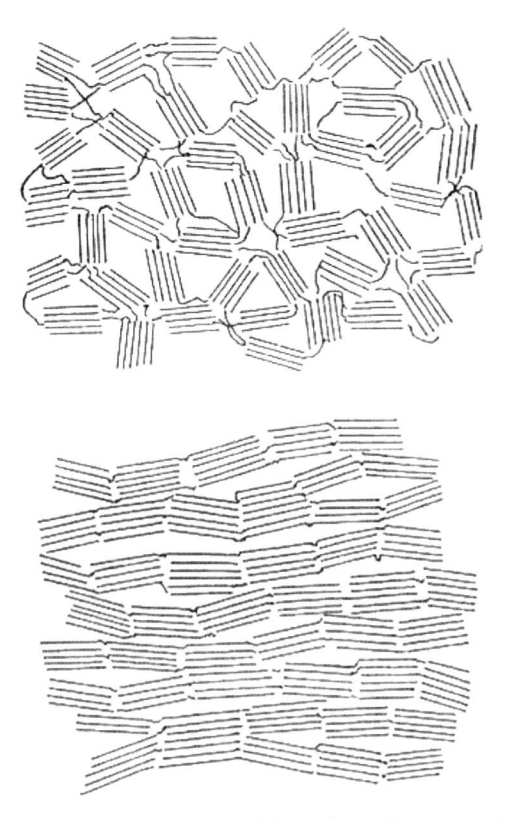

Figure 3.78 Structural model (Franklin) of (a) nongraphatizable carbon and (b) graphitizable carbon [10].

The L_a value of graphitizable carbon reaches 100 nm when carbonized at 3000 °C, but that of nongraphitizable carbon stops at 10 nm. Similarly, the L_c value of graphitizable carbon is 100 nm, whereas that of nongraphitizable carbon is 4 nm.

Figure 3.80 shows the results of Raman spectroscopy for crystalline graphite and various carbon materials of different crystallite sizes [12]. In crystalline graphite, the asymmetric C=C bond appears at about 1582 cm^{-1} (G mode) [13], while the 1355 cm^{-1} (D mode) overtone peaks are located at 2708 cm^{-1}. For smaller crystallite sizes (L_a) or amorphous carbon materials, peaks are observed at 1355 cm^{-1} (D mode), ~1622 cm^{-1} (D′ mode), and ~2950 cm^{-1} (D and D′ modes). Peaks at ~1355 and ~1622 cm^{-1} arise from the diamond structure. The width of the Raman peak decreases with larger crystallites, and the 1355 cm^{-1} peak intensity increases when the graphite crystal is reduced in size or at higher disorder. As shown in Figure 3.81, the ratio of peak intensity at 1355 cm^{-1} to 1575 cm^{-1} (1582 cm^{-1} of G mode) is inversely proportionate to the size of graphite crystallites [14]; that is, the smaller the peak intensity ratio, the larger the size of crystallites. Based on this inversely proportional relationship, it is possible to predict the size of graphite crystallites from the Raman peaks.

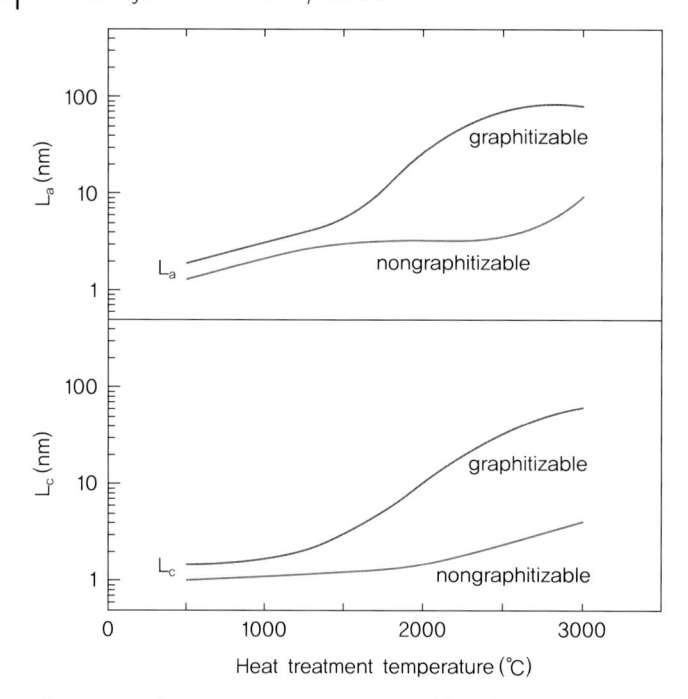

Figure 3.79 Changes in L_a and L_c of graphitizable and nongraphitizable carbon with varying heat treatment temperature [11].

The microstructure of graphitizable carbon and nongraphitizable carbon materials can be observed through transmission electron microscopy. Figures 3.82 and 3.83 show TEM photos of carbon precursors, anthracene and sucrose, which have undergone heat treatment at 1000 and 2300 °C [15].

Nongraphitizable carbon obtained from heat treatment at 1000 °C shows an isotropic structure with crosslinking between graphene layers. Graphitizable carbon that has been subjected to the same heat treatment temperature is arranged similar to graphite and possesses anisotropic characteristics. For carbon materials produced from heat treatment at 2300 °C, nongraphitizable carbon results in a disordered arrangement of small, crooked graphene layers whereas graphitizable carbon exhibits a well-developed graphite structure.

Electrochemical Reactions of Low-Crystalline Carbon Graphitic carbon is derived from heat treatment of soft carbon, but temperatures below 2000 °C lead to low crystallinity and greater disorder. Reactions of low-crystalline carbon at different heat treatment temperatures are described below.

When soft carbon is subjected to heat treatment at temperatures below 900 °C, the resulting carbon has a higher lithium ion storage capacity compared to crystalline graphite, and exhibits hysteresis, where the potential of lithium deintercalation is much higher than that of intercalation. As shown in Figure 3.84, the lithium ion

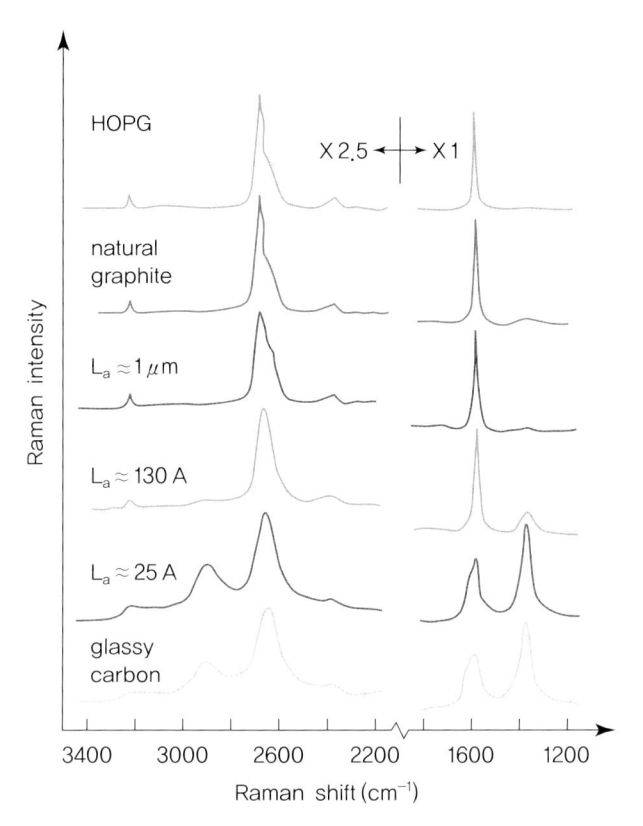

Figure 3.80 Raman spectroscopy for crystalline graphite and various carbon materials of different crystallite sizes [12].

storage capacity and extent of hysteresis increase with lower heat treatment temperatures [16].

The high capacity and hysteretic characteristics of soft carbon produced by heat treatment at low temperatures are related to the amount of hydrogen in carbon materials [17–19]. Figure 3.85 shows the amount of hydrogen expressed in hydrogen/carbon atomic ratio with varying heat treatment temperature. The amount of hydrogen in carbon (H/C atomic ratio) is inversely proportionate to the heat treatment temperature. We can assume that reactions between hydrogen and lithium ions in soft carbon contributed to the high capacity and hysteretic characteristics observed in low-crystalline carbon resulting from heat treatment below 1200 °C. From Figure 3.86, we can see that the 1 V potential plateau increases with an increase in the H/C atomic ratio. The increase in capacity arising from hydrogen is gradually reduced with repeated charge/discharge cycles, while the 1 V plateau decreases with higher heat treatment temperatures, thus causing less significant hysteretic characteristics.

Reactions between hydrogen and lithium ions arising from hysteretic characteristics are related to the hydrogen–lithium bonds. As the lithium atom bonds with

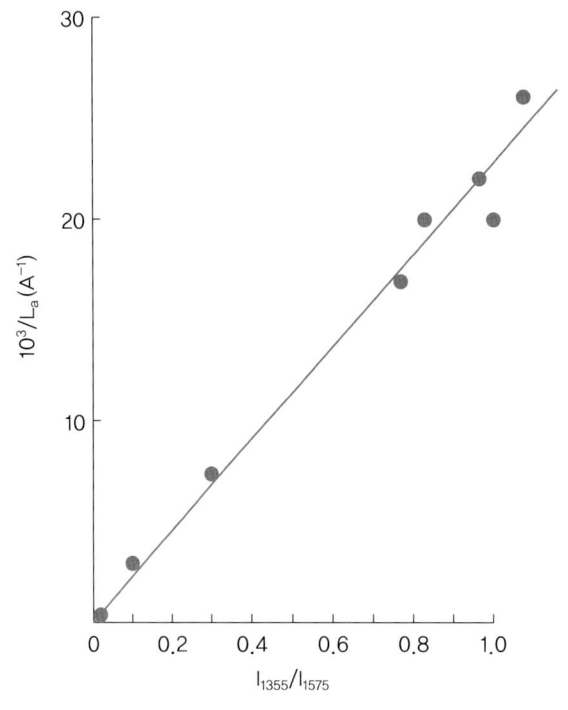

Figure 3.81 Peak intensity at 1355 cm^{-1} versus 1575 cm^{-1} (I_{1355}/I_{1575}) [14].

hydrogen existing on the edge of the hexagonal mesh, the carbon changes from an sp^2 to an sp^3 hybrid orbital, thus requiring a great amount of activation energy during lithium intercalation/deintercalation [20, 21]. This appears as hysteretic behavior in the charge/discharge curves, and cycle characteristics are reduced with a gradual decrease in capacity pertaining to hydrogen during repeated charge/discharge of

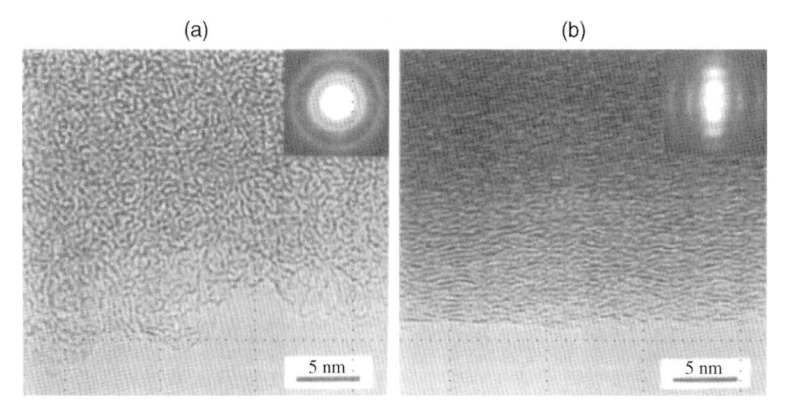

Figure 3.82 High-resolution TEM photos of (a) nongraphitizable carbon (sucrose precursor) and (b) graphitizable carbon (anthracene precursor) treated at 1000 °C under inert conditions [15].

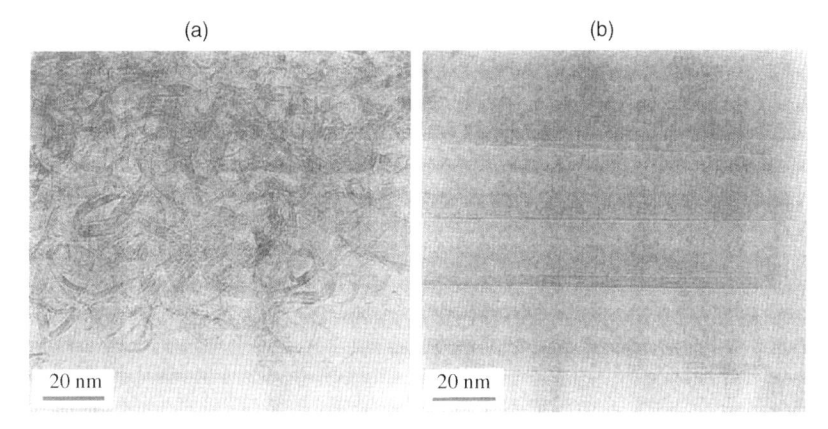

Figure 3.83 High-resolution TEM photos of (a) nongraphitizable carbon (sucrose) and (b) graphitizable carbon (anthracene) treated at 2300 °C under inert conditions [15].

carbon materials [18]. The high specific capacity of carbon materials produced from heat treatment at low temperatures is explained by various reaction mechanisms. Some representative examples are shown in Figure 3.87.

When soft carbon undergoes heat treatment at temperatures greater than 1000 °C, most hydrogen is released and graphene planes are packed in parallel. However, they participate in crosslinking and form a disordered arrangement in the c-axis. During

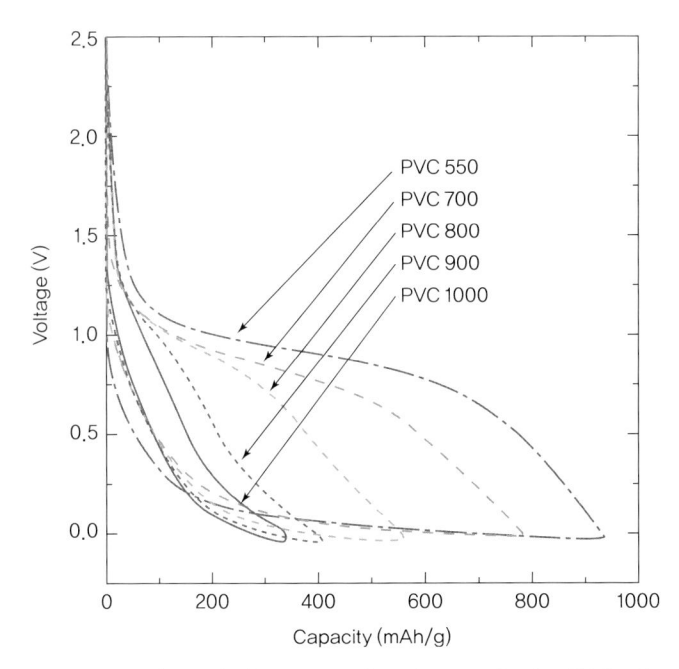

Figure 3.84 Charge/discharge curves from the second cycle of PVC with varying heat treatment temperature [17].

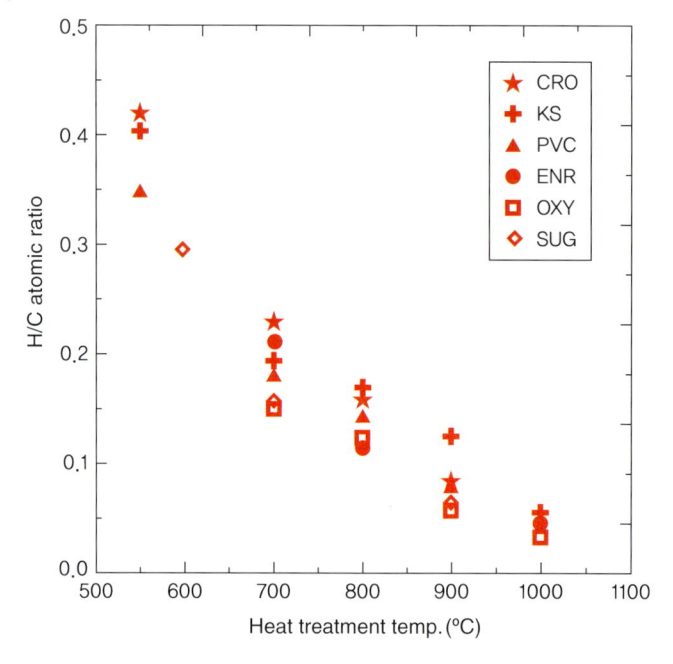

Figure 3.85 H/C atomic ratio with varying heat treatment temperature [16].

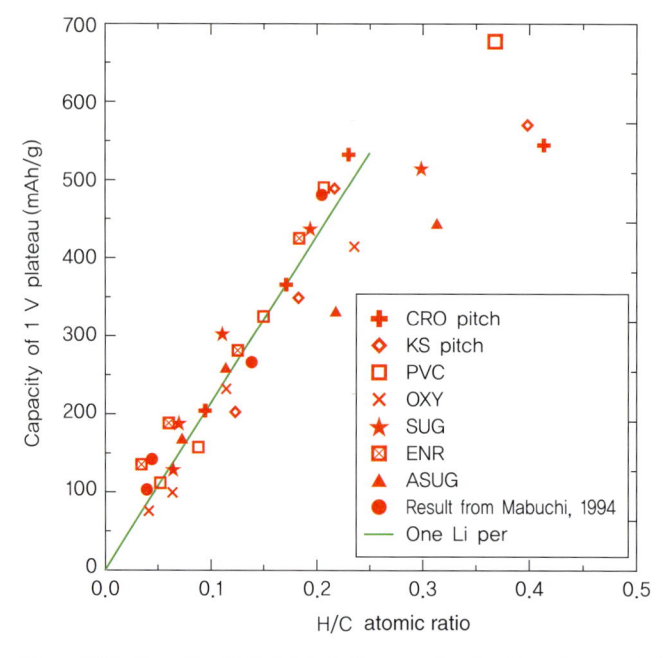

Figure 3.86 Capacity of 1 V plateau in the second cycle with an increase in H/C atomic ratio [16].

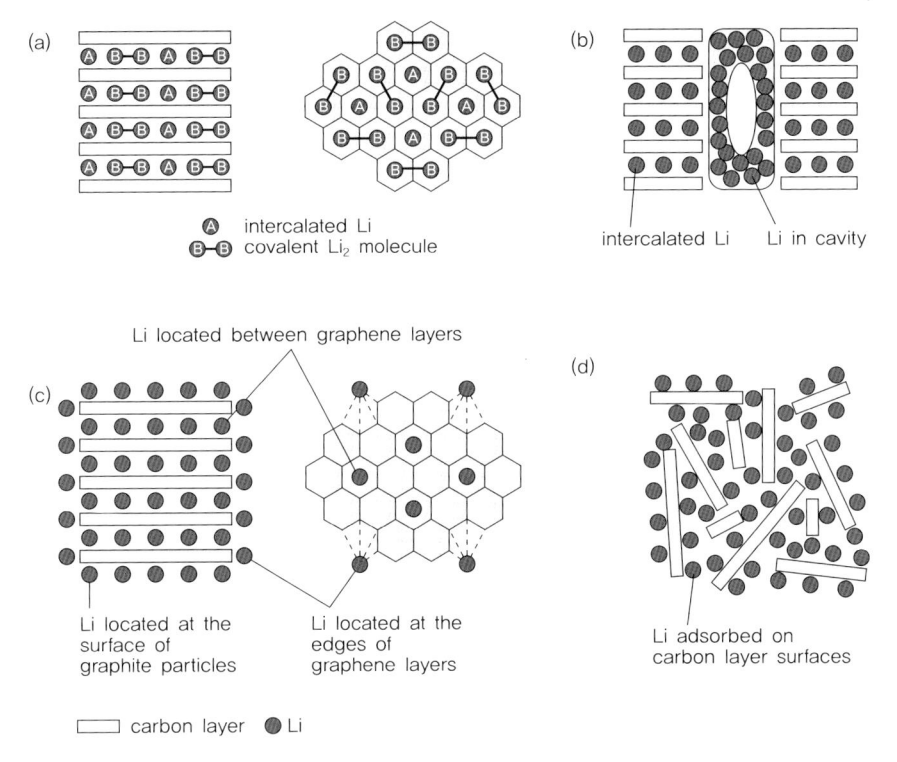

Figure 3.87 Lithium storage mechanism of a carbon material with high reversible capacity. (a) Li_2 covalent bond of intercalated lithium between graphene layers [22], From [22]. Reprinted with permission from AAAS. (b) lithium in nanosized cavities [23], (c) lithium located at the surface and around structural defects of graphene layers [24], Reprinted from [24] Copyright 1995, with permission from Elsevier. (d) lithium adsorbed on carbon layer surfaces. Reproduced by permission of ECS – The Electrochemical Society.

lithium intercalation between graphene layers in crystalline graphite, the surface of graphene layers shifts to allow stacking from ABAB to AAAA. However, lithium intercalation is difficult in graphene sheets with turbostratic disorder [25–27]. Because of the small crystallites in carbon materials produced from heat treatment at low temperatures, there is less space between graphene layers for lithium intercalation as compared to graphite [9, 28]. The reversible lithium storage capacity changes with the disorderly arrangement along the c-axis and the size of graphene layers. The chaotic structure opens various sites for lithium to be intercalated. Since lithium is distributed into various locations, the charge/discharge curves do not exhibit any plateau, in contrast to crystalline graphite. Figure 3.88 shows the charge/discharge characteristics for the first cycle of soft carbon. We can see that it has a large reversible capacity of 220 mAh/g.

Compared to graphite, soft carbon has a low capacity but high surface area and a stable crystal structure. Recently, it has been considered as an anode material for HEV lithium secondary batteries. First, electrodeposition of lithium metal hardly happens even at high-rate current due to the slope of charge/discharge curves. Second, rate capability is enhanced with an increase in surface area involved in lithium reactions.

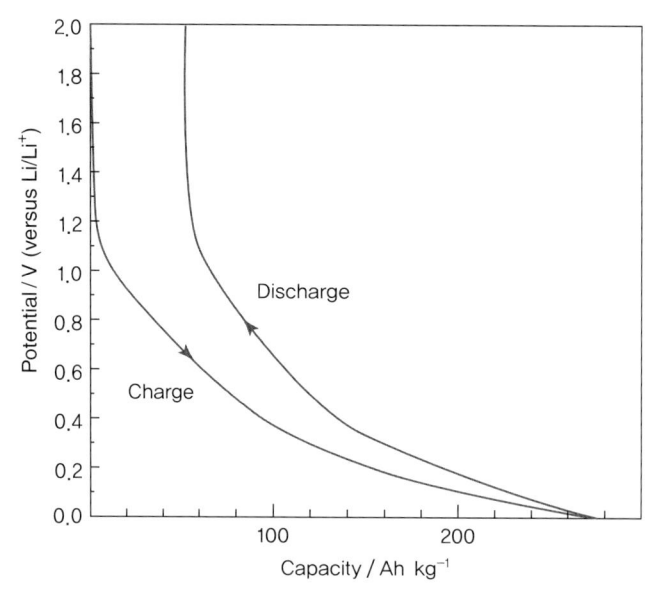

Figure 3.88 First charge/discharge curves of soft carbon (coke) [29].

Third, it is easy to adjust the depth of charging so as to control voltage changes. For the commercialization of soft carbon, the irreversible capacity arising from surface defects has to be reduced.

Electrochemical Reactions of Noncrystalline Carbon Carbon materials prepared from hard carbon do not have stacked graphene sheets and exhibit an amorphous structure with numerous micropores. Just like soft carbon, hard carbon subjected to heat treatment at temperatures below 800 °C contains a large amount of hydrogen and displays similar charge/discharge curves. Hard carbon processed at 1000 °C has a higher reversible capacity. With most hydrogen removed, the charge/discharge curves do not display hysteretic characteristics. As shown in Figure 3.89, a plateau is observed at a low potential of 0.05 V. The high reversible capacity of hard carbon can be explained by lithium adsorbed on carbon layer surfaces or the formation of lithium clusters within micropores. When the heat treatment temperature exceeds 1000 °C, the capacity of hard carbon is greatly reduced with a decrease in micropores and less space for lithium adsorption.

As shown in Figure 3.90 [30], the opening of micropores from heat treatment at high temperatures allows the electrolyte to penetrate, which leads to lowering of the reversible storage capacity since lithium cannot be stored [31, 32].

During lithium intercalation/deintercalation, graphite and soft carbon undergo 10% volume expansion or contraction, whereas hard carbon shows no change in volume due to the large size of micropores [33]. For electrodes using hard carbon, stable life characteristics can be obtained since there is no jelly-roll distortion from volume expansion. The rapid migration of carbon through surface pores results in

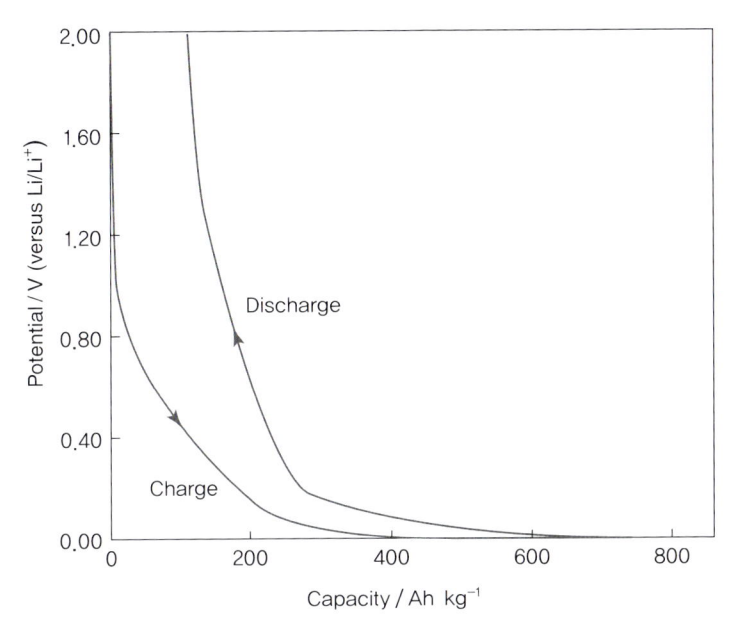

Figure 3.89 First charge/discharge curves of hard carbon [29].

superior rate capability compared to soft carbon. Hard carbon is suitable for batteries requiring high-rate capability but accompanies high manufacturing cost.

Figure 3.91 shows the change in reversible capacity per unit mass of soft/hard carbon with varying heat treatment temperatures. When soft carbon undergoes heat treatment at temperatures below 1000 °C, it attains an extremely high capacity. This drops to a minimum of around 1800–2000 °C. With an increase in heat treatment temperature, it reaches the theoretical value of 372 mAh/g. Meanwhile, hard carbon subjected to heat treatment at 1000 °C has a high capacity of 600 mAh/g.

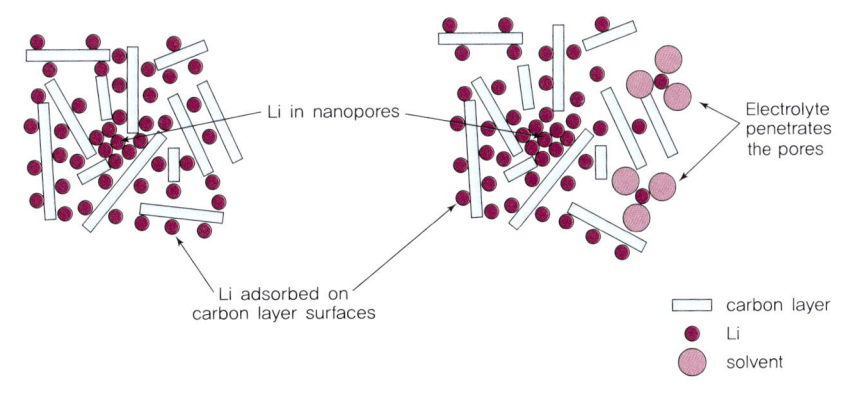

Figure 3.90 (a) Lithium storage mechanism of hard carbon and (b) penetration of electrolyte in pores [32].

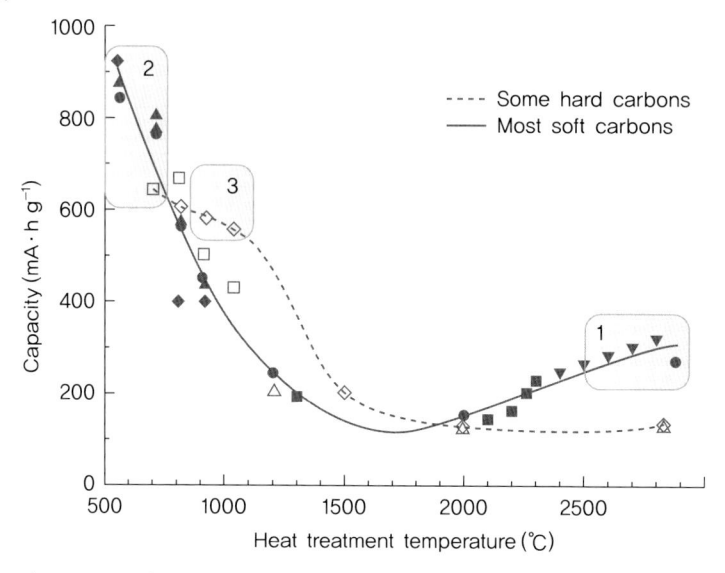

Figure 3.91 Change in reversible capacity per unit mass of soft/hard carbon with varying heat treatment temperature (dotted line: hard carbon; solid line: soft carbon) [17].

At temperatures above 2000 °C, it has a lower capacity than soft carbon, which has less space for lithium adsorption and fewer micropores. In Figure 3.91, region 1 shows graphite carbon produced from heat treatment of soft carbon at temperatures higher than 2400 °C. Region 2 corresponds to soft or hard carbon subjected to heat treatment at 500–700 °C and contains a large amount of hydrogen. Region 3 represents hard carbon with many micropores but almost no stacking of graphene layers.

Reactions Involving Electrolytes Figure 3.92 shows the potential curve during charging (lithium intercalation) and discharging (lithium deintercalation) of a graphite carbon material under a constant charge [34]. In theory, lithium intercalation and deintercalation into the carbon material is completely reversible. From the actual charge/discharge curves, we can see that more lithium is consumed than the theoretical capacity of graphite, 372 mAh/g, and only 80–95% is recovered upon discharge. In the second cycle, less lithium ions are intercalated during charging, and most are deintercalated during discharging.

The difference in capacity (C_{irr}) appearing in the first charge/discharge reaction is known as irreversible capacity loss, while the capacity (C_{rev}) for reversible intercalation/deintercalation within the carbon material is called reversible capacity. In commercialized lithium ion batteries, lithium ions are supplied from the cathode consisting of lithium metal oxides, and the carbon anode is produced without lithium. As such, it is important to minimize the loss of irreversible capacity during the initial stages of charge and discharge. This loss of irreversible capacity is known to contribute to electrolyte decomposition at the surface of carbon materials. In the range of potential where lithium ions are intercalated, the electrolyte exists in a

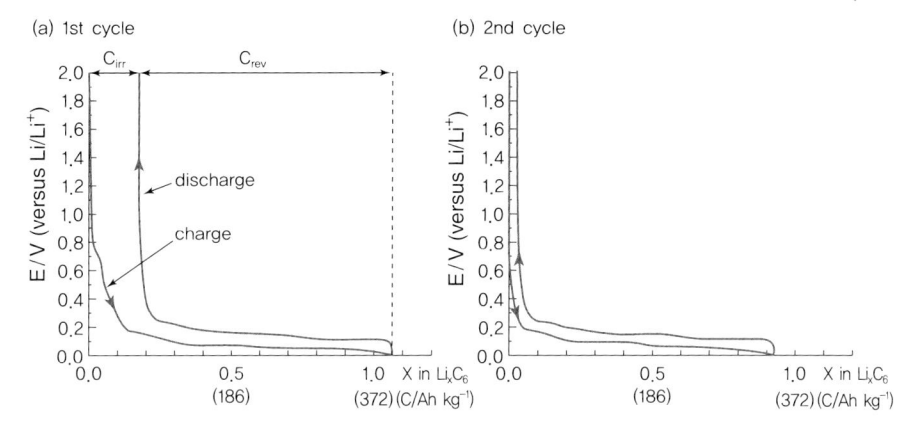

Figure 3.92 Charge/discharge curves of graphite Timrex KS 44 with LiN(SO$_2$CF$_3$)$_2$/ethylene carbonate/dimethyl carbonate as an electrolyte [34].

thermodynamically unstable state with the anode surface. During the first charging reaction, electrochemical reactions arising from electrolyte decomposition create an SEI layer on the carbon surface. In the case of crystalline graphite, this reaction occurs at around 0.8 V (Li$^+$/Li) and appears as a plateau when a constant current charge is applied. The size of the plateau is determined by the extent of electrolyte decomposition. The formation of this protective film requires a great amount of lithium ions and leads to a loss of reversible capacity. However, by selectively migrating lithium ions without transfer of electrons, further electrolyte decomposition is suppressed, and cycle characteristics are enhanced. This SEI formation is greatly influenced by electrolyte composition and characteristics, which will be dealt with in detail in the following sections. The protective layer takes on different characteristics depending on structural features of the material surface. In general, a larger specific surface area increases the absolute quantity of the SEI layer, thus causing a greater loss of irreversible capacity [35]. In graphitic carbon materials, more electrolyte decomposition takes place at the edge plane site as it is more electrochemically active than the basal plane [36].

Besides the formation of SEI layer, other factors that contribute to irreversible reactions are reduction reactions with impurities such as H$_2$O and O$_2$, reduction of surface functional groups containing oxygen at the material surface, and loss of reversibility from lithium that has undergone reduction but not oxidation [37].

In the case of crystalline graphite, lithium ions and polar molecules of the electrolyte can be intercalated between graphene layers. The intercalation of lithium ions and the polar solvent occurs simultaneously, expanding the space between graphene layers and destroying the structure [38]. Figure 3.93 shows the solvent intercalation, which can be understood as the exfoliation of graphene layers.

During the early stage of lithium intercalation, that is, when there is a low concentration of lithium ions in the crystalline graphite ($x \leq 0.33$ in Li$_x$C$_6$), lithium ions tend to be intercalated together with electrolyte solvents due to thermodynamic

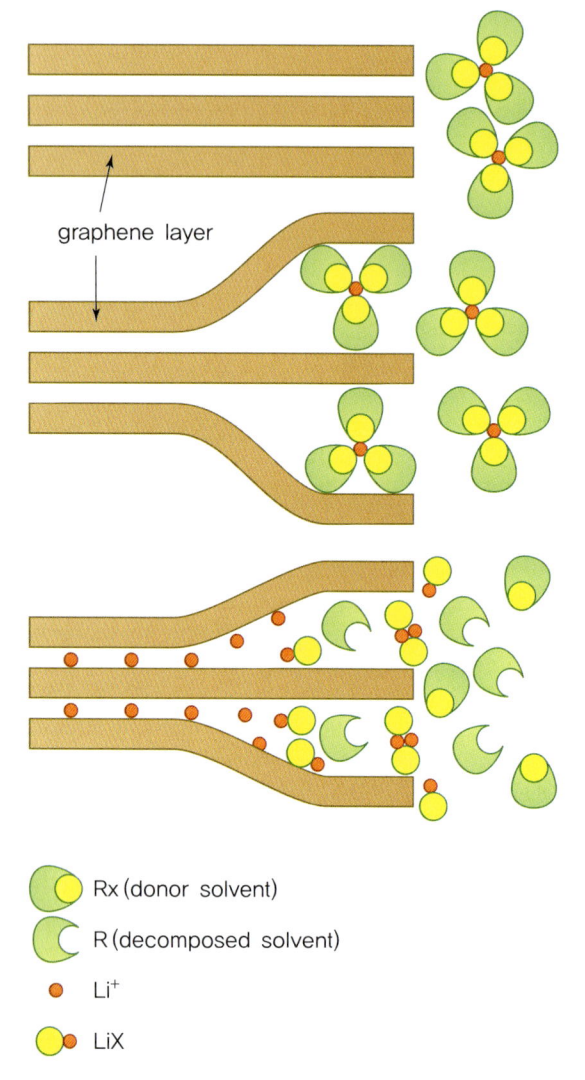

Figure 3.93 Schematic diagram of SEI film formation and intercalation between graphene layers from electrolyte decomposition [38]. Reprinted from [38] Copyright 1995, with permission from Elsevier.

reasons [39]. To prevent this from occurring, an electrolyte additive is used to create the SEI layer on the graphite surface before the simultaneous intercalation of lithium ions and the electrolyte solvent.

Since the simultaneous intercalation of lithium ions and the electrolyte solvent occurs between graphene layers of the edge plane, it is difficult for such reactions to occur in low-crystalline soft carbon or amorphous hard carbon, which are characterized by structural defects and chaotic arrangements [9, 40]. By coating the surface of low-crystalline soft carbon and amorphous hard carbon with graphite particles,

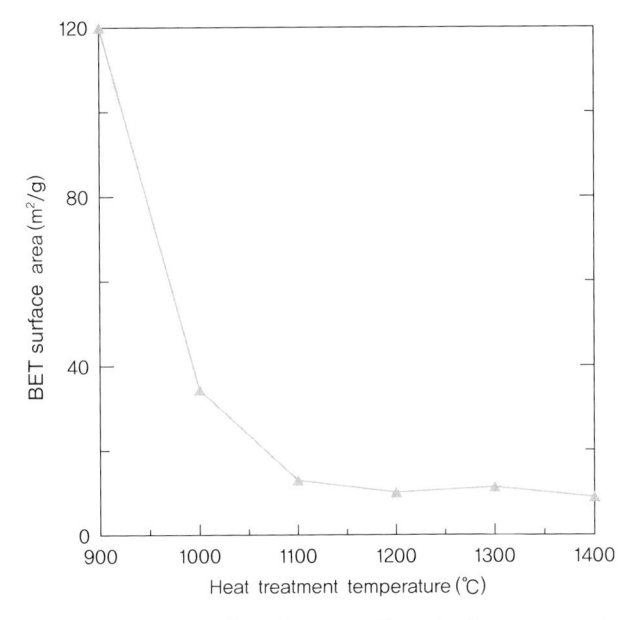

Figure 3.94 BET specific surface area with varying heat treatment temperature [41].

simultaneous intercalation is suppressed, and irreversible reactions arising from the formation of the SEI layer are greatly reduced [40, 42–43].

The structure and surface area of carbon materials are important factors that determine the characteristics of electrode–electrolyte reactions. Because the intercalation/deintercalation of lithium ions during charge/discharge takes place at the surfaces of particles, surface structure and surface area are crucial factors in determining the power characteristics of a battery.

Figure 3.94 shows the BET-specific surface area of hard carbon with varying heat treatment temperature. We can see that the BET-specific surface area decreases as the heat treatment temperature increases. This is caused by the closing of micropores within particles or at the surface upon exposure to high temperatures [44]. The decrease in specific surface area not only suppresses electrolyte decomposition at the surface but also reduces the rate capability by restricting the movement of lithium ions. These factors should be carefully considered in the design of materials to produce batteries with outstanding characteristics.

As summarized in Table 3.4, the surface of carbon materials consists of various chemical functional groups. These functional groups affect the chemical and electrochemical characteristics of carbon materials. Figure 3.95 shows the results of a X-ray photoelectron spectroscopy (XPS) analysis on various functional groups existing on the surface of carbon materials.

For a low level of oxidation, the peak intensity corresponding to the C–OH bond is strong while that of the C=O and HO–C=OOH bonds is relatively weak. For a high level of oxidation, the C–OH bond gives a low-peak intensity, while C=O and HOC=OOH bonds show high-peak intensities [46]. From this, we know

Table 3.4 XPS peaks for carbon materials [45].

Carbon	Binding energy for C_{1S}(eV)	Interpretation of spectra
Channel black	285.5	C bonded to H
	288.5	C bonded to O
Oxidized carbon fiber	286.0	Hydroxyl group
	287.0	Carbonyl group
	288.6	Carboxyl group
Carbon from PTFE	285.4–285.9	C atoms in the basic C skeleton
Reduced by Li	290.2–290.8	C atoms in surface COOH groups
Carbon fiber	285.0	Hydrocarbon group
	287.0	C singly bonded to O
	289.0	C doubly bonded to O
Graphite oxide	285.0	Carboxyl group
	287.2	Phenolic hydroxyl group
	289.0	Carbonly and ether groups
Air-oxidized carbon fiber	1.5^a	C–O– group
	2.5^a	C=O group
	4.5^a	Carboxyl group
Carbon fiber	1.6^a	C–O– group
	3.0^a	C=O group
	4.5^a	Carboxyl or ester group
Electrochemically	11.6^a	C–O– group
Oxidized carbon fiber	3.0^a	C=O group
	4.5^a	Carboxyl or ester-type group
Electrochemically oxidized carbon fiber	$\sim2.1^a$	C=O and/or quinine-type group
	$\sim4.0^a$	Ester-type group
	$>6.0^a$	CO_3^{2-}-type group
Fluorinated graphite	4.7^a	CF group
	6.7^a	CF_2 group
	9.0^a	CF_3 group

a) Chemical shift for C_{1S} peak (eV).

that most oxygen reacts with carbon at the surface of carbon particles during oxidation.

Thermochemical Characteristics The marketability of lithium ion batteries is affected by thermochemical characteristics and safety issues. In the case of commercialized lithium ion batteries, thermal safety is guaranteed until around 60° C. However, higher temperatures accelerate thermal decomposition of the active material or electrolyte from mutual interactions between components, thus greatly impeding battery performance due to a structural collapse. Thermal characteristics can be described using various complex reaction paths such as anode/electrolyte interfacial reactions, cathode/electrolyte interfacial reactions, and decomposition reactions of the electrolyte and electrodes. Such thermal decomposition leads to thermal runaway,

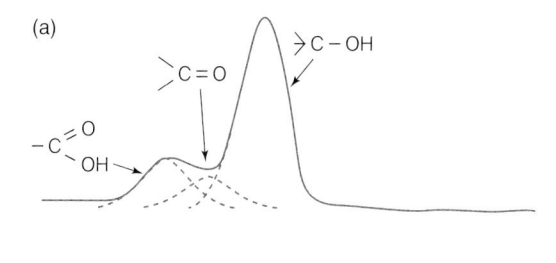

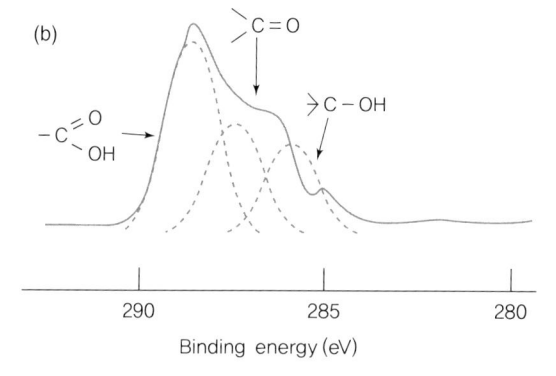

Figure 3.95 XPS pattern of carbon fibers after oxidation: (a) low level of oxidation and (b) high level of oxidation [46]. Reprinted from [46] Copyright 1984, with permission from Elsevier.

which causes serious safety problems such as fires and explosions. The fundamentals of thermal safety in batteries have been introduced in the previous section on cathode materials (refer to Section 3.1.5.1).

The decomposition of the SEI layer formed during the initial charging process at the anode/electrolyte is important in understanding the thermal characteristics of the anode and battery safety. As shown in Figure 3.96, the SEI layer begins to break down at around 80 °C and gets completely decomposed at 100–120 °C [47].

Components of the SEI layer differ according to electrolyte composition and commonly consist of metastable substances such as Li_2CO_3, LiF, or $(CH_2OCO_2Li)_2$. It is necessary to understand reactions between battery components in order to control thermal characteristics and prevent thermal runaway. In general, exothermic reactions at the anode that affect thermal characteristics are similar to those occurring at the anode. The following is a brief summary [48]:

1) As temperature rises, the SEI layer existing in the metastable state participates in decomposition through exothermic reactions at 90–120 °C.
2) At temperatures above 120 °C, exothermic reactions take place upon reactions between the electrolyte and the lithium ions intercalated in the anode.
3) Reactions between fluorine-containing binding agents (e.g., PVdF) and lithium ions intercalated in the anode proceed at temperatures similar to those above.
4) Thermal decomposition of the electrolyte occurs at temperatures above 200 °C.
5) Lithium metal may be deposited at the anode when overcharged and engaged in reactions with the electrolyte and binding agents.

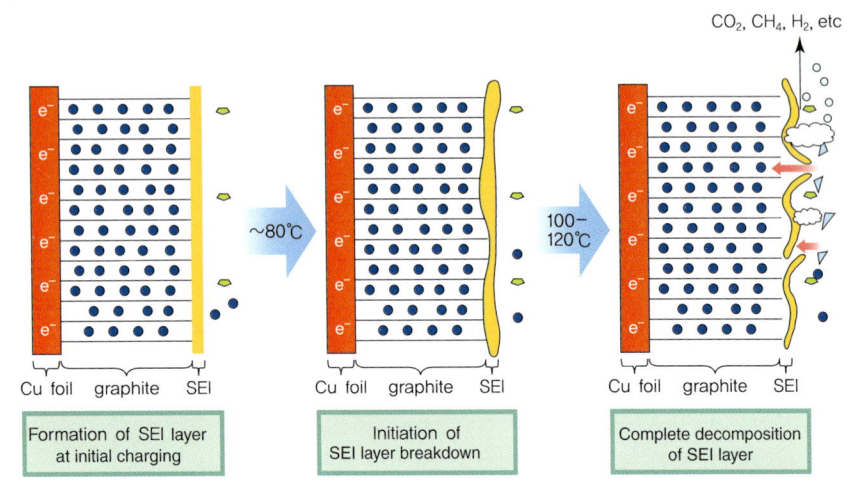

Figure 3.96 SEI layer decomposition reaction path. Reprinted with permission from American Chemical Society Copyright 2003.

Spotnitz and Franklin obtained DSC (differential scanning calorimetry) results for components of lithium secondary batteries based on kinetic variables such as exothermic onset temperature, exothermic peak temperature, thermal capacity, activation energy, and speed coefficient [48]. As shown in Figure 3.97, thermal characteristics of the anode are related to SEI decomposition and exothermic reactions of the carbon anode/electrolyte and carbon anode/binder. Similar to the cathode, thermal characteristics of the anode exhibit different calorimetric values depending on the charged state of lithium. Since thermal characteristics differ greatly with electrode and electrolyte types, it is important to understand thermal reaction paths and optimize battery design in order to produce thermally stable lithium ion batteries. Reaction enthalpy should be minimized in order to avoid thermal runaway

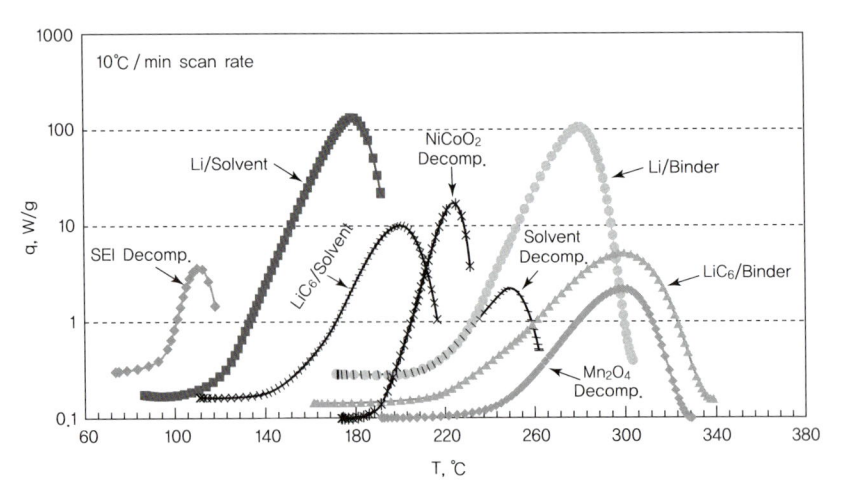

Figure 3.97 Typical DSC characteristics of lithium ion battery components [48].

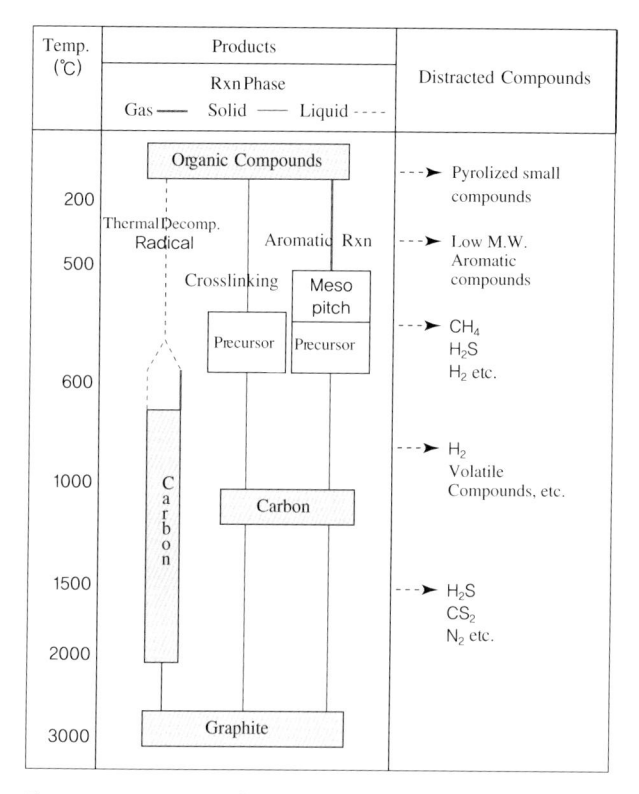

Figure 3.98 Overview of carbonization reactions [49].

caused by reactions between the anode and the electrolyte at relatively low temperatures (below 200 °C).

Carbon Raw Materials and Carbonization Depending on the type of precursor organic compound, carbon materials can be produced in gaseous, liquid, or solid states. Figure 3.98 shows the products resulting from heat treatment of these precursors. The resulting carbon differs with raw materials (gas, liquid, solid, etc.), intermediate precursors, and temperature. Each carbonization method is described in the following sections.

Gaseous Carbonization Carbon is produced from the gaseous precursor through thermal decomposition of volatile organic matter or hydrocarbons, followed by aromatization based on radical polymerization reactions. Examples of carbon materials obtained from the gaseous carbonization process are carbon black, pyrolytic carbon, and carbon whiskers.

Liquid-Phase Carbonization In the process of liquid-phase carbonization, low molecular compounds in raw materials such as fusible organic substances or fusible coals evaporate when subjected to heat, and this rise in temperature produces aromatic hydrocarbons through various reactions such as reduced viscosity, thermal decomposition, polymerization, dehydrogenation, and aromatization. Liquid viscosity increases when the components of carbon compounds undergo aromatization or an increase in molecular weight, eventually resulting in a solid carbon. Pores are formed within the material during solidification, and the increase in temperature releases hydrogen, hydrocarbons, and other impurities. This is followed by densification, and a graphite crystal structure is produced at temperatures above 1500 °C. Impurities (S, N, etc.) within the carbon material are released during graphitization.

In the case of liquid carbonization, viscosity begins to increase at around 400 °C and reaches a maximum point, and then drops to a minimum before rising again for solidification. Changes in viscosity with temperature differ according to carbon materials and carbonization conditions. The organic compound contains isotropic liquid and anisotropic liquid crystal, which gradually grow into anisotropic solid cokes. Since the solid configuration does not require rearrangement of large atoms, the order of hexagonal plates in the solid phase determines the progress of graphitization. The extent of graphitization is influenced by crystallite sizes (L_a and L_c) for carbon that is easily graphitized. However, the difference between graphitizable and nongraphitizable carbon arises from the regularity of arrangements, and not the size of crystallites. Nongraphitizable carbon takes the form of amorphous carbon since crystallites do not grow upon heat treatment.

Solid-Phase Carbonization Solid-phase carbonization occurs when raw materials (such as thermosetting polymers) do not produce liquid during heat treatment and remain in the solid state even with pyrolysis reactions. The initial solid structure determines the carbon microstructure since the movement of molecules is limited during carbonization reactions. When fusible organic substances become insoluble after oxidation and heat treatment, the process involves solid carbonization reactions. By the assumption of a specific shape before carbonization, it is possible to maintain the given shape in resulting carbon materials. However, without additional densification, this is likely to result in porous carbon materials, as volatile components are released during the carbonization process. Representative materials for solid carbonization are cellulose, phenol formaldehyde resin, furfuryl alcohol resin, and phenol furan resin. Carbon materials produced from solid phase carbonization include carbon fibers and activated carbon.

3.2.3.3 Noncarbon Materials

Following the commercialization of lithium secondary batteries in 1991, graphite and other carbon materials have been used as anode materials. Since the early period of development, battery performance has greatly improved, including a twofold increase in energy density. Along with an increase in charging voltage of the cathode, this has allowed the specific capacity of anode materials to increase from the level of noncrystalline hard carbon, at 170 mAh/g, to high-density graphite, at 360 mAh/g.

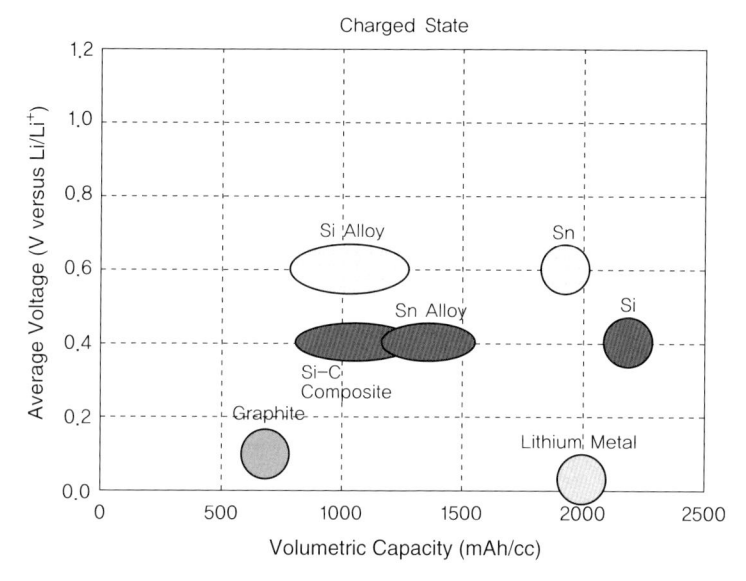

Figure 3.99 Voltage and capacity of metal elements that form alloys with lithium.

As mobile devices become increasingly lightweight, compact, and multifunctional, the energy density of lithium secondary batteries should be enhanced to achieve long operating time. For commercially available graphite, the lithium storage capacity (LiC$_6$) has been limited to 372 mAh/g (or 820 mAh/cm^3). This problem can be overcome by using anode materials with a larger lithium storage capacity. Besides graphite, Si and Sn are good examples of high-capacity materials that can react with lithium to form alloys. Research is being actively carried out on various alloys related to such metals.

Figure 3.99 shows the relationship between voltage and capacity density for representative anode materials. The working voltage of Sn and Sn alloy at 0.6 V is 0.2 V higher than that of Si and Si alloy. Si and Sn metals show a high specific capacity similar to that of lithium metal. The use of metals without alloying causes large volume expansion during charging; thus, more research is needed for commercial applications. Other elements such as Al, Ge, and Pb have low reversible reaction efficiency and high average working voltage, making them inadequate as anode materials.

Since metals and alloys including Si have a higher voltage than graphite, they lead to a lower potential difference between electrodes when applied to actual batteries, and thus a lower battery voltage. Figure 3.100 shows the discharge curves for a LiCoO$_2$ cathode and either graphite or Si–C as the anode. The curve for LiCoO$_2$/graphite exhibits a relatively higher average voltage, while the high capacity of LiCoO$_2$/Si–C varies greatly with cutoff voltage. When the cutoff voltage is set at 3.0 V, the capacity increase is about 10%. When the cutoff voltage is reduced to 2.5 V, the capacity improves to 15%. Si metal and other materials with a higher voltage than lithium may be used as high-capacity anodes but yield much smaller values than

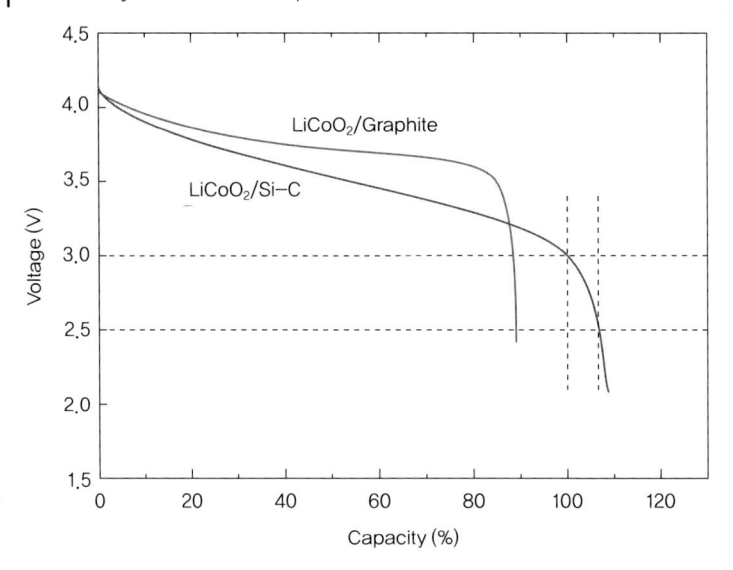

Figure 3.100 Discharge curves of batteries with LiCoO$_2$/graphite and LiCoO$_2$/Si-C.

expected in terms of energy density. These characteristics must be carefully considered in the design of batteries.

Alloys Li, Li–Al, and Li–Si alloys have been considered as cathode materials with the aim of increasing the capacity of lithium secondary batteries, but lithium metals have not been commercialized due to safety issues caused by dendrite formation. To overcome this problem, other metals (Si, In, Pb, Ga, Ge Sn, Al, Bi, S, etc.) that form alloys with Li have been proposed [50]. These metals react with lithium at a specific voltage range during charging to become alloys and return to their original state when discharged, thus allowing continuous reversible charge/discharge. Unlike intercalation/deintercalation reactions in graphite, reversible reactions occur upon alloying/dealloying with lithium. The charge/discharge reaction of the electrode involving metal elements is as follows:

$$x\text{Li}^+ + xe^- + \text{M} \leftrightarrow \text{Li}_x\text{M} \quad (\rightarrow : \text{Charge}, \leftarrow : \text{Discharge})$$

Alloying occurs when metals are neutralized by receiving electrons and lithium ions during charging, whereas dealloying is the opposite process of returning lithium ions to the original metal. The capacity per unit mass/volume of metal elements known to form alloys with lithium (Li$_x$M) is shown in Figure 3.101 [51].

Most metals show a higher specific capacity than graphite (LiC$_6$), while Si has a theoretical capacity greater than 4000 mAh/g. Here, we should take note of the working voltage of these materials. As shown in Figure 3.102, metal–lithium reactions occur at relatively low potential but not for graphite. If these metals are selected as anode materials of lithium secondary batteries, energy density can be reduced even with a high capacity per unit mass due to the decrease in battery voltage. However, these high-capacity alloys have become promising materials, as energy

3.2 Anode Materials | 121

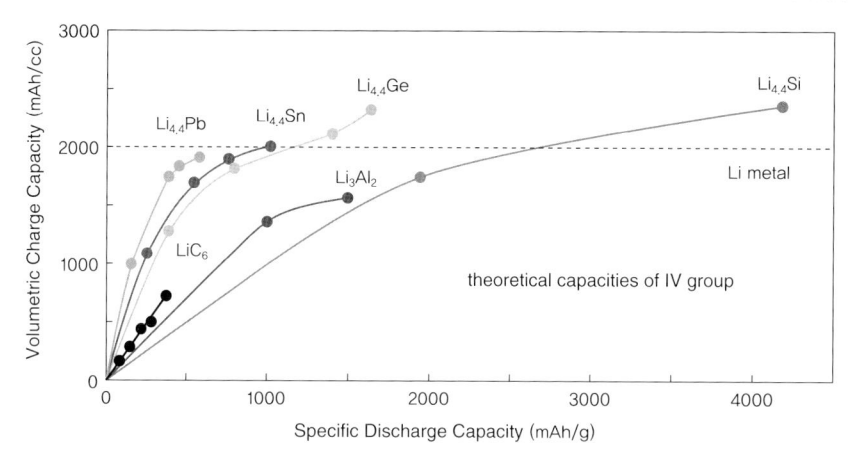

Figure 3.101 Discharge capacity of lithium metal alloys (*note*: charge capacity per unit volume includes the change in volume after alloying with lithium) [51].

consumption and working voltage have dropped with recent developments. Since past research has focused on Si and Sn, more studies should be conducted on Ge, Pb, and Al.

From the equilibrium phase diagram, we are able to infer the phase and composition of alloys derived from reactions between metals and lithium.

For instance, when Li is added to pure Sn, the resulting compounds are Li_2Sn_5, LiSn, Li_5Sn_2, $Li_{13}Sn_5$, and $Li_{22}Sn_5$. The phase of each compound can be conjectured from the phase diagram shown in Figure 3.103. These metals show a high specific capacity because alloys of various compositions are formed when lithium is added, and the reactions involve a great amount of lithium despite the capacity of one lithium ion accommodation for six carbon atoms. Under equilibrium conditions, the potential for lithium–metal reactions can be determined. Figure 3.104 compares the change in potential of Sn and Si with varying lithium composition.

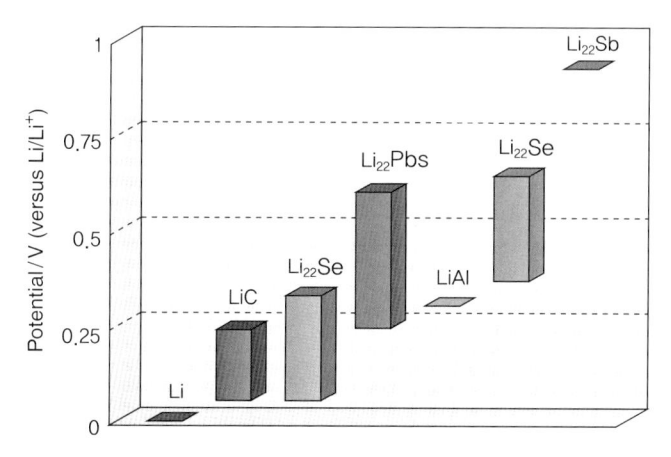

Figure 3.102 Range of potential for reactions with lithium [52]. Reprinted from [52] Copyright 1999, with permission from Elsevier.

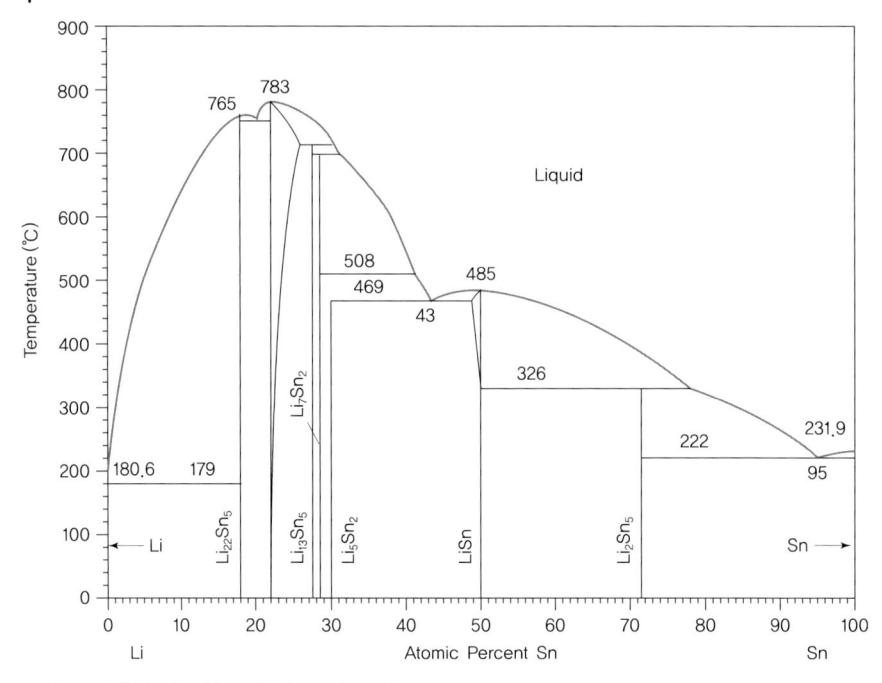

Figure 3.103 Sn–Li equilibrium phase diagram.

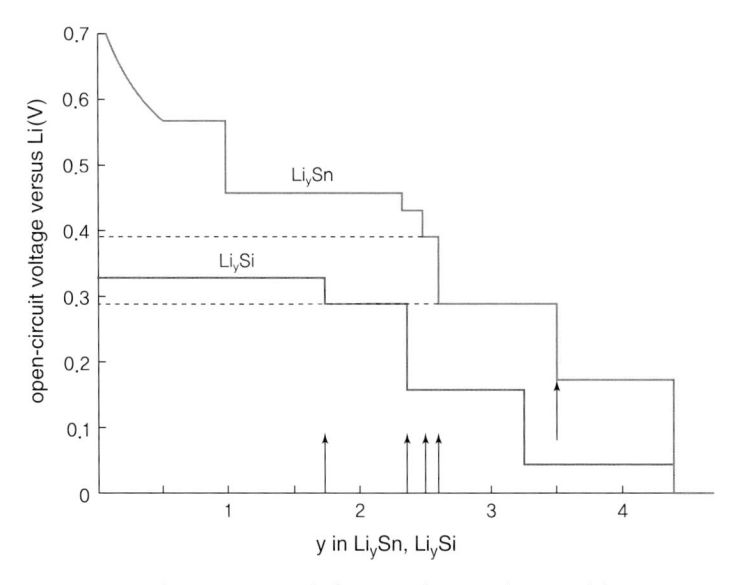

Figure 3.104 Change in potential of Sn–Li and Si–Li with varying lithium composition [50]. With kind permission from Springer Science and Business Media: [50]

Lithiation →

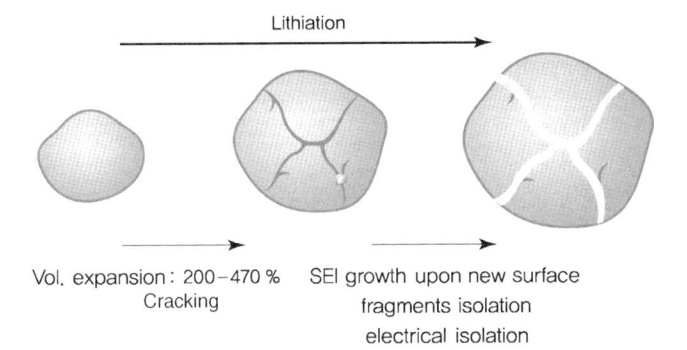

Vol. expansion: 200–470 %
Cracking

SEI growth upon new surface
fragments isolation
electrical isolation

Figure 3.105 Schematic diagram of cracking in metal alloys.

The potential of Sn–Li is higher than that of Si–Li but both maintain a constant potential when two phases coexist. These materials form alloys with lithium during charging, and such reactions are accompanied by a decrease in voltage and volume expansion. If the working voltage is high, the lowered voltage does not pose a problem in actual operation. However, the change in volume may be detrimental to battery performance.

Volume expansion of the metal anode contributes to a larger lattice constant, as the sites between metal atoms are filled during alloying. Since one Si atom can react with up to 4.4 lithium ions, this expansion in volume may reach 400%. The weak ionic bonds between lithium and metals are easily broken from the stress caused by volume expansion. For inorganic materials consisting of ionic bonds, the critical point is reached with a 5% volume increase. Figure 3.105 shows the cracking of metals from volume expansion during charging.

In the early stage of charging, the alloying of lithium and metal causes excessive volume expansion and cracking of particles. Further alloying reactions create a new surface layer, which eventually forms an SEI layer with electrolyte decomposition. Since the cracks are not always radial, certain areas within particles cannot be reached by the electrolyte. The isolated fragments do not participate in electrochemical reactions, resulting in a huge capacity loss. Figure 3.106 is a comparison of SEI layer formation at the surface of graphite and metal particles [51].

To prevent the cracking of metal particles, we should explore methods to suppress the volume expansion of anode materials.

Some methods of minimizing volume expansion are (1) refinement of metal particles involved in lithium reactions, (2) multiphase alloying with lithium, (3) use of active/inactive metal composites, and (4) formation of lithium alloy/carbon composites.

Micrometal Particles in Lithium Reactions When metal reacts with lithium, the extent of volume expansion is affected by the size of metal particles. Minimizing the size of particles is known to reduce stress caused by expansion. An excessive change in volume leads to the deterioration of active materials. During metal–lithium reactions, the large change in volume results in cracking within particles. More cracking occurs

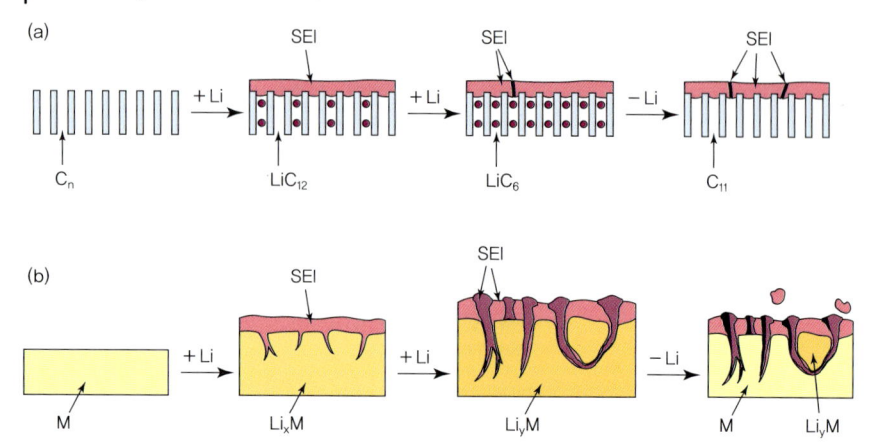

Figure 3.106 SEI layer formation in (a) graphite and (b) metal. Reproduced by permission of ECS – The Electrochemical Society.

with an increase in the amount of lithium ions, and a new SEI layer is formed. Eventually, battery performance is impeded by reduced capacity from electric isolation.

Figure 3.107 shows that a loose arrangement of metal microparticles at the electrode allows stable charging and discharging by alleviating the stress produced from volume expansion during lithium reactions. Maintaining the size of metal particles at a minimum not only avoids cracking but also accelerates reactions between metal and lithium. Cracking does not occur if the size of metal particles is kept below the critical point. The strain energy produced from volume expansion should be larger or equal to the surface energy during cracking. The critical size of metal particles is given by the following equation [53]:

$$d_{\text{crit}} = 32.2\gamma(1-2\nu)2V_0^2/E\Delta V^2$$

Here, d_{crit} is the critical particle size, γ the surface energy, ν Poisson's ratio, V_0 initial volume, and ΔV volume change.

In the above theoretical calculations, inaccuracies may arise if an estimate is used for each factor. However, the critical value obtained from actual calculations is smaller

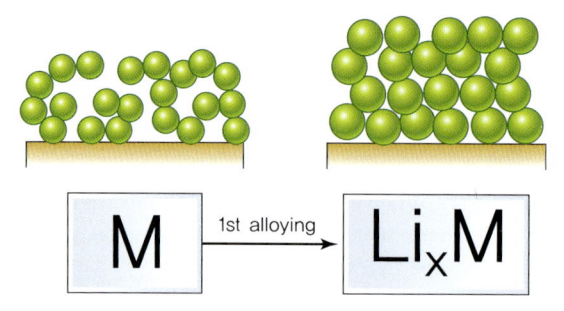

Figure 3.107 Lithium alloying of loosely arranged micrometal particles [52]. Reprinted from [52] Copyright 1999, with permission from Elsevier.

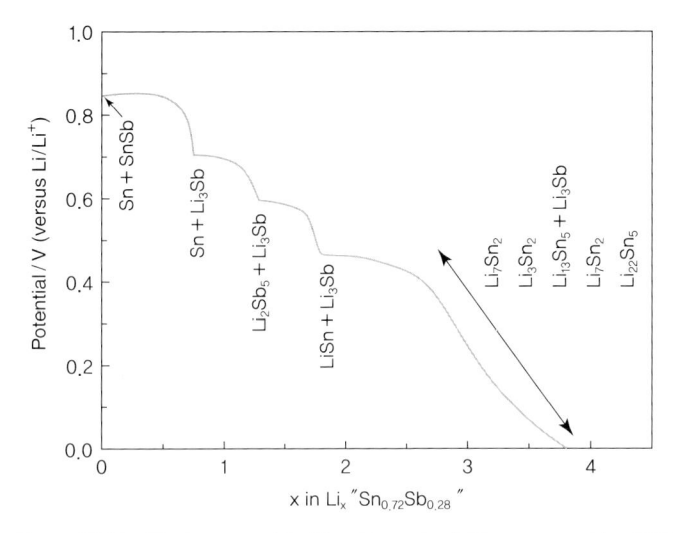

Figure 3.108 Alloying potential of Sn with varying lithium composition [52]. Reprinted from [52] Copyright 1999, with permission from Elsevier.

than the unit lattice size of metal, indicating that cracking cannot be avoided simply by minimizing the size of particles.

Multiphase Lithium Alloys Metals existing in a single phase (e.g., Sn) react with lithium ions at a specific potential, but other multiphase metals (e.g., multiphase Sn/SnSb) react with lithium across various ranges. As shown in Figure 3.108, the Sn–Sb alloy first reacts with lithium ions at 800–850 mV, while the remaining Sn reacts at a lower range of 650–700 mV. Cycle characteristics can be enhanced by alleviating volume expansion of the phase involved in reactions with lithium ions.

The above method helps to suppress volume expansion through progressive reactions with lithium. However, there is limited control over volume expansion since the resulting lithium metal alloy continues to react with a great amount of lithium.

Ag_3Sn is another substance that undergoes similar reactions. When existing as a metal, Ag provides electronic conductivity and forms LiAg through progressive reactions with lithium ions. LiAg reduces volume expansion by suppressing reactions between lithium ions and Sn. Subsequent phase changes are discussed as follows [54–56].

Composites of Lithium-Reactive/Nonreactive Metals Better performance can be expected when composites of lithium-reactive metals and nonreactive metals are used instead of pure metal. Volume expansion and contraction are alleviated as the active phase becomes surrounded by the inactive phase. This concept is illustrated in Figure 3.109.

To minimize volume expansion, metal particles of the active phase should be finely distributed within the inactive phase.

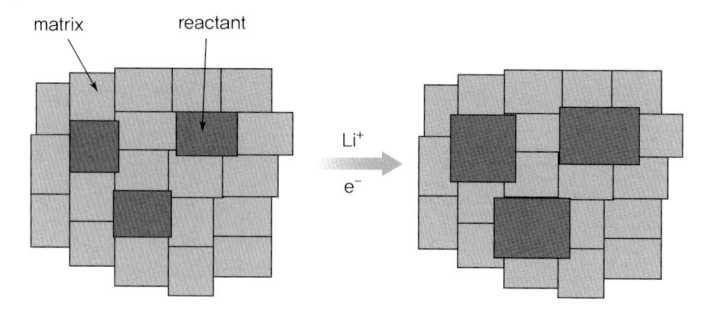

Figure 3.109 Volume expansion and contraction in composites of reactive and nonreactive metals [52]. Reprinted from [52] Copyright 1999, with permission from Elsevier.

In the early cycle of lithium reactions, tin oxides such as SnO, SnO_2, and $Sn_xAl_yB_zP_pO_n$ engage in the irreversible reaction shown below to form Li_2O and Sn. In the Li_2O continuous phase, nanosized Sn metal particles are found in the dispersed phase and form a composite of active (Sn) and nonactive phases (Li_2O). As the cycle continues, Sn and Li engage in reversible reactions.

Due to the huge capacity loss from irreversible reactions associated with Li_2O formation, it is difficult to use tin oxides (SnO) in actual lithium secondary batteries. This problem of irreversible capacity loss may be overcome by forming composites of lithium-reactive metals and nonreactive metals. Examples of active metals are Sn and Si, while nonactive metals refer to transition metals such as Fe, Ni, Mn, and Co. These composites react with lithium as follows:

$$SnM + xLi^+ + xe^- \rightarrow Li_xSn + M$$

The composite formation process follows the same principle as oxides. From the initial reaction between composite SnM and lithium ions, SnM is split into Li_xSn and M. Since the resulting active metal (Li_xSn) is distributed in the continuous phase of the transition metal M, the structural stability of the electrode is enhanced by suppressing volume changes during reactions with lithium. The nonactive metal (M) has better electronic conductivity compared to Li_2O but hinders the movement of lithium ions and lowers rate capability characteristics. Figure 3.110 represents the reactions between Sn–M and Li.

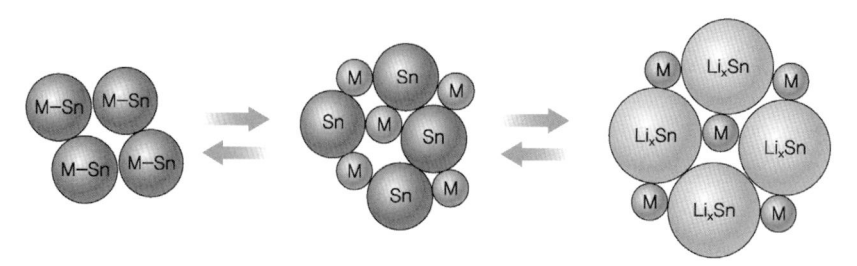

Figure 3.110 Reactions between Sn–M and Li [57].

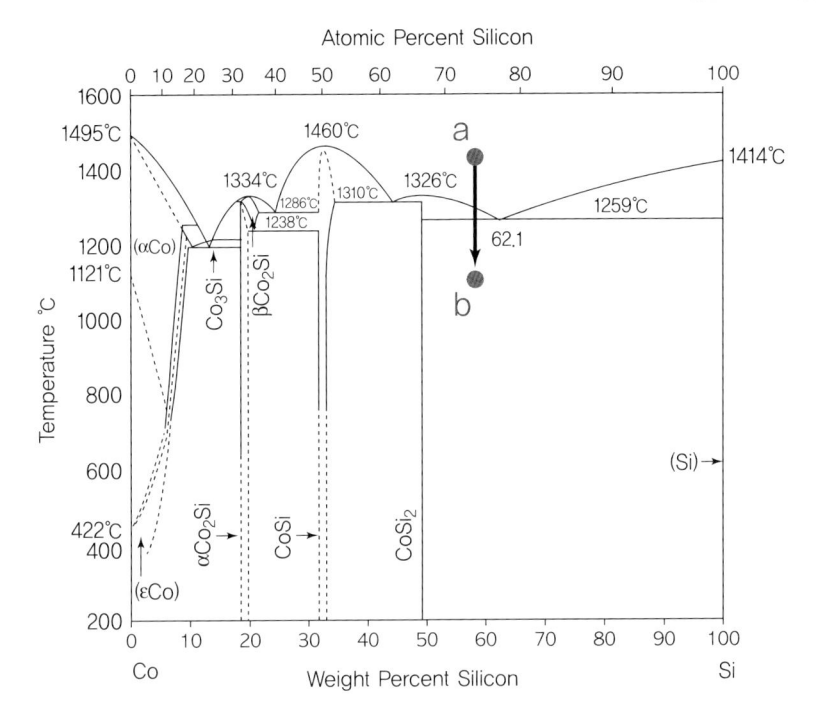

Figure 3.111 Phase diagram of Co–Si.

By finely distributing lithium-reactive metals (e.g., Si, Sn, etc.) in the continuous phase of nonreactive metals, change in volume during charging can be suppressed or alleviated. The nonreactive metal should have high mechanical strength and elasticity so as to withstand the stress from volume expansion, and superior electronic conductivity to facilitate the movement of electrons. More outstanding charge/discharge cycle characteristics can be obtained when the lithium-reactive Si forms a composite with nonreactive metals such as TiN, TiB_2, and SiC. However, the weight and volume of nonreactive metals may lead to a smaller lithium storage capacity in the composite and limit reactions between Li and Si by restricting the flow of lithium ions.

The phase diagram of a metal alloy can be used to design substances consisting of both active and nonactive phases. When an alloy at eutectic composition undergoes melting followed by rapid solidification, microstructures within the metal alloy can take on various forms. For example, in the phase diagram of Si–Co shown in Figure 3.111, Co–58Si is melted at temperature "a" and rapidly solidified by cooling to temperature "b," thus forming a metal composite with various microstructures. Figure 3.112 shows the cross section of a metal composite particle obtained in this manner. The particle is of globular form, which is characteristic of particles derived from atomization. Si particles are uniformly distributed in the matrix phase of Si–Co. This metal composite enhances cycle life by suppressing volume expansion of metal particles that react with nanosized lithium ions. However, this method requires

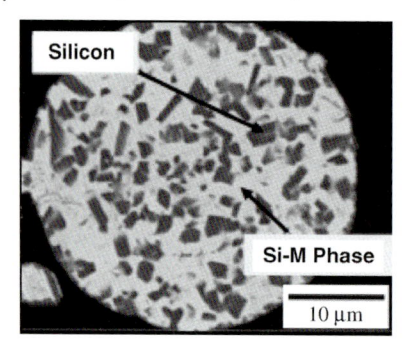

Figure 3.112 Cross section of rapidly solidified Co–Si alloy.

heating the metal to high temperatures and has not been commercialized due to the complex process and high manufacturing cost.

Despite various attempts being made to suppress volume expansion during charging of metals and alloys, successful commercialization has not been attained. However, the various methods introduced above are being actively studied. The following section describes composites of metals/alloys and carbon.

Metal/Alloy–Carbon Composites Metal/alloy–carbon composites resolve the problem of reduced electronic conductivity arising from the use of nonactive metals and provide outstanding electrochemical characteristics by minimizing volume expansion. The characteristics of metal/alloy–carbon composites as anode materials differ with microstructure and manufacturing methods. This section introduces basic concepts related to the microstructure and design of metal/alloy–carbon composites. While Si and Sn are common elements that constitute the metal or alloy of such composites, Sn may have limited uses in the design and manufacturing of carbon composites due to its low melting point.

Figure 3.113 shows the microstructure and schematic diagram of the Sn–Co–C composite. In this composite, Sn forms an alloy with both Co and C. Although alloying between three elements is difficult to achieve, Sn–Co–C alloying has been

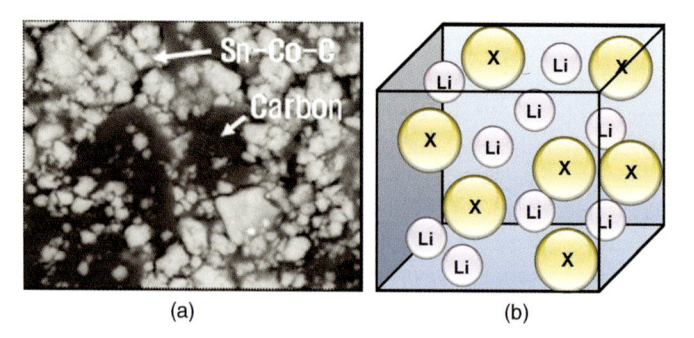

Figure 3.113 (a) Microstructure image and (b) schematic diagram of Sn–Co–C composite of alloy and carbon.

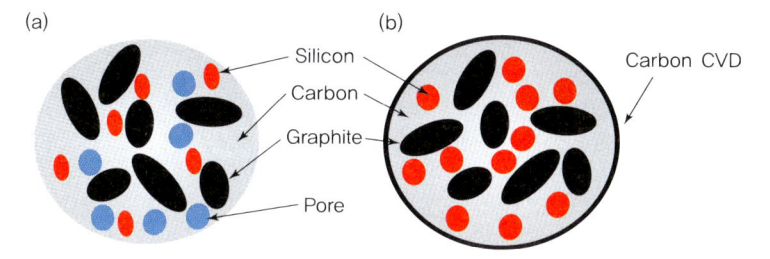

Figure 3.114 Schematic diagram of Si–carbon composite: apply carbon CVD to (a) to obtain (b).

experimentally proven. Volume expansion is suppressed both by the alloying of Sn with Co and C and by the presence of carbon. Batteries based on Sn–Co–C composites have been commercialized for camcorders [58].

However, complex synthesis and an inadequate increase in capacity have limited this material to special purposes.

In a general carbon composite, carbon coating is applied to Si and Si alloy, allowing Si to be dispersed within carbon. The carbon layer on the surface of active materials enhances electronic conductivity between particles and electrochemical characteristics of the electrolyte. This can be achieved through thermal decomposition of the carbon precursor on the surface of Si and Si alloys at temperatures higher than 1000 °C or simultaneous pyrolytic deposition of Si and the carbon precursor. Figure 3.114 shows a schematic diagram of a general composite of Si and carbon.

In the composite obtained using the above method, Si particles are dispersed within the carbon continuous phase. Electrode deterioration from volume expansion of Si particles is lessened but the use of carbon leads to reduced cycle life and irreversible reactions in the early stage. In Figure 3.114b, a carbon layer has been applied to the Si–C composite using CVD to reduce the number of irreversible reactions and improve cycle life. This is more efficient than the method shown in Figure 3.114a but entails a high manufacturing cost and does not contribute to a significant capacity increase due to the drop in energy density of the electrode.

Another method is to use composites of silicon and graphite, which are formed by milling followed by coating of graphite layers onto silicon particles. This method not only prevents an increase in irreversible capacity but also suppresses volume expansion of silicon particles from the presence of graphite layers, thus resulting in high electrode capacity and superior performance. Figure 3.115 shows a schematic diagram of a silicon and graphite composite.

Silicon particles are coated with graphite layers through ball milling and covered with an additional carbon layer to form silicon–graphite composites. The carbon coating allows a uniform distribution of silicon particles and alleviates volume expansion of Si particles when reacting with lithium during charging/discharging. By bonding graphite with Si particles, electronic and ionic conductivities can be improved. This carbon layer is formed by first combining the carbon precursor with the Si–graphite composite and is then followed by thermal decomposition at 1000 °C and carbonization. The use of Si alloys (Si/M, M: transition metal) instead of pure Si allows higher electronic conductivity and less change in volume [59].

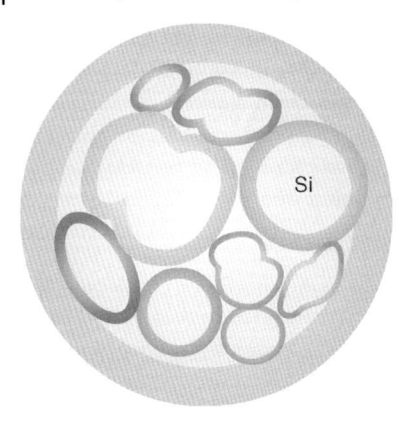

Figure 3.115 Schematic diagram of Si and graphite composite.

Metal Thin-Film Electrodes Metal active materials can be used as thin-film electrodes without binders [52]. Figure 3.116 shows the typical volume change in a thin-film electrode during charging (alloying with lithium) and discharging. The thickness increases from 6 to 11 μm when the battery is first charged and then discharged. After the second cycle, it further expands to 17 μm. The surface of copper was coated with cylindrical-shaped tin particles to absorb the stress from volume expansion. No cracking occurs, as there is sufficient space to accommodate volume expansion in the horizontal direction. It is helpful to design the geometrical shape of electrodes with consideration of such changes in volume. For more widespread commercialization, chemical methods should be developed to suppress volume expansion.

Compounds
Metal Oxides Anode materials consisting of metal oxides exhibit two different behaviors when reacting with lithium during the charging and discharging process. The first is lithium intercalation/deintercalation while maintaining the crystal structure and the second is oxide decomposition during lithium reactions. Some representative anode materials belonging to the latter are transition metal oxides with

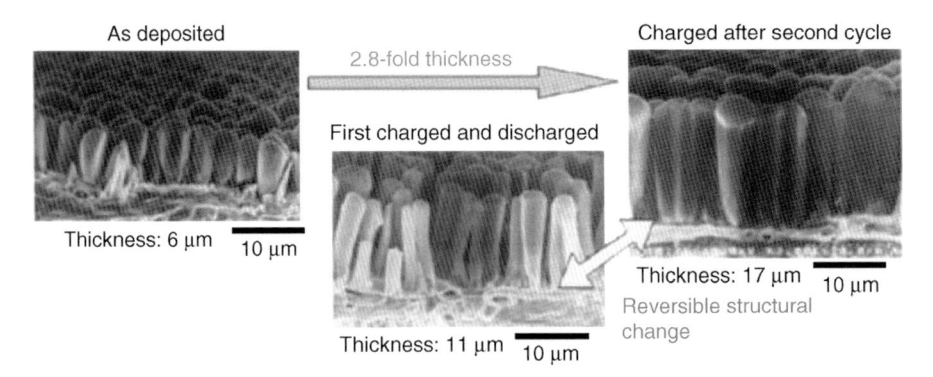

Figure 3.116 Volume change in metal (Si) thin-film during charge/discharge [52].

a rock-salt structure, such as CoO, NiO, FeO [60], and TiO_2 [61]. These oxides have a large irreversible capacity and a high average discharge voltage of 0.8–2.0 V. Depending on its structure, the TiO_2 oxide exists in various phases such as anatase, rutile, ramsdellite TiO_2, and TiO_2–B [62, 63].

Anatase titanium dioxide has a body-centered tetragonal ($I4_1/amd$) structure, lattice constants of $a = 3.782$ Å and $c = 9.502$ Å, and a density of 3.904 g/ml. Li_xTiO_2 is formed through electrochemical reactions with lithium, while x changes reversibly in the range of 0.0–0.5. The chemical equation is given as follows:

$$TiO_2 + xLi^+ + xe^- = Li_xTiO_2, (x = 0-0.5)$$

In the case of anatase TiO_2, intercalation and deintercalation reactions occur at the flat region, while charge/discharge proceeds in curved regions as TiO_2 and lithium react without electrolyte decomposition. Such intercalation and deintercalation reactions are two-phase equilibrium reactions by the tetragonal and the orthorhombic structure of $Li_{0.05}TiO_2$, and the electrochemical potential at this point is 1.8 V (Li/Li$^+$) [64, 65]. Despite having a high capacity of 200 mAh/g, this material has limited applications due to its large irreversible capacity arising from intercalation reactions between TiO_2 and lithium at the particle surface with an average voltage of 1.8 V. Unlike carbon-based materials, it shows a significant potential difference between charge and discharge curves due to its low electronic conductivity. A large polarization resistance can be observed in Figure 3.117. This material is expected to be useful in areas requiring low voltage and current density.

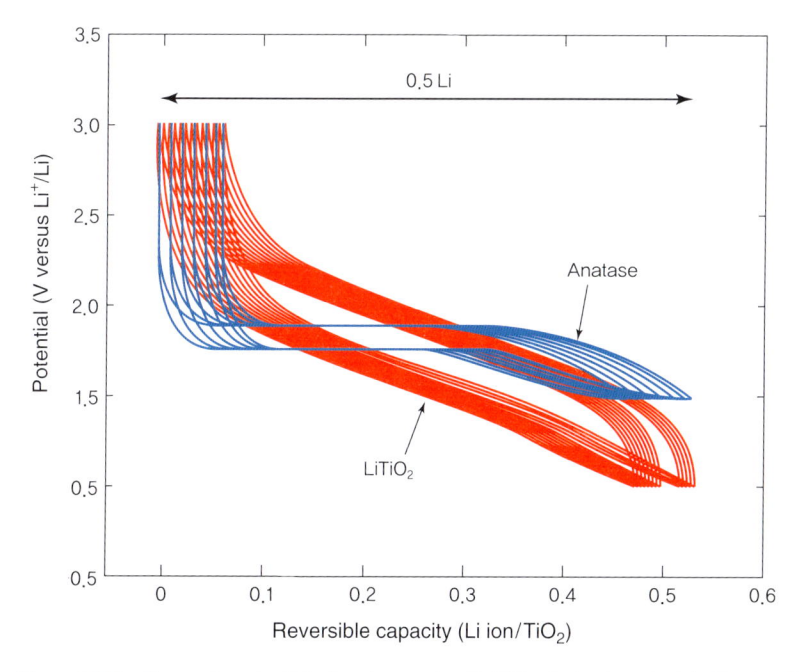

Figure 3.117 Change in potential of anatase and rutile TiO_2 electrode materials. Reprinted with permission from American Chemical Society Copyright 2007.

Rutile TiO_2 has the same rock-salt structure as $LiTiO_2$ and is electrochemically active. However, it is slow to react and its volume expands by 4.5% during charging. Its average working voltage is similar to that of anatase TiO_2 but it has a sloped potential curve [66]. Figure 3.117 shows the charge and discharge curves for anatase and rutile $LiTiO_2$. In the early stage of charging, TiO_2 initially shows an open-circuit voltage of 3.0 V but this rapidly drops to 1.8 V after reacting with lithium ions during charging. SEI layers are not formed since the voltage of lithium intercalation into the oxide is greater than that of electrolyte decomposition. Unlike carbon materials, the absence of coating on the surface of particles allows better output characteristics. Nanosized particles are suitable to achieve high-rate capability by increasing the surface area; nevertheless, the material itself has poor electronic conductivity and slow lithium diffusion.

The transition metal MO (M: Co, Ni, Fe, etc.) has a rock-salt structure and reacts with lithium to form Li_2O and a nanosized metal through oxide decomposition. The resulting nanosized metal is dispersed within Li_2O [60]. In the case of CoO, the equation is as follows:

$$CoO + 2Li \Leftrightarrow Li_2O + Co$$

The reverse reaction of forming CoO from a Li_2O continuous phase is based on the high surface energy of a finely distributed nanosized metal. Such reversible reactions are possible if transition metal oxides are nanosized. As shown in Figure 3.118 [67], the lithium reaction potential for the above reversible reaction is greater than 0.8 V,

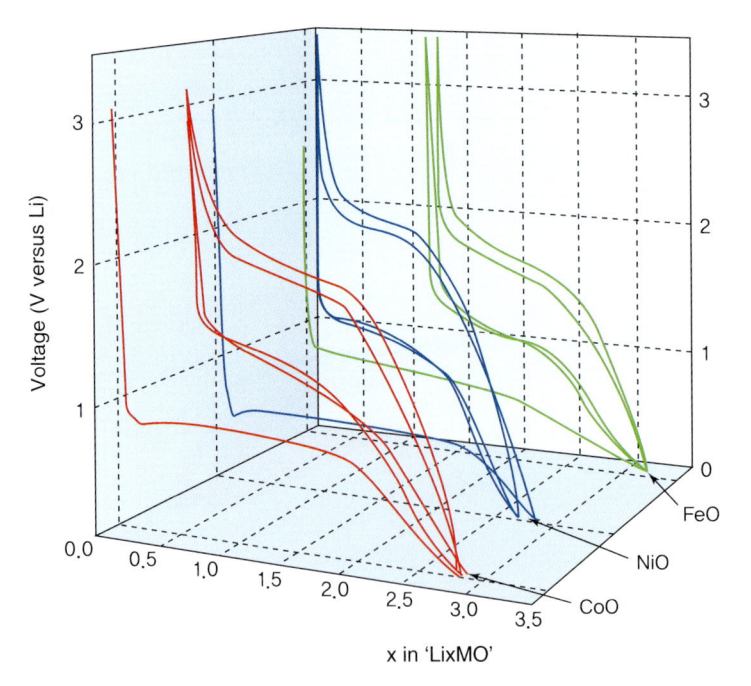

Figure 3.118 Charge/discharge curves of CoO, NiO, and FeO [67]. Reprinted by Permission from Macmillan Publishers Ltd: [67], copyright 2000.

with a large gap between charge and discharge curves. This is due to the rapid decrease in electronic conductivity, with the Li_2O continuous phase acting as an insulator. Even if nanosized metals within the Li_2O continuous phase are electrochemically reactive, electronic flow is disrupted by the insulator, which eventually causes a large polarization resistance. The gap between charge and discharge curves may generate excessive heat in batteries. While a high energy density can be obtained from such metal oxides, it is necessary to develop techniques to overcome the problem of low electronic conductivity.

Similar to the case of using metal oxides as cathodes, lithium–titanium oxides such as $Li(Li_{1/3}Ti_{5/6})O_4$ are electrochemically active toward lithium and have a specific capacity of 175 mAh/g and a high potential of 1.5 V (Li^+/Li) [60]. $Li(Li_{1/3}Ti_{5/6})O_4$ is known as a zero-strain material [68–71] that shows no change in the crystal lattice before and after charging. In its general form of $Li_4Ti_5O_{12}$, Li and Ti exist together at $16d$ octahedral sites, while the remaining Li occupies $8a$ tetrahedral sites. $Li_1(Li_{1/3}Ti_{5/6})O_4$ and $Li_2(Li_{1/3}Ti_{5/6})O_4$ have the same Fd-$3m$ (227) space group, and a lattice constant of 8.3595 Å and 8.3538 Å, respectively. The volume change accompanying reduction is very small at 0.0682%. The electrochemical reaction is given by the following equation:

$$Li_1(Li_{1/3}Ti_{5/6})O_4 + Li^+ + e^- = Li_2(Li_{1/3}Ti_{5/6})O_4$$

Figure 3.119 shows the typical charge and discharge curves for $Li_4Ti_5O_{12}$ as an anode materials. A flat potential range is exhibited due to two-phase reactions. The initial charge and discharge cycle efficiency is close to 100% because the high working voltage prevents SEI formation of the anode via electrolyte decomposition. This material may be utilized for its high output characteristics.

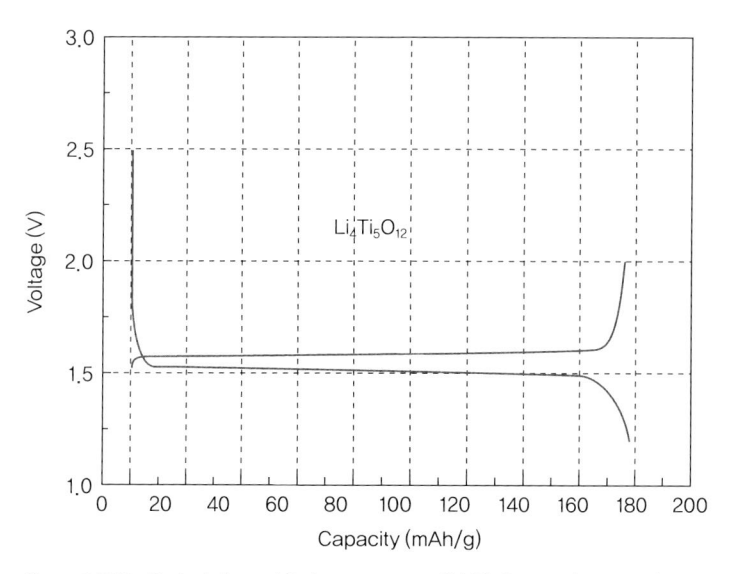

Figure 3.119 Typical charge/discharge curves of $Li_4Ti_5O_{12}$ anode material.

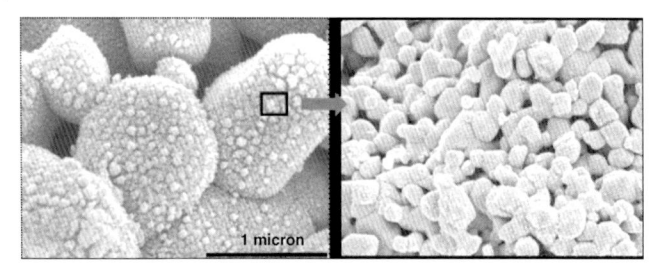

Figure 3.120 Particle shape of the $Li_4Ti_5O_{12}$ anode material.

As can be seen from the flat charge/discharge potential curves, nanosized particles should be used to minimize the migration path in order to overcome the low diffusivity of lithium ions. From the particles of commercial $Li_4Ti_5O_{12}$ in Figure 3.120, we can see that nanosized primary particles are aggregated to form secondary particles.

Electrode slurry manufacturing based on nanosized particles requires a great amount of solvent and thus lowers the productivity of the electrode. Furthermore, since nanosized particles are sensitive to moisture, excessive moisture is adsorbed when exposed to air. This not only impedes the electrode manufacturing process but also deteriorates battery characteristics. When the level of moisture increases in the electrode, hydrogen and oxygen decompose to release gases within the battery, thereby affecting battery performance. This problem can be resolved by introducing an additional process to control nanosized particles or by aggregating nanosized primary particles for use as electrode active materials.

$Li_4Ti_5O_{12}$, with relatively low specific capacity but high rate capability, is being actively studied for use in HEV batteries. Figure 3.121 shows the cycle life characteristics at a high C-rate for a battery consisting of a $Li_4Ti_5O_{12}$ anode and $LiMn_2O_4$ spinel cathode. Cycle life characteristics were found to be stable at a charge rate of 2 C and discharge rates of 10 C and 20 C. It was found that the use of $Li_4Ti_5O_{12}$ as an

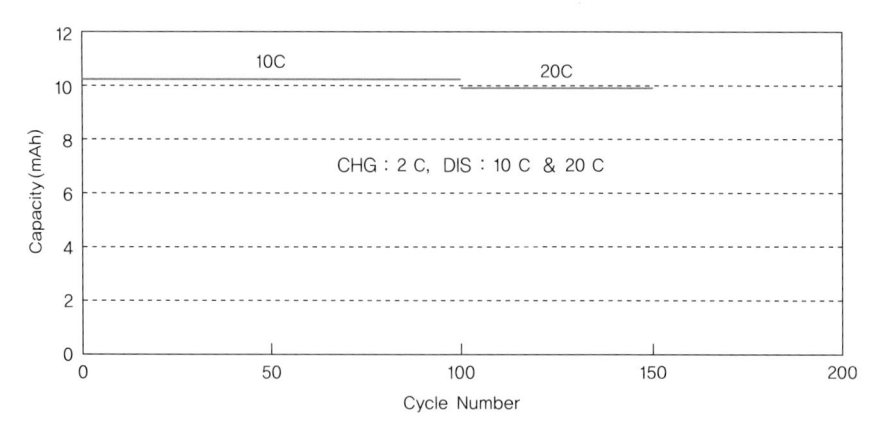

Figure 3.121 Rate capability characteristics of a battery with $Li_4Ti_5O_{12}$ anode and $LiMn_2O_4$ cathode.

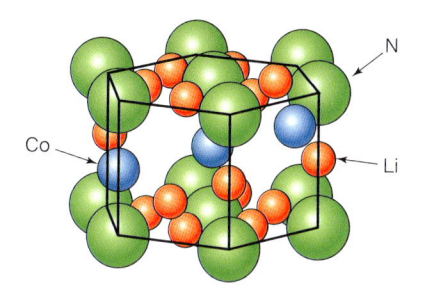

Figure 3.122 Crystal structure of $Li_{2.6}Co_{0.4}N$.

anode material greatly contributed to improved life characteristics. The formation of an SEI layer was avoided, as the electrolyte decomposition voltage was beyond the working voltage of the battery. In addition, high-rate capability can be achieved by the use of the nanosized particles displaying rate capability characteristics.

Nitride Anode Materials $Li_2(Li_{1-x}M_x)N$ anode (M = Co, Ni, or Cu), a representative nitride anode materials, has a layered structure and a high ionic conductivity. Figure 3.122 shows the crystal structure of $Li_{2.6}Co_{0.4}N$ with space group $P6/mmm$, lattice constants of $a = 3.68$ Å and $c = 3.71$ Å, and a density of 2.12 g/ml that is similar to that of graphite [72].

Figure 3.123 compares the charge/discharge curves of the above electrode with the existing graphite electrode [73]. Charge and discharge occur at a potential of 0–1.4 V, and a large reversible capacity of 800 mAh/g is shown. This material has a capacity two times larger than that of graphite and exhibits superior cycle life characteristics. In the half-cell, the discharge potential (0.7–0.8 V) versus lithium is much higher than that of graphite. When lithium is released in the early stage, the crystal changes into

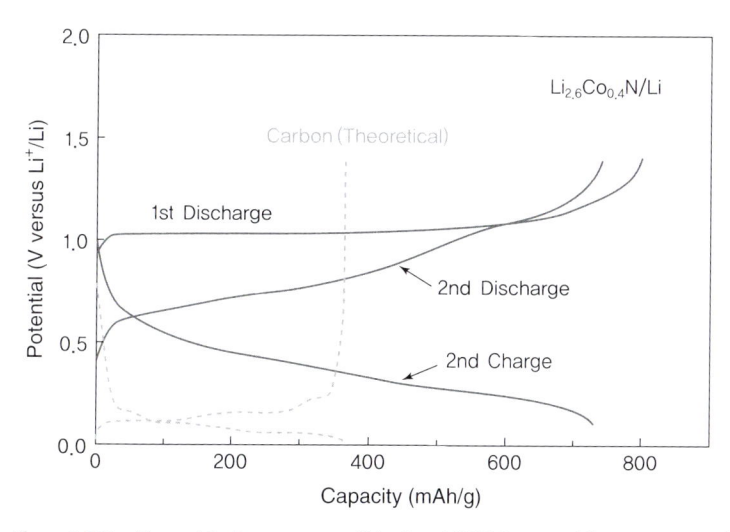

Figure 3.123 Charge/discharge curves of $Li_{2.6}Co_{0.4}N$ [73]. Reprinted from [73] Copyright 1999, with permission from Elsevier.

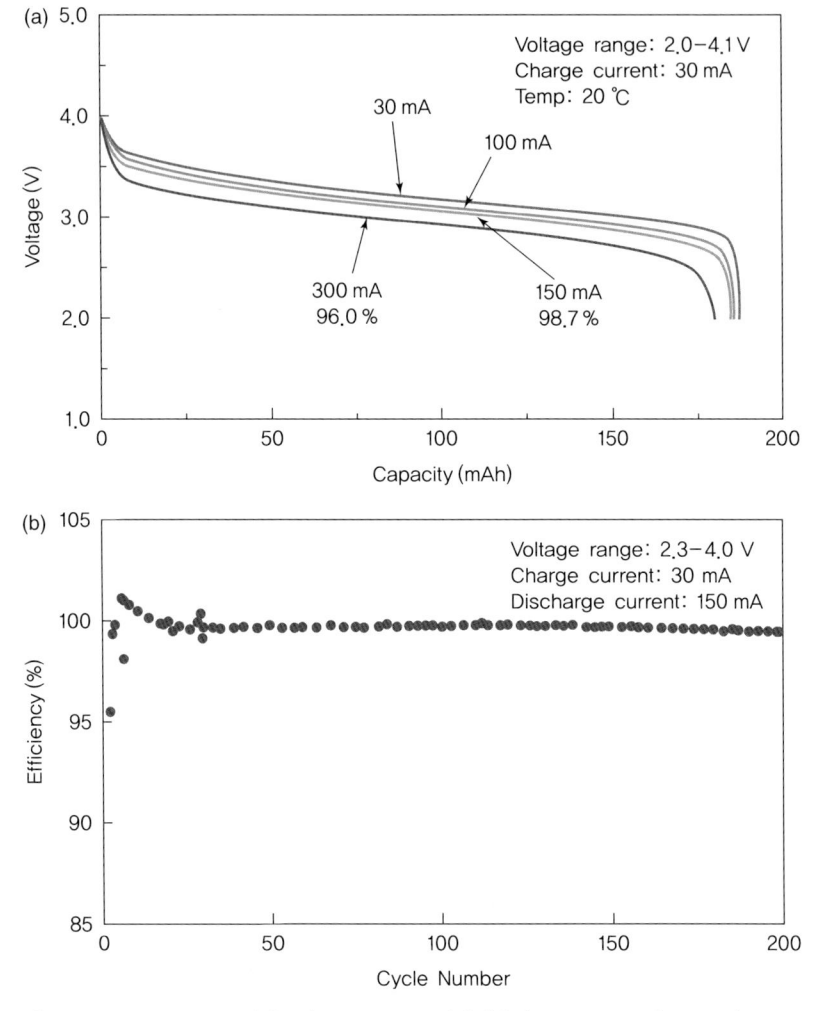

Figure 3.124 (a) Rate capability characteristics and (b) life characteristics of a square battery with a $Li_{2.6}Co_{0.4}N$ anode.

amorphous form and affects charge/discharge characteristics. Nitride materials face many constraints in various applications, as they are sensitive to moisture. In lithium secondary batteries, the cathode should accept lithium ions that are first released at the nitride anode. Unlike typical batteries, initial charging is not required if nitride anode materials are used.

Figure 3.124 shows (a) rate capability characteristics and (b) cycle life characteristics of a pristine lithium secondary battery with a $Li_{2.6}Co_{0.4}N$ anode. Li_xCoO_2 that has undergone lithium deintercalation is used as the cathode. The cell capacity of 1 C rate was 96% in comparison with C/10 (30 mA), and outstanding cycle life characteristics were shown, with a capacity retainability close to 100% even after

200 cycles. However, it is difficult to make a decision regarding use in commercial batteries based on the results, shown in Figure 3.124b, due to the low rate capability. This material should be subjected to discharge reactions first for the release of lithium, as it already contains lithium from the preparation process. By mixing this material with other anode materials having high irreversible capacity, the overall battery capacity may be improved.

3.2.4
Conclusions

Carbon-based materials are mainly used as anode materials of lithium secondary batteries. Artificial graphite was commonly employed in the past but is being replaced with natural graphite. New anode materials consisting of silicon and tin are being considered to overcome the low theoretical capacity of graphite and to enhance battery performance. Since the use of silicon and tin alone leads to excessive volume expansion and poor cycle life characteristics, they are being developed in the form of carbon composites [74].

Active research is being carried out on thermal stability, which is an important issue in lithium secondary batteries [48]. In particular, studies have focused on the mechanism of thermal reactions, the amount of heat generation, and the rate of heat release so as to prevent thermal runaway in batteries.

High-energy and high-power lithium secondary batteries are expected to be widely used for energy storage and in hybrid electric vehicles. Carbon materials with high stability and superior charge/discharge characteristics are being developed as anode materials. Meanwhile, noncarbon materials will initially be adapted to small-size batteries before advancing to application in high-capacity, high-power batteries.

References

1 Reynolds, W.N. (1968) *The Physical Properties of Graphite*, Elsevier.
2 Pierson, H.O. (1993) *Handbook of Carbon, Graphite, Diamond and Fullerenes*, Noyes Publications, Park Ridge, NJ.
3 Walker, P.L., Jr. (1969) *Chemistry and Physics of Carbon – A Series of Advances, 5: Deposition, Structure and Properties of Pyrolytic Carbon*, Marcel Dekker.
4 Winter, M. *et al.* (1998) Insertion electrode materials for lithium batteries. *Adv. Mater.*, **10**, 10.
5 Kambe, N., Dresselhaus, M.S., Dresselhaus, G., Basu, S., McGhie, A.R., and Fischer, J. (1979) *Mater. Sci. Eng.*, **40**, 1.
6 Song, X.Y., Kinoshita, K., and Tran, T.R. (1996) *J. Electrochem. Soc.*, **143**, L120.
7 Billaud, D., McRae, E., and Herold, A. (1979) *Mater. Res. Bull.*, **14**, 857.
8 van Schalkwijk, W.A. and Scrosati, Bruno (2002) *Advances in Lithium Ion Batteries*, Kluwer Academic Publishers.
9 Winter, M., Besenhard, J.O., Spahr, M.E., and Novak, P. (1998) *Adv. Mater.*, **10**, 725.
10 Mochida, I., Yoon, S.H., Korai, Y., Kanno, K., Sakai, Y., Komatsu, M., Marsh, H., and Rodriguez-Reinoso, F. (eds) (2000) *Science of Carbon Materials*, Publicaciones de la Universidad de Alicante, Alcante, Spain.
11 Otani, S. (1965) *Carbon*, **3**, 31.

12 Nemanich, R.J. and Solin, S.A. (1979) *Phys. Rev. B*, **20**, 392.
13 Mathew, S., Joseph, B., Sekhar, B.R., and Dev, B.N. (2008) *Nucl. Instrum. Methods Phys. Res. B*, **266**, 3241.
14 American Institute of Physics (1970) *J. Chem., Phys.*, **53**, 1126.
15 Alvarez, R., Diez, M.A., Garcia, R., Gonzalez de Andres, A.I., Snape, C.E., and Moinelo, S.R. (1993) *Energy Fuels*, **7**, 953.
16 Burchell, T.D. (1999) *Carbon Materials for Advanced Technologies*, Elsevier Science.
17 Dahn, J.R., Zheng, T., Liu, Y., and Xue, J.S. (1995) *Science*, **270**, 590.
18 Zheng, T., Liu, Y., Fuller, E.W., Tseng, S., Von. Sacken, U., and Dahn, J.R. (1995) *J. Electrochem. Soc.*, **142**, 2581.
19 Gao, Y., Myrtle, K., Mejji, Z., Reimers, J.N., and Dahn, J.R. (1996) *Phys. Rev. B*, **54**, 23.
20 Zheng, T., McKinnon, W.R., and Dahn, J.R. (1996) *J. Electrochem. Soc.*, **143**, 2137.
21 Claye, A. and Fischer, J.E. (1999) *Electrochim. Acta*, **45**, 107.
22 Sato, K., Noguchi, M., Demachi, A., Oki, N., and Endo, M. (1994) *Science*, **264**, 556.
23 Mabuchi, A. (1994) *Tanso*, **165**, 298.
24 Wang, S., Matsumura, Y., and Maeda, T. (1995) *Syn. Metals*, **71**, 1759.
25 Zheng, T. and Dahn, J.R. (1995) *Syn. Metals*, **73**, 1.
26 Zheng, T., Reimers, J.N., and Dahn, J.R. (1995) *Phys. Rev. B*, **51**, 734.
27 Zheng, T. and Dahn, J.R. (1996) *Phys. Rev. B*, **53**, 3061.
28 Dahn, J.R., Sleigh, A.K., Shi, H., Reimers, J.N., Zhong, Q., and Way, B.N. (1993) *Electrochim. Acta*, **38**, 1179.
29 Nazri, G.A. and Pistoia, Gianfranco (2004) *Lithium Batteries Science and Technology*, Kluwer Academic Publishers.
30 Nazri, G.A. and Pistoia, Gianfranco (2004) Chapter 5, in *Lithium Batteries Science and Technology*, Kluwer Academic Publishers.
31 Mabuchi, A., Fujimoto, H., Tokumitsu, K., and Kasuh, T. (1995) *J. Electrochem. Soc.*, **142**, 3049.
32 Surampudi, S. and Koch, V.R. (1993) *Lithium Batteries*, The Electrochemical Society, Pennington, NJ, Pv 93-24.
33 Wakihara, M. and Yamamoto, O. (1998) Chap. 8, in *Lithium Ion Batteries*, Kodansha/Wiley-VCH, Tokyo/Weinheim.
34 Peled, E. (1979) *J. Electrochem. Soc.*, **126**, 2047.
35 Fong, R., von Sacken, U., and Dahn, J.R. (1990) *J. Electrochem. Soc.*, **137**, 2009.
36 Yamamoto, O., Takeda, Y., and Imanishi, N. (1993) *Proceeding of the Symposium on New Sealed Rechargeable Batteries and Supercapacitors*, The Electrochemical Society, Inc., Pennington, NJ, p. 302.
37 Bittihn, R., Herr, R., and Hoge, D. (1993) *J. Power Sources*, **43–44**, 409.
38 Baseshard, J.O., Winter, M., Yang, J., and Biberacher, W. (1995) *J. Power Sources*, **54**, 228.
39 Winter, M., Basenhard, J.O., and Novak, P. (1996) *GDch Monographie*, **3**, 438.
40 Yamada, K., Tanaka, H., Mitate, T., and Yashikawa, M. (1997) U.S. Patent 5595938.
41 Shu, Z.X., McMillian, R.S., and Murray, J.J. (1993) *J. Electrochem. Soc.*, **140**, L101.
42 Kuribayashi, I., Yokoyama, M., and Yamashita, M. (1995) *J. Power Sources*, **54**, 1.
43 Yamasaki, M., Nohma, T., Nishio, N., Kusumoto, Y., and Shoji, Y. (1999) U.S. Patent 5888671.
44 Buil, E., George, A.E., and Dahn, J.R. (1998) *J. Electrochem. Soc.*, **145**, 2252.
45 Kinoshita, K. (1988) *Carbon: Electrochemical and Physicochemical Properties*, John Wiley & Sons, Inc.
46 Takahagi, T. and Ishitani, A. (1984) *Carbon*, **22**, 43.
47 Du Pasquier, A., Disma, F., Bowmer, T., Gozdz, A.S., Amatucci, G., and Tarascon, J.M. (1998) *J. Electrochem. Soc.*, **145**, 472.
48 Spotnitz, R. and Franklin, J. (2003) *J. Power Sources*, **113**, 81.
49 Isao Mochida, Chemistry and engineering of the carbon materials, Asakura (1990).
50 Huggins, R.A. (1999) *J. Power Sources*, **81–82**, 13–19.
51 Tamura, N., Ohshita, R., Fujimoto, M., Kamino, M., and Fujitani, S. (2003) *J. Electrochem. Soc.*, **150**, A679.
52 Winter, M. and Besenhard, J.O. (1999) *Electrochim. Acta*, **45**, 31.

53 Wolfenstine, J. (1999) *J. Power Sources*, **79**, 111.
54 Ronnebro, E., Yin, J., Kitano, A., Wada, M., and Sakai, T. (2005) *Solid State Ionics*, **176**, 2749.
55 Sreeraj, P., Wiemhofer, H.D., Hoffmann, R.D., Walter, J., Kirfel, A., and Pottgen, R. (2006) *Solid State Sciences*, **8**, 843.
56 Yin, J., Wada, M., Yoshida, S., Ishihara, K., Tanese, S., and Sakai, T. (2003) *J. Electrochem. Soc.*, **150**, A1129.
57 Nazri, G.A. and Pistoia, Gianfranco (2004) Chapter 4, in *Lithium Batteries Science and Technology*, Kluwer Academic Publishers.
58 http://www.sony.net/SonyInfo/News/Press/200502/05-006E/.
59 Xue, J.S. and Dahn, J.R. (1995) *J. Electrochem. Soc.*, **142**, 3668.
60 Kuhn, A., Amandi, R., and Garcia-Alvarado, F. (2001) *J. Power Sources*, **92**, 221.
61 Julien, C.M., Massot, M., and Zaghib, K. (2004) *J. Power Sources*, **136**, 72.
62 Ariyoshi, K., Yamato, R., and Ohzuku, T. (2005) *Electrochim. Acta*, **51**, 1125.
63 Ohzuku, T., Takeda, S., and Iwanaga, M. (1999) *J. Power Sources*, **81–82**, 90.
64 Armstrong, A.R., Armstrong, G., Canales, J., and Bruce, P.G. (2005) *J. Power Sources*, **146**, 501.
65 Brousse, T., Marchand, R., Taberna, P.L., and Simon, P. (2006) *J. Power Sources*, **158**, 571–577.
66 Kuhn, A., Amandi, R., and Garcia-Alvarado, F. (2001) *J. Power Sources*, **92**, 221–227.
67 Polzot, P., Laruelle, S., Grugeon, S., Dupont, L., and Tarascon, J.M. (2000) *Nature*, **407**, 496.
68 Ohzuku, T., Ueda, A., and Yamamoto, N. (1995) *J. Electrochem. Soc.*, **142**, 1431.
69 Oh, S.W., Park, S.H., and Sun, Y.K. (2006) *J. Power Sources*, **161**, 1314.
70 Yamada, H., Yamato, T., Moriguchi, I., and Kudo, T. (2004) *Solid State Ionics*, **175**, 195–198.
71 Baudrin, E., Cassaignon, S., Koelsch, M., Jolivot, J.P., Dupont, L., and Tarascon, J.M. (2007) *Electrochem. Commun.*, **9**, 337.
72 Shodai, T., Okada, S., Tobishima, S.I., and Yamaki, J.I. (1996) *Solid State Ionics*, **86–88**, 785.
73 Shodai, T., Sakurai, Y., and Suzuki, T. (1999) *Solid State Ionics*, **122**, 85.
74 Kasavajjula, U., Wang, C., and Appleby, A.J. (2007) *J. Power Sources*, **163**, 1003.

3.3
Electrolytes

An electrolyte acts as a medium for the movement of ions and commonly consists of a solvent and salt. Molten electrolytes are also possible. Liquid electrolytes are formed from organic solvents and solid electrolytes are derived from inorganic compounds or polymers, while polymer electrolytes are prepared from polymers and salts. Polyelectrolytes are also considered polymer electrolytes. In general, the term "electrolyte solution" is used to refer to liquid electrolytes.

The electrodes of lithium ion batteries use materials capable of lithium intercalation/deintercalation and are separated by a membrane before being immersed in a liquid electrolyte. The liquid electrolyte transports lithium ions from anode to cathode during charging, and vice versa. Porous electrodes with a transition metal oxide and carbon as an active material are used for the cathode and anode of lithium secondary batteries. As such, the electrolyte not only supplies lithium ions by permeating into the micropores but also exchanges lithium ions at the surface of active materials. The working voltage and energy density of lithium secondary batteries are determined by the cathode and anode materials. The choice of electrolyte is also important since high ionic conductivity between electrodes is essential for high-performance batteries.

Table 3.5 shows the characteristics of electrolytes in lithium secondary batteries. (1) Liquid electrolytes with lithium salt dissolved in an organic solvent have been widely used since the 1970s when lithium primary batteries were first developed. Most lithium secondary batteries available today use organic electrolytes. (2) Ionic liquid electrolytes are comprised of molten salts with a melting point below room temperature and used together with lithium salts. Due to the absence of a combustible and flammable organic solvent, they are known to produce safer batteries. (3) Solid polymer electrolytes are manufactured by dissolving lithium salts in polymers of a high polarity and have not been applied to actual batteries due to their low conductivity. (4) Gel polymer electrolytes consist of a polymer matrix and liquid electrolyte, and exhibit transitional characteristics between liquid and polymer electrolytes. Lithium ion batteries with gel polymer electrolytes are called lithium ion polymer batteries. This section discusses the characteristics of liquid electrolytes, ionic liquid electrolytes, polymer electrolytes, and gel polymer electrolytes. We will also look at other components that contribute to battery performance and safety, namely, membranes, binders, conductive agents, and current collectors.

Principles and Applications of Lithium Secondary Batteries, First Edition. Jung-Ki Park.
© 2012 Wiley-VCH Verlag GmbH & Co. KGaA. Published 2012 by Wiley-VCH Verlag GmbH & Co. KGaA.

Table 3.5 Electrolytes for lithium secondary batteries.

	Liquid electrolytes	Ionic liquid electrolytes	Solid polymer electrolytes	Gel polymer electrolytes
Composition	Organic solvents + lithium salts	RT ionic liquids + lithium salts	Polymer + lithium salts	Organic solvents + polymer + lithium salts
Ion conductivity	High	High	Low	Relatively high
Low-temp. performance	Relatively good	Poor	Poor	Relatively good
Thermal stability	Poor	Good	Excellent	Relatively good

3.3.1
Liquid Electrolytes

3.3.1.1 Requirements of Liquid Electrolytes

Typical liquid electrolytes for lithium secondary batteries are lithium salts dissolved in organic solvents. While there are many types of organic solvents and liquid salts, not all are suitable for lithium secondary batteries. In order for liquid electrolytes to be used in lithium secondary batteries, they should have the following characteristics:

1) **The electrolyte should have high ionic conductivity.** Electrolytes with higher ionic conductivity have superior battery performance. The movement of lithium ions at the electrodes and diffusion within the electrolyte are especially important when lithium secondary batteries are rapidly charged or discharged. At room temperature, liquid electrolytes of lithium secondary batteries should have an ionic conductivity higher than 10^{-3} S/cm.

2) **The electrolyte should exhibit high chemical and electrochemical stability toward electrodes.** Since lithium ion batteries engage in electrochemical reactions at the cathode and anode, the electrolyte should be electrochemically stable within the potential range of redox reactions at the two electrodes. In addition, the electrolyte should be chemically stable toward various metals and polymers constituting the cathode, anode, and battery.

3) **The electrolyte should be used over a wide temperature range.** Lithium ion batteries with liquid electrolytes are usually employed in mobile devices and must satisfy the above requirements in the temperature ranging from -20 to $60°C$. At higher temperatures, electrochemical stability drops, while ionic conductivity increases.

4) **The electrolyte should be highly safe.** Organic solvents used in electrolytes are flammable and may cause fires or explosions when heated to high temperatures during short circuits. Higher ignition points or flash points are favored, and nonflammable materials should be used if possible. The electrolyte should have a low toxicity in case of leakage or disposal.

5) **The electrolyte should be low cost.** High-performance electrolytes may be difficult to commercialize if they come at a high cost. Given the fierce market competition for lithium ion batteries, expensive materials are unlikely to be adopted.

As mentioned above, electrolytes in lithium secondary batteries should exhibit high ionic conductivity over a wide temperature range, and remain stable over a wide potential with a working voltage higher than the battery. Electrolyte characteristics are determined by properties of the solvent and lithium salt, and vary according to combinations.

3.3.1.2 Components of Liquid Electrolytes

Organic Solvents Since lithium secondary batteries have a high working voltage, organic solvents are used instead of aqueous electrolytes. The greatest disadvantage of organic solvents is the low dielectric constant. Electrolytes should possess high ionic conductivity to dissolve lithium salts and exist as polar aprotic solvents to avoid chemical reactions with lithium. Representative physicochemical characteristics of electrolytes in lithium batteries are summarized in Table 3.6 [1]. The specific dielectric constant of the solvent affects ionic dissociation and association of lithium salt. A higher dielectric constant (ε) results in greater dissociation as it is inversely proportional to the Coulombic force between cations and anions of the lithium salt.

Table 3.6 Physicochemical properties of organic solvents in lithium batteries.

Solvent	T_m (°C)	T_b (°C)	Dielectric constant	Viscosity (cP)	Donor number (DN)	Acceptor number (AN)	E_{ox} [b] (V versus Li/Li$^+$)
Ethylene carbonate (EC)	39	248	89.6	1.86[a]	16.4	—	6.2
Propylene carbonate (PC)	−49.2	241.7	64.4	2.53	15.1	18.3	6.6
Dimethyl carbonate (DMC)	0.5	90	3.11	0.59	—	—	6.7
Diethyl carbonate (DEC)	−43	126.8	2.81	0.75	—	—	6.7
Ethylmethyl carbonate (EMC)	−55	108	2.96	0.65	—	—	6.7
1,2-Dimethoxyethane (DME)	−58	84.7	7.2	0.46	24.0	—	5.1
γ-Butyrolactone (GBL)	−42	206	39.1	1.75	—	—	8.2
Tetrahydrofuran	−108.5	65	7.3	0.46	20.0	8.0	5.2
1,3-Dioxolane (DOL)	−95	78	6.8	0.58	—	—	5.2
Diethylether (DEE)	−116.2	34.6	4.3	0.22	19.2	3.9	—
Methyl formate (MF)	−99	31.5	8.5	0.33	—	—	5.4
Methyl propionate (MP)	−88	79	6.2	0.43	—	—	6.4
Sulfolane (S)	28.9	287.3	42.5	9.87	14.8	19.3	—
Dimethylsulfoxide (DMSO)	18.4	189	46.5	1.99	29.8	19.3	—
Acetonitrile (AN)	−45.7	81.8	38	0.35	14.1	18.9	—

a) Measured at 40 °C.
b) E_{ox}: Oxidative potential (scan rate: 5 mV/s; reference electrode: Li).

In general, dielectric solvents with a dielectric constant larger than 20 are recommended because lithium dissociation is difficult to achieve with smaller dielectric constants. According to Stokes' law, the movement of ions within liquid electrolytes is inversely proportional to the solvent viscosity. As such, solvents should have a low viscosity of 1 cP or less. The donor number (DN) and acceptor number (AN) represent the nucleophilicity and electrophilicity of the solvent, respectively. These numbers provide information on the mutual interaction between cations and anions, or the solvation strength of the salt. The dissociation of lithium increases with a higher DN. The working temperature is influenced by the melting and boiling points of the organic solvent, which should remain in liquid form at room temperature and be able to dissolve lithium at a temperature as low as $-20\,°C$. The organic solvent should exhibit a high boiling point and low vapor pressure. While a high dielectric constant and low viscosity are required for electrolytes to possess high ionic conductivity, a higher dielectric constant leads to increased polarity and viscosity. This can be resolved by mixing solvents with a high dielectric constant and those with a low viscosity. For example, cyclic carbonates such as ethylene carbonate (EC) and propylene carbonate (PC) have a high dielectric constant and high viscosity due to large mutual interactions with solvent particles. Lithium dissociation is possible in EC but it cannot be used alone given its high melting point. In contrast, linear carbonates such as dimethyl carbonate (DMC) and diethyl carbonate (DEC) have a low dielectric constant and low viscosity. Thus, cyclic carbonates and linear carbonates are combined to obtain desirable characteristics as organic solvents for lithium secondary batteries. Table 3.7 shows ionic conductivities of liquid electrolytes prepared from dissolving 1 M of $LiPF_6$ in organic solvents. From Table 3.7, we can see that mixed solvents tend to have higher ionic conductivity.

Lithium Salts Table 3.8 shows the physicochemical properties of lithium salts commonly used in lithium secondary batteries. Anions with larger radii are favored since lithium salts having delocalized anions tend to dissociate more readily. In general, the dissociation of lithium salts takes the following order [2]:

Table 3.7 Ionic conductivities of organic electrolytes prepared with 1 M $LiPF_6$.

Organic solvents	Ionic conductivity (mS/cm at 25°C)
EC	7.2
PC	5.8
DMC	7.1
EMC	4.6
DEC	3.1
EC/DMC (50/50, vol%)	11.6
EC/EMC (50/50, vol%)	9.4
EC/DEC (50/50, vol%)	8.2
PC/DMC (50/50, vol%)	11.0
PC/EMC (50/50, vol%)	8.8
PC/DEC (50/50, vol%)	7.4

Table 3.8 Physicochemical properties of representative lithium salts.

Lithium salts	T_m (°C)	Anion diameter (nm)	Λ_o[a] in PC (Scm^2/mol)	E_{ox} in PC (V versus SCE)
LiBF$_4$	>300	0.229	28.9	3.6
LiClO$_4$	236	0.237	27.4	3.1
LiPF$_6$	194	0.254	26.3	3.8
LiAsF$_6$	>300	0.260	26.0	3.8
LiCF$_3$SO$_3$	>300	0.270	2.3	3.0
Li(CF$_3$SO$_2$)$_2$N	228	0.325	22.8	3.3
LiC$_4$F$_9$SO$_3$	>300	0.339	21.5	3.3
Li(CF$_3$SO$_2$)$_3$C	263	0.375	20.2	3.3
LiBPh$_4$	—	0.419	17.0	1.0

a) Λ_0: Limiting molar conductivity.

$$\text{Li(CF}_3\text{SO}_2)_2\text{N} > \text{LiAsF}_6 > \text{LiPF}_6 > \text{LiClO}_4 > \text{LiBF}_4 > \text{LiCF}_3\text{SO}_3$$

On the other hand, an increase in ionic radius leads to less movement of anions. Ionic mobility (μ_0) follows Stokes' law and is given by Eq. (3.16), which can also be expressed in terms of the diffusion coefficient (D).

$$\mu_0 = \frac{\lambda_0}{zF} = \frac{ze}{6\pi\eta_0 r} = \frac{zFD}{RT}$$

Here, λ_0, z, F, e, r, η_0, R, and T, respectively, represent the limiting molar conductivity, charge number, Faraday constant, elementary electric charge, ionic radius, viscosity, gas constant, and absolute temperature. As shown above, the size of anions is an important factor that determines the properties of lithium salts. Figure 3.125 shows the ionic radii of lithium salts based on space-filling models and van der Waals radius [3].

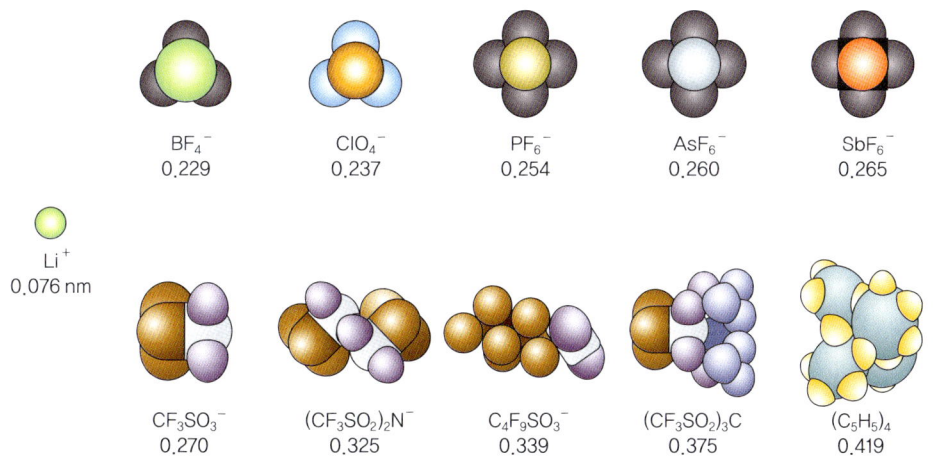

Figure 3.125 Space-filling models and ionic radii of lithium salts [3]. Reproduced by permission of ECS – The Electrochemical Society.

LiClO$_4$ is used as a lithium salt in lithium primary batteries but not in lithium secondary batteries due to safety issues arising from the oxidation environment during charging. Salts such as LiBF$_4$ or LiPF$_6$ that contain fluorinated Lewis acids are commonly used in lithium ion batteries because of their solubility and chemical stability. Other substances being considered are inorganic lithium salts, organic sulfonates, and imide salts. The lower ionic conductivity of electrolytes with LiBF$_4$ than those containing LiClO$_4$ or LiPF$_6$ negatively affects high-rate characteristics. In contrast, LiPF$_6$ electrolytes possess high ionic conductivity but low thermal stability even without side reactions at electrodes. Battery performance deteriorates with side reactions and electrolyte decomposition when HF is produced from exposure to moisture. Table 3.9 compares the various characteristics of representative lithium salts.

Meanwhile, perchloro alkyl sulfonate (LiR$_f$SO$_3$, R$_f$=C$_n$F$_{2n-1}$) has not been commercialized as it has limited solubility and low ionic conductivity. Lithium sulfonylimide (Li[R$_f$SO$_2$]$_2$N) is chemically stable but less tolerant to oxidation at the cathode and causes corrosion of the conducting aluminum, thus making it inadequate for use in actual batteries.

Application of Molecular Orbital Theory in Solvent Design A molecular orbital (MO) function describes the wave-like behavior of an electron using the Schrodinger equation $H\psi = E\psi$ (H: Hamiltonian operator; E: sum of potential and kinetic energy). The wavefunction ψ is the probability of finding an electron within a particular molecule, and the actual probability is obtained through $|\psi|^2$. While the molecular orbital function is similar to the atomic wavefunction, electrons are distributed across molecules instead of atoms. Electrons occupy the wavefunction beginning with low energy levels. Here, the highest occupied molecular orbital and lowest occupied molecular orbital are referred to as the HOMO and LUMO. Figure 3.126 illustrates the HOMO and LUMO for different energy levels.

Such energy levels are unique for different substances and vary according to the type of solvent used in the electrolyte. The potential window of a given electrolyte can be calculated based on the HOMO–LUMO theory.

Having a high HOMO energy level facilitates oxidation with strong electron–donor properties, while a low LUMO energy contributes to reduction by accepting electrons. Solvents with a low HOMO and high LUMO are suitable for use in electrolytes.

Table 3.9 Comparison of lithium salt characteristics.

	LiPF$_6$	LiBF$_4$	LiCF$_3$SO$_3$	Li(CF$_3$SO$_2$)$_2$N	LiClO$_4$
Solubility	◉	○	○	◉	◉
Ionic conductivity	◉	○	△	◉	◉
Low-temp. performance	○	△	△	○	○
Thermal stability	X	○	○	○	X
Stability toward Al	○	○	X	X	○
Stability toward Cu	○	○	○	○	○

◉: Excellent; ○: good; △: normal; X: poor.

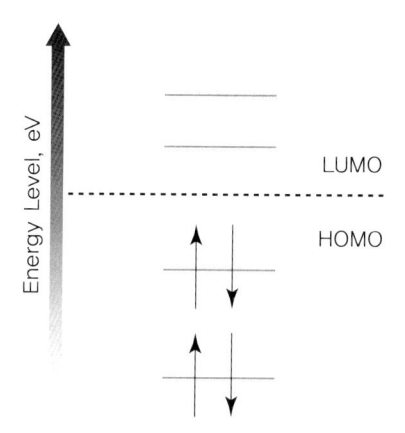

Figure 3.126 Comparison of HOMO and LUMO of various solvents with different energy levels.

Figure 3.127 shows the relationships of various solvents with different HOMO and LUMO energy levels [4].

3.3.1.3 Characteristics of Liquid Electrolytes

Ionic Conductivity The ionic conductivity of electrolytes is an important property that is directly related to rate the characteristics and other measures of battery performance. As shown in Eq. (3.17), ionic conductivity is proportional to the charge number z of ionic species i, concentration c, and mobility μ_i.

$$\sigma = N_A e \sum |z_i| c_i u_i$$

Here, N_A and e are the Avogadro constant and elementary electric charge, respectively. Ionic conductivity becomes higher with an increase in the number of

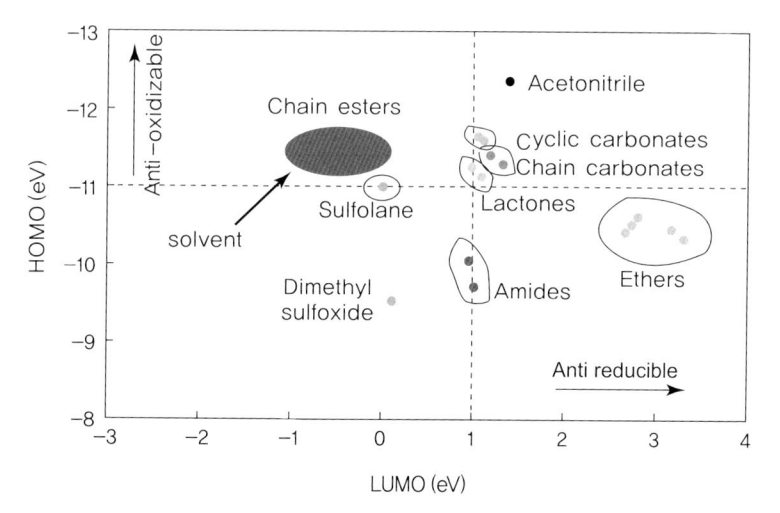

Figure 3.127 Comparison of HOMO and LUMO energy levels by solvent.

dissociated free ions and more rapid migration of such ions. Normal battery operation is difficult for low ionic conductivity since lithium ions produced at one electrode cannot be easily transported to the other. The ionic conductivity of electrolytes in lithium ion batteries at room temperature should be higher than 10^{-3} S/cm. Electrode active materials may not achieve sufficient capacity if lithium ions are not properly transported between the two electrodes due to low ionic conductivity. While ionic conductivity can be measured using a conductivity meter, it can also be calculated based on the resistance of an electrode having a known cell constant. Another method is to first obtain the solvent resistance of the electrolyte and then calculate it using the following formula: ionic conductivity = distance between electrodes/(solvent resistance × electrode surface area).

Electrochemical Stability The electrochemical stability of an electrolyte is determined by the range of potential that does not participate in redox reactions. A potentiostat is used to scan the potential of a working electrode at a constant rate with respect to the reference electrode. A rapid increase or decrease in current corresponds to the decomposition voltage. This value can also be decided by the potential at a point when the redox current reaches a certain value. The working electrode uses an electrode made of platinum (Pt), carbon (C), or stainless steel, while the reference electrode consists of a lithium metal or Ag/AgCl. This method is known as linear sweep voltammetry (LSV). Since the decomposition voltage differs according to various measurement conditions, the reference potential and scan rate must be noted. A slow scan rate (below 1 mV/s) allows more accurate measurement of electrochemical stability.

Table 3.6 shows the oxidative decomposition voltage (E_{ox}) and various physical properties of representative organic solvents. These voltages were scanned at 5 mV/s and converted against Li/Li$^+$ for a current density higher than 1 mA/cm^2. The quaternary ammonium $(C_2H_5)_4NBF_4$ was used instead of lithium to prevent a drop in potential from desolvation of lithium ions or intercalation into electrodes. As shown in the table, alkyl carbonates or ester solvents are able to withstand oxidation at 1 V higher than ether solvents. For the same reason, low-viscosity ether-based solvents are widely used in 3 V lithium primary batteries, while carbonate solvents are common in 4 V lithium secondary batteries. Using similar methods to those described above, we can compare electrochemical stability by dissolving different lithium salts into a given solvent. Table 3.8 shows the oxidative decomposition voltage of various salts in PC. The oxidative stability of lithium salt can be arranged in the following order [5–7]:

$$LiAsF_6 > LiPF_6 > LiBF_4 > Li(CF_3SO_2)_2N > LiClO_4, LiCF_3SO_3$$

Interfacial Characteristics of Electrodes/Electrolytes Reactions between liquid electrolytes and electrode active materials form an SEI layer, which greatly affects the charge/discharge cycle characteristics of lithium secondary batteries [8–11]. As lithium diffuses with cycle progression, the SEI layer has a direct influence over additional reactions between the electrolyte and the electrodes. For a graphite cathode, ethylene carbonate is favored over propylene carbonate, which destroys graphene layers and hinders SEI formation. Lithium salts also contribute to creation of a protective layer. SEI characteristics at the anode can be enhanced by adding compounds such as vinylene carbonate (VC) into the organic electrolyte to facilitate reduction. While various studies on surface reactions between carbon cathodes and organic electrolytes have been carried out, the relationship between additives and SEI formation has not been determined. A more fundamental understanding is thus necessary despite the ongoing research on suppressing electrolyte reactivity through the use of additives [12–14].

Operating Temperature Since lithium ion batteries operate in the temperature range from −20 to 60°C, the melting and boiling points of solvents must be carefully considered. For instance, solvents with a low melting point such as DEC, DME, and PC are mixed with EC, DMC, or other salts that exist as a solid at 0°C. When lithium salt is precipitated due to weak dissociation by the solvent, this temperature becomes the lower limit for the electrolyte. In addition, solvents with a low boiling point have limited use, as packing materials such as aluminum laminates may expand when vapor pressure rises. At higher temperatures, thermal and electrochemical stability drops while ionic conductivity increases.

Cation Transport Number As shown in Eq. (3.17), ionic conductivity is the sum of conductivities of cations and anions. In lithium secondary batteries, lithium cations engage in electrochemical reactions to produce current at the electrodes. As such, the conductivity of cations within the electrolyte is critical. The contribution of cations to overall ionic conductivity is represented by the cation transport number (t^+), and this is given in Eq. (3.18).

$$t^+ = \frac{\sigma_+}{\sigma_+ + \sigma_-} = \frac{\mu_+}{\mu_+ + \mu_-}$$

In the above equation, the conductivity (σ) ratio can also be expressed in terms of mobility (μ) because lithium salt dissociates into the same number of cations and anions. When the cation transport number is small, the overall resistance of the cell increases due to concentration polarization of the anion in the electrolyte. The cation transport number can be calculated using various methods such as alternating current impedance, direct current polarization measurement, Tubandt's method, Hittorf's method, and pulsed field gradient NMR (PFG NMR) [15, 16]. This number is influenced by numerous factors including temperature, salt concentration in the electrolyte, ionic radius, and electric charge [16].

3.3.1.4 Ionic Liquids

An ionic liquid refers to a salt in the liquid state. In particular, those found in the liquid state at room temperature are called room-temperature ionic liquids (RTILs). With the discovery that compounds of pyridinium or imidazolium and aluminum chlorides lead to ionic liquids, ionic liquids began to be actively studied in the 1950s [17, 18]. Compared to liquid electrolytes, ionic liquids have the following advantages:

1) They exist in liquid form over a wide temperature range and have a low vapor pressure.
2) They are flammable and thermally resistant.

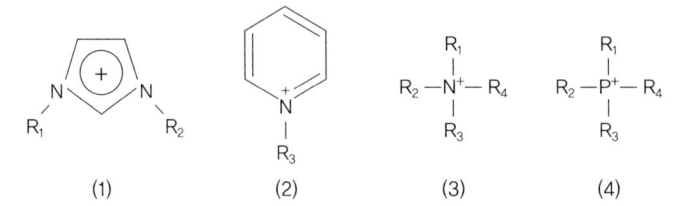

Figure 3.128 Representative cations used in ionic liquids: (1) 1,3-dialkylimidazolium cation, (2) N-alkylpyridinium cation, (3) tetra-alkylammonium cation, and (4) tetra-alkylphosphonium cation.

3) They are chemically stable.
4) They have a relatively high polarity and ionic conductivity.

However, ionic liquids exhibit poor battery performance, as ionic bonds result in a low viscosity while lithium diffusion is hindered by the presence of other cations.

Structure of Ionic Liquids Ionic liquids consist of organic cations and inorganic anions. From Figure 3.128, we can see that ionic liquids are centered on N or P and take on various structures such as alkylimidazolium, alkylpyridinium, alkylammonium, and alkyl-phosphonium. Even with the same cation, ionic liquids may or may not be in a liquid state at room temperature depending on the type of anion. For instance, an ionic liquid containing 1-ethyl-3-methylimidazolium (EMI) as a cation can have different melting points according to the anion. If the anion is Br^-, the ionic liquid is found as a white crystal solid (melting point of 78 °C) at room temperature. For BF_4^- and $TFSI^-$ anions, it becomes a colorless, transparent liquid, and the melting point drops to 15 and −16°C respectively. Flourine-based anions such as BF_4^-, PF_6^-, $CF_3SO_3^-$, and $(CF_3SO_2)_2^-$ are commonly used in ionic liquids. As shown in Figure 3.128, ionic liquids have a low melting point due to the following structural characteristics:

1) Large size of cations and anions.
2) Charge delocalization in ions.
3) Considerable conformational freedom of cations and anions, and high melting entropy.
4) Asymmetric cation structure.

Characteristics of Ionic Liquids As mentioned above, ionic liquids have unique properties such as high ionic conductivity, nonvolatility, nonflammability, and superior thermal stability. Furthermore, they have a high polarity that allows dissolution of inorganic and organic metal compounds, and they can exist in a liquid state over a wide temperature range. From the physicochemical properties of imidazolium ionic liquids shown in Table 3.10, we can see that varying the structure of cations and anions leads to diverse characteristics.

Viscosity and Ionic Conductivity Ionic liquids are special liquids composed solely of ions. Due to the high concentration of ions, ionic liquids possess high ionic conductivity. The viscosity of ionic liquids differs according to the combination of

Table 3.10 Physicochemical properties of representative imidazolium ionic liquids.

	T_m (°C)	Density (g/cm^3)	Viscosity (cP)	Ionic conductivity (mS/cm)	Water solubility
a) EMI-(CF$_3$SO$_2$)$_2$N	−15	1.53	26.1	8.8	Insoluble
b) BMI-(CF$_3$SO$_2$)$_2$N	< −50	1.44	41.8	3.9	Insoluble
c) HMI-(CF$_3$SO$_2$)$_2$N	−9	1.37	44.0	—	Soluble
EMI-PF$_6$	60	—	—	—	Insoluble
BMI-PF$_6$	6.5	1.37	272.1	—	Insoluble
HMI-PH$_6$	−73.5	1.30	497	—	Soluble
EMI-BF$_4$	15	1.24	37.7	14	Soluble
BMI-BF$_4$	−71	1.21	118.3	—	Soluble
HMI-BF$_4$	−82	1.15	234	—	Insoluble
EMI-CF$_3$SO$_3$	−9	1.39	45	9.2	Soluble
BMI-CF$_3$SO$_3$	15	1.29	99	3.7	Insoluble
HMI-CF$_3$SO$_3$	21	—	—	—	Insoluble

a) EMI: 1-Ethyl-3-methylimidazolium.
b) BMI: 1-Butyl-3-methylimidazolium.
c) HMI: 1-Hexyl-3-methylimidazolium.

cations and anions. This value is often 10 times higher than those of organic solvents. Ionic interactions increase with the addition of lithium salt, which raises viscosity but reduces ionic conductivity.

As an example, LiTFSI was added to an ionic liquid consisting of trimethylpropyl ammonium (TMPA) and bis(trifluoromethyl sulfonylimide) (TFSI). The change in viscosity and ionic conductivity according to salt concentration is shown in Figure 3.129. From the graph, adding 1.0 M of lithium salt leads to a threefold increase in viscosity and a one-fourth decrease in ionic conductivity.

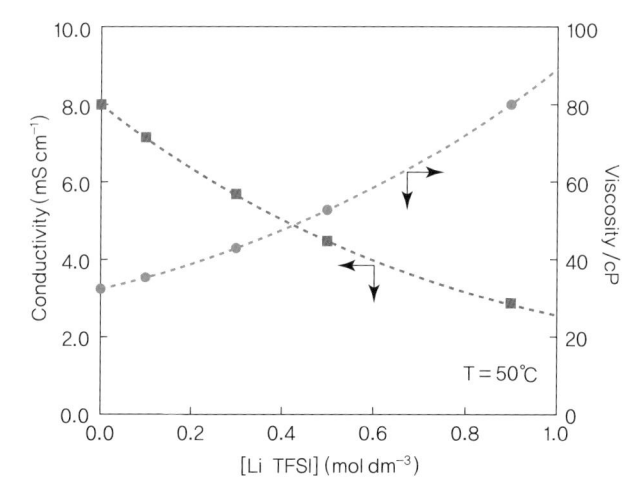

Figure 3.129 Viscosity and ionic conductivity with varying salt concentration in ionic liquid electrolytes.

Density and Melting Point Similar to organic liquid electrolytes, the density of electrolytes containing ionic liquids increases when lithium salt is added. Since the melting point of most lithium salts is higher than 200 °C, this value increases with higher concentrations of salt. For example, the electrolyte freezes at room temperature when 1.2 M of LiTFSI is added to TMPA-TFSI. As such, ionic liquids in secondary batteries should have a low melting point below room temperature. The melting point of ionic liquids takes on various values depending on the combination of cations and anions. As shown in Table 3.10, even for liquids having the same cation, the type of counterion contributes to different melting points.

Electrochemical Stability The electrochemical stability of ionic liquids can be measured using three-electrode cells in a cyclic voltammetry (CV) experiment. Figure 3.130 shows the CV measurement results for ionic liquids containing (trifluoromethylsulfonyl)imide (TFSI) as an anion and different types of cations [19]. In Figure 3.130, anodic current flows at $+2.5$ V, whereas cathodic current ranges from -1.5 to 3.0 V depending on the cation. The resistance to reduction and oxidation of ionic liquids is determined by the type of cations and anions, respectively.

Most known ionic liquids satisfy the conditions for oxidative stability required of lithium secondary batteries. However, they are not as resistant to reduction. Considering that the reduction potential of EMI is $+1.1$ V (versus Li/Li$^+$) [20], a compound with a higher reduction potential than that of lithium should be added to form an SEI layer, or the battery should have an anode with a higher potential than the lithium electrode [21, 22]. In ionic liquids, aliphatic quaternary ammonium cations are known to be more stable toward electrochemical reduction than aromatic cations such as EMI.

Electrolytes in Secondary Batteries High EMF is one of the greatest advantages of lithium secondary batteries. This is a result of an anode or metal anode having a very

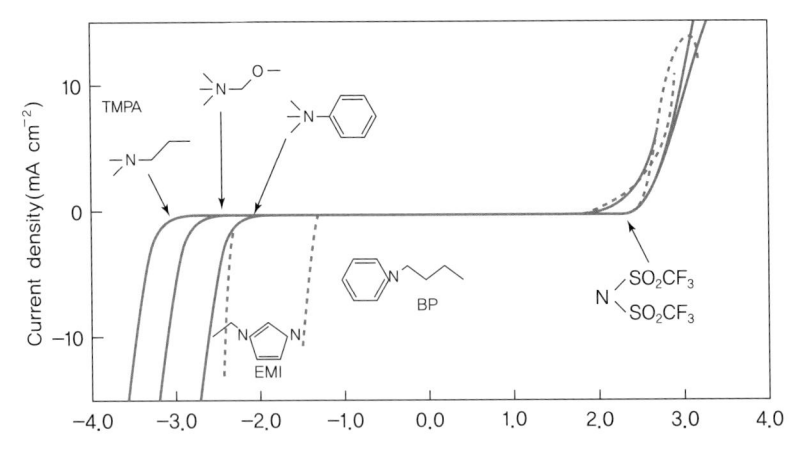

Figure 3.130 Cyclic voltammograms of ionic liquids comprised of TFSI$^-$ ions (working electrode: glassy carbon; reference electrode: Pt wire deposited on EMI-TFSI with dissolved iodide redox pair; scan rate: 50 mV/s).

low potential combined with a transition metal oxide cathode having a very high potential. Conventional organic liquid electrolytes cannot be easily applied to lithium secondary batteries for they do not satisfy safety requirements such as flame retardancy and nonvolatility. On the other hand, ionic liquids do not burn easily, have low volatility, and exhibit relatively high ionic conductivity. One of the most common ionic liquids includes EMI as cation. The EMI cation can be used with various anions to form ionic liquids with a low melting point and viscosity. However, one drawback of EMI is its low cathodic stability. As such, quaternary ammoniums that do not require additives for SEI formation are being considered as electrolyte materials for lithium secondary batteries. Aliphatic quaternary ammoniums with fluorine-containing anions tend to have low viscosity and high oxidation resistance. These ammoniums may have a methoxy side chain, or exist as systems containing BF_4^- or ClO_4^- [23], TFSI [24], or 2,2,2-trifluoro-N-(trifluoromethylsulfonyl)acetamide (TSAC) [25].

Figure 3.131 shows the charge/discharge results for laminated lithium metal secondary batteries with ionic liquids containing lithium salt [26]. Among the three types of ionic liquids, N-methyl-N'-propylpiperidinium (PP13)-TFSI shows the best charge/discharge cycle characteristics. Liquids with a higher reduction potential than PP13-TFSI, namely, TEA-TSAC and EMI-TSAC, rapidly decrease in capacity as cycling proceeds. From this, we can confirm the relationship between reduction stability and cycle characteristics of ionic liquids. In order to attain outstanding charge/discharge characteristics, it is important to use ionic liquids with high cathodic stability. Ionic liquids such as TFSI and TSAC have a viscosity of 10–150 cP, which is 10 times higher than that of organic solvents such as PC. This viscosity rises further when lithium salt is added to ionic liquids at room temperature. However, at temperatures higher than 80 °C, the viscosity of ionic liquids is slightly higher or similar to organic liquid electrolytes. If we take into account the nonvolatile characteristics of ionic liquids, they may be used as electrolytes for lithium secondary batteries that operate at high temperatures.

3.3.1.5 Electrolyte Additives

Function of Additives The main function of electrolyte additives is to enhance ionic conductivity, battery life, or safety [27]. Most liquid electrolytes used in batteries contain small quantities of additives, which affect battery performance and safety. The additives engage in different reactions with electrodes and the electrolyte. The global market is dominated by a few electrolyte manufacturing companies and only limited information is available on the technology.

Characteristics of Different Additives Additives can be grouped by function into SEI formation additives, overcharge inhibitors, ionic conductivity enhancers, and flame retardants.

SEI Formation Additives Vinylene carbonate (VC) is a common additive used to form and maintain an SEI layer at the surface of a carbon anode [28, 29]. VC produces a

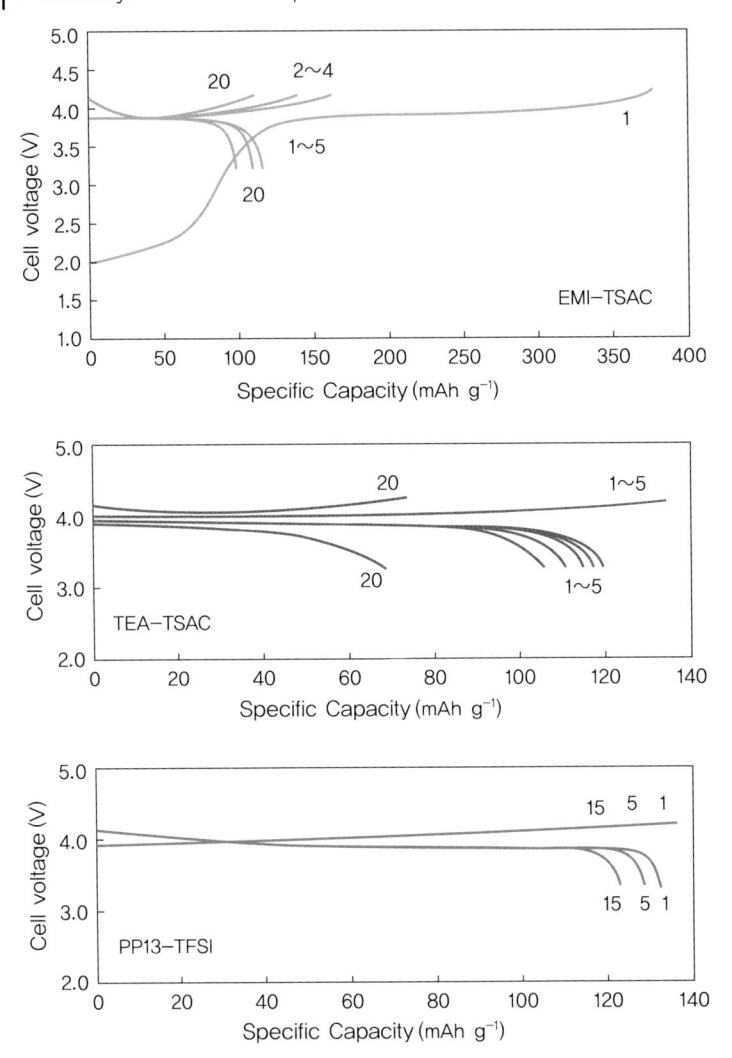

Figure 3.131 Charge/discharge curves of Li/LiCoO$_2$ cell with various ionic liquids.

stable SEI layer during the initial charging process, and enhances battery life by preventing carbon exfoliation and direct reactions with the electrolyte. Figure 3.132 shows the chemical structure of VC.

Due to the unstability from ring strains of carbon (sp^2 mixed orbital), VC participates in ring-opening reactions to take on a more stable structure.

Figure 3.132 Chemical structure of VC.

The presence of a vinyl group allows polymerization and produces a stable protective layer. Just by adding a small amount of VC, the irreversible capacity of the carbon anode is reduced with the SEI layer. This is especially effective for PC electrolytes. For example, when 1 M LiPF6/PC/VC electrolyte is used for the $LiMn_2O_4$ cathode, it exhibits a stable reversible capacity up to 4.3 V [30]. Battery performance remains unaffected even when an excess amount is added, and this facilitates process control during manufacturing. VC does not affect the cathode and maintains SEI stability at high temperatures. Despite the high performance of VC, other additives are being developed as VC is difficult to synthesize and is expensive.

Overcharge Inhibitors Since battery safety is a top priority, lithium ion batteries are assembled with various safety devices such as a positive temperature coefficient (PTC), protection circuit modules (PCMs), and safety vents. With these devices, however, batteries become more expensive. Batteries should also have internal safety devices to suppress chemical reactions [31, 32]. Overcharge inhibitors have been proposed as an additive to solve safety issues caused by overcharging. The redox shuttle type restricts high voltage by allowing the excess charge to dissipate within the cell, while the cathode layer-forming type creates a protective layer to block current flow and ion diffusion.

Redox Shuttle Type The addition of n-butylferrocene to a 2 V Li/TiS_2 battery was the first case of a redox shuttle additive [33–35]. When the cell exceeds its cutoff voltage, as shown in Figure 3.133, n-butylferrocene at the cathode surface is oxidized and transferred to the anode. This cycle repeats as n-butylferrocene at the anode is reduced and transported back to the cathode. Increases in voltage are suppressed, and battery overcharge is prevented.

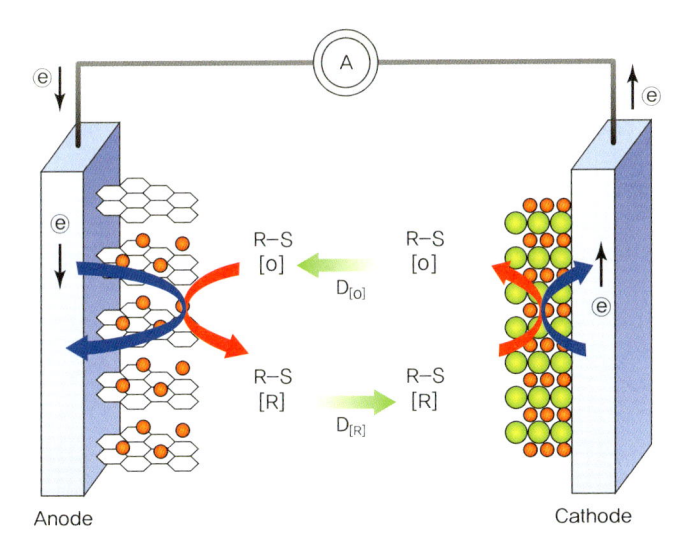

Figure 3.133 Operating mechanism of redox shuttle additive [36].

Recently, anisole structures of halogen elements substituted with benzene are being used as additives in 3 V batteries. At a voltage higher than 4.3 V, the excess current is eliminated through repeated oxidation and reduction [37–39]. These redox additives are effective for a small excess current, but should be highly concentrated in order to prevent overcharging during larger overcurrents. When the battery is subjected to an excess current large enough to cause damage to the battery, the use of redox additives may still lead to overcharging.

Cathode Layer-Forming Type Cathode layer-forming additives are more stable methods than redox shuttle additives. When the cutoff voltage is exceeded, a high polymer insulating layer is created at the cathode to block current and Li^+ diffusion. As shown in Figure 3.134, polymerizable monomers such as biphenyl (BP) are decomposed at the cathode and polymerized to form a polymer protective layer. This blocks the movement of ions within the battery and restricts external current flow [40, 41]. These additives help to prevent overcharging by deactivating the battery, which can also be regarded as a disadvantage.

Ionic Conductivity Enhancers To improve ionic conductivity, the dissociation of lithium salt must be promoted and the dissociated ions should remain in the ionic state. At the same time, ions within the electrolyte must have high mobility. Crown ether additives allow greater dissociation of lithium salt by isolating Li^+ based on the ion dipole interaction and separating from the anion. Examples of crown ether additives are 12-crown-4 and 15-crown-5 [42, 43]. As a result of application of these additives, the ionic conductivity of organic solvents with a low dielectric constant and polymer electrolytes is slightly improved. When these additives were used in polymer electrolytes, the glass transition temperature (T_g) was lowered. However, cation receptors of crown ether compounds impede battery performance by slowing the movement of Li^+. Furthermore, the use of crown ether additives is restricted, as they are highly toxic and harmful to the environment [44, 45]. Anion receptors are able to compensate for the weakness of cation receptors by increasing the number of transferred cations through isolation of anions such as PF_6^- or BF_4^- and by preventing recombination with lithium cations. Cation decomposition is suppressed at the electrode, and the battery cycle is further stabilized. Cation receptors consist of substituents that maximize mutual interactions with cations by attracting electrons to boron [46]. When boron is combined with poly(ethylene glycol) (PEG), ionic conductivity is enhanced with greater dissociation of lithium salt within the electrolyte. Figure 3.135 shows the chemical structure of a representative boron-based additive.

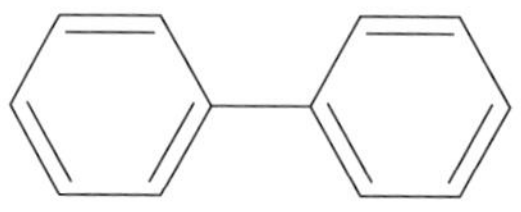

$$BP + (BP)_n \rightarrow (BP)_{n+1} + 2H^+ + 2e^-$$
$$2H^+ + 2e^- \rightarrow H_2$$

Figure 3.134 Chemical structure of biphenyl and layer-forming mechanism [12]. Reprinted from [12] Copyright 2006, with permission from Elsevier.

R_F — B — R_F, R_F

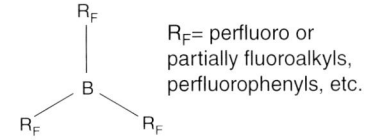

R_F = perfluoro or partially fluoroalkyls, perfluorophenyls, etc.

Figure 3.135 Boron-based additives.

Flame Retardants Liquid electrolytes consist mostly of organic solvents that combust readily once ignited even if the external current supply is cut off. To suppress the flammability of electrolytes, organic solvents should have a high boiling point and form a protective layer during thermal decomposition to block out oxides and combustible gases. Among lithium salts, $LiPF_6$ is known as an effective flame retardant in liquid electrolytes. In the case of polymer electrolytes, polyacrylonitrile (PAN) gel polymer electrolytes have some flame retardancy [47]. The CN triple bond in PAN is broken at around 200 °C with carbonization reactions arising from thermal decomposition, and the ladder arrangement stiffens to form a graphene structure where carbon layers serve as protection against combustible gases.

Since it is difficult for liquid electrolytes comprised of organic solvents and salts to acquire flame retardancy, they require flame-retardant additives. These additives should be compatible with the electrolyte without affecting electrochemical performance and also be reasonably priced. Most flame-retardant additives are phosphates, such as trimethyl phosphate (TMP), tris(2,2,2-trifuloroethyl) phosphate (TFP), and hexa-methoxycyclotriphosphazene (HMTP). While the working mechanism of these additives remains unclear, they are highly effective in reducing heat generation from the battery. Similar to PC, TMP causes exfoliation when inserted between graphene layers. A small amount of HMTP can prevent thermal runaway, and fluorine-containing TFP enhances both electrochemical stability and cycle performance [48–50].

We have examined the functions of representative electrolyte additives used in lithium secondary batteries. Table 3.11 provides a summary of common additives.

3.3.1.6 Enhancement of Thermal Stability for Electrolytes

A prerequisite for successful commercialization of lithium secondary batteries is to secure battery safety in all environments. Lithium batteries involve the risk of thermal runaway caused by high temperatures, smoke, explosions, and flames. As shown in Figure 3.136, the three main factors of thermal runaway are oxygen in the cathode material, organic liquid electrolyte serving as fuel, and heat generated by the battery.

Critical safety issues may arise from overheating when an electrolyte is heated above its flash point. Thermal runaway occurs at high temperatures through chemical reactions with oxygen released from cathode material and organic liquid electrolytes as fuel. To resolve this problem, a positive temperature coefficient (PTC) produces electrical resistance when the temperature exceeds a certain point, while a protection circuit module (PCM) cuts off current when overcharged. Other devices include a safety vent and a separating film. In particular, the PCM is highly priced compared to other battery components and more expensive for larger size batteries.

Table 3.11 Characteristics of representative electrolyte additives by type.

Additive types		Additive chemical structures	Remarks
Anode SEI additives		Vinylene carbonate (VC); Catechol carbonate (CC)	VC: Prior reduction to other electrolyte components, forming stable SEI on anodes. Irreversible cap. ↓ and cycle life ↑, little effects on cathodes. No negative effects on cell performance with excessive addition. Effective for PC suppression. CC: Effective for PC suppression and reversible capacity enhancement
Overcharge protectors	Redox-shuttle	n-Butylferrocene (BF); Substituted benzene (SB)	BF: When overcharging, BF is oxidized upon cathodes and moves to anode, and is reduced repeatedly; thereby, the cell voltage is maintained SB: Used for 2 V batteries when not substituted by halogens. With halogens, it can be used for 3 V-type batteries
	Cathode layer forming	Cyclohexylbenzene (CHB); Biphenyl (BP)	When overcharging, polymerizable monomers such as CHB and BP can form insulating layer upon cathode surface, which blocks ion conduction permanently More effective in overcharging protection than in redox-shuttle types

Ionic conductivity enhancers

Cation receptors: Enhanced Li salts dissociation due to the interaction b/w ether groups and Li ion. Effective for organic solvents with low dielectric const. and polymer electrolytes. Not used any more due to toxicity

Anion receptors: Electron-deficient characteristics of boron can impart strong interaction b/w AR and anions, resulting in better dissociation and electrochemical stability of anions

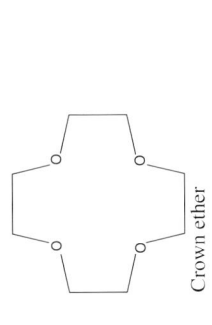

Crown ether

Boranes
R_F = perfluoro or partially fluoro alkyls, perfluorophenyls, etc.

Flame retardants

Phosphorus: Combustion reaction can be stopped by carbonized layer. Or selfheating can be suppressed by reaction with radicals. Flame retardancy increases with adding more amount; however, cell performance can be degraded

Halogen-substituted phosphorus: Flame retardancy and cycle enhancement due to phosphorus and halogen, respectively

Hexamethocycyclotriphosphazene (HMTP)

Trimethyl phosphate (TMP)

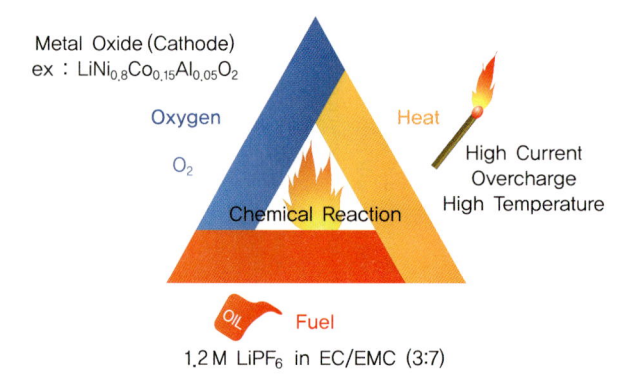

Figure 3.136 Main factors of thermal runaway.

Despite the high costs of protective devices, their use is considered mandatory according to present technological standards. However, some alternatives have been proposed. For instance, we can consider selecting a suitable solvent for the synthesis of a thermally stable salt. As shown in Figure 3.137, different lithium salts result in varying levels of chemical reactivity, which in turn affects heat generation and the self-heating rate. Another method is to introduce functional or flame-retardant additives. Functional additives can create and maintain an SEI layer at the surface of electrodes during the initial cycle. They can also be used to delay thermal runaway by preventing battery overcharge.

As shown in Figure 3.138, hexamethoxycyclotriphosphazene (HMTP), a flame-retardant additive, suppresses thermal reactivity and self-generated heat, thus avoiding thermal runaway [48, 49]. In summary, additives used to enhance thermal

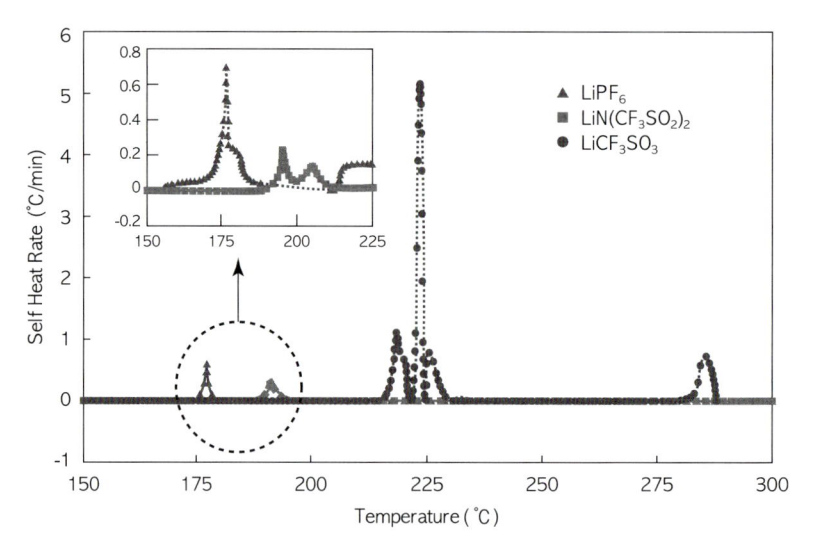

Figure 3.137 Self-heat rate of electrolytes containing different lithium salts based on ARC measurements.

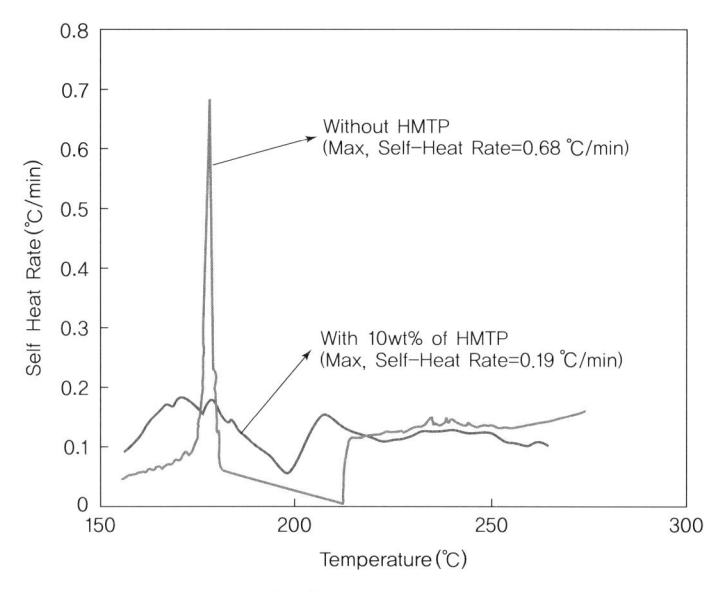

Figure 3.138 Comparison of self-heat rates between liquid electrolytes with and without flame-retardant additives [51].

stability of liquid electrolytes should satisfy various requirements such as high solubility, a large potential window, high ionic conductivity, and low viscosity. Reactivity with electrodes is closely related to battery safety. By securing thermal stability of liquid electrolytes using flame retardants, we will be able to enter new markets not only for mobile devices such as cell phones, laptops, and digital cameras but also for large-scale applications such as electric vehicles.

3.3.1.7 Development Trends of Liquid Electrolytes

Organic Solvents New organic solvents are being developed to satisfy the following three requirements. They should be able to enhance the reversibility of electrochemical reactions at the graphite anode, achieve high ionic conductivity for improved battery characteristics at low temperatures, and have excellent resistance to flame at high temperatures. Chemically modified solvents such as PC and EC are being proposed to allow more reversible chemical reactions in the graphite anode. While the PC solvent does not show good reversibility toward the graphite anode, treating the solvent with fluorine slows down decomposition reactions and facilitates lithium intercalation/deintercalation. It can be used together with partially chlorinated EC, which suppresses reduction and decomposition reactions between graphite and PC [52]. Linear ester solvents such as methyl formate (MF) and isopropyl acetate are also being considered for their low melting point and low viscosity at low temperatures. Due to their low solubility in lithium salts when used alone, they are usually mixed with EC. As for nonflammable solvents, phosphoric esters, fluoroesters, and fluoroethers are available. Combining asymmetrical phosphoric ethers such as ethyldimethyl phosphate (EDMP) or butyldimethyl phosphate (BuDMP) with EC/DEC is known to produce flame retardancy. Ethylnonafluorobutylether (EFE), a representative example of fluoroethers, is a flame-retardant and has good charge/discharge reversibility. Safe electrolytes can be made by adding EFE in excess, despite some problems in high-rate capability. New organic solvents must be further explored to satisfy various requirements including safety, low cost, and high performance.

Lithium Salts Among lithium salts used in lithium ion batteries, $LiPF_6$ lacks thermal stability and easily undergoes hydrolysis when exposed to moisture, while $LiBF_4$ has low ionic conductivity and leads to SEI formation on the surface of electrodes. To overcome these weaknesses and fulfill the necessary requirements, lithium salts are being developed in various structures through molecular design. One example is organic cations containing weakly coordinated fluorine such as perfluoroalkyl sulfonate salt or imide salt [53, 54]. These cations have been substituted with electron acceptors such as fluorine and CF_3, thus resulting in less interaction between molecules and electrostatic interaction with lithium ions. However, they have limited use in lithium secondary batteries due to low stability, high cost, and corrosion of the conductive agent. Other salts being considered are borates and chelates. There have been attempts to dissolve insoluble salts such as LiF together with complex ions in organic solvents and to use existing lithium salts for greater added value [55–57].

Ionic Liquids With the growing concern of safety in lithium secondary batteries, there has been increased interest in nonflammable and flame-resistant ionic liquid electrolytes. Most studies have concentrated on enhancing reductive stability through new cations or additives, selecting new anode materials, and lowering viscosity to improve battery performance at room temperature. Recently, a reversible capacity higher than 360 mAh/g was achieved in a graphite anode with an ionic liquid consisting of 1-ethyl-3-methylimidazolium bis(fluorosulfonyl)imide (EMI-FSI) and LiTFSI [58, 59]. In addition, research is being conducted to overcome the high viscosity of ionic liquids by lowering current density or reducing the thickness of electrodes.

3.3.2
Polymer Electrolytes

3.3.2.1 Types of Polymer Electrolytes

For lithium ion batteries available on the market, liquid electrolytes are used along with polyolefin separators that are 10–20 μm thick to facilitate the movement of lithium ions. When polymer electrolytes are used in place of liquid electrolytes, it becomes easier to produce compact batteries since metal packaging is not necessary. Some examples of polymer electrolytes are solid polymer electrolytes, gel polymer electrolytes, and polyelectrolytes. The transport of lithium ions in polymer electrolytes depends on the segment motion of polymer chains, while gel polymer electrolytes are influenced by liquid electrolytes incorporated into polymers and their ionic conductivity. Among polymer electrolytes, gel polymer electrolytes have high ionic conductivity and mechanical strength at room temperature. Thus, they are more commonly used in lithium secondary batteries than solid polymer electrolytes or polyelectrolytes. In order to resolve safety issues in lithium secondary batteries, solid polymer electrolytes are likely to be further studied. The next section introduces the characteristics, applications, and development of various polymer electrolytes.

Solid Polymer Electrolytes Solid polymer electrolytes have been actively studied following Wright's discovery that ions can be transported through polymers [60] and Armand's observation that they can be applied to electrochemical devices including batteries [61]. The advantages of all solid-state batteries with solid polymer electrolytes are as follows:

1) Higher energy density of the battery with a lithium metal anode.
2) Highly reliable and no risk of leakage.
3) Can be manufactured into various shapes and designs.
4) Possibility to produce ultrathin batteries.
5) No release of combustible gases at high temperatures.
6) Realization of a low-cost battery that does not require a separator or protection circuit.

Solid polymer electrolytes consist solely of a polymer and salt, and studies have mainly focused on molecular design and synthesis of polymers. In solid polymer electrolytes, polymers should be amorphous and include polar elements such as oxygen (O), nitrogen (N), and sulfur (S) to enhance both the movement of polymer chains at room temperature and the salt dissociation. Among the research conducted in the past on derivatives such as poly(ethylene oxide) (PEO), poly(propylene oxide) (PPO), polyphosphazene, and polysiloxane, studies on PEO-based polymers are the most active.

Matrix polymers of solid polymer electrolytes are closely related to ionic conductivity. In order for solid polymer electrolytes to have high ionic conductivity, matrix polymers should have the following characteristics:

1) Contain a polar group with ionic coordination ability, and have polar groups in neighboring chains to participate in joint coordination.
2) Adequate spatial conformation to allow salt dissociation.
3) Polar groups must include an electron donating group such as ether, ester, and amine for cation solvation.
4) Low T_g for greater flexibility of polymer chains.

In the preceding list, (1)–(3) are requirements for salt dissociation, while (4) relates to the transport of ions.

Poly(ethelyne oxide) consists of repeated units of $-CH_2CH_2O-$ (ethylene oxide, EO), in which oxygen atoms form coordinate bonds with alkali metal ions. This is because oxygen atoms have stronger donor properties than other polar elements (N, S, etc.) with coordination abilities. As such, lithium salts undergo dissociation when alkali metal cations coordinate with oxygen in ethers. The movement of

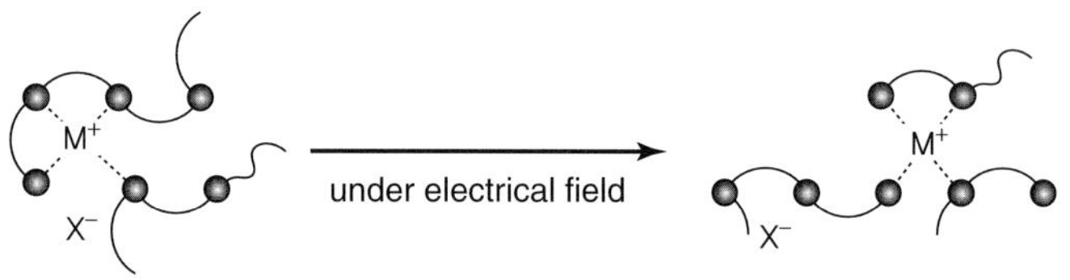

Figure 3.139 Conduction mechanism of lithium ions in a solid polymer electrolyte.

lithium ions in solid polymer electrolytes follows the mechanism illustrated in Figure 3.139 [62].

Due to the low rotational energy barrier between methylene and oxygen in PEO, the flexible polymer chains can easily take a conformation for cation coordination. Similar to crown ether, ion-dipole interactions between the cation and the oxygen lone electron pair in ether create a complex, and lithium salt undergoes dissociation within PEO. To obtain high ionic conductivity, the dissociated ions must have high mobility in polymers. Polymer chains engage in active thermal motion around room temperature, which is much higher than the glass transition temperature of (T_g) flexible polymers. Lithium ions thus change their local position through coordination exchange with polymer chains. As such, lithium cations move freely within the free volume and participate in the local structural change of polymers. Meanwhile, with the redistribution of free volume from the relaxational motion of polymers, anions are able to move freely without much restraint. PEO matrix complexes have high crystallinity from strong $O - Li^+$ interactions and thus insufficient ionic conductivity below room temperature. Most research has focused on synthesizing new polymers to enhance ionic conductivity. Various approaches are being explored [62–64], including grafting a short EO unit into the side chain. This method lowers T_g while maintaining an amorphous structure, thus allowing a high conductivity of 10^{-4} S/cm at room temperature. Another method is to introduce a cross-linking structure to expand the amorphous region while improving the mechanical structure. Ionic conductivity, mechanical strength, and electrode/electrolyte surface characteristics can be enhanced by adding inorganic particles such as aluminum oxide (Al_2O_3), silica (SiO_2), and titanium dioxide (TiO_2) to solid polymer electrolytes. This is because inorganic fillers suppress the crystallization of polymers, and excess moisture or impurities are adsorbed onto the surface of inorganic particles. The ferroelectric inorganic material also promotes salt dissociation. Despite such active research, lithium polymer batteries with solid polymer electrolytes have yet to be fully commercialized. Compared to liquid electrolytes, solid polymer electrolytes have low ionic conductivity at room temperature, weak mechanical properties, and poor interfacial characteristics. As such, they are being studied as large-scale secondary batteries for high-temperature operation in electric vehicles and energy storage devices.

Gel Polymer Electrolytes Comprised of polymers, organic solvents, and lithium salts, gel polymer electrolytes are produced by mixing organic electrolytes with solid polymer matrices. Despite existing in a solid film state, gel polymer electrolytes have an ionic conductivity of ~10^{-3} S/cm due to the electrolyte encapsulated in the

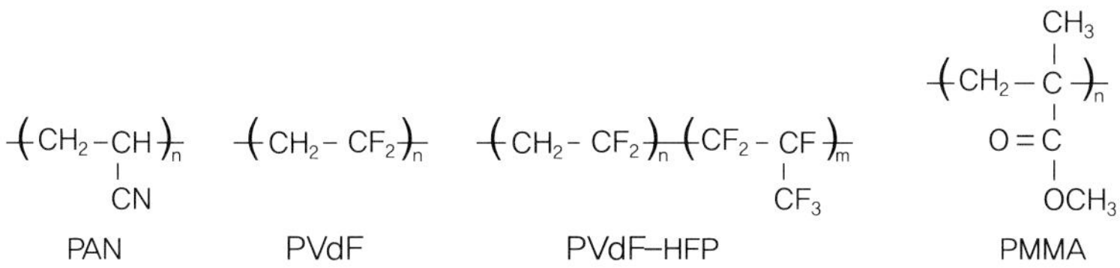

Figure 3.140 Representative matrix polymers of gel polymer electrolytes.

polymer chains. Encompassing the advantages of both solid and liquid electrolytes, they have been actively studied for use in lithium secondary batteries [65, 66]. As shown in Figure 3.140, representative polymers used as matrix polymers of gel polymer electrolytes are polyacrylonitrile, poly(vinylidene fluoride) (PVdF), poly(methyl methacrylate) (PMMA), and PEO.

Polyacrylonitrile System The highly polar CN side chain of PAN attracts lithium ions and solvents, thus making it appropriate as a matrix polymer of gel polymer electrolytes. In general, PAN is formed by infusing organic solvents such as EC or PC with $LiPF_6$ dissolved in an electrolyte, and shows a high ionic conductivity of $\sim 10^{-3}$ S/cm at room temperature. PAN-based gel polymer electrolytes have a wide potential window, strong mechanical properties, and low reactivity with cathode materials. Electrolyte preparation, which will be described later, involves the heating of PAN dissolved in EC or PC up to 100 °C. When PAN and salt are completely dissolved, the solution is cast and then left to cool at room temperature.

Polyvinylidene Fluoride System Gel polymer electrolytes made from PVdF polymers can be dissolved at high temperatures, and are capable of physical gelation with phase separation and crystallization during cooling. Since the PVdF-based gel polymer electrolyte film has micropores from phase separation, the liquid electrolyte in the micropores allows high ionic conductivity. P(VdF-co-HFP), a copolymer of vinylidene fluoride and hexafluoropropylene, has been used as a matrix of gel polymer electrolytes. These electrolytes are obtained by preparing a porous membrane created by extracting the plasticizer from a polymer film [67]. In this case, the electrolyte film is produced under atmospheric conditions and activated only at the final stage.

Poly(Methyl Methacrylate) System A type of PMMA gel polymer electrolytes can be derived from polymerization of MMA monomers and dimethacrylate difunctional monomers. PMMA-based gel polymer electrolytes comprised of transparent films can be used in electrochromic devices. Chemically crosslinked gel polymer electrolytes prepared with $LiClO_4$ and EC/PC nonaqueous electrolytes exhibit an ionic conductivity of 10^{-3} S/cm at room temperature and a wide potential window of 4.5 V toward lithium metals. In addition, the interfacial resistance of lithium metal is maintained at a constant level even after a period of time.

Polyethylene Oxide System Gel polymer electrolytes consist of a PEO polymer matrix with the EO structure either in the main chain or side chain. For a polymer matrix that carries the EO structure in the main chain, the end hydroxyl group (−OH) of PEO uses isocyanate for chemical cross-linking. For gel polymer electrolytes having a polymer matrix with an oligo(ethylene oxide) in the side chain, methacrylate and acrylate are common derivatives. Side chains can also include polyurethane, polyphosphazene, or a grafted PEO chain of molecular weight 2000.

Within gel polymer electrolytes, ions are transported in a liquid medium while the polymer matrix maintains the mechanical strength of the film and stores liquids. Depending on the amount of liquid electrolyte, gel polymer electrolytes exhibit different mechanical properties. As previously mentioned, the ionic conductivity of gel polymer electrolytes is close to that of liquid electrolytes. Gel polymer electrolytes can be grouped into two categories according to their physical and chemical cross-links. Physically cross-linked gel polymer electrolytes take on a physically cross-linked structure when polymer chains become entangled with one another or through cross-links formed by the partial molecular orientation of polymer chains. These electrolytes gain mobility as the polymer chains become untangled upon heating and then turn into gel form when cooled. By making use of such properties, the liquid electrolyte can be inserted into the battery before being cooled into gel form. However, at high temperatures, gel polymer electrolytes may be fluidified or become prone to leakage. Among physically cross-linked gel polymer electrolytes, the most actively studied gel polymer electrolyte consists of poly(vinylidene fluoride-hexafluoro propylene) copolymers.

To overcome the weak mechanical properties of physical cross-linking, lithium ion polymer batteries are manufactured by coating a polyolefin separator or electrode with gel polymer electrolytes. These lithium ion polymer batteries have the same structure as lithium ion batteries. Polymers used for physical gelation are PEO, PAN, PVdF, and PMMA. The performance of lithium ion polymer batteries depends on the thickness of the coating applied to the polyolefin separator [68]. The gel coating on the porous membrane compensates for the weak mechanical strength of the electrolyte, enhances the adhesive property of the electrodes, and improves battery safety. Based on good interfacial contact between the electrode and the electrolyte, batteries can take on the form of an aluminum pouch. Meanwhile, cylindrical batteries maintain good interfacial contact through winding pressure and the use of metal can covers. Recently, polyethylene (PE) particles with a low melting point have been dispersed in gel polymer electrolytes to serve as a fuse. Battery performance was thus greatly improved with a rapid increase in resistance when these particles melted at around 100 °C [69]. On the other hand, structural change is more difficult in chemically cross-linked gel polymer electrolytes since the networked structure is based on chemical bonds instead of van der Waals forces. Also, battery performance may be affected by the presence of nonreactive monomers or cross-linking agents. To produce lithium ion polymer batteries from chemically cross-linked gel polymer electrolytes, a polymer precursor capable of chemical cross-linking is dissolved in an electrolyte and injected into a battery. This is followed by thermal polymerization to attain uniform gelation of the electrolyte [70]. This process of battery fabrication is shown in Figure 3.141. Polymer batteries produced from this method show similar performance to lithium ion batteries and can be applied in various mobile devices.

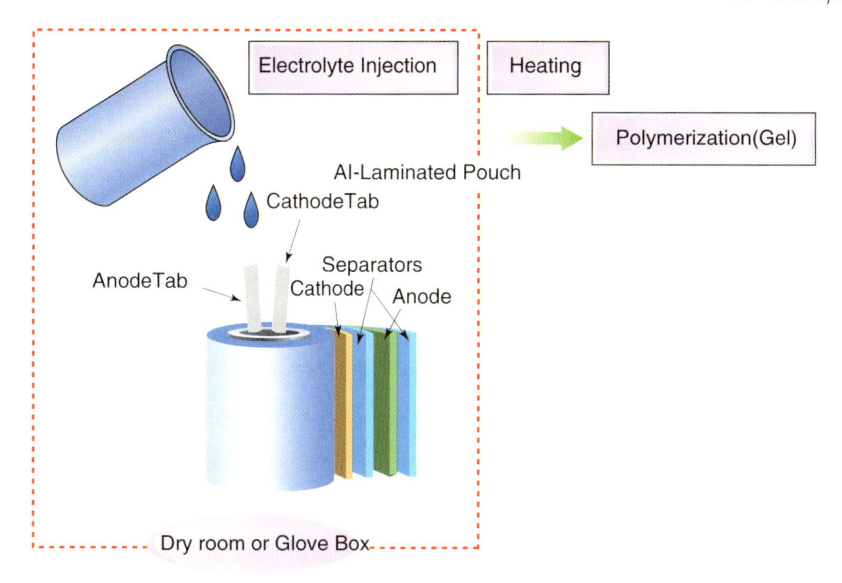

Figure 3.141 Manufacturing of chemically cross-linked gel polymer electrolytes in lithium ion polymer batteries.

Even with an aluminum-laminated film as packaging, such batteries are leakage-free due to the complete gel-type electrolyte.

Polyelectrolytes: Single-Ion Conductor Polyelectrolytes are conductive substances formed from the dissociation of cations and anions in polymers. They are also called single-ion conductors as cations or anions can be transported independent of the other. In polymer electrolytes containing lithium salts, dissociated anions migrate without interacting with polymer chains. The cation transport number, which indicates the ratio of cation conductivity, is usually below 0.5. When used in secondary batteries, both lithium ions and counter anions migrate during charge/discharge. Lithium ions are able to flow through both electrodes while anions accumulate on the surface of electrode active materials. This results in concentration polarization between the two electrodes and increases the electrolyte resistance over time. In polyelectrolytes, the lithium cation transport number is close to 1.0 due to the lack of anion movement. When polyelectrolytes are used in secondary batteries, a stable discharge current can be attained since no concentration polarization occurs and resistance remains the same over time.

For polyelectrolytes comprised of lithium cations, the anions must be fixed to polymer chains. The molecular design should enhance the degree of dissociation and provide a migration path for lithium ions. To improve the ionic conductivity of polyelectrolytes with anions attached to polymer chains through shared bonding, ion pair formation should be weakened to allow greater dissociation. This can be achieved either by lowering the charge density of cations or by using a substituent to block access to anions. PEO chains with repeated EO units or oligo(ethylene oxide) and other ether chains are used as the migration path of lithium ions. In addition to

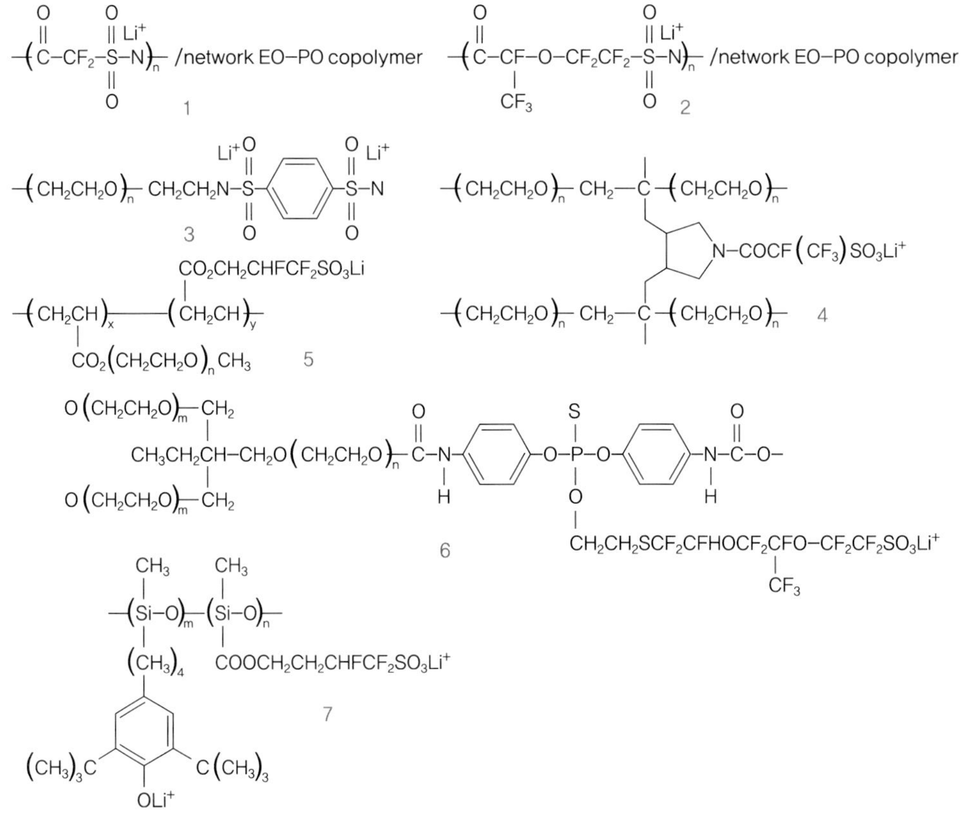

Figure 3.142 Structures of representative polymer salts [71–76].

fixing polymers to ether chains, polyelectrolytes may be mixed with polyethers. Representative polyelectrolytes developed to date are shown in Figure 3.142.

In the networked EO–PO copolymer, polymers (1 and 2) with the fluoroalkane sulfonamide structure have a higher degree of dissociation, thus resulting in significant ionic conductivity [71]. Here, ionic conductivity is greatly affected by the distance between anions. Polyelectrolyte 1 has low ionic conductivity due to the short distance between anions whereas polyelectrolyte 2 has a conductivity of 10^{-6} S/cm at room temperature. The insufficient dissociation of polyelectrolyte 3, which contains benzene sulfonamide, leads to low ionic conductivity at room temperature [72]. Polyelectrolyte systems 4, 5, and 6 with fluoroalkanesulfonate and ether chains exhibit high ionic conductivity above 10^{-6} S/cm [73–75]. On the basis of the percolation calculations, we can see that polymer electrolytes should contain an ionic group in the side chain and an oligoether chain as an ionic conduction path. In other words, structures with an oligoether and relatively mobile cations in the side chain are suitable polyelectrolytes. Polyelectrolyte 7, in which a phenolate group is introduced through the *t*-butyl substituent in 2,6-, shows high ionic conductivity in the range of 10^{-6}–10^{-5} S/cm at room temperature [76]. However, most polyelectrolytes have low ionic conductivity due to strong interactions between cations and anions, and a low concentration of ions. Ionic conductivity may be improved by adding a highly polar plasticizer or a small amount of lithium salt.

3.3.2.2 Preparation of Polymer Electrolytes

Polymer electrolytes can be prepared using various methods, which vary according to material characteristics and battery manufacturing techniques. Some examples are solution casting, immersion of a porous membrane, *in situ* crosslinking, and hot melting [66]. Advantages and disadvantages of each method are introduced below.

Solution Casting The easiest method to prepare polymer electrolytes is solution casting, wherein a film is produced by evaporating the volatile solvent from a casting solution of polymers and lithium salts. Figure 3.143 shows the process of obtaining a polymer electrolyte film using solution casting. A polymer solution is poured onto a glass slide, and the film thickness is controlled using a doctor blade. The film becomes thinner as the solution evaporates and is removed from the glass plate. For the film to be easily peeled off, teflon or polytetrafluoroethylene may be used instead of glass. In addition, the rate of evaporation may be adjusted by covering the tray to restrict air circulation. This simple and low-cost method is widely used in preparing polymer electrolytes. In preparing gel polymer electrolytes, linear carbonate solvents such as DMC and DME cannot be used since they may be removed during the evaporation process. As such, this method is limited to the preparation of gel polymer electrolytes that include organic electrolytes with solvents having high boiling points, such as EC and PC. Examples of polymer electrolytes prepared by solution casting are PEO and PMMA.

Immersion of Porous Membrane Microporous membranes being used in lithium ion batteries are produced and immersed in an electrolyte to obtain gel polymer electrolytes. Unlike hydrophobic polyolefin porous membranes in lithium ion batteries, the microporous membranes created using this method have a good affinity to liquid electrolytes and high polarity. Examples of representative polymers are P(VdF-*co*-HFP) and other PVdF copolymers. The preparation of gel polymer electrolytes using P(VdF-*co*-HFP) copolymers by immersion of porous membranes

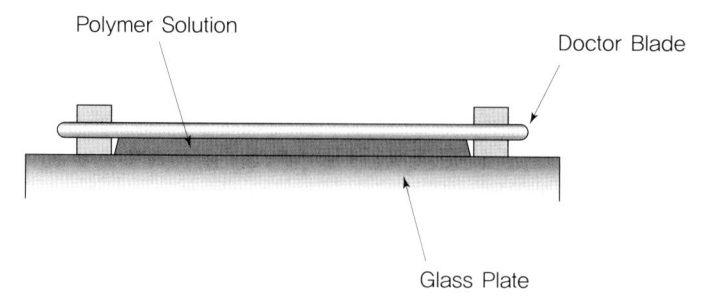

Figure 3.143 Preparation of polymer electrolytes by solution casting.

can be described as follows. First, a mixture containing a polymer, plasticizer, and filler is dissolved in an acetone solvent. Polar solvents with high boiling points such as dibutylphthalate (DBP) are suitable plasticizers, while silica or alumina is used as filler. The inorganic material added at this point not only enhances the mechanical strength of the porous membrane but also allows a greater amount to be immersed into the membrane. After casting the polymer solution on a glass plate at an appropriate thickness, the acetone is volatilized and removed. With acetone removed, the film contains a large amount of plasticizer. When this film is placed in a nonsolvent such as water, methanol, or ether, the DBP plasticizer is eliminated from the film and is replaced with micropores. This film is vacuum dried and then immersed in the electrolyte to eventually form a gel polymer electrolyte. Instead of direct impregnation, gel polymer electrolytes can be obtained by creating a cell comprised of a cathode/porous membrane/anode before injecting the electrolyte. Since the electrolyte is introduced in the final stage, it is easier to maintain a hydrophobic atmosphere. One advantage over solution casting is that various types of electrolytic solutions may be used. However, aging is required for uniform impregnation within the gel polymer electrolyte.

In situ Crosslinking As shown in Figure 3.144, *in situ* crosslinking can be used to prepare gel polymer electrolytes having a three-dimensional networked structure by filling a cell comprised of a cathode/separator/anode with reactive oligomers, cross-linking agents, initiators, and linear polymers dissolved in an electrolyte, followed by heating or ultraviolet irradiation. A commonly used cross-linking agent is polyethylene gylcol dimethacrylate (PEGDMA), which contains two functional groups. Thermal and light initiators are azobisisobutylonitrile (AIBN) and aromatic ketone, respectively.

Due to the networked structure of chemically cross-linked gel polymer electrolytes, they undergo little structural change compared to those of physically cross-linked electrolytes. In addition, the supporting polyolefin separator produces outstanding mechanical properties. However, this process affords low productivity since cross-linking requires exposure to high temperatures over a certain period of time. Another weakness is the difficulty of removing nonreactive monomers from the polymer precursor.

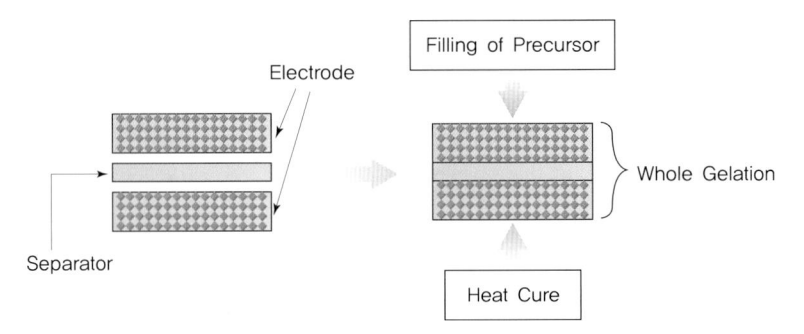

Figure 3.144 *In situ* preparation of polymer electrolytes.

Hot Melting Hot melting is a method of obtaining gel films through direct dissolution of polymers in an electrolyte, followed by film casting and cooling at room temperature. There is hardly any loss involved since gel polymer electrolytes can be prepared from organic electrolytes and polymers without using cosolvents. Organic solvents with low boiling points cannot be used because polymers are dissolved directly in the electrolyte at high temperatures. The hot melting method is commonly used to prepare PAN-based polymer electrolytes with EC or PC solvents as an electrolyte.

3.3.2.3 Characteristics of Polymer Electrolytes

Ionic Conductivity Low ionic conductivity compared to liquid electrolytes is the main reason for the little advancement in the development of lithium secondary batteries based on polymer electrolytes. For instance, 1 M $LiPF_6$-EC/DMC has an ionic conductivity of $\sim 10^{-2}$ S/cm at room temperature whereas thin-film polymer electrolytes require this value to be above 10^{-3} S/cm for practical applications. Figure 3.145 shows the change in ionic conductivity with varying temperature for liquid electrolytes, gel polymer electrolytes, and solid polymer electrolytes. Solid polymer electrolytes have low ionic conductivity at ambient temperatures, and we can see that conductivity sharply declines as temperature decreases. As such, solid polymer electrolytes are being considered for lithium secondary batteries operating at higher temperatures. Meanwhile, gel polymer electrolytes and liquid electrolytes have high ionic conductivity (above 10^{-3} S/cm) at room temperature and over a wide temperature range, thus permitting greater applications.

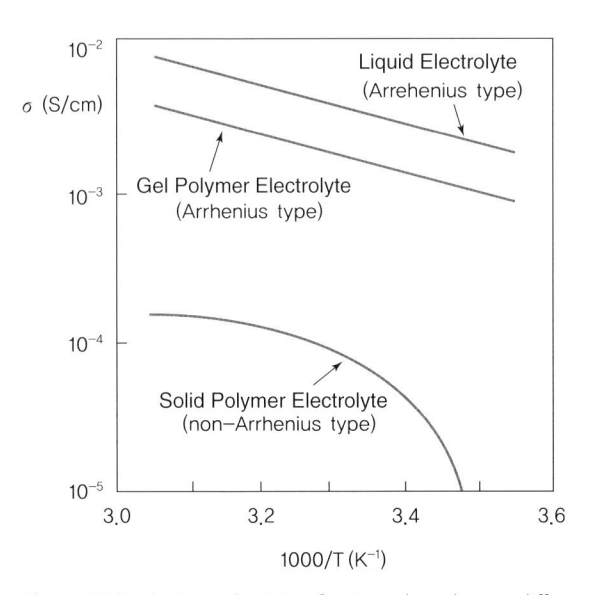

Figure 3.145 Ionic conductivity of various electrolytes at different temperatures.

Electrochemical Stability The electrochemical stability of polymer electrolytes affects determination of the operating voltage range. Polymer electrolytes must not be involved in decomposition arising from oxidation or reduction within the given working voltage range of the cathode and anode. Polymer electrolytes should be electrochemically stable until 4.5 V because lithium secondary batteries with 4 V metal oxide cathodes such as $LiCoO_2$, $LiNiO_2$, and $LiMn_2O_4$ have a voltage of 4.3 V when fully charged. Since the development of high-voltage lithium secondary batteries is needed for improved energy density, polymer electrolytes must have oxidative stability and be free of side reactions with electrode active materials.

Cation Transport Number Lithium cations are the charge carrier in lithium secondary batteries. As such, cations should have a higher mobility than anions and thus make a more significant contribution to ionic conductivity. Cation transport number for polymer electrolytes is usually in the range of 0.2–0.5. The transport number is close to 1.0 for hydrogen ion conductors such as Nafion, which is used as a polyelectrolyte in fuel cells. Polyelectrolytes are thus strong candidates being developed for use in lithium secondary batteries.

Electrode–Electrolyte Interfacial Reactions Major problems caused by using polymer electrolytes include low ionic conductivity and poor interfacial contact between electrodes. Unlike liquid electrolytes, polymer electrolytes can be divided into areas having sufficient contact with electrodes or lacking contact. The former is actively involved in charge transfer reactions but the latter fails to utilize electrodes due to a nonuniform current distribution. For example, when charging or discharging proceeds at high current, active materials within electrodes take on different distributions depending on localized sections. The nonuniform expansion and contraction result in rapid heating of graphite electrodes or transition metal oxide electrodes. To form a uniform surface, electrodes in secondary batteries may consist of anode materials or cathode materials mixed with polymer electrolytes. For instance, a monomer is inserted into pores at the anode or cathode and then polymerized to achieve close contact between the polymer electrolyte and the electrodes. Such electrochemical reactions between polymer electrolytes and electrodes are known to produce similar products to those of liquid electrolytes. Compared to liquid electrolytes, polymer electrolytes have a higher molecular weight, less mobility, and lower surface reactivity. In particular, the gradual increase in electrode–electrolyte interfacial resistance of lithium metals and polymer electrolytes implies that surface reactions continue to occur due to instability.

Mechanical Properties The greatest advantage of polymer electrolytes is the large surface area provided by thin films. Thinner films allow higher energy density since a greater amount of active materials can be packed into the battery. Film thickness is an important factor that affects battery performance, which is also determined by the mechanical strength of polymer electrolytes. The mechanical strength of electrolyte films is related to the defect rate and productivity. Raising the glass transition temperature of polymers or increasing the amount of organic solvent improves ionic conductivity but leads to a decrease in mechanical strength. Considering this relationship, ionic conductivity and mechanical properties should be optimized within a suitable range. For enhanced surface characteristics with electrodes, polymer electrolytes should possess adequate adhesive properties and flexibility.

3.3.2.4 Development Trends of Polymer Electrolytes

Gel polymer electrolytes with ionic conductivity close to that of liquid electrolytes can be used in lithium secondary batteries operating at room temperature. While not many polymer electrolytes are suitable for use in lithium secondary batteries, research is being actively carried out to improve their ionic conductivity and mechanical properties. In particular, solid polymer electrolytes are being considered for lithium secondary batteries in electric vehicles, which operate at relatively high temperatures. For instance, polymer electrolytes based on polyethylene glycol boric ester compound have been applied to lithium polymer batteries. The boric atom was introduced as a boric ester compound to prevent disruption of polymer movement by the polyethylene oxide group, thus allowing battery operation both at high temperatures and at room temperature. In addition, research is underway to lower costs by reducing salt content and to achieve high power of lithium secondary batteries by improving ionic conductivity. Recently, there have been attempts to develop ionic gels consisting of ionic liquids and polymers for lithium secondary batteries [77]. Flexible, transparent thin films with free-standing characteristics can be obtained by dissolving a vinyl monomer in an ionic liquid followed by radical polymerization. There are also active studies on the bulk and surface structure of polymer electrolytes to enhance surface characteristics. Despite these trends, polymer electrolytes have various disadvantages in terms of cost competitiveness and mass production. At temperatures below $-10\ °C$, battery performance rapidly deteriorates and energy density decreases. Polymer electrolytes may gain competitiveness with developments in production technology and higher production volume. Additives are being developed to improve performance even at low temperatures.

3.3.3 Separators

Separators are nonactive materials that do not participate in electrochemical reactions. They provide a pathway for ion transport that is essential for battery operation and separate physical contact between the anode and the cathode. Together with anode, cathode, and electrolyte, separators play an important role in determining battery performance and safety. In order to obtain a better understanding of electrolyte characteristics, we should examine separator functions, structure, and characteristics.

3.3.3.1 Separator Functions

Separators in lithium secondary batteries are microporous polymer films with pores ranging from nanometers to micrometers. The most commonly used separators are polyolefins such as polyethylene (PE) and polypropylene (PP), and these materials have various advantages including outstanding mechanical strength, chemical stability, and low cost [78–80]. Commercialized separators have pores that are 0.03–1 μm large, a porosity of 30–50%, and low thermal shutdown temperature (PE: ∼135 °C; PP: ∼165 °C). If the temperature rises during internal short circuits, the melted separator blocks pores and restricts ion movement, thus improving battery safety by delaying thermal reactions. When making batteries, thin separators are used to maximize battery capacity. In particular, separators with a thickness of 16 μm are used in high-capacity cylindrical batteries (18 650 type, 3 Ah).

Figure 3.146 shows the surface of a PE separator prepared by the dry process. The PE crystal consists of a networked structure (light areas) and air gaps are interconnected to form micropores (black areas) that are ∼μm in size. Depending on the size of micropores, separators can prevent contact between the electrodes and block the passing of substances released from the electrodes. Ionic conduction is possible by filling the micropores with a liquid electrolyte.

3.3.3.2 Basic Characteristics of Separators

The basic characteristics of separators should be optimized for improved battery safety and to prevent problems with mechanical strength that may occur during production. Table 3.12 shows the main physical properties of separators.

The basic characteristics of separators can be categorized as follows [36]:

1) **Thickness**: Since the ionic conductivity of organic liquid electrolytes is 100 times lower than that of aqueous electrolytes, it is important to maximize electrode surface area while reducing the distance between electrodes to achieve high output and energy density. As such, film thickness should be no more than 25 μm. The most commonly used separators have a film thickness of

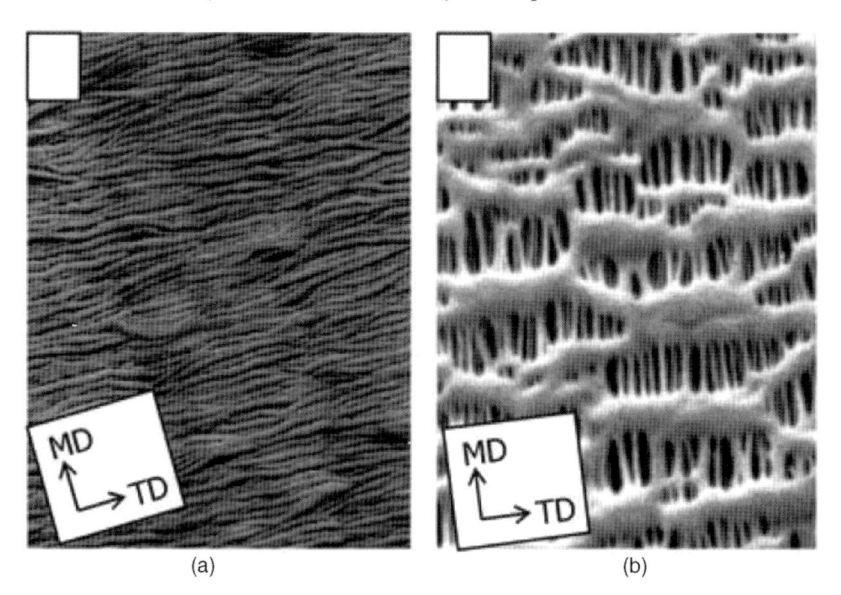

Figure 3.146 Microstructure of a PE separator prepared from the dry process [36]: (a) Before elongation and (b) after elongation along one axis.

Table 3.12 Characteristics of separators: measurement parameters and physical properties.[81]. Adapted with permission from [81] Copyright 2004 American Chemical Society.

Parameter	Value
Thickness[a),b)]	<25 μm
Electrical resistance	<8 (MacMullin, dimensionless)[c)]
Electrical resistance	<2 Ohms cm^2
Gurley[d)]	~25 s/mil
Pore size[e)]	<1 μm
Porosity	~40 %
Puncture strength[f)]	>300 kgf/mil
Mix penetration strength	>100 kgf/mil
Shrinkage[g)]	<5% in both md and TD
Tensile strength[h)]	<2% offset at 1000 psi
Shutdown temperature	~130°C
High-temp. melt integrity	>150°C
Wettability	Completely wet in typical battery electrolytes
Chemical stability	Stable in battery for a long period of time
Dimensional stability	Separator should lay flat; be stable in electrolyte
Skew	<0.2 mm/m

a) ASTM D5947-96.
b) ASTM D2103.
c) D.L. Caldwell, Pouch, U.S. Patent 4,464,238 (1984)
d) ASTM D726.
e) ASTM E128-99.
f) ASTM D3763.
g) ASTM D1204.
h) ASTM D882.

    20, 16, or 10 μm. Thin separators increase the discharge capacity of electrodes by increasing the concentration of surrounding liquid electrolyte and facilitating the movement of substances. However, thin separators may cause pinholes and are prone to tearing. Battery safety is also reduced from the increased risk of short circuits between electrodes.

2) **MacMullin number**: The MacMullin number is the resistance of the separator filled with an electrolyte divided by the resistance of the electrolyte alone, and is usually as high as 10–12.
3) **Electrical resistance**: The separator serves as an insulator and should have a low electrical resistance when filled with electrolyte. A high electrical resistance negatively affects battery characteristics including discharge capacity.
4) **Permeability**: Permeability, expressed in Gurley units, measures the time taken for air to flow through under uniform conditions (uniform pressure, uniform area, etc.). It is one of the characteristics of separators that affect battery performance.
5) **Pore size and porosity**: Porosity is usually at around 40%. Pore size should be below tens of micrometers and smaller than particle size to prevent internal short circuits from dendritic growth and impurities.
6) **Puncture strength**: Internal short circuits may be caused by impurities released from electrodes, the surface state of the anode and cathode, and dendritic

growth of lithium. Puncture strength represents the resistance of the separator against such threats and is measured by compressing the separator with a probe. A higher value lowers the risk of internal short circuits caused by separators.

7) **Thermal shrinkage**: While thermal shrinkage differs according to manufacturer, 1 h of drying at 90 °C in vacuum should result in less than 5% shrinkage.
8) **Tensile strength**: Like winding, tensile strength is a property that has a significant effect on the manufacturing process. Separators have high tensile strength in the direction of elongation. A separator with a thickness of 25 μm has a tensile strength above 1000 kgf/cm^2 in the machine direction (MD). In the case of uniaxial elongation, tensile strength in the transverse direction is as low as 1/10 of the machine direction. For two-axes elongation, tensile strength in the transverse direction is roughly the same as that in the machine direction.
9) **Shutdown**: Shutdown is a safety function that cuts off the circuit by blocking micropores during excess current caused by internal or external short circuits. PE separators are commonly used in lithium secondary batteries as micropores are shutdown to prevent the temperature from rising in case of early short-circuiting.
10) **Melt integrity**: Melt integrity is a characteristic that maintains the form of the separator for long periods above the meltdown temperature. Along with shutdown, it is an important factor in acquiring battery safety.
11) **Wettability**: A fast wetting rate and sufficient wettability are required.
12) **Chemical stability**: Chemical stability refers to stability under redox conditions. Separators should exhibit corrosion resistance to electrolytes at high temperatures.
13) **Average molecular weight and MW distribution**: This is an important factor that determines thermal and mechanical characteristics of polyolefin substances. Outstanding mechanical properties and narrow melting range can be achieved with a large $\overline{M}_W$ and small $\overline{M}_W/\overline{M}_n$.

3.3.3.3 Effects of Separators on Battery Assembly

Electrode active materials are coated on current collectors and a separator is inserted between them followed by winding. This is then inserted into a can and sealed after filling it with electrolyte to form lithium ion batteries. A mandrel is used during the winding process. To ensure high density without exfoliation of active materials or twisting of the separator, it is necessary to use a thin separator with strong mechanical properties. Sufficient tensile break strength and elongational break strength in the machine direction are required to avoid winding damage and necking. In addition, high puncture strength is needed for protection against impurities or damage. These characteristics are determined by porosity and film thickness and affect the manufacturing process. Other requirements include sampling of the wound jelly-roll, wettability during electrolyte insertion, and chemical stability to electrolytes.

3.3.3.4 Oxidative Stability of Separators

Separators in contact with the anode and cathode of a battery experience oxidation and reduction at the surface of each electrode. Polyolefin separators undergo oxidative decomposition due to their low resistance to oxidation. This oxidative decomposition worsens at higher operating temperature and eventually reduces cycle characteristics. Oxidation resistance differs according to material, and PP separators are known to be more resistant than PE. Some manufacturers produce three-layer products consisting of PP/PE/PP, which has a higher oxidative resistance compared to a single layer of PE [36]. The resistance of separators to oxidation has become increasingly important with the demand for high-capacity batteries. This is accomplished by improving the oxidative resistance of the electrolyte and separator. However, the resistance of separators to oxidation has not been actively studied. One method of improving the oxidative resistance of separators is through the coating of PVdF-based polymer electrolytes on the anode surface [82]. Figure 3.147 shows the results of an FTIR analysis on a disassembled battery stored for 4 h at 4.4 V voltage and 90°C. The C=C double bond peak observed in the range of 700–1900 cm^{-1} shows that oxidation occurred at the surface of the PE separator. This is confirmed by the color change in the PE separator in the FTIR analysis in Figure 3.147: (a) existing LIB and (b) PVdF on anode surface. As shown in Figure 3.148, when a PVdF-based gel polymer electrolyte is coated on the cathode surface, the battery maintains high performance at 4.4 V without any oxidation. This is because the PVdF-based gel polymer electrolyte on the cathode surface of the PE separator prevents direct contact with the cathode [82].

In addition to electrode surface modification using gel polymer electrolytes, oxidative resistance may be improved by introducing organic/inorganic composite layers into the surface of electrodes or separators [83–86].

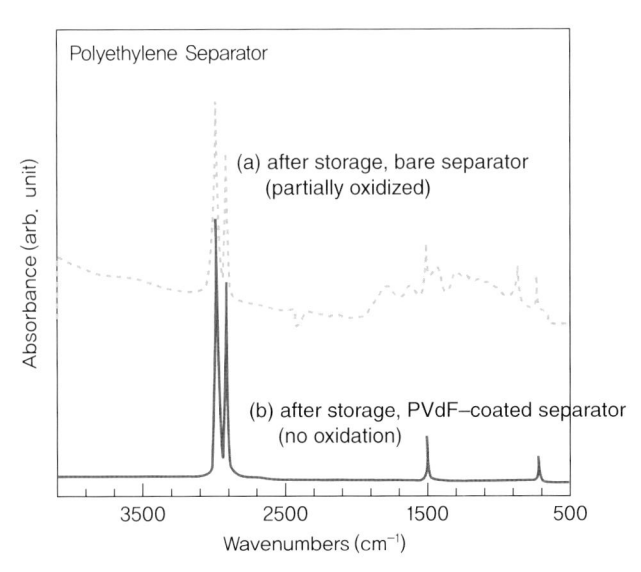

Figure 3.147 FTIR spectra of PE separators stored at 90 for 4 h: (a) bare separator and (b) PVdF-coated separator [83]. Reprinted from [83] Copyright 2007, with permission from Elsevier.

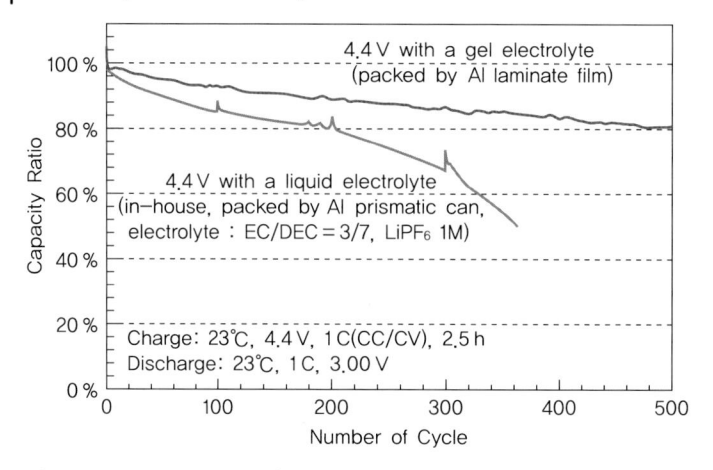

Figure 3.148 Comparison of cycle performance between 4.4V LIPB and 4.4V LIB [83]. Reprinted from [83] Copyright 2007, with permission from Elsevier.

3.3.3.5 Thermal Stability of Separators

When the temperature of a battery rises from an excess of current caused by external or internal short circuits, reactions between electrodes and the electrolyte or electrolyte decomposition may trigger the release of gases or liquids and result in ignition. Here, the separator contributes to improving battery safety. As the battery temperature increases, the separator melts to block micropores and limits ion conductivity, thus delaying ignition with time consumed in thermal diffusion even when the battery stops reacting at shutdown temperature. Since the internal temperature of the battery continues to rise, the separator should have a high meltdown temperature. Along with basic characteristics such as porosity and permeability, the shutdown temperature and meltdown temperature are important factors in securing battery safety. In consideration of both short-circuit and meltdown characteristics, attempts have been made to enhance meltdown characteristics by using adequate molecular weight and distribution of polyethylene to suppress mobility above the meltdown temperature. Another method is to combine materials with different meltdown temperatures, such as PE and PP, to obtain a low shutdown temperature and high meltdown characteristics.

PE Separator By using ultrahigh molecular weight polyethylene, which has almost no mobility above the meltdown temperature, the separator serves as an insulator by maintaining a low shutdown temperature and high meltdown temperature.

PE/PP Multilayer Separator By stacking PE and PP layers having different meltdown temperatures, battery safety is enhanced with insulation near the core during short circuits in lithium secondary batteries. For separators to effectively maintain battery safety, the method of combining layers with different meltdown temperatures is known to provide insulation across a wide range of temperature [83, 87]. For the three-layer separator shown in Figure 3.149, we can see that a short-circuit occurs at around 130°C and insulation remains stable at 180°C without any meltdown.

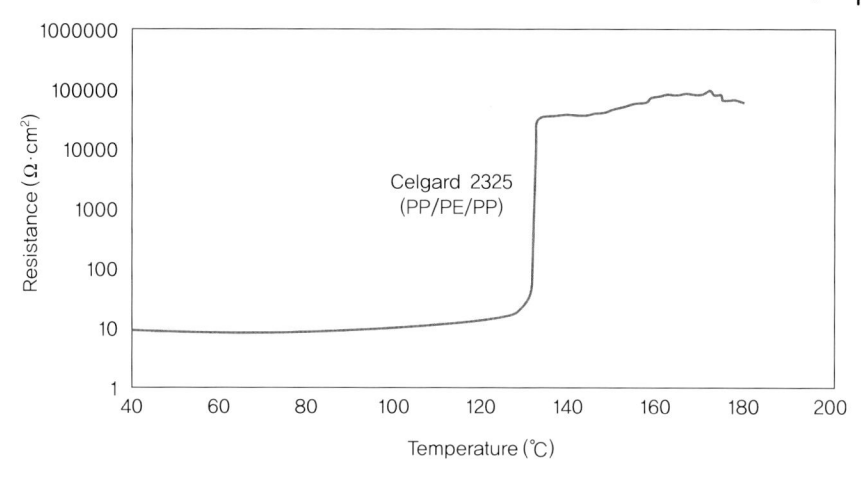

Figure 3.149 Shutdown characteristics of a PP/PE/PP three-layer separator (Celgard 2325) [82]. Adapted with permission from [82] Copyright 2004 American Chemical Society.

3.3.3.6 Development of Separator Materials

Microporous Polyolefin Film   In the early days, separators were made of PE with micropores. At temperatures above the meltdown temperature of 120°C, there is limited movement of ions and organic solvents through pores and the battery becomes deactivated. Since PE remains mobile at high temperatures, it is difficult to keep electrodes apart or proceed with meltdown during ignition. To overcome this weakness, PE is combined or stacked with PP, which has a meltdown temperature greater by at least 40°C. However, PE separators have been more actively developed due to the complexity and high manufacturing cost of multilayers. Research is underway to replace PP with ultrahigh molecular weight PE to resolve mobility issues that occur above the PE meltdown temperature.

Porous PVdF Film   PVdF has been used for binders in electrodes of lithium ion batteries. Compared to polyolefins, flourinated polymers contain fluorine atoms with high electronegativity in the main chain and thus have a greater affinity to liquid electrolytes due to strong interactions with polar solvents. P(VdF-*co*-HFP) separators were used in the plastic lithium ion battery (PLIB) developed by Bellcore in the early 1990s. Through copolymerization of VdF and HFP, the crystallization temperature and crystallinity of PVdF were lowered. Despite having high electrolyte uptake and ion conductivity, PVdF was not utilized for lithium ion batteries due to its weak mechanical strength compared to polyolefin. Other developments include composite-type separators, which combine the high mechanical strength of polyolefins and superior qualities of PVdF polymers.

Inorganic Coating on Separator   Polyolefin separators show extreme thermal shrinkage above 100°C due to material characteristics and stretching during film production. They can be easily ruptured due to metal particles and other impurities in the battery. This is also known as the main cause of internal short circuits between the anode and the cathode [82]. To overcome this weakness, research is being actively

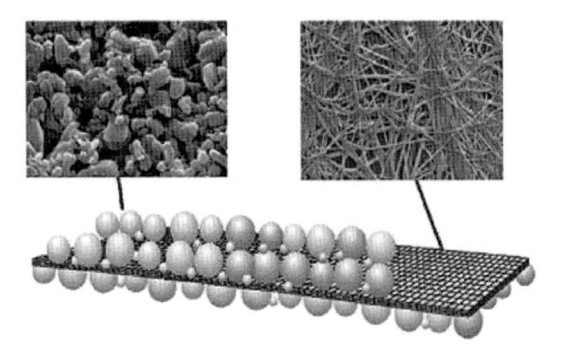

Figure 3.150 Mimetic diagram of Separion, a ceramic separator membrane developed by Evonik Degussa of Germany.

conducted on new separators having a layer of inorganic nanoparticles (SiO_2, TiO_2, Al_2O_3, ZrO_2, etc.) and binder (polymer or inorganic) coated on the surface of a polyolefin [83], nonwoven [88], or electrodes. In particular, advanced technology has been applied to separators made of a polyester nonwoven and inorganic coating (Figure 3.150) instead of the typical polyolefin [88].

Inorganic coating allows improved thermal/mechanical properties over polyolefin separators and enhances battery safety by suppressing internal short circuits. Furthermore, battery performance is better compared to gel polymer electrolytes due to greater ion conductivity obtained from inorganic nanoparticles, binder characteristics, and microstructure control of the inorganic layer. While various approaches are being adopted depending on the manufacturer in the development of separators with inorganic coatings, results have not been released in detail. More specifically, fundamental research is being carried out on a free-standing inorganic composite separator comprised of $CaCo_3$ and teflon [89], and a composite separator made of Al_2O_3 and PVdF binder [90].

3.3.3.7 Separator Manufacturing Process

Film Technology Film technology consists of extrusion and stretching. While extrusion is usually carried out by a twin-screw extruder, a single-screw extruder may be used if manufacturing does not involve mixing of polymers and solvents. Sheets extruded from a T-die are stretched in the machine direction by uniaxial elongation or both the machine and the transverse directions by two-axes elongation. Two-axes elongated films are more suitable as separators due to their high strength and isotropic characteristics. Another method is to carry out extrusion using a cylindrical die followed by tubular elongation.

Technology for Making Pores Technology for making pores can be classified into the dry and wet process.

Dry Process The dry process forms pores by elongating an extruded film at low temperature to produce small cracks on the lamellar crystal surface. Figure 3.151a shows a SEM photo of a microporous film prepared by using this method.

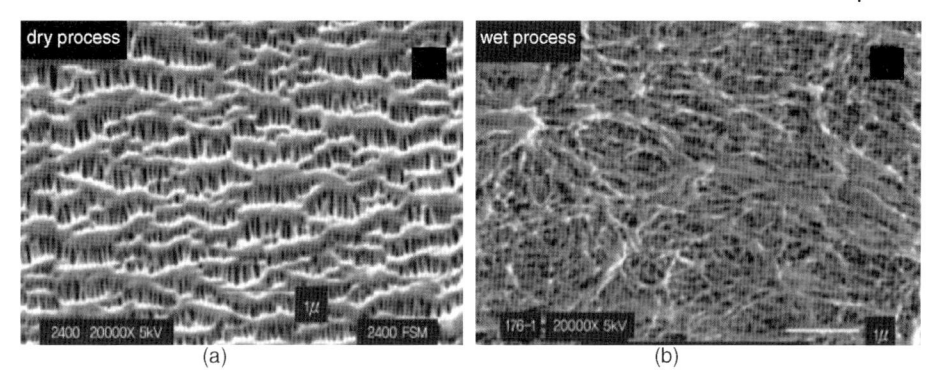

Figure 3.151 SEM photo of a separator prepared from (a) dry process and (b) wet process [82]. Adapted with permission from [82] Copyright 2004 American Chemical Society.

Wet Process In the wet process, a polymer and plasticizer are mixed uniformly at high temperature followed by cooling for phase separation. The plasticizer is then removed to create pores (Figure 3.151b). An inorganic powder may also be added and then removed together with the plasticizer. The latter method is able to produce larger pore diameter and higher porosity.

3.3.3.8 Prospects for Separators

As lithium ion batteries gain high energy density and high power density while becoming more compact, separators should be thin, high strength, and less prone to contraction. In particular, dimensional stability is critical when PE-laminated aluminum film is used as a packaging material since external force may cause bending or twisting of the battery. Thinner separators facilitate impregnation of the electrolyte but may reduce the impregnated amount or electrolyte retention. As such, miscibility between the separator and the electrolyte must be improved. In order to satisfy high energy density and high-power characteristics, new developments are needed to enhance pore closing and meltdown in separators.

3.3.4
Binders, Conducting Agents, and Current Collectors

In this section, we examine other battery components such as binders, conducting agents, and current collectors. They are important components in addition to those introduced above and considered as priorities in production. The overall battery performance depends on the characteristics of these components.

3.3.4.1 Binders

Binder Functions Binders play an essential role in mechanical stabilization of electrodes. For instance, when lithium ions are intercalated, as shown in Figure 3.152, the graphite electrode expands by 10% in the c-axis and contracts when the ions are removed.

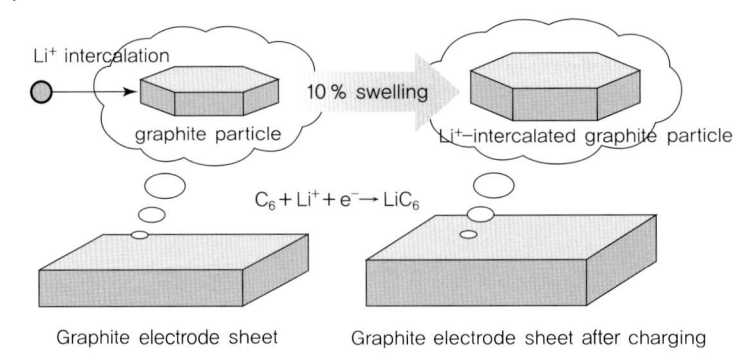

Figure 3.152 Volume change of a graphite electrode sheet after charging.

This process repeats with charging and discharging of the battery. When such changes occur, the interfacial contacts between active materials or conductive agents are weakened. This is accompanied by an increase in contact resistance between particles. As a result, battery characteristics deteriorate due to the greater ohmic resistance of electrodes. Such problems arising from the mechanical instability of electrodes may be resolved through the use of binders. It is necessary to maintain strong binding when the electrode temperature is 200 °C during the dry process of electrode manufacturing. The structural control of polymers used as binders is important to forming bonds between active materials and current collectors.

Requirements of Binders

i) Battery Performance and Safety Binders must maintain a binding property without being dissolved in the highly polar carbonate organic solvent used as an electrolyte in lithium ion batteries. They should remain stable in electrochemical environments. Fluorinated polymers containing TFE (tetrafluoroethylene, $-CF_2-CF_2-$) or HFP (hexafluoropropylene, $-CF_2-CF(CF_3)-$) units are more easily reduced compared to partially fluorinated PVdF [91]. As such, PVdF is the most stable material for binders known today. Binders should also exhibit oxidative resistance, as cathode materials are usually metal oxides and may produce reactive oxygen when overcharged. The oxidation potential values derived from theoretical calculations (molecular orbital method) are -12.12 eV (PE), -14.08 eV (PVdF), and -15.47 eV (PTFE). The oxidation potential of organic solvents such as EC and PC are -12.46 eV and -12.33 eV, respectively. This shows that fluorinated polymers have greater resistance to oxidation. In addition, binders serve as a buffer for expansion and contraction of active materials caused by lithium intercalation/deintercalation at the electrodes. Given the binding pattern of binders to active materials, as shown in Figure 3.153, elastic polymers have yet to be commercialized due to poor adhesive properties and swelling at high temperatures.

In the manufacturing of lithium ion batteries, battery performance is negatively affected by moisture. Considering that electrolytes have a moisture level of 10 ppm, electrodes should ultimately be free of water. Furthermore, binders should have

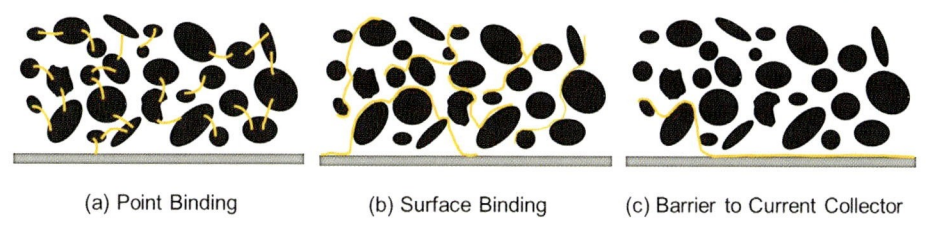

(a) Point Binding   (b) Surface Binding   (c) Barrier to Current Collector

Figure 3.153 Binding model of active materials.

excellent heat resistance since the temperature can rise as high as 200°C during the manufacture of electrodes. These conditions restrict the type of polymers suitable for use as binders.

ii) Manufacturing Process Binders should allow uniform, stable preparation of an active material paste followed by rapid coating. Active materials must maintain strong binding even with increase in temperature or throughout charge/discharge cycles. Good adhesion between the active materials and the current collector prevents the powder from blowing during slitting and contributes to battery safety. Binder solutions with high viscosity are required for active materials having a high specific gravity.

Figure 3.154 shows the relationship between concentration and viscosity for three different binders. As shown in Figure 3.155, all binders must be adjusted to appropriate values of true density and apparent density. Battery performance is

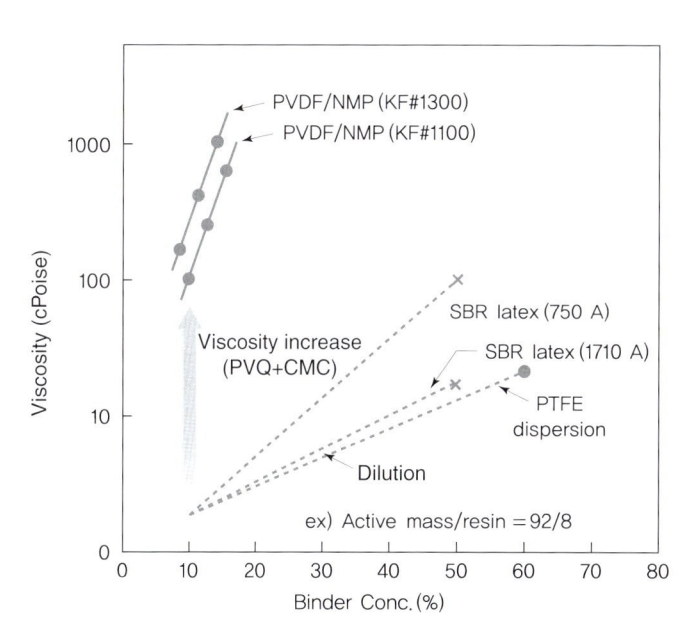

Figure 3.154 Viscosity of binding solutions.

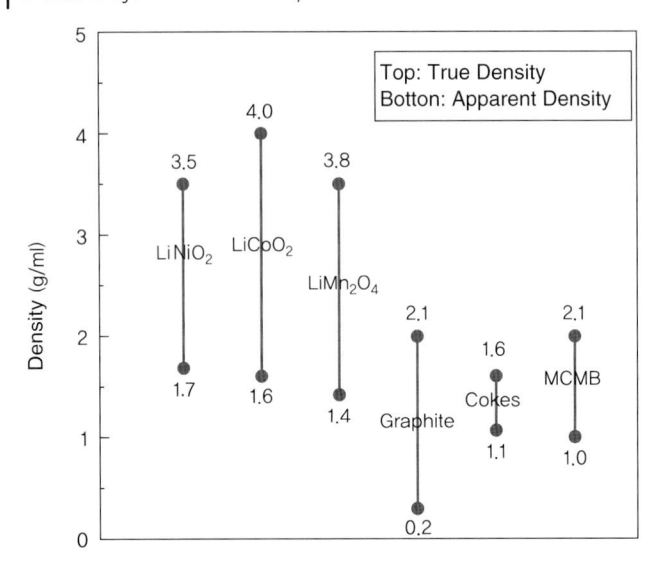

Figure 3.155 True density and apparent density of electrode active materials.

mostly determined by the performance of electrode active materials and is unaffected by binders, which show only lithium ion conductivity in the presence of PVdF.

Suitable Polymers for Binders Suitable polymers for binders include PVdF, highly adhesive SBR (styrene-butadiene rubber)/CMC(carboxy methyl cellulose), chemically stable and heat-resistant PTFE, polyolefin, polyimide, polyurethane, and polyester. Polyimide, polyester, and other polymers obtained from condensation reactions have strong adhesive properties and excellent heat resistance. However, amide and ester bonds do not satisfy the conditions mentioned above. Epoxy resins cannot be used due to their long curing time.

A polymer binder can be seen as a mixture of a polymer and dispersion medium, usually in the form of an organic solvent or water. Effects (solubility, amount of residue, etc.) of the dispersion medium on active materials and reuse of the solvent are issues to be carefully considered. For aqueous dispersion mediums, cycle characteristics may be affected by surface activators remaining in active materials after drying. Flame retardancy is another important property of binders, and can be arranged in descending order as PTFE > PVdF > polyimide > polyethylene.

Ideal Binders The role of binders in terms of manufacturing process and battery characteristics can be summarized as follows. Various technical problems are present in the coating and drying of active materials contained in a conductive solution. However, this method will continue to be used for its high productivity. Existing wet binders will no longer be needed if active material layers can be directly and continuously formed on collector plates based on self-binding from the dry process or CVD. While it is unclear how binders bind to active materials in batteries, atomic force microscopy is being used to examine the state of graphite and PVdF.

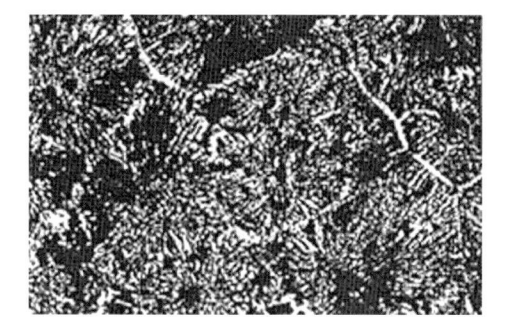

Figure 3.156 Nucleating activity of active materials.

In Figure 3.156, the result of adding a nongraphite anode to PVdF binder is seen through a polarizing microscope. We can observe the nucleating activity of active materials with carbon acting as the nucleus.

PVdF Binders

i) Basic Characteristics   PVdF(2F), a type of fluorinated polymer resin similar to PTFE (4F), is a vinyl copolymer having thermoplasticity and solubility in polar solvents. Fluorinated polymers higher than 3F are insoluble, while 2F and 6F (hexafluoropropylene) copolymers are highly soluble due to their low crystallinity.

Compared to other polymers, PVdF has an unusually high relative dielectric constant (9–10 at 102–103 Hz) and exhibits a lithium ion conductivity of 10^{-6}–10^{-5} S/cm after swelling in a polar solvent. PVdF is a crystalline polymer with a specific gravity of 1.78 g/ml, glass transition temperature (T_g) of approximately $-35°C$, melting temperature (T_m) of 174°C, and crystallization temperature (T_c) of 142°C. Compared to emulsion polymerization, low-temperature suspension polymerization produces PVdF with a higher melting point and crystallinity, thus allowing greater resistance to swelling.

ii) Solubility   In order to be used as binders, PVdF must first be dissolved or dispersed in an appropriate medium. The solubility of PVdF differs greatly according to its polymer structure and crystallinity, which are properties determined by the method of polymerization.

Figure 3.157 shows the solubility of PVdF with the solubility parameter of the solvent as the transverse axis. We can see that PVdF shows solubility higher than 10% at 35°C in DMAc (dimethylacetate) and NMP solvents. PVdF hardly dissolves in commonly used organic solvents such as carbonates, esters, and lactones. Emulsion polymerization facilitates swelling or dissolution in most organic solvents by lowering PVdF crystallinity through flexible chains. However, PVdF prepared from suspension polymerization cannot be dissolved in widespread solvents such as MEK (methyl ethyl ketone) or THF (tetrahydrofuran). Among these, NMP is a stable solvent with a flash point of 95–97°C, ignition point of 346°C, and an established method of solvent recovery.

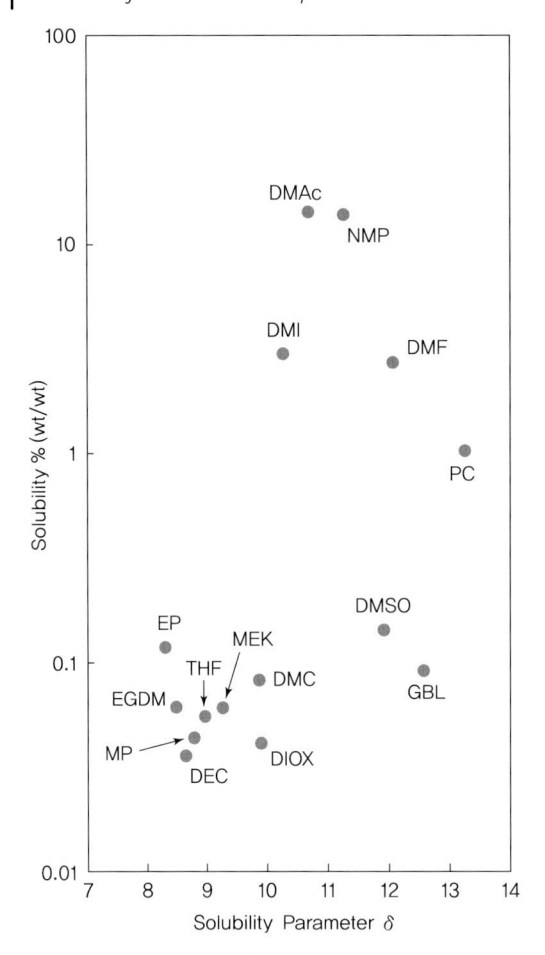

Figure 3.157 Solubility of PVdF (KF#1100, 35°C).

iii) Solvent Characteristics Solvents can have a wide range of viscosity that is appropriately set to the specific gravity (true specific gravity and apparent specific gravity) of active materials. If a cathode material paste having high true specific gravity (3.5–4.5 g/ml) is not prepared under high viscosity, it may readily cause sedimentation and other problems. The mass ratio of active materials and PVdF binder is usually in the range of 96: 4–88: 12. If viscosity is increased by adding more polymer, the amount of binder becomes excessive. As such, polymers with a high degree of polymerization are used during the preparation of a paste with high viscosity. The binder is filtered after dissolution with the microgel completely removed, and examined for metal impurities including alkali metals, Fe, Zn, and Cu prior to its use. Once the appropriate viscosity is determined for the paste, it is transferred to a coating device and coated on current collectors. The paste should maintain a uniform viscosity across a broad range of shear speeds.

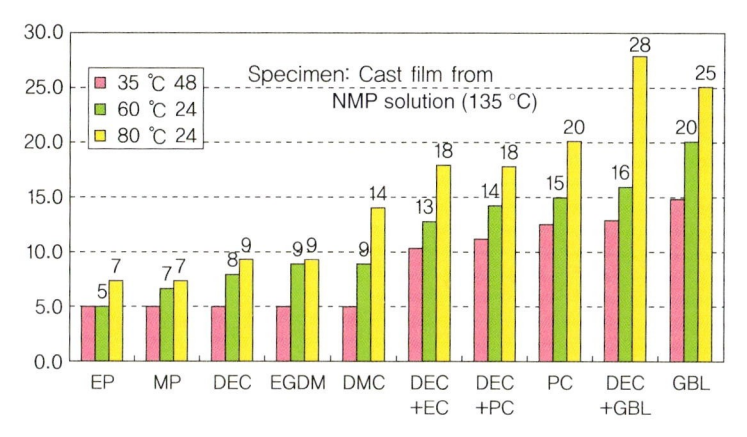

Figure 3.158 Swelling ratio of PVdF.

iv) Swelling and Stability toward Organic Solvents Figure 3.158 presents the swelling ratio of PVdF binder as the weight percentage of the organic solvent in the film after swelling. High swelling ratios are displayed by solvents consisting of or mixed with cyclic carbonates having a high dielectric constant. Most linear carbonates, with the exception of dimethyl carbonate, show a low dielectric constant below 10. When the swelling ratio is under 20%, impregnated electrolytes do not affect the crystal structure but remain distributed at the boundary of crystal particles and noncrystalline areas.

PTFE power is a suitable binder for harsh conditions as there is almost no swelling.

The swelling ratio of SBR latex and thermosetting resins, which will be introduced in later sections, are affected by cross-linking induced by the curing reaction of the polymers. In practice, the cross-linking density is determined by the choice of catalyst and curing time. Lowering the cross-linking density leads to more significant swelling depending on the type of organic solvent.

SBR/CMC Binders Recently, the use of SBR/CMC (styrene-butadiene rubber/carboxymethyl cellulose) binders has been started, instead of PVdF binders, in carbon and noncarbon anodes (Si/C composite, Sn, or Si group). To maximize the capacity and energy density, it is important to increase the amount of electrode active materials while reducing binder content. However, PVdF is restrictive in terms of lowering the amount of binder used. This problem becomes more serious when active material particles are nanosized. SBR/CMC is being considered for its superior adhesive qualities that allow strong binding of active materials and conductive particles with minimal contact. Without tight binding, the large volume change in noncarbon anodes may result in increased resistance. Since SBR/CMC binder can be dispersed into the aqueous solution, it is more environment-friendly compared to PVdF in NMP solvents (see Table 3.13). In addition to being nonflammable, SBR/CMC lowers manufacturing cost by enabling slurry formation and coating under atmospheric conditions.

Table 3.13 Characteristics of SBR/CMC binders.

	Category	[a)]SBR/CMC
Binder	Rubber type	Diene-type rubber
	Glass transition temp.	$-5°C$
	Diameter	130 nm (dried)
	Onset temp. for thermal decomp.	248°C (air)
		342°C (N_2)
Binder solution	Diluent	Water
	Concentration	40 wt%
	Viscosity	12 mPa s
	pH	6
Binder performance with electrolyte	Swelling	1.6-fold (wt)
	Chemical stability	No color change
	Electrochemical stability	Antireduction
Slurry composition (mass based)	SBR	15
	CMC	1.0
	MCMB (0.9 m^2/g)	100
	Water	52.25
Slurry properties	Total solid	66.7 wt%
	Viscosity (60 rpm)	3000 mPa s
	Storage stability (7 days)	No sedimentation
Cell performance	Binding ability toward current collectors (unpressed)	3 g/cm
	Stiffness ($H = 10$ mm)	2 g
	Cracking point	1 mm
	Electrode surface roughness (Ra, unpressed)	3 μm

a) SBR: BM-400B, Zeon, Japan.

Basic Characteristics SBR is a diene-based synthetic rubber with better heat resistance compared to PVdF. In Na-CMC (sodium carboxy methylcellulose), the hydroxyl group of cellulose is partially substituted with sodium carboxylate ($-CH2COONa^+$) with a substitution degree of 0.6–0.95. It is highly soluble in water. SBR and CMC formerly were treated separately but are now combined to maximize adhesion while minimizing binder content. PVdF is still widely used as cathode binders for its high oxidative stability.

Performance SBR in SBR/CMC binders consists of fine rubber particles uniformly dispersed in water. It is lower in content compared to PVdF and exhibits both good mechanical flexibility and stability at high temperatures. In addition to strong binding, binder performance is determined by electrochemical stability and various manufacturing conditions.

Future Trends of Binders As lithium secondary batteries become more compact, slim, high capacity, and stable, improvements should be made to battery production such as efficient preparation of the electrode slurry, faster electrode manufacturing, more rapid impregnation of the electrolyte, and high-speed winding of electrodes. The development of binders is a priority since they are essential in the enhancement of battery performance and productivity.

Cathode materials are moving away from cobalt to nickel or manganese, which face less severe limitations in resources. Nickel-based cathode materials are water-soluble, highly alkaline, and able to absorb water at a rate of 2000–3000 ppm. Aqueous binders are not very effective on cathode materials with high alkali content. Even if nonaqueous binders are used, the slurry does not gain mobility after mixing.

While carbon anodes are chemically inactive, various characteristics (hydrophobic/hydrophilic) are exhibited by the structure and surface of carbon. For example, natural graphite requires a greater amount of dispersion medium due to its low specific gravity, and further developments are needed to improve the fluidity of the slurry.

3.3.4.2 Conducting Agents

Fine carbon powder added to improve electronic conductivity between active material particles at electrodes or to the metal current collector is known as an electronic conducting agent. This is necessary to prevent the binder from acting as a nonconductor and to compensate for the lack of electronic conductivity in electrode active materials.

Types of Conducting Agents Carbon-based materials are commonly used as conducting agents. Examples of carbon powder are carbon black, acetylene black, and ketjen black. Carbon powder takes the form shown in Figure 3.159, with fine particles intertwined into a fibrous web. The basic crystal structure is the same as that of a graphite anode in lithium ion batteries, but the graphitic structure of carbon powder is only partially developed.

Different manufacturing methods lead to differences in the degree of graphitization and microstructure between acetylene black and ketjen black. In general, carbon particles of ketjen black show higher conductivity due to the graphite structure being more developed. While the resistance is below 0.5 Ωcm for both

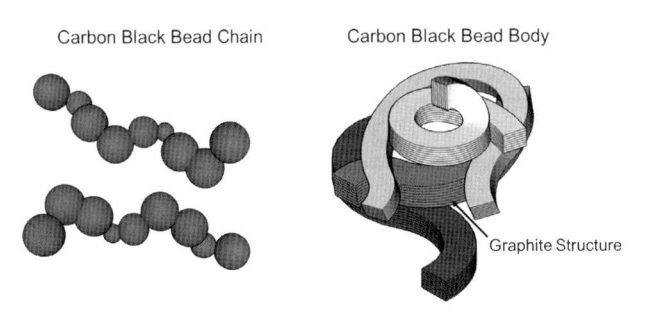

Figure 3.159 Structural model of carbon powder.

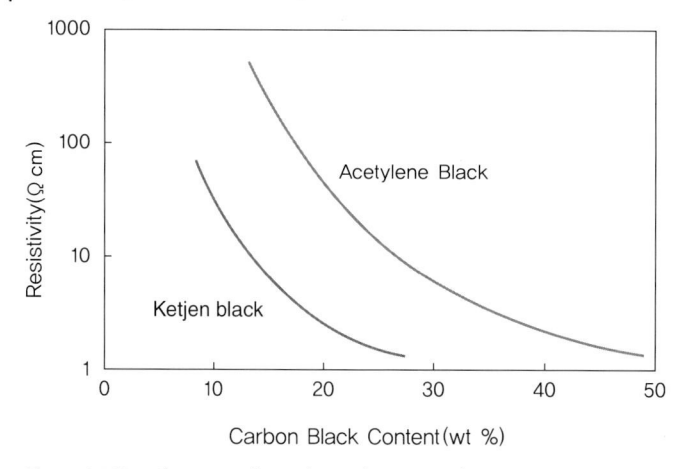

Figure 3.160 Change in electrode conductivity with varying carbon content.

cases, the two carbon types produce a higher resistance when mixed compared to using ketjen black alone. Carbon powder in acetylene black is smaller (about 30 nm) than ketjen black and connected in a manner akin to prayer beads. Although ketjen black has larger particles, its specific surface area of 800 m^2/g is greater than that of acetylene black at 100 m^2/g. This is because the fine powder in ketjen black contains many pores. Both conducting agents show excellent uniformity with strong adhesion to anode and cathode materials. Figure 3.160 shows the change in electrode resistance by varying the amount of carbon black. We can see that resistance is greatly reduced by adding carbon. Since the pattern of resistance reduction is influenced by carbon type, it is important to select an appropriate carbon type according to the active materials and particle state.

Dispersibility of Conducting Agents   When using carbon powder as a conducting agent, carbon, active materials, and binders should be uniformly combined, as shown in Figure 3.161.

Here, it is important to uniformly mix electrode active materials and carbon powder. If dispersed as shown in Figure 3.162a, a greater amount of carbon powder is required. Figure 3.162b is an example of good carbon dispersion. While carbon powder exists as a chain structure, the chains should be spaced out instead of clumping together.

Conducting Agents and Wettability of Electrodes   As a hydrophobic substance, carbon powder has no wettability. Commonly used organic solvents have high permittivity and better wettability than water. However, they may not become wet depending on the state of carbon. As such, the surface of acetylene black and ketjen black should be modified to secure wettability. For instance, the energy density of electrodes with the same pore size can be adjusted by lowering the affinity of carbon powder to electrolytes. Carbon is able to retain electrolytes due to the presence of pores within particles.

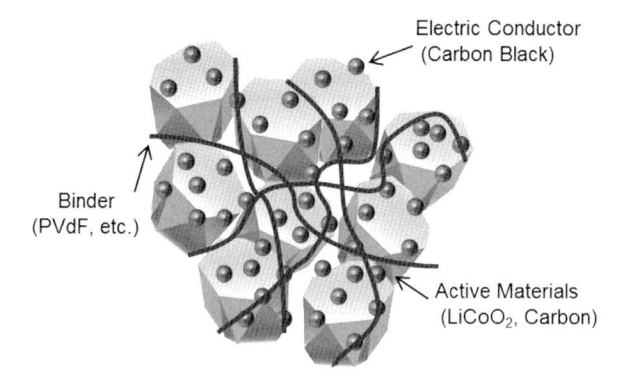

Figure 3.161 Structural model required of actual electrodes.

Modification of Conducting Agents While there has been little research on carbon as a conducting agent, boron doping has been used to improve carbon crystallinity or control the surface state of carbon powder. Battery performance can be improved if carbon itself is able to contribute to charging and discharging. The development of conductive polymer materials as conducting agents is not being emphasized due to material costs and characteristics.

3.3.4.3 Current Collectors

Role of Current Collectors Current collectors act as a medium in supplying electrons to electrode active materials from external circuits or delivering resulting electrons of electrode reactions to the external circuit. They are also important materials that constitute electrodes. In consideration of electronic conductivity, electrochemical stability, and electrode manufacturing process, metals are commonly used as material for current collectors.

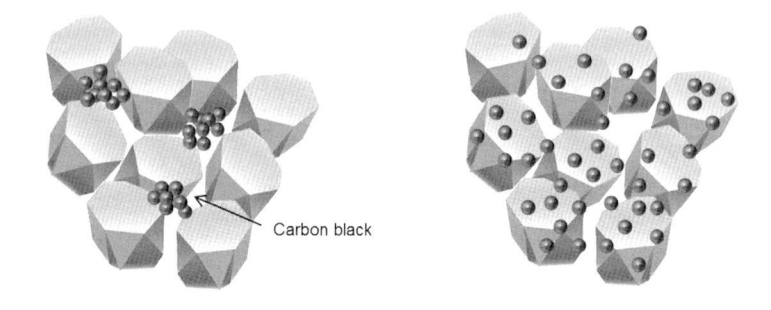

(a) Poor dispersion caused by carbon black aggregation

(b) Well dispersion of carbon black

Figure 3.162 Electrodes with (a) poor carbon dispersion and (b) good dispersion.

Requirements In current collectors of lithium ion batteries, aluminum and copper are used at the cathode and anode, respectively. Electrode active materials are coated on current collectors followed by drying to create electrodes. Despite being comprised of a very thin foil (thickness: 10–20 μm), current collectors show sufficient mechanical strength. Their surface state is critical to the preparation of electrode slurry. The actual foil surface exhibits high wettability to the paste, and strong binding is possible between the binder and the current collector after removing the solvent.

Anode Current Collectors Anode current collectors consist of materials, such as copper or nickel, which are electrochemically inactive within the working potential (0.01–3.0 V versus Li) of carbon electrodes. In particular, copper is relatively stable toward reduction, while nickel is not affordable.

Cathode Current Collectors For cathodes, it is important to avoid oxidation of metal current collectors at high potential. Copper cannot be used for cathodes as oxidation occurs at 3 V. Considering various factors such as cost and electrochemical stability in the operating range, aluminum is the most appropriate material.

References

1 Blomgren, B.E. (1983) *Lithium Batteries*, Academic Press, New York.
2 Ue, M., Murakami, A., and Nakamura, S. (2002) *J. Electrochem. Soc.*, **149**, A1385.
3 Ue, M. (1994) *J. Electrochem. Soc.*, **141**, 3336.
4 Yoshio, M. and Kozawa, A. (1996) *Lithium-Ion Secondary Battery: Materials and Application*, 3rd edn, Nikkan Kougyo Shinbunsya.
5 Ue, M., Ida, K., and Mori, S. (1994) *J. Electrochem. Soc.*, **141**, 2989.
6 Ue, M., Takeda, M., Takehara, M., and Mori, S. (1997) *J. Electrochem. Soc.*, **144**, 2684.
7 Ue, M., Murakami, A., and Nakamura, S. (2002) *J. Electrochem. Soc.*, **149**, A1572.
8 Mori, S., Asahina, H., Suzuki, H., Yonei, A., and Yokoto, K. (1997) *J. Power Sources*, **68**, 59.
9 Kominato, A., Yasukawa, E., Sato, N., Ijuuin, T., Asahina, H., and Mori, S. (1997) *J. Power Sources*, **68**, 471.
10 Wang, Y., Nakamura, S., Ue, M., and Balbuena, P.B. (2001) *J. Am. Chem. Soc.*, **123**, 11708.
11 Wang, Y., Nakamura, S., Tasaki, K., and Balbuena, P.B. (2002) *J. Am. Chem. Soc.*, **124**, 4408.
12 Abe, K., Ushigoe, Y., Yoshitake, H., and Yoshio, M. (2006) *J. Power Sources*, **153**, 328.
13 Xiao, L., Ai, X., Cao, Y., and Yang, H. (2004) *Electrochim. Acta*, **49**, 4189.
14 Shima, K., Shizuka, K., Ue, M., Ota, H., Hatozaki, T., and Yamaki, J. (2006) *J. Power Sources*, **161**, 1264.
15 Bockris, J.O'M. and Reddy, A.K.N. (1970) Chapter 4, in *Modern Electrochemistry*, vol. 1, Plenum, New York.
16 MacInnes, D.A. (1961) Chapter 4, in *The Principles of Electrochemistry*, Dover, New York.
17 Devynck, J., Mossina, R., Pingarron, J., and Tremillon, B. (1984) *J. Electrochem. Soc.*, **131**, 2274.
18 Scordilis-Kelly, C. and Carlin, R.T. (1994) *J. Electrochem. Soc.*, **141**, 873.
19 Matsumoto, H., Yanagida, M., Tanimoto, K., Kojima, T., Tamiya, Y., and Miyazaki, Y. (2000) *Molten Salts XII*, The Electrochemical Society, Pennington, p. 186.

20 Fuller, J., Carlin, R.T., and Osteryoung, R.A. (1997) *J. Electrochem. Soc.*, **144**, 3881.

21 Fuller, J., Osteryoung, R.A., and Carlin, R.T. (1995) *J. Electrochem. Soc.*, **142**, 3632.

22 Kim, D.W., Sivakkumar, S.R., MacFarlane, D.R., Forsyth, M., and Sun, Y.K. (2008) *J. Power Sources*, **180**, 591.

23 Cooper, E.I. and Angell, C.A. (1986) *Solid State Ionics*, **18–19**, 570.

24 MacFarlane, D.R., Meakin, P., Sun, J., Amini, N., and Forsyth, M. (1999) *J. Phys. Chem. B*, **103**, 4164.

25 Matsumoto, H., Kageyama, H., and Miyazaki, Y. (2002) *Chem. Commun.*, **16**, 1726.

26 Hajime, M. (2002) *Polym. Prep. Jpn.*, **51**, 2758.

27 Xu, K. (2004) *Chem. Rev.*, **104**, 4303.

28 Barker, J. and Gao, F. (1998) U.S. Patent 5712059.

29 Fujimoto, M., Shouji, Y., Nohma, T., and Nishio, K. (1997) *Denki Kagaku*, **65**, 949.

30 Simon, B. and Boeuve, J.P. (1997) U.S. Patent 5626981.

31 Holleck, G.L., Harris, P.B., Abraham, K.M., Buzby, J., and Brummer, S.B. (1982) Technical Report No. 6, Contract N00014-77-0155, EIC Laboratories, Newton, MA.

32 Abraham, K.M. and Brummer, S.B. (1983) *Lithium Batteries* (ed. J. Ganabo), Academic Press, New York.

33 Behl, W.K. and Chin, D.T. (1988) *J. Electrochem. Soc.*, **135**, 16.

34 Behl, W.K. and Chin, D.T. (1988) *J. Electrochem. Soc.*, **135**, 21.

35 Behl, W.K. (1989) *J. Electrochem. Soc.*, **136**, 1305.

36 Lee, Y.G. and Kim, K.M. (2008) *J. Korean Electrochem. Soc.*, **11**, 242.

37 Adachi, M., Tanaka, K., and Sekai, K. (1999) *J. Electrochem. Soc.*, **146**, 1256.

38 Chen, J., Buhrmester, C., and Dahn, J.R. (2005) *Electrochem. Solid State Lett.*, **8**, A59.

39 Moshurchak, L.M., Buhrmester, C., and Dahn, J.R. (2005) *J. Electrochem. Soc.*, **152**, A1279.

40 Yoshino, A. (2002) Proceedings of the 4th Hawaii Battery Conference, Jan 8–11, ARAD Enterprises, Hilo, HI, p. 102.

41 Tobishima, S., Ogino, Y., and Watanabe, Y. (2003) *J. Appl. Electrochem.*, **33**, 143.

42 Izatt, R.M., Bradshaw, J.S., Nielson, S.A., Lamb, J.D., and Christensen, J.J. (1985) *Chem. Rev.*, **85**, 271.

43 Salomon, M. (1990) *J. Solution Chem.*, **19**, 1225.

44 Lee, H.S., Yang, X.Q., McBreen, J., Choi, L.S., and Okamoto, Y. (1996) *J. Electrochem. Soc.*, **143**, 3825.

45 Lee, H.S., Sun, X., Yang, X.Q., McBreen, J., Callahan, J.H., and Choi, L.S. (2000) *J. Electrochem. Soc.*, **146**, 9.

46 Tasaki, K. and Nakamura, S. (2001) *J. Electrochem. Soc.*, **148**, A984.

47 Prakash, J. (2005) Battery Technology Symposium, KERI, Korea.

48 Lee, C.W., Joachin, H., Hui, Y., and Prakash, J. (2000) Rechargeable Lithium Batteries, The Electrochemical Society Proc. PV 2000-21, p. 297.

49 Lee, C.W., Venkatachalapathy, R., and Prakash, J. (2000) *Electrochem. Solid State Lett.*, **3** (2), 63.

50 Prakash, J., Lee, C.W., and Amine, K. (2002) U.S. Patent 6455200.

51 Lee, C.W., Venkatachalapathy, R., and Prakash, J. (2000) *Electrochem. Solid State Lett.*, **3** (2), 63.

52 Shu, Z.X., McMillan, R.S., Murray, J.J., and Davidson, J. (1996) *J. Electrochem. Soc.*, **143**, 2230.

53 Arai, J., Katayama, H., and Akahoshi, H. (2002) *J. Electrochem. Soc.*, **149**, A217.

54 Zhang, S.S., Xu, K., and Jow, T.R. (2002) *J. Electrochem. Soc.*, **149**, A586.

55 Barthel, J., Wuhr, M., Buestrich, R., and Gores, H.J. (1995) *J. Electrochem. Soc.*, **142**, 2527.

56 Barthel, J., Schmidt, M., and Gores, H.J. (1998) *J. Electrochem. Soc.*, **145**, L17.

57 Handa, M., Suzuki, M., Suzuki, J., Kanematsu, H., and Sasaki, Y. (1999) *Electrochem. Solid State Lett.*, **2**, 60.

58 Ishikawa, M., Sugimoto, T., Kikuta, M., Ishiko, E., and Kono, M. (2006) *J. Power Sources*, **162**, 658.

59 Sugimoto, T., Kikuta, M., Ishiko, E., Kono, M., and Ishikawa, M. (2008) *J. Power Sources*, **183**, 436.

60 Wright, P.V. (1975) *Br. Polym. J.*, **7**, 319.

61 Vashishta, P., Mundy, J.N., and Shenoy, G.K. (1979) *Fast Ion Transport in Solids*, North-Holland, Amsterdam.

62 MacCallum, J.R. and Vincent, C.A. (1987) *Polymer Electrolyte Reviews*, vol. **1**, Elsevier Applied Science, London.; MacCallum, J.R. and Vincent, C.A. (1989) *Polymer Electrolyte Reviews*, vol. **2**, Elsevier Applied Science, London.
63 Scrosati, B. (1993) *Applications of Electroactive Polymers*, Chapman and Hall, London.
64 Gray, F.M. (1997) *Polymer Electrolytes*, The Royal Society of Chemistry, Cambridge.
65 Song, J.Y., Wang, Y.Y., and Wan, C.C. (1999) *J. Power Sources*, **77**, 183.
66 Kim, D.W. and Park, J.K. (2000) *IEEK J.*, **27** (8), 803.
67 Gozdz, A.S., Tarascon, J.M., Gebizlioglu, O.S., Schmutz, C.N., Warren, P.C., and Shokoohi, F.K. (1995) PV 94-28 The Electrochemical Society Proceedings Series, Pennington, NJ, p. 400.
68 Kim, D.W. and Jeong, Y.B. (2004) *J. Power Sources*, **128**, 256.
69 Gee, M. and Olsen, I. (1996) U.S. Patent 5534365.
70 Narukawa, S. and Nakane, I. (2000) 10th IMLB, June 1, Como, Italy, Abstract 38.
71 Watanabe, M., Suzuki, Y., and Nishimoto, A. (2000) *Electrochim. Acta*, **45**, 1187.
72 Tominaga, Y. and Ohno, H. (2000) *Electrochim. Acta*, **45**, 3081.
73 Benrabah, D., Sylla, S., Alloin, F., Sanchez, J.Y., and Armand, M. (1995) *Electrochim. Acta*, **40**, 2259.
74 Cowie, J.M.G. and Spence, G.H. (1999) *Solid State Ionics*, **123**, 233.
75 Bayoudh, S., Pavizel, N., and Reibel, L. (2000) *Polym. Intern.*, **49**, 703.
76 Okamoto, Y., Yeh, T.F., Lee, H.S., and Skotheim, T.A. (1993) *J. Polym. Sci.*, **31**, 2573.
77 Noda, A. and Watanabe, M. (2000) *Electrochim. Acta*, **45**, 1265.
78 Sakaebe, H. and Matsumoto, H. (2003) *Electrochem. Commun.*, **5**, 594.
79 Mao, Z. and White, R.E. (1993) *J. Power Sources*, **43**, 181.
80 Tye, F.L. (1983) *J. Power Sources*, **9**, 89.
81 Caldwell, D.L.,Pouch, U.S. Patent 4,464,238 (1984).
82 Arora, P. and Zhang, Z. (2004) *Chem. Rev.*, **104**, 4419.
83 Yamamoto, T., Hara, T., Segawa, K., Honda, K., and Akashi, H. (2007) *J. Power Sources*, **174**, 1036.
84 Lee, S., Kim, J., Park, P., Shin, B., Hong, J., Kim, I., and Ahn, S. (2007) Lithium Mobile Power, San Diego.
85 Augustin, S., Hennige, V., Hoerpel, G., and Hying, C. (2002) *Desalination*, **146**, 23.
86 Kim, J., Han, W., and Min, J. Korea Patent 2007-0005341.
87 Zhang, S.S. and Xu, K. (2005) *J. Power Sources*, **140**, 361.
88 Augustin, S., Hennige, V., Hoerpel, G., and Hying, C. (2002) *Desalination*, **146**, 23.
89 Zhang, S.S. and Xu, K. (2005) *J. Power Sources*, **140**, 361.
90 Takemura, T., Aihara, S., Hamano, K., Kise, M., Nishimura, T., Urishibata, H., and Yoshiyasu, H. (2005) *J. Power Sources*, **146**, 779.
91 Park, J.K. (1988) Proceedings of the 33rd International Power Sources Symposium, 97, p. A0063366.

3.4
Interfacial Reactions and Characteristics

The deterioration of lithium secondary battery performance and cycle life is often caused by overcharging of a battery, subsequent oxidative decomposition of the electrolyte at the cathode and reductive decomposition of the electrolyte at the anode, self-discharge, change in phase or dissolution of electrode materials, and corrosion of current collectors and current leads. Those reactions originate from interfacial reactions between electrodes/current collectors and electrolytes. To prevent these drawbacks, it is necessary to have a fundamental understanding of chemical and electrochemical reactions occurring at the interface and find a path to improved battery performance.

Electrode–electrolyte interfacial reactions differ depending on the type of electrode materials and electrolytes. This section summarizes well-established interfacial reactions of anode and cathode materials and aluminum current collector.

3.4.1
Electrochemical Decomposition of Nonaqueous Electrolytes

Electrolytes in lithium secondary batteries mainly serve as a medium in providing lithium ions and transferring lithium ions between the anode and the cathode. Since large surface area active materials can provide enlarged interfacial contact area with the electrolyte, the electrochemical reaction kinetics can be enhanced.

Nonaqueous electrolytes used in lithium secondary batteries mostly consist of lithium salts and organic carbonate solvents. In order to understand the interfacial reaction behavior of electrolytes, we have to first look at the electrochemical decomposition of lithium salts and carbonate solvents. Since lithium ions (Li^+) have a great degree of solvation by carbonate solvent molecules [1], the mobility of solvated Li^+ is lower than that of anions such as AsF_6^-, PF_6^-, ClO_4^-, BF_4^-, and $N(SO_2CF_3)_2^-$. The highly mobile anions reach the surface of electrodes earlier than Li^+ and are thus more likely to participate in electrochemical decomposition. Among anions, AsF_6^- is the most reactive and easily decomposed. However, decomposition for other less reactive anions tends to begin when the potential becomes close to the Li/Li^+ potential [2].

In reduction reactions of propylene carbonate (PC), the reduction potential is below $+1.0\,V$ versus Li/Li^+, while tetrahydrofuran (THF) and 2-methyl THF are

Table 3.14 Comparison of calculated [5] and experimental potential values of solvent reduction. Reproduced by permission of ECS – The Electrochemical Society.

Solvent	Calculated (V)	Experimental (V)
EC	1.46	1.36
DEC	1.33	1.32
PC	1.24	1.00–1.60
DMC	0.86	1.32
VC	0.25	1.40

reduced at potential lower than −2.0 V [3]. Table 3.14 shows the cyclic voltammetry results on a gold (Au) electrode with THF/LiClO$_4$ electrolyte. The electrochemical reduction potential of ethylene carbonate (EC), diethylene carbonate (DEC), dimethyl carbonate (DMC), PC, and vinylene carbonate (VC) was all greater than 1 V versus Li/Li$^+$. In particular, EC is reduced at around +1.46 V versus Li/Li$^+$. In most cases, the values were similar to theoretical estimates of the density functional theory (DFT) [4]. This implies that electrolyte decomposition takes place within the range of the discharge potential.

Reduction reactions of carbonate solvents by one-electron transfer are shown in Figure 3.163. Electron transfer drives the ring-opening reaction of cyclic carbonate molecules [4].

In addition to single molecular reactions arising from one-electron transfer, chain reactions of various carbonate molecules can produce various compounds with higher molecular weight. According to the DFT calculations for EC, continuous reduction reactions lead to Li$^+$(EC)$_n$ ($n = 1$–4) (see Scheme 3.1). The main product of EC decomposition is (CH$_2$OCO$_2$Li)$_2$, and other derivatives include lithiumalkyl bicarbonates, (CH$_2$CH$_2$OCO$_2$Li)$_2$, LiO(CH$_2$)$_2$CO$_2$(CH$_2$)$_2$OCO$_2$Li, Li(CH$_2$)$_2$OCO$_2$Li, and Li$_2$CO$_3$ [5].

On the other hand, oxidative decomposition reactions of carbonate solvents begin at a relatively higher potential of approximately +4.0 V versus Li/Li$^+$. Table 3.15 presents the oxidation and reduction potential of solvents at inert electrodes [6].

In the table above, E_a is the oxidation potential and E_c is the reduction potential.

Based on theoretical calculations, the oxidation potential of each solvent is EC (5.58 V), DEC (5.46 V), DMC (5.62 V), and VC (4.06 V). In the 0–5 V operation voltage range of lithium batteries, some of these organic solvents operate beyond the thermodynamically stable region of electrolytes. The oxidation of electrolytes occurs near 4.0 V versus Li/Li$^+$ (close to 1.0 V versus NHE). During the charging process, decomposition takes place at the anode and cathode surface due to solvent reduction and oxidation, respectively. Actual reduction and oxidation potentials of organic solvents may differ depending on electrode type.

Lithium cells consisting of inert metal electrodes and electrolytes are used to determine the oxidation potential. In PC and other cyclic carbonates, electron transfer to the cathode occurs at around 4.2 V. Beginning with PC ring-opening reactions shown in Figure 3.164, oxidative decomposed compounds are absorbed on

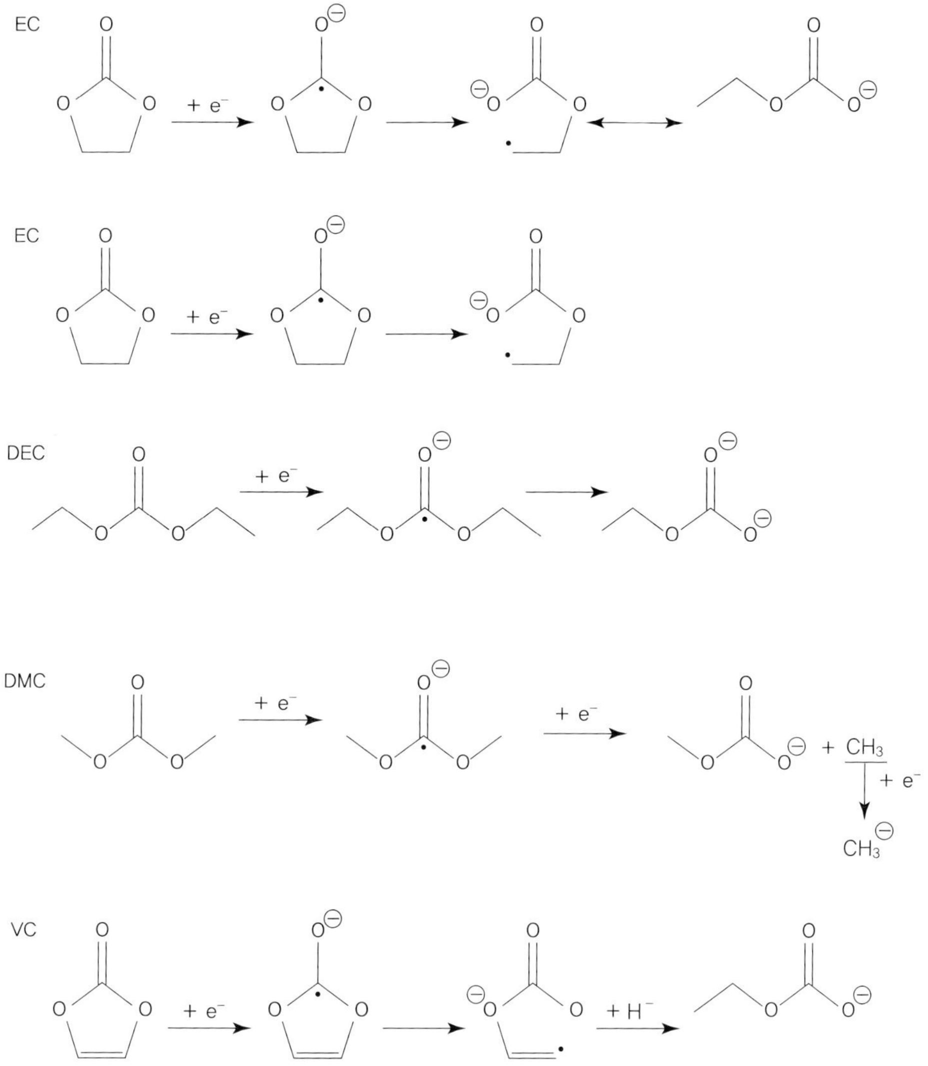

Figure 3.163 Reduction from one-electron transfer of nonaqueous organic solvents [4].

the electrode surface [7]. Part of the adsorbed compounds may either dissolve in the electrolyte or form a passive layer containing organic matter such as −COOR and −COOH at the anode surface.

According to the DFT calculations, continuous oxidative decomposition reactions occur in the free radical state when an electron is delivered from the organic solvent to the electrode [8]. After a positive charge containing C=O is formed in linear carbonates (DMC, DEC, and EMC), charge is dispersed through the resonance structure of O−C$^+$O−O, as shown in Figure 3.165.

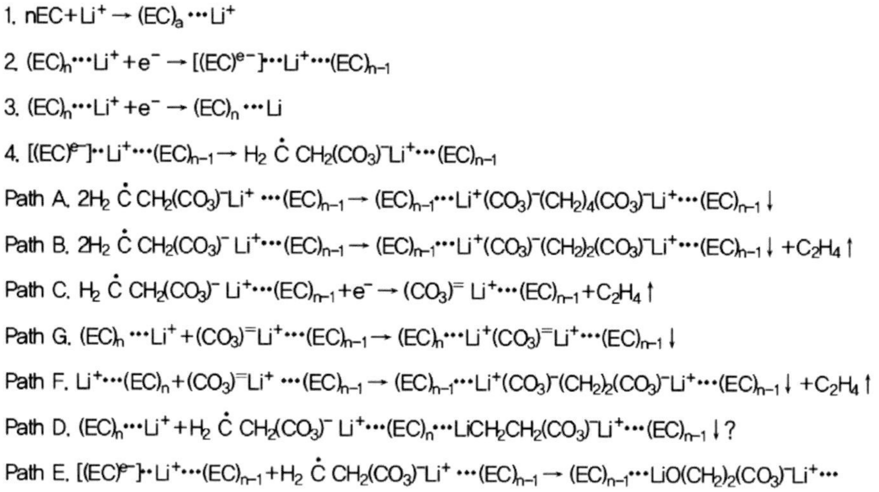

1. $nEC + Li^+ \rightarrow (EC)_a \cdots Li^+$

2. $(EC)_n \cdots Li^+ + e^- \rightarrow [(EC)^{e-}] \cdots Li^+ \cdots (EC)_{n-1}$

3. $(EC)_n \cdots Li^+ + e^- \rightarrow (EC)_n \cdots Li$

4. $[(EC)^{e-}] \cdot Li^+ \cdots (EC)_{n-1} \rightarrow H_2\dot{C}CH_2(CO_3)^- Li^+ \cdots (EC)_{n-1}$

Path A. $2H_2\dot{C}CH_2(CO_3)^- Li^+ \cdots (EC)_{n-1} \rightarrow (EC)_{n-1} \cdots Li^+(CO_3)^-(CH_2)_4(CO_3)^- Li^+ \cdots (EC)_{n-1} \downarrow$

Path B. $2H_2\dot{C}CH_2(CO_3)^- Li^+ \cdots (EC)_{n-1} \rightarrow (EC)_{n-1} \cdots Li^+(CO_3)^-(CH_2)_2(CO_3)^- Li^+ \cdots (EC)_{n-1} \downarrow + C_2H_4 \uparrow$

Path C. $H_2\dot{C}CH_2(CO_3)^- Li^+ \cdots (EC)_{n-1} + e^- \rightarrow (CO_3)^= Li^+ \cdots (EC)_{n-1} + C_2H_4 \uparrow$

Path G. $(EC)_n \cdots Li^+ + (CO_3)^= Li^+ \cdots (EC)_{n-1} \rightarrow (EC)_n \cdots Li^+(CO_3)^= Li^+ \cdots (EC)_{n-1} \downarrow$

Path F. $Li^+ \cdots (EC)_n + (CO_3)^= Li^+ \cdots (EC)_{n-1} \rightarrow (EC)_{n-1} \cdots Li^+(CO_3)^-(CH_2)_2(CO_3)^- Li^+ \cdots (EC)_{n-1} \downarrow + C_2H_4 \uparrow$

Path D. $(EC)_n \cdots Li^+ + H_2\dot{C}CH_2(CO_3)^- Li^+ \cdots (EC)_n \cdots LiCH_2CH_2(CO_3)^- Li^+ \cdots (EC)_{n-1} \downarrow ?$

Path E. $[(EC)^{e-}] \cdot Li^+ \cdots (EC)_{n-1} + H_2\dot{C}CH_2(CO_3)^- Li^+ \cdots (EC)_{n-1} \rightarrow (EC)_{n-1} \cdots LiO(CH_2)_2(CO_3)^- Li^+ \cdots$

Scheme 3.1 EC/Li^+ Reductive Decomposition Mechanism.

Decomposition is caused by β-decomposition of the C–O bond with electron movement within the free radical group. The C–O$^+$=C cation is formed and CO_2 is produced (Figure 3.166). The release of CO_2 gas has a critical effect on the safety of lithium cells. The oxidation potential of linear carbonates is known to be 5.5–5.6 V versus Li/Li^+.

Table 3.15 Electrochemical stability of electrolyte solvents (nonactive electrodes). Adapted with permission from American Chemical Society Copyright 2004.

Solvent	Salt/Concentration (M)	Working electrode	E_a (V)	E_c (V)
PC	Et_4NBF_4/0.65	GC	6.6	
	None	Pt	5.0	~1.0
	Bu_4NPF_6	Ni		0.5
	$LiClO_4$/0.1	Au, Pt		1.0-1.2
	$LiClO_4$/0.5	Porous Pt	4.0	
	$LiClO_4$	Pt	4.7	
	$LiClO_4$	Au	5.5	
	$LiAsF_6$	Pt	4.8	
EC	Et_4NBF_4/0.65	GC	6.2	
	Bu_4NPF_6	Ni		0.9
	$LiClO_4$/0.1	Au, Pt		1.63
DMC	Et_4NBF_4/0.65	GC	6.7	
	$LiClO_4$/0.1	Au, Pt		1.32
	$LiPF_6$/1.0	GC	6.3	
	LiF	GC	5.0	
DEC	Et_4NBF_4/0.65	GC	6.7	
	$LiClO_4$/0.1	Au, Pt		1.32
EMC	Et_4NBF_4/0.65	GC	6.7	
	$LiPF_6$/1.0	GC	6.7	
BL	$LiAsF_6$/0.5	Au, Ag		1.25

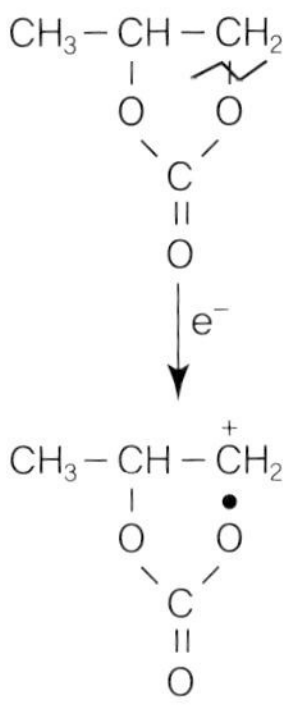

Figure 3.164 Oxidation reactions from electron transfer of PC [7].

Cyclic carbonates such as EC and PC have a five-membered ring structure. Mass spectrum analysis shows that CO_2 is produced from thermal decomposition of EC [9] (Figure 3.167). When CO_2 is released, hydrogen ions are transferred to another carbon.

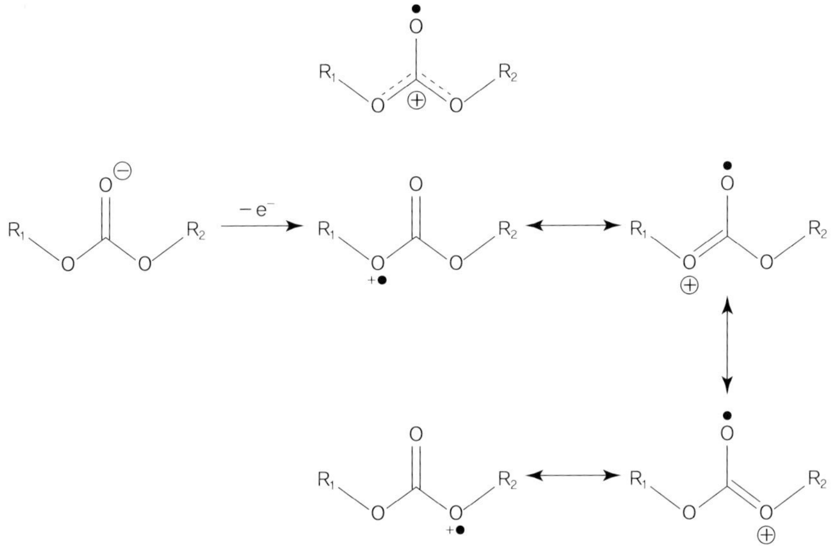

Figure 3.165 Resonance structure of the free radical state in a linear carbonate [8].

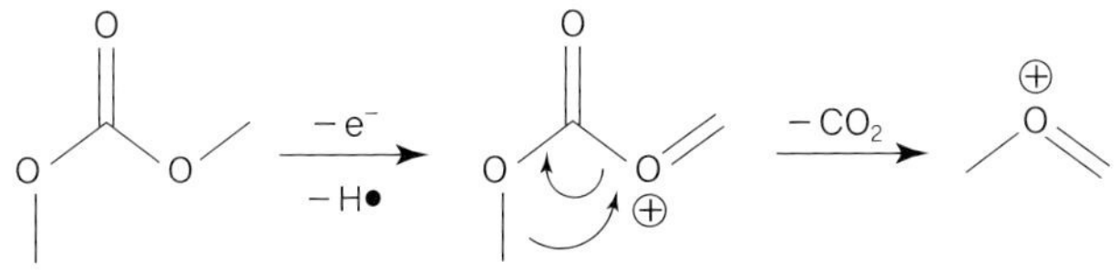

Figure 3.166 Oxidative decomposition reactions of a linear carbonate [8].

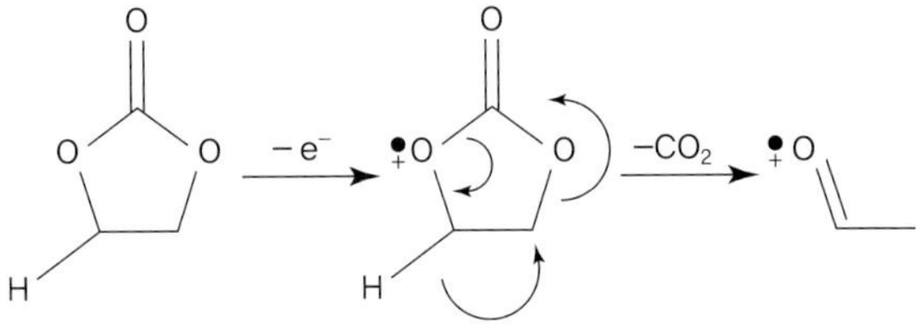

Figure 3.167 Oxidative decomposition reactions of a cyclic carbonate [8].

Vinylene carbonate, a type of unsaturated cyclic carbonate, goes into the free radical state when one electron is transferred to the electrode (Figure 3.168). At this point, the C—O bond of the carbonyl group is shortened, while that of the ether group becomes longer. These oxidative decomposition reactions occur at 4.6 V versus Li/Li$^+$. Dione is formed with the release of CO gas.

Decomposed ionic components may develop ionic bonds with Li$^+$ ions to create lithium organic salts or react with one another for polymerization.

3.4.2
SEI Formation at the Electrode Surface

With redox decomposition reactions of electrolyte components at the interface with metals, carbons, or oxide electrodes, a new layer is formed from the deposition of electrolyte decomposition products. Some of these products become permanent deposits and form a passive layer on the electrode surface. This insulating layer is known as the solid electrolyte interphase (SEI) layer and is characterized by low electronic conductivity, high ionic conductivity, and behavior similar to solid electrolyte (Figure 3.169) [10, 11]. The porosity of the SEI layer allows Li$^+$ ions to pass through but not other electrolyte components.

In addition to transferring Li$^+$ ions at the electrode–electrolyte interface, the SEI layer facilitates Li$^+$ ion transport under a uniform current distribution by reducing concentration polarization and overvoltage, and securing equal grain size and chemical composition. To obtain a long cycle life of lithium cells, the SEI layer should have strong adhesion to electrodes and possess physical/chemical stability.

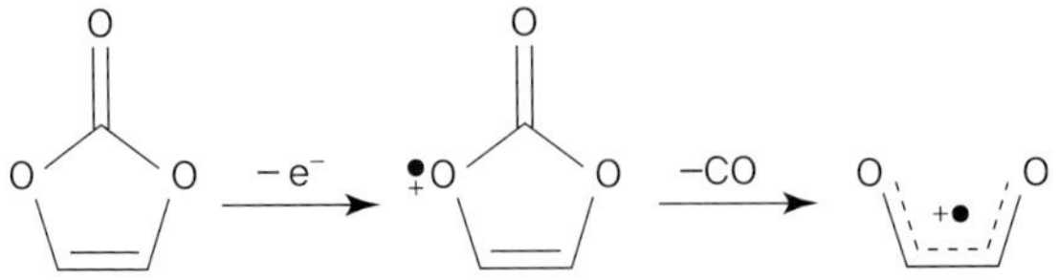

Figure 3.168 Oxidative decomposition reactions from electron transfer of an unsaturated cyclic carbonate [8].

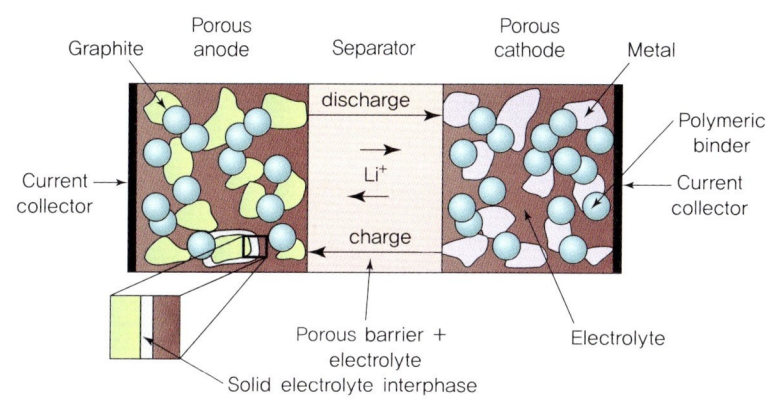

Figure 3.169 SEI layer formation at the electrode–electrolyte interface.

A basic understanding of the SEI formation mechanism is essential in order to enhance cycling performance, stability, and cycle life of lithium cells.

When the electrolyte comes into contact with lithium metals, its reduction might occur since the standard reduction potential of lithium is much lower than that of the solvent [2]. To form a stable SEI layer with low dissolubility, it is important to choose electrolyte components with high standard reduction potential and charge density.

Electrochemical reactions through the SEI layer proceeds in the three stages described as follows (Figure 3.170):

1) Electron transfer occurs through the electrolyte–SEI layer interface.
2) Li^+ ions pass through the SEI layer.
3) Charge transfer occurs at the SEI layer active material interface.

In general, the second stage is the rate-determining step. In this stage, an additional layer is formed above the first formed layer at the lithium metal surface. This is because the solvent component of the electrolyte first fills up the pores in the first SEI layer and continuous reductive reactions occur with electron transfer from lithium to the solvent during direct contact. The same form of oxidative reactions occurs at the interface with the cathode. When the electrolyte comes in contact with the surface of electrode active materials, SEI layers are rapidly formed and then filled up, which gradually reduces the concentration of SEI species. The first SEI layer is thin and has a low degree of filling. Since the subsequent components exist together with the components of the first layer, the overall system can be described as porous, open, and disordered. Figure 3.171 presents an SEI layer having a mosaic-type polyhetero microphase structure [12].

In actual lithium cells, the SEI layer formation occurs during the early cycles. The formation of the SEI layer gradually slows down as it is electrochemically stable and prevents further electrolyte decomposition reactions.

$$M^+_{M/SE} + e(M) \rightleftarrows M^e_M \quad (\Delta\emptyset_{M/SE})$$

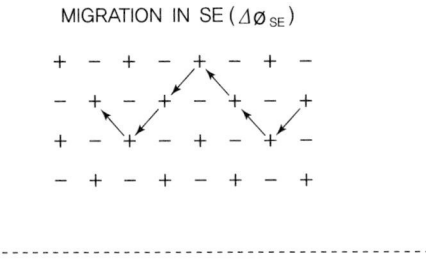

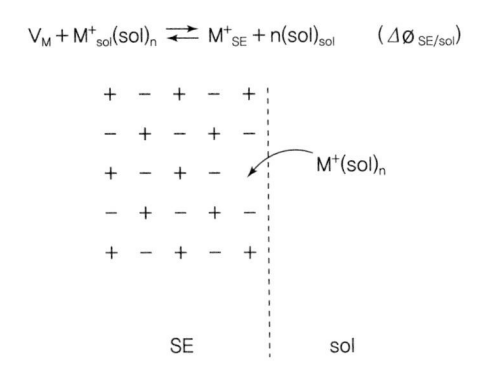

MIGRATION IN SE ($\Delta\emptyset_{SE}$)

$$V_M + M^+_{sol}(sol)_n \rightleftarrows M^+_{SE} + n(sol)_{sol} \quad (\Delta\emptyset_{SE/sol})$$

Figure 3.170 Li$^+$ ion transport through the SEI layer [10]. Reproduced by permission of ECS – The Electrochemical Society.

The composition and formation behavior of the SEI layer differ depending on the types of electrolyte and electrode. An SEI layer consists of a thick organic layer and thin inorganic layer. Various *in situ* and *ex situ* analysis techniques such as FTIR, XPS, AFM, DSC-TGA, DEMS, EDS, and EQCM have been used to examine the composition of SEI layers. Depending on the method used, SEI layers were found to have slightly different compositions. This is due to the different sensitivity of each device to the SEI components. It is thus important to carry out compositional analyses using various methods.

The three general interface types in secondary lithium batteries are anode–electrolyte, cathode–electrolyte, and current collector–electrolyte. The SEI layer is

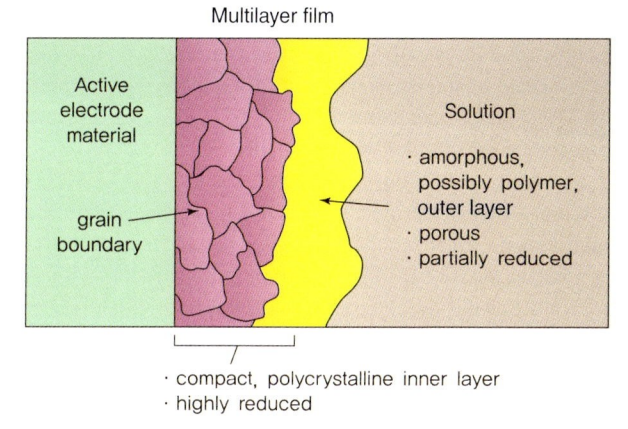

Figure 3.171 Structure of the SEI layer at the electrode surface.

more easily formed at the surface of the anode, but less common or thinner at the surface of the cathode [6]. The formation of the SEI layer is considered to be the main cause responsible for irreversible lithium intercalation/deintercalation (initial Coulombic efficiency) and capacity of the anode and cathode. Electrode–electrolyte interfacial reactions also lead to self-discharge.

The SEI layer forms at the current collector–electrolyte interface. Since most active materials are porous, the electrolyte passes through the pores between particles and directly comes in contact with the current collector to form surface layers.

This book describes interfacial reactions according to the three aforementioned types: cathode–electrolyte, anode–electrolyte, and current collector–electrolyte.

3.4.3
Anode–Electrolyte Interfacial Reactions

Before examining anode–electrolyte interfacial reactions of lithium metal, we should look at the native layer that is naturally formed at the surface of lithium. Although water and oxygen are limited to small amounts in the glove box, passive layers can be formed from the reactions between water, oxygen and CO_2, and lithium. Layers further away from lithium consist of Li_2CO_3 or $LiOH$, while layers closer to lithium are comprised of Li_2O. These porous films have no significant effect on the formation of the SEI layer [13].

When the anode is carbon (e.g., graphite), organic layers are produced from the decomposition of carbonate solvents, while inorganic layers result from lithium salt decomposition. To understand the SEI formation mechanism occurring during charge/discharge of lithium batteries, it is important to study chemical reactions (i.e., no electricity is applied) at the interface of lithium metals and electrolytes. Unlike chemical reactions, electrochemical reactions by applying a suitable current or potential force interfacial reaction, promote reaction kinetics, and serve as a catalyst in the formation and growth of passive layers.

3.4.3.1 Lithium Metal–Electrolyte Interfacial Reactions

Contact between lithium metals and PC is sufficient to create a surface layer. In other words, PC decomposition is caused by spontaneous chemical reactions even without electrochemical reactions. The surface layer consists mainly of Li_2CO_3 [14, 15], and *ex situ* FTIR analysis points to lithium alkyl carbonate ($ROCO_2Li$) as the main surface group [16]. When in contact with lithium metals, the HOMO and LUMO of the electrolyte undergo changes. In lithium metals, electrons move freely from the valence band to the conduction band, and from the conduction band to the electrolyte LUMO. A positive charge may be produced at the surface of lithium metal, while a negative charge is induced for the electrolyte. These chemical reactions create a layer at the lithium metal surface. When a lithium cell is charged, the valence band level becomes higher and approaches the LUMO energy level as lithium ions are reduced and electrons are delivered to the anode through an external circuit. Reductive decomposition reactions of the electrolyte are stimulated due to easier electron transfer from the anode to the electrolyte.

Surface layers are also formed when lithium metals and graphite are placed in other organic solvents such as EC, DEC, or EC + DEC. In the case of EC, chemical electron transfer takes place at the lithium anode to form passive layers on lithium metals and the lithium–graphite interface. When DEC is used as solvent, layers are not produced at the lithium metal surface. Instead, lithium ions are dissolved out to the electrolyte and the organic solvent turns into dark brown. FTIR analysis of the brown solution detected a mixture of $CH_3CH_2OCO_2Li$ and CH_3CH_2OLi [16]. By studying such electrolyte reductive decomposition reactions, we can obtain a better understanding of electrochemical interfacial reactions.

From a thermodynamic perspective, solvents with a low LUMO level have greater interfacial reactivity since they are able to better accept electrons from lithium metals. While most cyclic carbonate solvents have similar LUMO energy levels, double bonds or substitution may result in a lowering of LUMO levels. Electrolyte additives containing various functional groups are used for the early formation of SEI layers and stabilization. The relatively lower LUMO levels of these additives allow electron transfer at higher reduction potentials, thus producing more stable SEI layers.

When lithium metal is placed in DMC, reductive decomposition reactions of the electrolyte take place with the electron transfer described below, and resulting products create a passive layer on the lithium metal surface [16].

$$CH_3OCO_2CH_3 + e^- + Li^+ \rightarrow CH_3OCO_2Li\downarrow + CH_3 \text{ or } CH_3OLi\downarrow + CH_3OCO\cdot$$

As shown in Figure 3.172, an FTIR analysis was conducted for a separately synthesized lithium methyl carbonate and lithium metal surface that had been soaked in DMC solution. By comparing these results with spectra derived from Hartree–Fock calculations, the passive layer was found to contain lithium metal carbonate, lithium oxalate, and lithium methoxide [17]. This implies that an interface is formed between the lithium and the organic solvent, and reductive decomposition reactions begin with electron transfer from the lithium metal to the organic solvent.

When lithium makes contact with DMC, electron transfer occurs from the lithium to DMC and the ester bond of DMC is broken. This process is shown in the

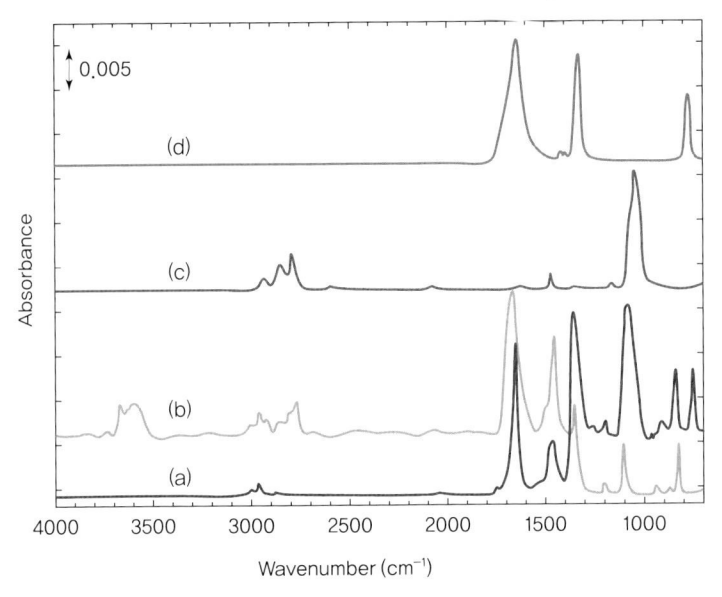

Figure 3.172 FTIR spectrum of (a) lithium methyl carbonate, (b) passive layer formed by lithium metal in DMC, (c) lithium methoxide, and (d) lithium oxalate.

chemical equation below. As a result, unstable CH3• radicals and methyl carbonate anions are produced. Ethane gas is released as CH3• radicals interact with other radicals.

$$\text{Li} + \text{H}_3\text{COC(O)OCH}_3 \longrightarrow \text{Li}^+ + \text{CH}_3\bullet + {}^-\text{OC(O)OCH}_3$$

$${}^-\text{OC(O)OCH}_3 + \text{Li}^+ \longrightarrow \text{LiOCH}_3 + \text{CO}_2$$

$$\text{CH}_3\bullet + \text{CH}_3\bullet \longrightarrow \text{C}_2\text{H}_6$$

$$\text{Li} + \text{H}_3\text{COC(O)OCH}_3 \longrightarrow \text{Li}^+ + \text{CH}_3\bullet + {}^-\text{OC(O)OCH}_3$$

$$ {}^-\text{OC(O)OCH}_3 + \text{Li}^+ \longrightarrow \text{LiOCH}_3 + \text{CO}_2$$

$$\text{CH}_3\bullet + \text{CH}_3\bullet \longrightarrow \text{C}_2\text{H}_6$$

Methyl carbonate anions react with lithium ions to form CO_2 and $LiOCH_3$, along with a release of gas. At the same time, the acyl bond of the ester group is broken, thus producing stable CH_3O^- anions and acyl radicals. Although lithium methyl carbonate is considered a thermodynamically stable compound, it produces oxylate and $CH_3O\bullet$ radicals when attacked by acyl radicals and eventually forms methoxide [16].

The resulting lithium methyl carbonate is stable up to 400 K but begins the following thermal decomposition process, which results in CO_2 gas and Li_2CO_3 [18]:

$$CH_3OCOOLi \rightarrow CH_3OLi + CO_2 \quad \text{or} \quad 2CH_3OCOOLi \rightarrow CO_2 + CH_3OCH_3 + Li_2CO_3$$

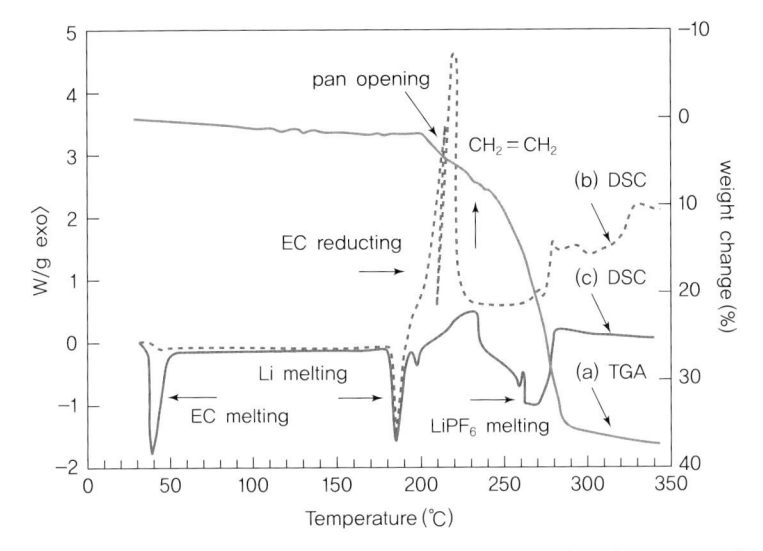

Figure 3.173 (a) Thermal analysis curve of graphite in electrolyte, (b) DSC curve for EC at the surface of lithium metal, and (c) DSC curve of LiPF$_6$ at the surface of lithium metal [18].

Similar observations were made for a graphite anode in 1 M LiPF$_6$/EC:DEC (2:1) electrolyte. According to the results of a thermal analysis shown in Figure 3.173, the passive layer is decomposed at approximately 220 °C and converted to Li$_2$CO$_3$ [17].

EC further amplifies the production of lithium alkyl carbonate. From the FTIR spectrum of lithium soaked in LiAsF$_6$/EC:DEC (1:1), we can see that (CH$_2$OCO$_2$Li)$_2$ is the main component of the SEI layer [16]. With the following two-electron transfer reductive reaction, EC produces lithium ethylene dicarbonate (CH$_2$OCO$_2$Li)$_2$ and releases ethylene gas:

$$2\,\text{EC} \xrightarrow{2e^-\ Li^+} (CH_2OCO_2Li)_2 \downarrow + CH_2{=}CH_2 \uparrow$$

Spectra for the synthesized lithium ethylene dicarbonate and metal surface compound were first obtained. These were compared with theoretical calculations to confirm the presence of lithium ethylene dicarbonate in the actual SEI layer [18]. The SEI layer containing lithium ethylene dicarbonate is produced from electrochemical reduction of the electrolyte at 1.8–1.9 V versus lithium and shows a large reduction peak. EC is known to be reduced by accepting electrons from lithium according to the steps as follows:

Lithium ethylene dicarbonate is the main component among other surface groups due to strong interactions within O...Li−O of $(CH_2OCO_2Li)_2$. The intermediate radical anions are conjugated with the side carbonyl, and neutral ethylene gas is released at the same time. However, lithium ethylene dicarbonate also reacts with water at the ppm level and is converted to Li_2CO_3.

$$2\ R\text{-OCO-OLi} + H_2O \longrightarrow Li_2CO_3 + 2ROH + CO_2$$

As indicated in the results of the experimental thermal analysis, ethylene gas was found to be a common product in the above EC reductive decomposition reactions. The release of ethylene and other gases is observed in reductive decomposition of all organic solvents. When electrolyte decomposition is accelerated within a lithium cell, the rapid increase in the amount of gases released causes the battery to expand.

$$LiF_6 \longrightarrow Li_2CO_3 + 2ROH + CO_2$$
$$PF_5 + H_2O \longrightarrow PF_3O + 2HF$$

The small amount of water existing as an impurity in the electrolyte or adsorbed on electrodes leads to hydrolysis of lithium anions (e.g., BF_4^-, PF_6^-) and produces HF [16], which is used to decompose lithium alkyl carbonate into LiF.

$$R\text{-OCO-OLi} \xrightarrow{HF} LiF + ROH + CO_2$$

As an unstable compound, lithium alkyl carbonate may transform into another compound due to continuous reductive reactions induced by long-term contact with lithium metals.

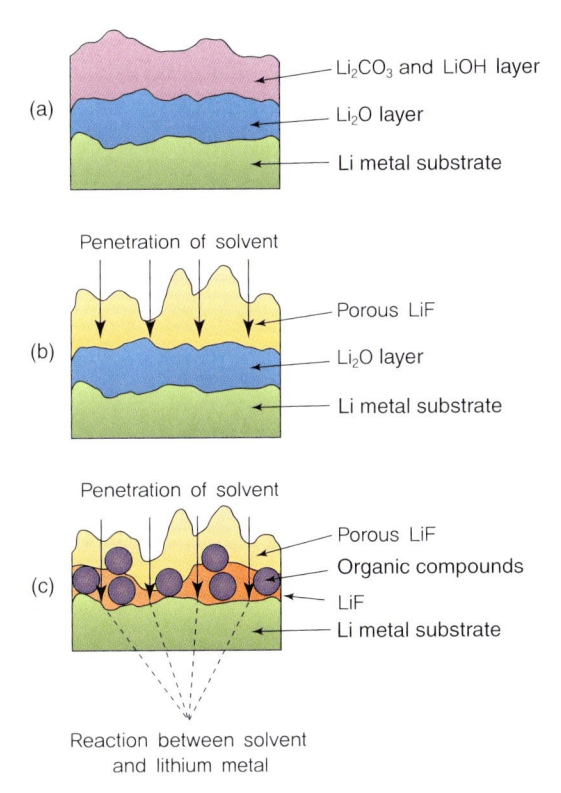

Figure 3.174 Reactions between electrolyte and lithium metal: (a) naturally produced passive layer, (b) acid–base reaction between HF and passive layer produced from electrolyte decomposition, and (c) direct reaction between solvent and lithium metal. [19]. Reproduced by permission of ECS – The Electrochemical Society.

Compounds produced from the above reductive decomposition of the electrolyte create an SEI layer on the surface of electrodes. An XPS analysis including depth profiling was used to detect components of the SEI layer by soaking lithium metal in 1 M LiBF$_4$ PC or γ-butyrolactone or after an electrochemical cycle. Lithium alkyl carbonate and other organic matter were found in the outer layer, while stable inorganic substances such as Li$_2$O or Li$_2$CO$_3$ were closer to lithium metal. LiOH, Li$_2$O, and Li$_2$CO$_3$ were all transformed into LiF with HF generated from LiBF$_4$ hydrolysis and located in the inner SEI layer. The resulting lithium-based inorganic compounds are known to coexist with naturally produced Li$_2$O, LiOH, and Li$_2$CO$_3$ at the lithium surface [19]. This means that SEI is a multilayered structure, as shown in Figure 3.174.

3.4.3.2 Interfacial Reactions at Graphite (Carbon)

Similar to SEI layer formation from reductive decomposition of organic solvents and lithium salts at the lithium–electrolyte interface, interfacial reactions take place between graphite and electrolyte, which is different from that at the lithium–electrolyte interface due to its different surface property.

Depending on the method of synthesis, graphite varies in surface structure, chemical composition, particle size and shape, pore size and distribution, degree of open pores, surface area, and types of impurities. When carbon is used as the cathode, a common factor is the large decrease in initial irreversible capacity. Since SEI layers are formed with electrolyte decomposition due to electron transfer from the graphite to electrolyte during charging, this decrease is caused by a consumption of electrons in addition to lithium intercalation into graphite [6, 19]. As such, the carbon–electrolyte interface increases for carbon with larger surface areas, thus leading to further decreases in initial irreversible capacity [20].

$LiBF_4 + H_2O \rightarrow 2HF + LiF + BOF$ (in bulk solution)
$LiOH + HF \rightarrow LiF + H_2O$ (at the lithium surface)
$Li_2O + 2HF \rightarrow 2LiF + H_2O$ (at the lithium surface)
$Li_2CO_3 + 2HF \rightarrow 2LiF + H_2CO_3$ (at the lithium surface)

Ideally, lithium ions dissolved in the electrolyte will penetrate the SEI layer for carbon intercalation while in the solvent-free state and thereby suppress additional reactions by the organic solvent. However, PC is intercalated together with lithium ions into the graphite crystal structure, thus destroying the layered structure of graphite and causing exfoliation [21].

With electrolyte decomposition, SEI layers begin to form on the graphite surface at 1.7–0.5 V and up to 0.0 V [6, 19]. The potential at which SEI layers are produced depend on various conditions such as the carbon lattice plane, basal-to-edge ratio, temperature, type of electrolyte solvent, lithium salt concentration, and applied charge density. These factors must be considered in order to maximize battery performance [21–24]. Impedance results for graphite including the SEI layer are shown in the equivalent circuit in Figure 3.175 [25].

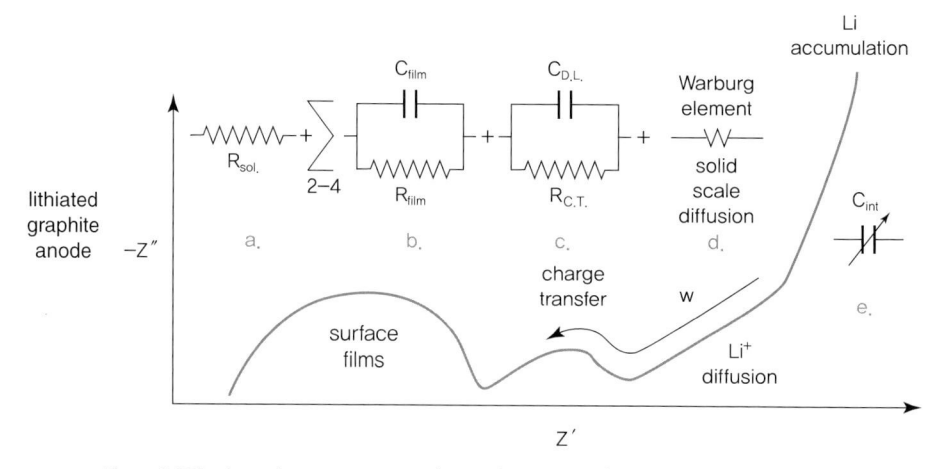

Figure 3.175 Impedance spectrum and equivalent circuit of graphite with SEI layer in the charged state [27]. Reprinted from [27] Copyright 2000, with permission from Elsevier.

The semicircle in the high-frequency range represents the resistance of lithium ions penetrating the surface layer, while the middle semicircle is the resistance during charge delivery through the surface–graphite interface. The Warburg element in the low-frequency range shows the resistance of lithium diffusion within graphite, and resistance in the lowest frequency range corresponds to the accumulation of lithium ions in the bulk. The impedance diagram illustrates the overall intercalation of lithium ions into the graphite anode.

3.4.3.3 SEI Layer Thickness

It is difficult to measure the actual thickness of SEI layers at the interface between electrodes and electrolyte. In the electrochemical method of measuring SEI layer thickness shown in Figure 3.176, CV is obtained using an electrolyte and nonactive metal electrode, and cathodic charge is given by the SEI layer on the metal surface. Measurements were then taken under the assumption that the SEI layer consists of a single compound (example: lithium ethylene carbonate) [17].

As can be seen in Figure 3.176, a reduction charge of $0.01\,C/cm^2$ for lithium ethylene dicarbonate is derived from the CV reduction peak (1.7 V). From this, we know that two electron transfers occur for each EC molecule. Assuming four electron transfers per dimer, this can be expressed as 1.56×10^{16} dimer/cm^2. If lithium ethylene dicarbonate is in the cylindrical form of 3 Å (diameter) × 20 Å (length) over the electrode surface, this figure becomes 1.67×10^{14} (dimer/cm^2)/layer, giving the surface layer a thickness of 300 Å (= 30 nm).

Recent advancements have allowed direct measurement of SEI layer thickness based on *in situ* AFM technology and observations of grain size change at the electrode surface with SEI formation [27]. From Figure 3.177, we are able to determine the SEI layer thickness by observing any change in the AFM image. In

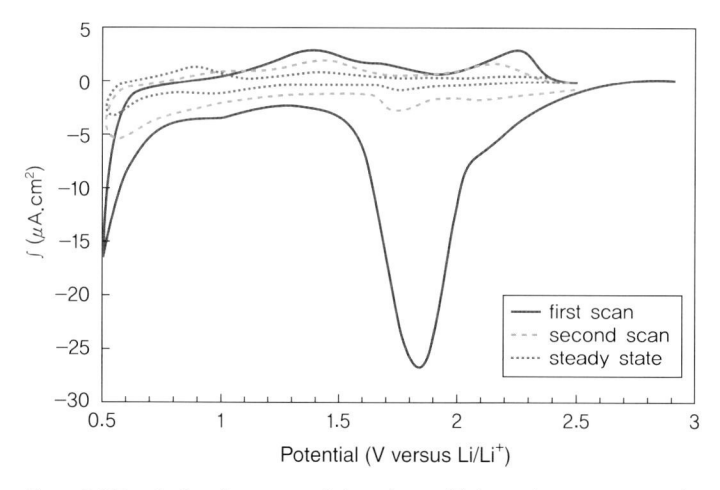

Figure 3.176 Cyclic voltammetry of electrolyte/gold electrode to measure SEI layer thickness [26]. Adapted with permission from [26] Copyright 2005 American Chemical Society.

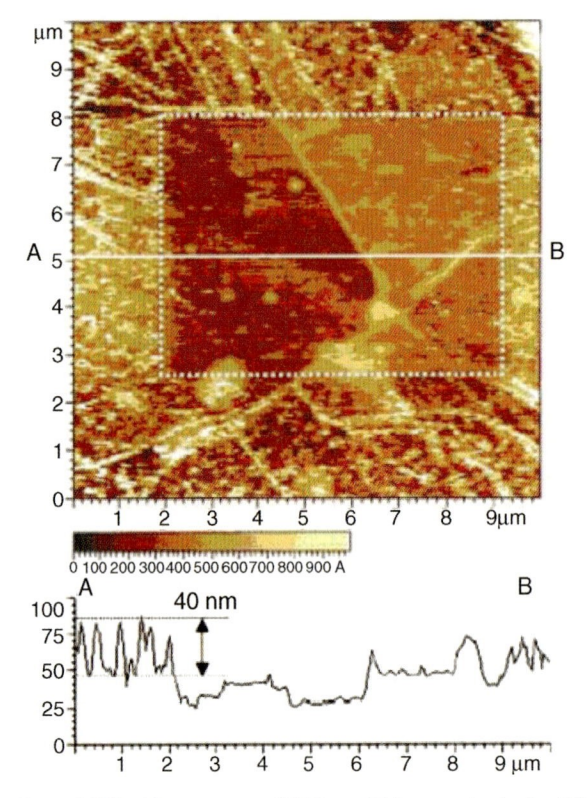

Figure 3.177 Measurement of SEI layer thickness using *in situ* AFM analysis [28]. Reproduced by permission of ECS – The Electrochemical Society.

this case, the SEI has a thickness of 40 nm after the first cycle, and expands to 70 nm after the second cycle.

The above values are similar to measurements of SEI layer thickness obtained from a different study. The SEI has a thickness of approximately 10–40 nm after the first cycle, and another layer of similar thickness is added after the second cycle [28].

3.4.3.4 Effect of Additives

PC has a limited use as an electrolyte since it is intercalated into graphite along with lithium ions and causes exfoliation during lithium deintercalation. These problems can be resolved by replacing PC with EC to induce SEI layer formation, but it results in a decrease in ionic conductivity at low temperatures. Another method to prevent graphite exfoliation was by adding small amounts of VC or lithium bisoxalato borate (LiBOB) in PC.

If PC alone is used as a solvent, LiBOB decomposition from one-electron transfer reactions can create SEI layers. The resulting compound containing carbonyl groups and lithium oxylate form passive layers on the graphite surface, thus protecting graphite from PC–graphite interfacial reactions and suppressing exfoliation [4].

The addition of LiBOB accelerates electrolyte decomposition and forms SEI layers on the surface. These layers are mainly comprised of boron and components similar to semicarbonates. Adding 1–5 mol% of LiBOB to $LiPF_6$ is known to prevent the problem of early exfoliation in graphite [29].

One of the most commonly used additives is VC. SEI characteristics of the carbon anode can be changed by adding 1% of VC to PC. A reversible capacity of 67% is achieved in the initial cycle and 93–95% in the next [30]. This is a significant improvement compared to the initial reversible capacity of 12% for the carbon cathode without VC. The decomposition of VC begins at 1.2 V versus lithium, which is before the intercalation of lithium ions. Passive layers are formed at the carbon surface with decomposition occurring at a potential higher than for PC and other organic solvents. This prevents carbon exfoliation by blocking the intercalation path of PC and lithium ions.

According to thermal and spectroscopic analyses, SEI layer components include polymer substances such as poly-VC, oligomers of VC, ring-opening polymers of VC, and polyacetylene; and lithium organic salts such as lithium vinylene carbonate $(CHOCO_2Li)_2$, lithium divinylene dicarbonate $(CH=CHOCO_2Li)_2$, lithium divinylene dialkoxide $(CH=CHOLi)_2$, and lithium carboxylate (RCO_2Li) [4].

As shown below, the decomposition of EC/DMC electrolyte produces ethylene, CO, and methane gas, while that of VC results in acetylene and CO.

$$2EC + 2e^- + 2Li^+ \rightarrow (CH_2OCO_2Li)_2 + C_2H_4$$

$$EC + 2e^- + 2Li \rightarrow (CH_2OLi)_2 + CO$$

$$DMC + e^- + Li^+ \rightarrow CH_3OCO_2Li + CH_3$$

$$DMC + 2e^- + 2Li^+ \rightarrow 2CH_3OCO_2Li + CO$$

$$(CH_3) + 2H \rightarrow CH_4$$

$$2VC + 2e^- + 2Li^+ \rightarrow LiO_2COC = COCO_2Li + C_2H_2$$

$$2VC + 2e^- + 2Li^+ \rightarrow LiOC = C - C = COLi + 2CO$$

VC is not only reduced before EC at a high potential of 1.0 V versus lithium but also raises the reduction potential of EC from 0.7 V to 0.8 V and facilitates ring-opening reactions. Thus, stable SEI layers are effectively formed on the graphite surface, and battery performance is enhanced at high temperatures.

3.4.3.5 Interfacial Reactions between a Noncarbonaceous Anode and Electrolytes

One disadvantage of commercialized graphite is its limited capacity (theoretical capacity of 372 mAh/g) before the formation of LiC_6. When a lithium cell having lithium metal oxide cathode is overcharged, lithium metal is deposited on the graphite surface. At the same time, electrolyte decomposition is stimulated and gases are released within the cell. These safety issues greatly reduce its practical applicability. Alternative anode materials are needed to resolve various problems including the decrease in irreversible capacity during initial charging and discharging. Now, Si/Sn/Sb-based metals, alloys, and carbon compounds are being studied as anode materials to replace carbon. These materials exhibit a high theoretical capacity up to several thousands of mAh/g. However, they are affected by a large decrease in initial irreversible capacity and difficulties in maintaining capacity during long-term cycling.

While various types of alloys and compounds containing Si, Sn, and Sb are being developed, there is little research on electrolyte interfacial chemistry and SEI layer analysis. Lithium alloying and dealloying are irreversible reactions accompanied by large changes in lattice volume and considered more disorderly than the intercalation/deintercalation of lithium ions into carbon. Unlike carbon, reactions between alloy substances and organic solvents will have a catalytic effect that varies with alloy components. Extreme changes in grain size are caused by reactions with lithium in alloy electrodes and may result in cracking. As such, it is important to secure mechanical flexibility of SEI layers.

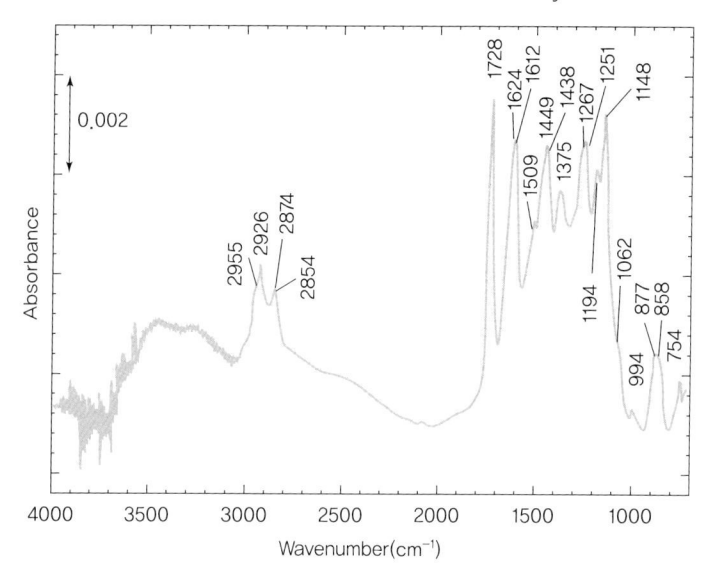

Figure 3.178 FTIR spectrum of SEI layer formed on the Cu_2Sb electrode surface after cycling.

Si/Sb-based alloys are prone to self-discharge because of the larger electrode–electrolyte interfacial area induced by volume change and particle milling during the initial cycle [31, 32]. Self-discharge involves gradual decomposition of the electrolyte and release of lithium ions from the anode due to electrode–electrolyte electron transfer even without any applied current or voltage.

By examining the surface components after cycling, we can see that the interfacial chemistry between metals and electrolytes is very different from graphite (Figure 3.178). Surface groups of Li_xCu_2Sb include $-CH_2CH_3$, $-COO-$ ester groups, saturated ester $-COOR$ group (with R alkyl chain), $-CO_2-$ carboxylate groups recognized as lithium oxalate $Li_2(O_2C)_2$ and lithium succinate $Li_2(O_2CCH_2)_2$, and $-C-O$ ether groups recognized as $LiOCH_3$ [33]. Other components are inorganic materials such as Li_2CO_3 and $-P-F-$ groups. While some of these surface groups are the same as those on the surface of graphite electrodes, additional decomposition products are formed from the catalytic activity of alloys.

LiF, Li_2CO_3, and Li_2O and small quantities of polymers were detected from an XPS analysis on a Sn–Sb–Cu–graphite alloy anode after cycling [34].

$Li_4Ti_5O_{12}$ is a representative oxide-based anode material. It is known as an ideal zero-strain material with a stable crystal structure that remains unaffected by lithium intercalation/deintercalation. Such a stabilized structure is very different from the extreme volume changes in metals and alloys. However, cathode materials and electrolytes have to be replaced to match its high working voltage of 1.5 V. Despite this disadvantage, the use of $Li_4Ti_5O_{12}$ is expected to enhance battery stability and safety. Since $Li_4Ti_5O_{12}$ operates at a potential that is 0.9 V higher than the reductive decomposition potential of $LiPF_6$/EC/PC electrolytes, SEI layers are not formed due to the lack of interfacial reactions. SEI layers may be produced when the

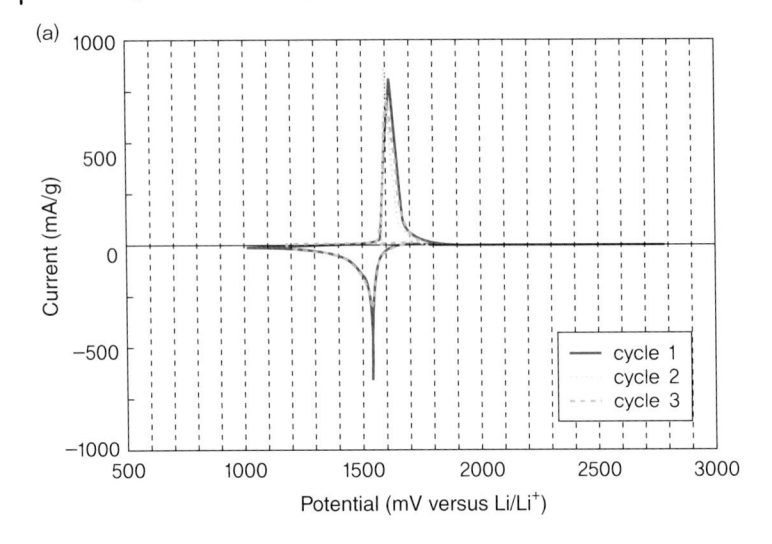

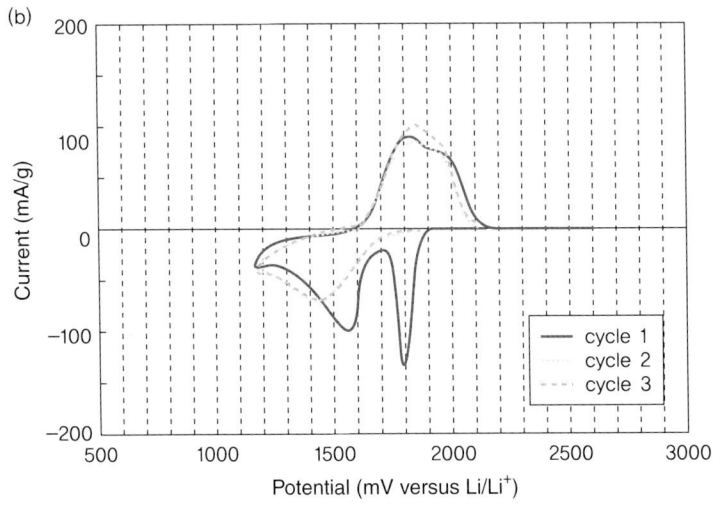

Figure 3.179 Reductive decomposition of electrolytes by lithium salt (a) LiPF$_6$ and (b) LiBOB [35]. With kind permission from Springer Science and Business Media: [35].

electrolyte containing LiBOB is decomposed at around 1.75 V. As presented in Figure 3.179, the reduction potential peaks at 1.5 V for LiPF$_6$ but increases to 1.75 V for LiBOB [35]. This shows that SEI layers may be formed on the surface of Li$_4$Ti$_5$O$_{12}$ with the use of LiBOB or other additives that decompose above 1.5 V.

3.4.4
Cathode–Electrolyte Interfacial Reactions

While there has been extensive research on the formation of SEI layers at the anode–electrolyte interface, cathode–electrolyte interfacial reactions have not been widely studied. This is because cathode materials maintain a stable crystal structure throughout lithium intercalation/deintercalation, whereas the intercalation of the electrolyte along with lithium results in exfoliation of graphite anodes. However, passive layers may be formed at the cathode from electrolyte oxidation reactions. This is similar to SEI layer formation by reductive reactions at the lithium metal and graphite surface, with the only difference being electrolyte oxidation. It is important to produce stable anode SEI layers in order to control and enhance battery performance.

3.4.4.1 Native Surface Layers of Oxide Cathode Materials

Before examining the formation of SEI layers at the cathode, we should look at the native layer existing on the cathode surface in order to differentiate between the native surface layer components and SEI components produced from chemical or electrochemical reactions (Figure 3.180. Common cathode materials such as $Li_{1-x}Ni_{1+x}O_2$, $LiCoO_2$, and $LiMn_2O_4$ are found to have a Li_2CO_3 at the surface. Li_2CO_3 is formed as a result of reaction with CO_2 absorbed from the air during electrode material synthesis [36]. It may also be soluble in the electrolyte and form native surface layers at the anode. Li_2CO_3 is produced according to the equation below and has a thickness of 10 nm [37].

When the cathode surface is covered by the insulating Li_2CO_3, some cathode particles may be electrically isolated, thus causing a decrease in battery output and capacity.

Li_2CO_3 at the cathode surface remains in a stable state when placed in nonaqueous pure solvents, but begins to be soluble in electrolytes containing HF [38]. This is due

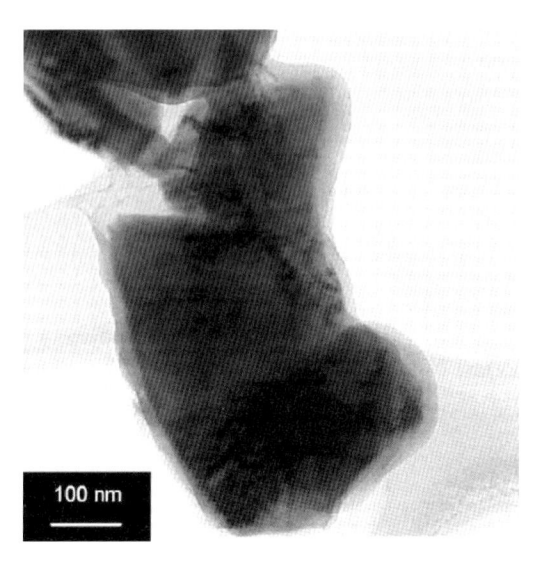

Figure 3.180 TEM image showing native layer on the cathode surface [37]. Reprinted from [37] Copyright 2004, with permission from Elsevier.

to acidic nature of lithium salts such as $LiPF_6$ and $LiAsF_6$ and the production of HF acid from reactions with small amounts of water in the electrolyte.

$$2HF + Li_2CO_3 \rightarrow 2LiF + H_2O + CO_2$$

3.4.4.2 SEI Layers of Oxide Cathodes

While the cathode shows a relatively small decrease in initial irreversible capacity compared to the graphite anode, an EIS analysis confirmed the presence of SEI layers at the cathode [39]. Cycling of a lithium cell using $Li_{1-x}CoO_2$ as an anode material shows a gradual increase in the cathode surface thickness and internal resistance of the cell [40]. Co^{4+} with high oxidizing power is formed at the $Li_{1-x}CoO_2$ ($x=0.34$) anode surface around 4.5. Co^{4+}–electrolyte electron transfer occurs with the mixing of PC and $Li_{1-x}CoO_2$ particles. This accelerates the oxidative decomposition of PC and produces a thick polymer surface area. Figure 3.181 shows the Randles circuit of the anode–electrolyte interface having an SEI layer.

Similar to the anode, additional SEI layers are formed after the first layer from electron transfer between lithium ions passes through the first SEI layer and makes contact with the cathode surface. Repeated cycling increases the internal resistance of the cell and may affect battery performance, battery life, and thermal stability.

3.4.4.3 Interfacial Reactions at Oxide Cathodes

Electrochemical reactions are not triggered by $LiMn_2O_4$ and other cathode materials, and SEI layers can be easily formed by chemical reactions during contact with the electrolyte [41, 42]. Making contact with the electrolyte changes the HOMO and LUMO levels of oxides, as was the case of the anode. Electron transfer takes place from the electrolyte HOMO to the conduction band of oxides, thus forming a negative charge on the oxide surface and a positive charge for the electrolyte. Compounds produced from oxidative decomposition of the electrolyte are deposited on the cathode surface as a passive layer. When a lithium cell is charged, lithium ions are deintercalated from the cathode and delivered to the anode through an external circuit. The valence band level is lowered and comes closer to the HOMO energy level, which facilitates electron transfer from the electrolyte to the cathode and accelerates both oxidative decomposition and SEI layer formation. SEI layers resulting from chemical reactions with the electrolyte are comprised of polyether, alkyl carbonate, LiF, Li_xPF_y, and $Li_xPF_yO_z$.

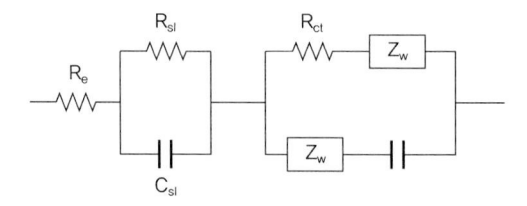

Figure 3.181 Randles circuit of the cathode–electrolyte interface with SEI layer [40].

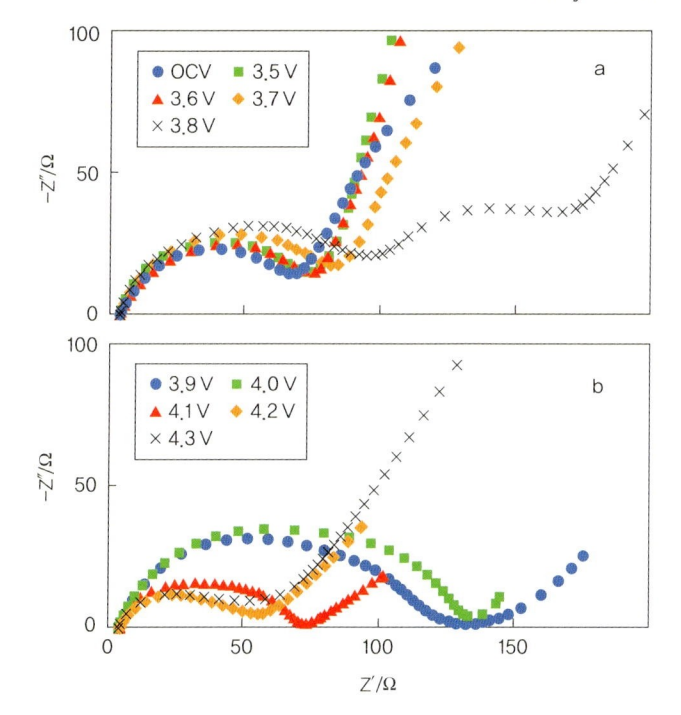

Figure 3.182 Nyquist plot measured (under constant voltage) for a charged Li/LiMn$_2$O$_4$ cell in 1 M LiPF$_6$/EC + EMC electrolyte [42].

In general, the initial irreversible capacity is reduced by 10% with the formation of SEI layers. In addition, the growth of SEI layers during charging and discharging leads to an increase in internal resistance. With repeated cycling, resistance is increased while capacity continues to shrink. As shown in Figure 3.182, the semicircle in the high-frequency range corresponding to the passive layer and interfacial resistance grows larger until a charge potential of 3.8 V, but subsequently drops up to 4.3 V. This indicates that SEI layer formation changes according to charge potential.

SEI layers on the surface of LiMn$_2$O$_4$ are less stable compared to that of the graphite anode. The stability of LiMn$_2$O$_4$ affects interfacial reactions and contributes to the disproportionation of Mn^{3+} in LiMn$_2$O$_4$. If LiMn$_2$O$_4$ is placed in pure DMC solvent, DMC is oxidized by LiMn$_2$O$_4$, and part of LiMn$_2$O$_4$ is reduced to Mn$_2$O$_3$ as follows [43]:

$$LiMn_2O_4 + CH_3OCO_2CH_3 \rightarrow Mn_2O_3 + CH_3OCO_2Li + CH_3OLi$$

$$LiMn_2O_4 + CH_3OCO_2CH_3 \rightarrow Mn_2O_3 + [CH_3OCH_2] + [OLi]^- CO_2$$

Reactions with DMC produce intermediate substances that engage in polymerization to form lithium alkoxide, and DMC is continuously attacked.

$$[CH_3OCH_2] + [OLi]^- \rightarrow CH_3OCH_2OLi$$

$$CH_3OCH_2OLi + CH_3OCO_2CH_3 \rightarrow (CH_3O)_2(CH_3OCH_2O)COLi$$

While LiMn$_2$O$_4$ oxidizes the electrolyte solvent, the resulting LiMn$^{3+}_2$O$_4^-$ from reductive reactions may be decomposed into MnO and Li$_2$MnO$_3$, MnO, Li$_2$O, or λ-MnO$_2$. Due to the porosity of SEI layers, λ-MnO may be dissolved out to the electrolyte according to the equation below. In practice, the dissolution of λ-MnO was observed at temperatures above 60 °C.

$$\text{LiMn(III)Mn(IV)O}_4 + e^- \rightarrow \text{LiMn(III)}_2\text{O}_4^-$$
$$\rightarrow 0.5\text{Li}_2\text{O} + 1.5\lambda\text{-Mn(II)} + 0.5\,\text{Mn(II)O}$$

Besides chemical reactions leading to SEI layer formation and interfacial reactions at the cathode, electrochemical reactions also have a significant impact on battery performance. As mentioned above, PC undergoes ring-opening oxidative reactions [40] and begins to be electrochemically oxidized at 4.1 V. It is deposited on the surface of LiCoO$_2$ as an organic material containing carboxylics, dicarboxylic acid anhydride, $-$CH$_2-$, and $-$CH$_3$ groups [44]. This effect becomes more prominent in LiMn$_2$O$_4$. Cycling the lithium cell in the range of 3.8–4.4 V significantly reduces the reversible capacity. This is due to increased cell resistance caused by SEI layers at the cathode surface produced from oxidative decomposition of the electrolyte [45]. As described in the equation below, self-discharge occurs when the electrolyte transfers electrons to electrodes while engaging in oxidative decomposition, and lithium ions are intercalated into LiMn$_2$O$_4$ to achieve charge balance. Decomposition products of the electrolyte are then deposited on the surface of LiMn$_2$O$_4$ to form SEI layers.

$$\text{E1} \rightarrow e^- + \text{E1}^+ \rightarrow e^- + \text{reaction products}$$

$$\text{Li}_x\text{Mn}_2\text{O}_4 + \gamma Li^+ + \gamma e^- \rightarrow \text{Li}_{x+y}\text{Mn}_2\text{O}_4$$

Similar reactions to LiMn$_2$O$_4$ occur in Li$_{1-x}$Ni$_{1+x}$O$_2$ and LiCoO$_2$ for SEI layer formation. As shown in Figure 3.183, SEI layers are comprised of polycarbonates, ROCO$_2$Li, ROLi, LiF, and P$-$F/As$-$F groups. Interfacial reactions between metal oxides and organic solvents follow the equation below [46].

As shown in the following equation, EC engages in nucleophilic reactions to produce lithium alkyl carbonates, which are then polymerized to yield polycarbonates [6].

LiNiO$_2$ + (ethylene carbonate) ⟶ NiO$_2-$CH$_2$CH$_2$O$-$C(=O)$-$OLi

LiNiO$_2$ + (dimethyl carbonate, H$_3$CO$-$C(=O)$-$OCH$_3$) ⟶ NiO$_2-$CH$_3$ + CH$_3$OCO$_2$Li
⟶ NiO$_2-$CH$_2$Li + CH$_3$OLi

Adapted with permission from American Chemical Society Copyright 2004.

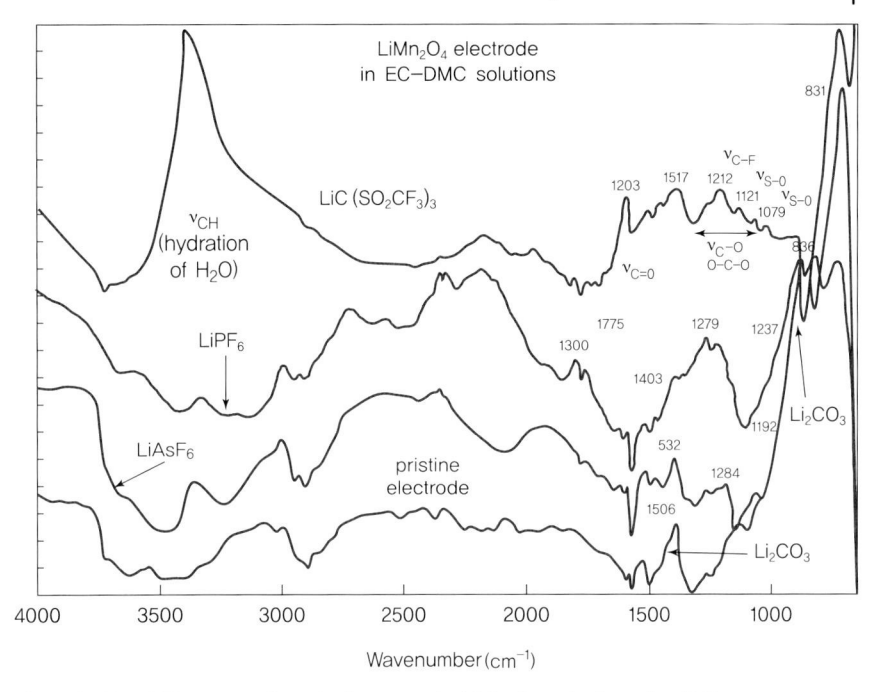

Figure 3.183 FTIR spectra of passive layers on the $LiMn_2O_4$ cathode surface [39].

$$RO^- + EC \rightarrow ROCH_2CH_2OCO_2^{-(EC)}$$
$$\rightarrow ROCH_2CH_2OCO_2CH_2CH_2OCO_2^{-(EC)} \rightarrow$$

Adapted with permission from American Chemical Society Copyright 2004.

LiF is commonly found in SEI layers because small amounts of water react with $LiPF_6$ to form HF, which reacts with lithium ions in the electrolyte or at the electrode surface to produce LiF, as shown in the equation as follows:

$$LiCoO_2 + 2x\,HF \rightarrow 2x\,LiF + Li_{1-2x}CoO_2$$

Since LiF acts as an insulator, an increase in the concentration of LiF leads to a higher EIS of the cathode. In addition to SEI layer formation, Mn^{3+} in $LiMn_2O_4$ undergoes disproportionation reactions [47].

$$2Mn^{3+}_{solid} \rightarrow Mn^{4+}_{soild} + Mn^{2+}_{solution}$$

Electrolyte salts and solvents must be carefully selected as disproportionation is closely related to electrolyte acidity. For lithium salts, the extent of decomposition induced by disproportionate manganese can be arranged in ascending order as shown below [48].

$$LiCF_3SO_3 < LiPF_6 < LiClO_4 < LiAsF_6 < LiBF_4$$

Besides disproportionation reactions, $Mn^{3+}/^{4+}$ of $LiMn_2O_4$ is reduced to Mn^{2+} by electron transfer from the electrolyte to $LiMn_2O_4$. Mn^{2+} dissolution then occurs. This not only affects the cathode structure but also impacts the anode performance since Mn^{2+} passes through the electrolyte and is adsorbed on the anode in the form of Mn metal.

$$Mn^{2+} + 2LiC_6 \rightarrow Mn + 2Li^+ + 2C_6$$

The potential for oxidative decomposition of the electrolyte and various reactions differ according to electrolyte components and cathode type. When PC is used, $LiNiO_2$ releases gas from 4.2 V, while $LiCoO_2$ and $LiMn_2O_4$ begin at 4.8 V. In PC/DMC solvents, CO_2 is released only from $LiNiO_2$ at 4.2 V [49].

Lithium intercalation/deintercalation takes place to stabilize the cathode structure, and this can be observed in $Li_{1-x}Ni_{1+x}O_2$. The nonstoichiometric $Li_{1-x}Ni_{1+x}O_2$ exists in a semilayered form due to cation mixing of atoms from Li and Ni layers. When charged up to 4.2 V (or when more than 0.6 lithium ions are intercalated), cation mixing occurs (Ni^{2+} is formed) in the cathode of $Li_{1-x}Ni_{1+x}O_2$, and the unstable oxide released from the cathode causes oxidative decomposition of the electrolyte. This results in SEI layers comprised of organic matter such as dicarbonyl anhydride and polyester [37]. As can be seen in Figure 3.184, the decrease in the initial irreversible capacity contributes to an irreversible structural change and SEI layer formation.

Active cation mixing in the initial cycle causes a decrease in the concentration of $Ni^{3+}/^{4+}$, an increase in Ni^{2+}, and continuous release of oxygen to maintain charge balance. The release of oxygen in turn generates heat and may lead to ignition of other components (e.g., electrolyte, binder, and organic matter). Since these interfacial reactions are triggered by electron or oxygen transfer, changes in Ni electron structure in the cathode are directly related to oxidative decomposition of the electrolyte and SEI layer formation. As such, the decrease in initial irreversible capacity observed in $Li_{1-x}Ni_{1+x}O_2$ can be traced to structural change of the cathode and interfacial reactions with the electrolyte [51].

To obtain a more accurate understanding of the electrode interface in lithium batteries, we need to look at the surface of metal foils that are being used as current collectors. The metal surface is always coated with Al_2O_3 because cathode current collectors made of Al are easily oxidized. As shown in Figure 3.185, the surface of

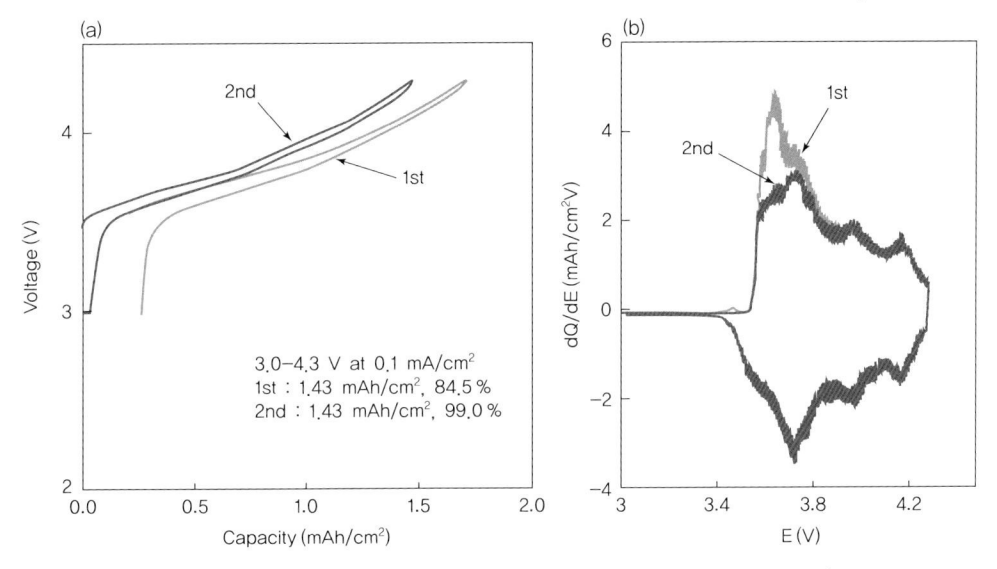

Figure 3.184 (a) Initial potential–capacity curve and (b) differential capacity–potential curve of a Li/LiNiO$_2$ cell obtained at a charge density of 0.1 mA/cm^2 and OCV 3.0–4.3 V [50].

commercialized Al metal is coated with polyamides or other substances to prevent oxidation [37].

When conducting FTIR and organic analyses of SEI layers, coating layers on current collectors and binder polymers must be analyzed beforehand to obtain more accurate information on SEI composition.

3.4.4.4 Interfacial Reactions of Phosphate Cathode Materials

A new cathode material under study, LiFePO$_4$ is usually coated with carbon due to its low electronic conductivity. Its low working voltage results in different interfacial characteristics compared to other metal oxides. Although naturally present on the surface of other cathode materials, Li$_2$CO$_3$ cannot be seen on LiFePO$_4$. This implies that phosphate groups do not undergo reactions in air [50–54]. Instead, lithium iron oxide (Li$_x$Fe$_y$O$_z$) produced during synthesis can be found in small amounts (<2wt%) on the surface of LiFePO$_4$ [50, 53]. The lithium iron oxide existing on the surface releases lithium when first charged and increases the charge capacity of LiFePO$_4$. Since this is an irreversible reaction that deactivates iron oxide, discharge capacity is obtained from LiFePO$_4$ only. The initial irreversible charge capacity is thus greatly reduced. Meanwhile, LiFePO$_4$ that has been synthesized using the sol–gel method is contaminated with surface impurities such as FeP and Li$_3$PO$_4$, which are likely to have been produced during thermal treatment [55]. Compound types and concentration differ according to the method of synthesis, thereby affecting the initial electrochemical cycle characteristics of LiFePO$_4$.

HF acid is formed from small quantities of water in the electrolyte, and may impede LiFePO$_4$ electrode performance, as was the case for other lithium metal oxides. For example, HF is produced when LiPF$_6$ is decomposed with water, and Fe is

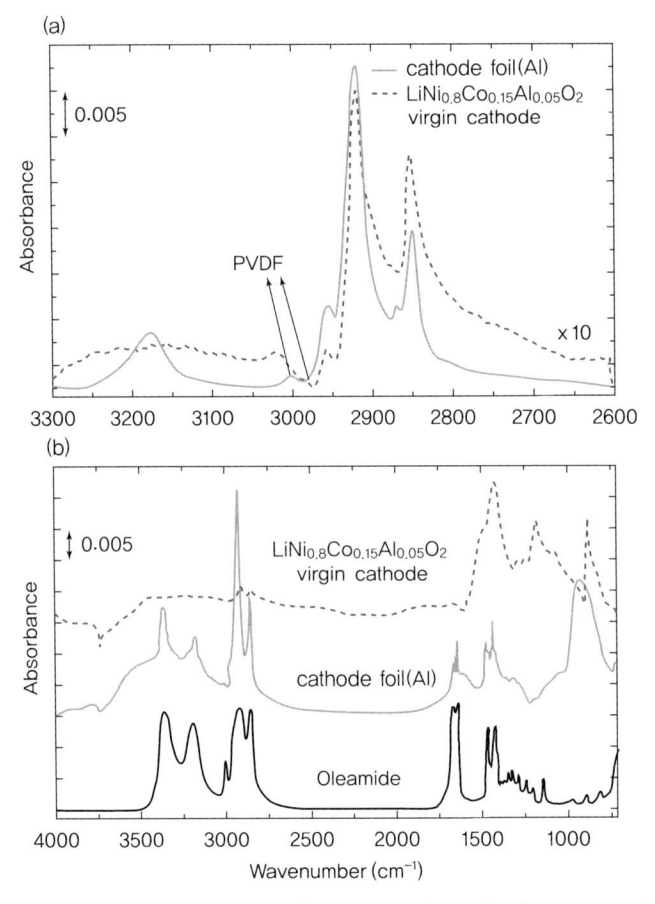

Figure 3.185 (a) Comparison of C−H groups observed in the spectrum of the cathode surface and coating layer on the Al current collector foil. (b) Comparison of reference spectra for the cathode, Al current collector foil, and oleamide.

eluted by attacking LiFePO$_4$. The Fe dissolution not only decreases the irreversible capacity of LiFePO$_4$ but Fe may also be adsorbed on the anode surface and result in a loss of irreversible capacity at the anode [56].

As shown in Figure 3.186, an FTIR analysis of the LiFePO$_4$ surface during electrochemical reactions shows that SEI layers consist of −OCOCO−, −CO$_2$M (M = metal) organic groups, and small quantities of −PF$_x$ or CO$_3^{2-}$ inorganic salts [57]. From the relatively weak organic bands compared to other oxide-based cathodes, we can expect thin SEI layers or a low concentration of compounds. By conducting an XPS analysis, SEI layers were found to have organic compounds containing CH, C=O, C−O groups, and phosphorus compounds having Li−F, P−F, or O−P−F bonds [58]. Electron transfer between the anode and the electrolyte is less active in the low working voltage (2.5/3.0–4.0 V) of LiFePO$_4$, thus leading to less significant oxidative decomposition of the electrolyte on the LiFePO$_4$ surface. Besides

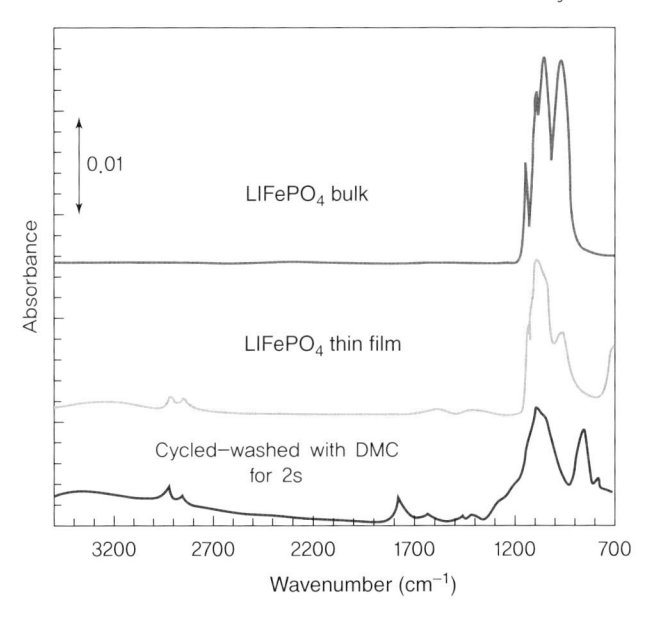

Figure 3.186 Comparison of FTIR spectra for LiFePO$_4$ bulk and LiFePO$_4$ film surface obtained after cycling in the 2.5–4.0 V range followed by DMC washing.

its structural stability, LiFePO$_4$ exhibits stable cycle performance based on the aforementioned interfacial structure and characteristics.

3.4.5
Current Collector–Electrolyte Interfacial Reactions

To fabricate the anode and cathode of lithium batteries using active materials, aluminum (Al) and copper (Cu) metal foil current collectors must be each covered with a mixture of bulk active material, binder, and carbon. When a lithium cell is charged, copper metal coated with anode materials approaches a low potential (close to the potential of lithium metal). Despite being chemically stable, repeated charging and discharging over a long period of time may result in physical cracking [59].

On the other hand, aluminum metal experiences chemical corrosion at high potentials. This becomes more severe with long-term cycling and causes an increase in cell internal resistance.

3.4.5.1 Native Layer of Aluminum
Since aluminum metal is thermodynamically unstable with an oxidation potential of 1.39 V [60], its surface is usually covered with thermodynamically stable Al$_2$O$_3$, oxyhydroxide, or hydroxide [61, 62]. Due to this native layer, aluminum metal is able to attain stability and resistance to corrosion. The oxidation stability limit of aluminum metal in 1.0 M LiClO$_4$/EC/DME electrolyte is 4.2 V [63].

Active materials in bulk form are porous substances, and aluminum metal exhibits porous characteristics despite its Al$_2$O$_3$ surface layer. When the cathode coated on aluminum is placed in an electrolyte, the electrolyte passes through pores between cathode particles and makes direct contact with the metal, thus forming the aluminum–electrolyte interface. When a lithium cell is charged, the anode and current collector must be maintained at high potential. Furthermore, repeated cycling (with increased charging at high potential) over long periods may lead to corrosion of aluminum. Aluminum–electrolyte interfacial reactions are directly related to battery safety, as pitting corrosion breaks down the aluminum and weakens contact between electrodes and the current collector. This not only shortens anode life but may also induce short circuits.

3.4.5.2 Corrosion of Aluminum

Aluminum corrosion is known to be affected by electrolyte components. PC/DEC solvent causes less corrosion than EC/DMC, but it is lithium salts that have a greater impact [64]. Compared to other salts, lithium bis-perfluoroalkylsulfonylimide (LiN$(SO_2CF_3)_2$) is superior in terms of ionic conductivity, cycle characteristics, thermal stability, and hydrolytic stability. However, it tends to accelerate aluminum corrosion [65] and leaves behind carbonate compounds such as LiF and Li_2CO_3 on the aluminum surface. To determine aluminum corrosion or oxidation stability, linear sweep voltammetry is performed over a wide potential range including the cathode working potential (approximately, 2.5–4.5 V). Figure 3.187 shows the stability of aluminum in various imide-based electrolytes.

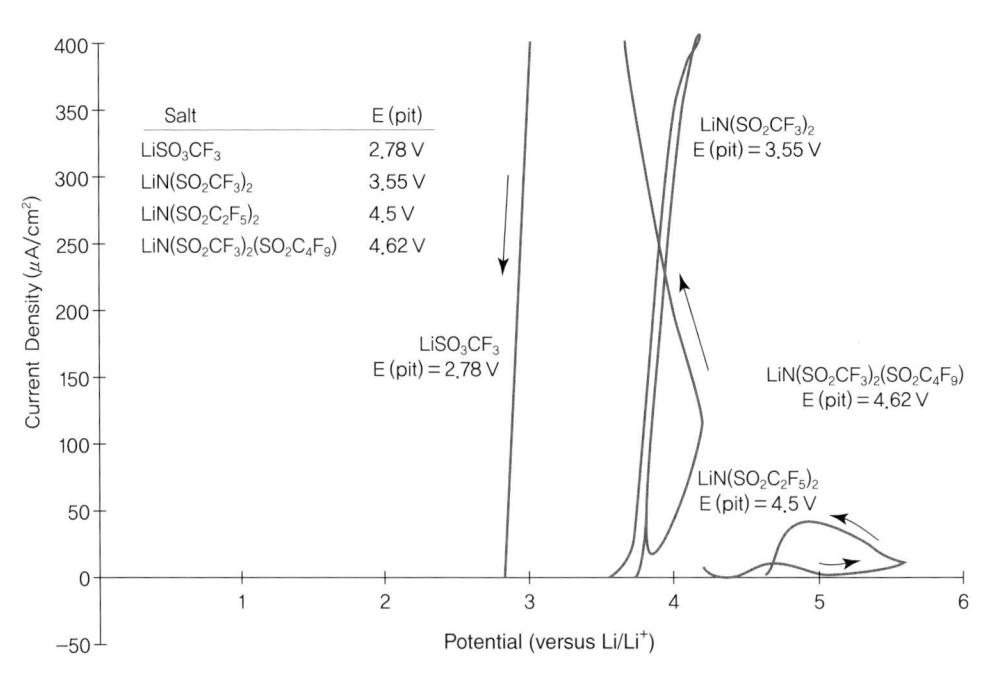

Salt	E (pit)
LiSO$_3$CF$_3$	2.78 V
LiN(SO$_2$CF$_3$)$_2$	3.55 V
LiN(SO$_2$C$_2$F$_5$)$_2$	4.5 V
LiN(SO$_2$CF$_3$)$_2$(SO$_2$C$_4$F$_9$)	4.62 V

Figure 3.187 Corrosion potential of various sulfonate, sulfonyl amide lithium salts in 1.0 M EC–PC solvent obtained using linear sweep voltammetry [65]. Reprinted from [65] Copyright 1997, with permission from Elsevier.

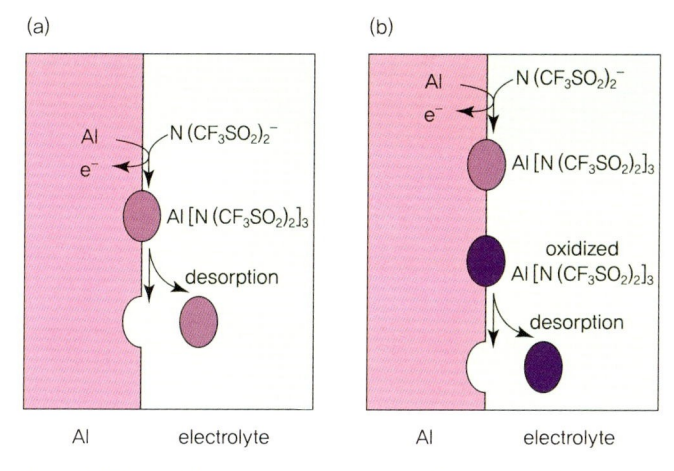

Figure 3.188 Possible corrosion mechanism of aluminum in LiN(CF$_3$SO$_2$)$_2$/PC electrolyte [6, 67]. Adapted with permission from [6, 67] Copyright 2004 American Chemical Society.

From Figure 3.187 above, we can see that the stability of aluminum toward oxidative corrosion can be arranged in ascending order as follows [65]:

$$LiCF_3SO_3 < LiN(SO_2CF_3)_2 < LiClO_4 < LiPF_6 < LiBF_4$$

Imide-based lithium salts such as LiCF$_3$SO$_3$ and LiN(SO$_2$CF$_3$)$_2$ induce aluminum corrosion at potentials below 3.8 V whereas Li(C$_4$F$_9$SO$_2$)CF$_3$SO$_2$N creates a passive layer [66].

As for imide-based electrolytes, N(SO$_2$CF$_3$)$_2^-$ anions remain stable up to 4.5 V without oxidation and react with Al^{3+} eluted from aluminum metal to form a Al[N(CF$_3$SO$_2$)$_2$] complex. As shown in Figure 3.188, Al[N(CF$_3$SO$_2$)$_2$] is adsorbed on the aluminum surface, and part of this deposition may dissolve in the electrolyte and cause corrosion by producing pits on the surface of aluminum metal [67].

In addition to reactions with imide-based electrolyte components, Table 3.16 presents other oxidative reactions that occur on the aluminum surface [60].

Table 3.16 Possible anodic reactions in organic electrolyte solutions and the expected mass change. Reprinted with permission from American Chemical Society Copyright 2002.

Anodic reaction	Mass change (g F^{-1})
Al $\rightarrow$ Al^{3+} + 3e$^-$	-9
2Al + 3H$_2$O $\rightarrow$ Al$_2$O$_3$ + 6H$^+$ + 6e$^-$	$+8$
Al + 3H$_2$O $\rightarrow$ Al(OH)$_3$ + 3H$^+$ + 3e$^-$	$+17$
Al(OH)^3 + HF $\rightarrow$ AlOF + 2H$_2$O	$+12$
Al + 3HF $\rightarrow$ AlF$_3$ + 3H$^+$ + 3e$^-$	$+19$
Al + 3F$^-$ $\rightarrow$ AlF$_3$ + 3e$^-$ {Al + (CF$_3$SO$_2$)$_2$N$^-$ $\rightarrow$ AlF$_3$ + 3e$^-$ + [(CF$_{1.5}$SO$_2$)2N$^-$]}	$+19$
Al + 3[(CF$_3$SO$_2$)$_2$N$^-$ $\rightarrow$ Al[(CF$_3$SO$_2$)$_2$N]$_3$ + 3e$^-$	$+283$

1F = 96485 Coulomb.

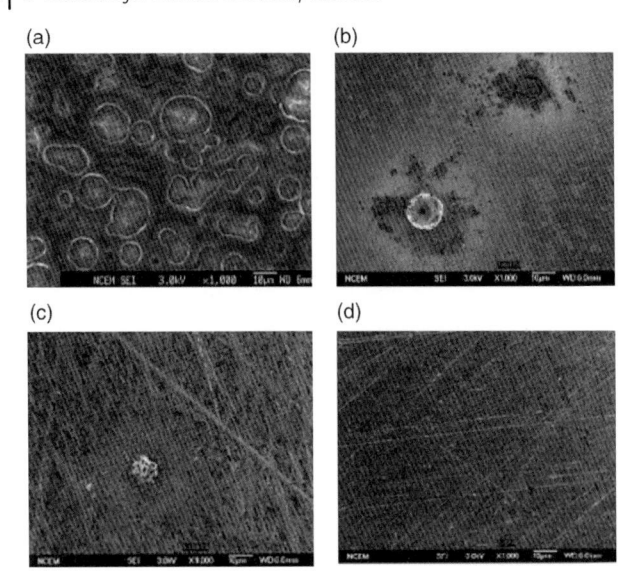

Figure 3.189 Effect on corrosion and passive layer formation with varying composition of $LiBF_4$ and LiTFSI salts: (a) 1.0 M LiTFSI/EC + DMC electrolyte, (b) LiTFSI:$LiBF_4$ = 8: 2, (c) LiTFSI: $LiBF_4$ = 5: 5, and (d) 1 M $LiBF_4$/EC + DMC electrolyte [45]. Reprinted from [45] Copyright 2004, with permission from Elsevier.

3.4.5.3 Formation of Passive Layers on Aluminum Surface

As can be seen in Figure 3.189, $LiPF_6$ and $LiBF_4$ form passive layers on the surface of aluminum, whereas imide-based salts induce corrosion. In both cases, deposits are produced on the aluminum surface [60, 67, 68]. In particular, $LiBF_4$ creates a more stable passive layer on aluminum and enhances cycle characteristics even under unfavorable conditions such as high temperature and several hundred ppm of water [69, 70]. $LiBF_4$ creates a passive layer while $LiN(CF_3SO_2)_2$ causes pitting corrosion, but these characteristics can be combined to form $LiN(CF_3SO_2)_2$. As shown in Figure 3.189, corrosion is suppressed and internal resistance is improved with the formation of a stable passive layer [71].

The corroded metal surface was found to contain $-SO_2-$ and $-CF_3$ groups of TFSI and $-OH$ groups. Passive layer components include $-CH_2CH_3$, $-OH$, carboxyl groups, ester groups, and B−F groups. Organic matter such as $CH_3CH_2-CO_2M$, $-COOR$, and lithium oxalate ($Li_2C_2O_4$), as well as LiOH and inorganic matter having B−F bonds were detected. Since these surface compounds are similar to SEI layer components of the cathode, the formation of passive layers is likely to have occurred through a similar mechanism. LiBOB is also known to create passive layers but their composition is still largely unknown [72].

The aforementioned changes in interfacial reactions by lithium salts apply only to aluminum with an uncoated cathode. For coated cathodes such as $LiMn_2O_4$ and $LiFePO_4$ in an electrolyte containing $LiPF_6$, oxidation occurs at a high potential of 5.0–6.5 V versus lithium and eventually results in corrosion [73].

References

1 Winter, M., Besenhard, J.O., Spahr, M.E., and Novak, P., (1998) *Adv. Mater.*, **10**, 725.
2 Peled, E., Golodnitsky, D., Menachem, C., and Bar Tow, D., (1998) *J. Electrochem. Soc.*, **145**, 3483.
3 Campbell, S.A., Bowes, C., and McMillan, R.S., (1990) *J. Electroanal. Chem.*, **284**, 195.
4 Zhang, X., Kostecki, R., Richardson, T.J., Pugh, J.K., and Ross, P.N., Jr., (2001) *J. Electrochem. Soc.*, **148**, A1341.
5 Wang, Y., Nakamura, S., Ue, M., and Balbuena, P.B., (2001) *J. Am. Chem. Soc.*, **123**, 11708.
6 Xu, K., (2004) *Chem. Rev.*, **104**, 4303.
7 Kanamura, K., Toriyama, S., Shiraishi, S., and Takehara, Z., (1995) *J. Electrochem. Soc.*, **142**, 1383.
8 Zhang, X., Pugh, J.K., and Ross, P.N., Jr., (2001) *J. Electrochem. Soc.*, **148**, E183.
9 Thompson, J.B., Brown, P., and Djerassi, C., (1966) *Tetrahedron*, **1**, 241.
10 Peled, E., (1979) *J. Electrochem. Soc.*, **126**, 2047.
11 Gabano, J.P., (1983) *Lithium Batteries*, Academic Press, New York, p. 43.
12 Peled, E., Golodnitsky, D., and Ardel, G., (1997) *J. Electrochem. Soc.*, **144**, L208.
13 Kanamura, K., Shiraishi, S., and Takehara, Z., (1995) *Chem. Lett.*, **24**, 209.
14 Selim, R., and Bro, P., (1974) *J. Electrochem. Soc.*, **121**, 1467.
15 Day, A.N. and Sullivan, B.P., (1970) *J. Electrochem. Soc.*, **117**, 222.
16 Aurbach, D., Markovsky, B., Shechter, A., and Ein-Eli, Y., (1996) *J. Electrochem. Soc.*, **143**, 3809.
17 Zhuang, G.V., Yang, H., Ross, P.N., Jr., Xu, K., and Richard Jow, T., (2006) *Electrochem. Solid State Lett.*, **9**, A64.
18 Du Pasquier, A., Disma, F., Bowmer, T., Gozdz, A.S., Amatucci, G., and Tarascon, J.M., (1998) *J. Electrochem. Soc.*, **145**, 472.
19 Kanamura, K., Tamura, H., Shiraishi, S., and Takehara, Z., (1995) *J. Electrochem. Soc.*, **142**, 340.
20 Fong, R., Sacken, U., and Dahn, J.R., (1990) *J. Electrochem. Soc.*, **137**, 2009.
21 Peled, E., Menachem, C., Bar-Tow, D., and Melman, A., (1996) *J. Electrochem. Soc.*, **143**, L4.
22 Besenhard, J.O., Winter, M., Yang, J., and Biberacher, W., (1993) *J. Power Sources*, **54**, 228.
23 Guyomard, D., and Tarascon, J.M., (1993) US Patent 5192629.
24 Shu, Z.X., McMillan, R.S., and Murray, J., (1993) *J. Electrochem. Soc.*, **140**, 922.
25 Dahn, J.R. et al. (1994) *Lithium Batteries: New Materials, Development and Perspectives*, Elsevier, p. 22.
26 Zhuang, G.V., Xu, K., Yang, H., Richard Jow, T., and Ross, P.N., Jr., (2005) *J. Phys. Chem. Soc.*, **109**, 17567.
27 Aurbach, D., (2000) *J. Power Sources*, **89**, 206.
28 Jeong, S.K., Inaba, M., Abe, T., and Ogumi, Z., (2001) *J. Electrochem. Soc.*, **148**, A989.
29 Xu, K., Lee, U., Zhang, S., Wood, M., and Richard Jow, T., (2003) *Electrochem. Solid State Lett.*, **6**, A144.
30 Xu, K., Zhang, S., and Richard Jow, T., (2005) *Electrochem. Solid State Lett.*, **8**, A365.
31 Ota, H., Sakata, Y., Inoue, A., and Yamaguchi, S., (2004) *J. Electrochem. Soc.*, **151**, A1659.
32 Song, S.W., Striebel, K.A., Reade, R.P., Roberts, G.A., and Cairns, E.J., (2003) *J. Electrochem. Soc.*, **150**, A121.
33 Song, S.W., Reade, R.P., Cairns, E.J., Vaughey, J.T., Thackeray, M.M., and Striebel, K.A., (2004) *J. Electrochem. Soc.*, **151**, A1012.
34 Ulus, A., Rosenberg, Y., Burstein, L., and Peled, E., (2002) *J. Electrochem. Soc.*, **149**, A635.
35 Wachtleri, M., Wohlfahrt-Mehrensi, M., Bele, S.S., Panitz, J.C., and Wietelmann, U., (2006) *J. Appl. Electrochem.*, **36**, 1199.
36 Aurbach, D., Levi, M.D., Levi, E., Markovsky, B., Salitra, G., Teller, H., Heider, U., and Hilarius F V., (1997) Batteries for Portable Applications and Electric Vehicles, PV 97-18, The Electrochemical Society Proceedings Series, Pennington, NJ, p. 941.
37 Zhuang, G.V., Chen, G., Shim, J., Song, X., Ross, P.N., and Richardson, T.J., (2004) *J. Power Sources*, **134**, 293.

38. Song, S.W., Zhuang, G.V., and Ross, P.N., (2004) *J. Electrochem. Soc.*, **151**, A1161.
39. Aurbach, D., Gamolsky, K., Markosky, B., Salitra, G., Gofer, Y., Heider, U., Oesten, R., and Schmidt, M., (2000) *J. Electrochem. Soc.*, **147**, 1322.
40. Thomas, M.G.S.R., Bruce, P.G., and Goodenough, J.B., (1985) *J. Electrochem. Soc.*, **132**, 1521.
41. Levi, M.D., Salitra, G., Markovsky, B., Teller, H., Aurbach, D., Heider, U., and Heider, L., (1999) *J. Electrochem. Soc.*, **146**, 1279.
42. Zhang, S.S., Xu, K., and Jow, T.R., (2002) *J. Electrochem. Soc.*, **149**, A1521.
43. Eriksson, T., Andersson, A.M., Bishop, A.G., Gejke, C., Gustafsson, T., and Thomas, J.O., (2002) *J. Electrochem. Soc.*, **149**, A69.
44. Matsuo, Y., Kostecki, R., and McLarnon, F., (2001) *J. Electrochem. Soc.*, **148**, A687.
45. Kanamura, K., Toriyama, S., Shiraishi, S., Ohashi, M., and Takehara, Z., (1996) *J. Electroanal. Chem.*, **419**, 77.
46. Guyomard, D., and Tarascon, J.M., (1992) *J. Electrochem. Soc.*, **140**, 3071.
47. Aurbach, D., Gamolsky, K., Markosky, B., Salitra, G., Gofer, Y., Heider, U., Oesten, R., and Schmidt, M., (2000) *J. Electrochem. Soc.*, **147**, 1322.
48. Hunter, J.C., (1981) *J. Solid State Chem.*, **39**, 142.
49. Jang, D.H. and Oh, S.M., (2002) *J. Electrochem. Soc.*, **144**, 3342.
50. Zhang, S.S., Xu, K., and Jow, T.R., (2002) *Electrochem. Solid State Lett.*, **5**, A92.
51. Imhof, R. and Novak, P., (1999) *J. Electrochem. Soc.*, **146**, 1702.
52. Song, S.W., Reade, R.P., Kostecki, R., and Striebel, K.A., (2006) *J. Electrochem. Soc.*, **153**, A12.
53. Herstedt, M., Stjerndahl, M., Nyten, A., Gustafsson, T., Rensmo, H., Siegbahn, H., Ravet, N., Armand, M., Thomas, J.O., and Edstrom, K., (2003) *Electrochem. Solid State Lett.*, **6**, A202.
54. Striebel, K., Shim, J., Sierra, A., Yang, H., Song, X., Kostecki, R., and McCarthy, K., (2005) *J. Power Sources*, **146**, 33.
55. Rho, Y.H., Nazar, L.F., Perry, L., and Ryan, D., (2007) *J. Electrochem. Soc.*, **154**, A283.
56. Koltypin, M., Aurbach, D., Nazar, L., and Ellis, B., (2007) *Electrochem. Solid State Lett.*, **10**, A40.
57. Song, S.W., Reade, R.P., Kostecki, R., and Striebel, K.A., (2004) Private Report.
58. Herstedt, M., Stjerndahl, M., Nyten, A., Gustafsson, T., Rensmo, H., Siegbahn, H., Ravet, N., Armand, M., Thomas, J.O., and Edstrom, K., (2003) *Electrochem. Solid State Lett.*, **6**, A202.
59. Koltypin, M., Aurbach, D., Nazar, L., and Ellis, B., (2007) *Electrochem. Solid State Lett.*, **10**, A40.
60. Morita, M., Shibata, T., Yoshimoto, N., and Ishikaw, M., (2002) *Electrochim. Acta*, **47**, 2787.
61. Lide, D.R. (2005) *CRC Handbook of Chemistry and Physics*, CRC Press, Boca Raton, FL.
62. Lopez, S., Petit, J.P., Dunlop, H.M., Butruille, J.R., and Tourillon, G.J., (1998) *J. Electrochem. Soc.*, **145**, 823.
63. Chen, Y., Devine, T.M., Evans, J.W., Monteiro, O.R., and Brown, I.G., (1999) *J. Electrochem. Soc.*, **146**, 1310.
64. Zhang, S. and Jow, T.R., (2002) *J. Power Sources*, **109**, 458.
65. Krause, L.J., Lamanna, W., Summerfield, J., Engle, M., Korba, G., Loch, R., and Atanasoski, R., (1997) *J. Power Sources*, **68**, 320.
66. Zhang, S.S., and Jow, T.R., (2002) *J. Power Sources*, **109**, 458.
67. Yang, H., Kwon, K., Devine, T.M., and Evans, J.E., (2000) *J. Electrochem. Soc.*, **147**, 4399.
68. Kanamura, K., Umegaki, T., Shiraishi, S., Ohashi, M., and Takehara, Z., (2002) *J. Electrochem. Soc.*, **149**, A185.
69. Wang, X., Yasukawa, E., and Mori, S., (2000) *Electrochim. Acta*, **45**, 2677.
70. Zhang, S.S., Xu, K., and Jow, T.R., (2002) *J. Electrochem. Soc.*, **149**, A586.
71. Song, S.W., Richardson, T.J., Zhuang, G.V., Devine, T.M., and Evans, J.W., (2004) *Electrochim. Acta*, **49**, 1483.
72. Zhang, X. and Devine, T.M., (2006) *J. Electrochem. Soc.*, **153**, B365.
73. Zhang, X., Winget, B., Doeff, M., Evans, J.W., and Devine, T.M., (2005) *J. Electrochem. Soc.*, **152**, B448.

4
Electrochemical and Material Property Analysis

4.1
Electrochemical Analysis

4.1.1
Open-Circuit Voltage

Open-circuit voltage [1–3] is the value measured when an electrochemical cell is in equilibrium without any difference in potential between electrode locations. In other words, it is the voltage of stabilized electrodes when there is no current flow. This value reflects the Gibbs free energy in a state of thermodynamic equilibrium. On the other hand, closed-circuit voltage (CCV) measures the voltage of electrodes when connected to an external circuit with electric current flowing between the terminals. To accurately represent the thermodynamic state of electrodes, it is important to derive the open-circuit voltage in relation to chemical composition.

The open-circuit voltage provides information on the voltage of electrode materials, the occurrence of internal short circuits, and initial interfacial chemical reactions. By measuring the open-circuit voltage of electrodes throughout charging and discharging cycles, we can examine the reversibility of charging and discharging. Changes in open-circuit voltage over time are useful in studying electrochemical reactions, such as charge transfer in electrode materials and self-discharge.

Figure 4.1 shows the open-circuit voltage measured from a cell consisting of single-walled carbon nanotubes (SWCNTs) and lithium metal electrodes with 1 M $LiPF_6$/ethylene carbonate (EC):diethylcarbonate (DEC) (1 : 1 by volume) as the electrolyte. Voltage was measured for the range of 0–3 V by applying a constant current of 50 mA/g for 1 h followed by no current for the next hour. Closed-circuit voltage measured during 1 h of constant current is represented by the thick dotted line while open-circuit voltage obtained under zero current is indicated by the solid line.

Principles and Applications of Lithium Secondary Batteries, First Edition. Jung-Ki Park.
© 2012 Wiley-VCH Verlag GmbH & Co. KGaA. Published 2012 by Wiley-VCH Verlag GmbH & Co. KGaA.

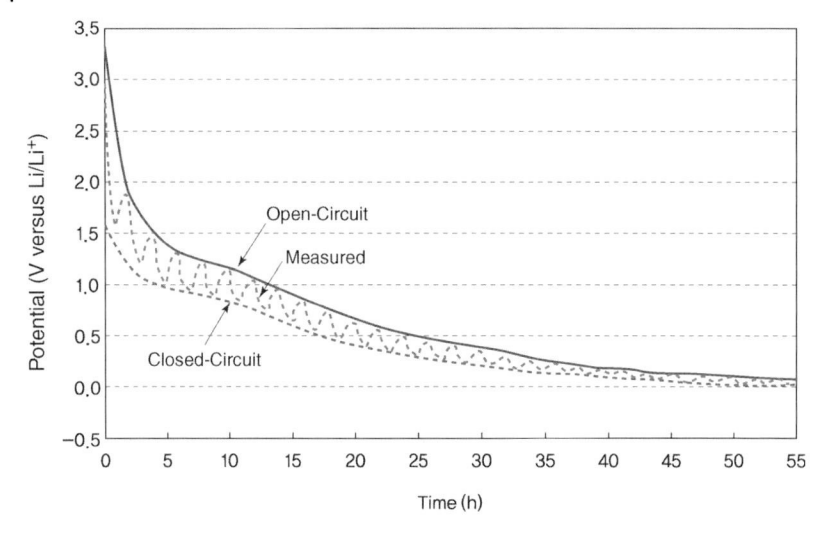

Figure 4.1 Open-circuit voltage and closed-circuit voltage of a carbon (SWCNT)/lithium battery.

4.1.2
Linear Sweep Voltammetry

Linear sweep voltammetry is an electrochemical analysis method that scans voltage with scan rate v (V/s) for a given potential range of an electrochemical cell and presents results in the form of current–voltage plots. When the cell undergoes oxidation or reduction within the measured voltage range, sharp changes are observed in the current. We can predict and analyze reactions occurring within the cell by measuring the amount of current and voltage at these points. On the basis of these characteristics, linear sweep voltammetry is widely used in evaluating the electrochemical stability of electrolytes.

Figure 4.2 is an example of linear sweep voltammetry that shows changes in the current with a constant scanning rate of voltage while varying the applied voltage. The cell consists of lithium metal and platinum electrodes with 1 M $LiPF_6$/dimethyl carbonate (DMC) as the electrolyte. In this case, negligible changes in current are seen for voltages below 4.5 V. The current changes more significantly in the range of 4.6–5.2 V and rapidly rises from 5.2 V. This implies that redox reactions take place within the electrochemical cell at voltages higher than 4.6 V and the electrolyte exhibits stable electrochemical characteristics up to 4.6 V.

4.1.3
Cyclic Voltammetry

Cyclic voltammetry [1, 2] is an electrochemical analysis method that scans voltage with a constant scan rate of voltage for a given potential range of an electrochemical cell. Similarly to linear sweep voltammetry, cyclic voltammetry observes changes in current by applying voltage under a constant scan rate. However, cyclic voltammetry

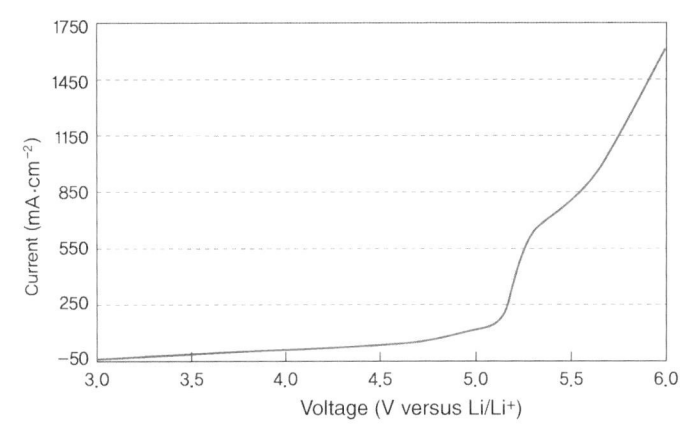

Figure 4.2 Current–voltage profile of Li/(1 M LiPF$_6$/DMC)/Pt cell under linear sweep voltammetry.

repeats the same experiment for each cycle. Current–voltage curves obtained from cyclic voltammetry are different from linear plots of linear sweep voltammetry. Cyclic voltammetry provides information on redox reactions occurring within the cell, including (1) potential, (2) quantity of electricity (3) reversibility, and (4) continuity (sustainability of reversible electrochemical reaction). While the scan rate differs depending on the purpose of experiments, a low scan rate is recommended for detailed analysis of electrochemical reactions.

Figure 4.3 is a typical cyclic voltammogram showing changes in current with varying applied potential. Oxidative reactions are induced by the anodic current when the scan direction is (+), while reductive reactions occur for (−). Figure 4.4 gives the results of cyclic voltammetry on a natural graphite anode in the 0–3 V range.

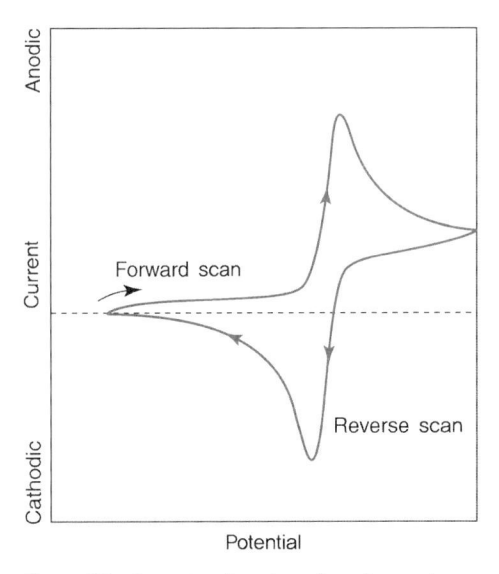

Figure 4.3 Current–voltage in cyclic voltammetry.

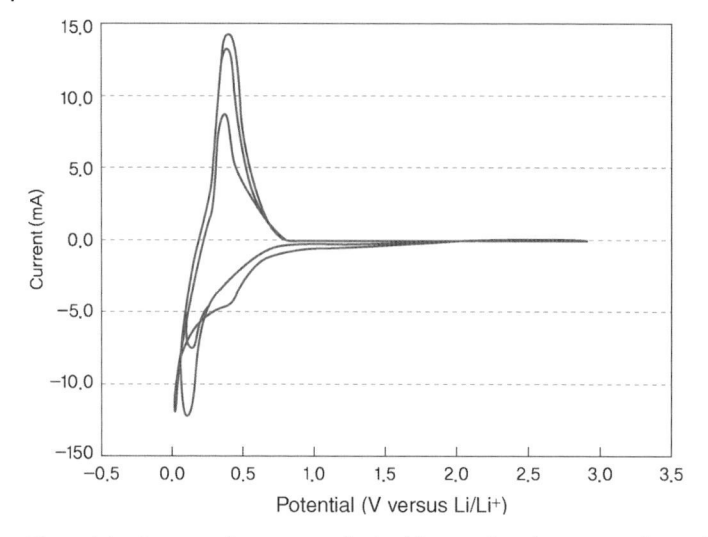

Figure 4.4 Current–voltage curves obtained from cyclic voltammetry of natural graphite/ [1 M LiPF$_6$/(PC:EC:DEC)]/Li cell.

The cell is comprised of natural graphite/[1 M LiPF$_6$/(PC:EC:DEC)]/lithium. The cathodic current is observed with electrochemical reactions arising from lithium intercalation, while the cathodic current corresponds to lithium deintercalation.

4.1.4
Constant Current (Galvanostatic) Method

The constant current method [1, 2] examines battery characteristics by measuring changes in voltage over time under a constant current. Electrochemical properties that can be obtained from this method include capacity, reversibility, resistance, and rate of diffusion. Depending on termination conditions, the constant current method can be divided into two types.

4.1.4.1 Cutoff Voltage Control
Continuous charging and discharging experiments are conducted to measure changes in voltage under a constant current for a given potential range. It is a method of electrochemical analysis that derives the quantity of electrical charge under continuous charging and discharging with varying voltage over time.

Figure 4.5 shows changes in voltage over time with the lower limit set to 0 V and upper limit to 3 V. From the graph, we can see that there are small changes in voltage over time per cycle. This implies that electrode materials are involved in reversible reactions with lithium ions. Figure 4.6 presents charge or discharge capacities derived from charging or discharging per cycle. This allows calculations of coulombic efficiency for each charge and discharge steps.

A differential capacity curve is a plot of dQ/dV against voltage based on time and voltage values obtained from constant current tests. The unit of differential capacity curves may also be represented as dt/dV using the equation below.

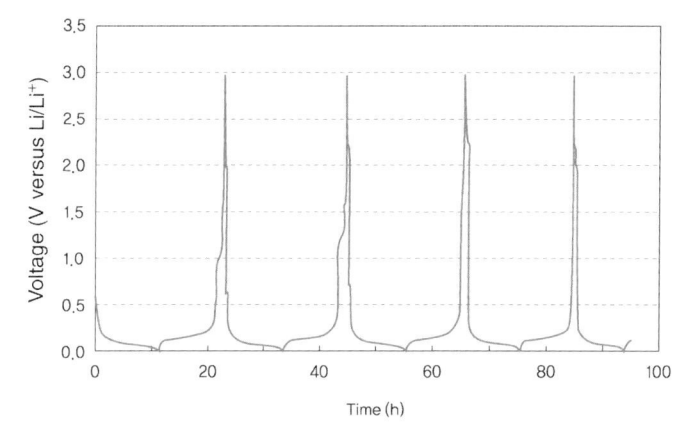

Figure 4.5 Voltage controlled constant current charge–discharge curve of a graphite/lithium secondary battery.

$$\frac{dQ}{dV} = \frac{dQ}{dt}\frac{dt}{dV} = I\frac{dx}{dV}$$

Figure 4.7 shows the differential capacity curves (dQ/dV) from voltage cutoff controlled constant current tests on an electrochemical cell consisting of graphite and lithium metal. The data are derived from the first cycle given in Figure 4.5. Thus, we can accurately measure the specific voltage of electrochemical reactions.

Differential capacity curves are similar to cyclic voltammetry curves with the only difference being a constant overpotential. In cyclic voltammetry, the scan rate should be lowered in order to separate electrochemical reactions by type. However, the voltage at which electrochemical reactions occur can be more easily detected in differential capacity curves.

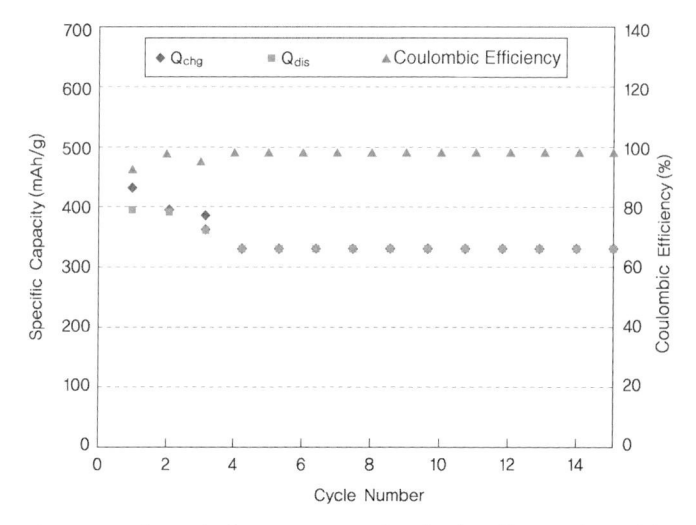

Figure 4.6 Charge/discharge capacity and coulombic efficiency obtained from voltage cutoff controlled constant current charge/discharge tests on a graphite/lithium cell.

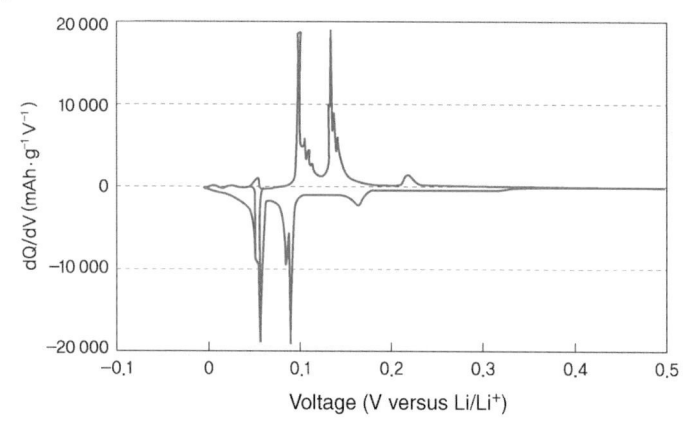

Figure 4.7 Differential capacity curves of a graphite/lithium cell.

4.1.4.2 Constant Capacity Cutoff Control

Constant capacity cutoff control is a constant current method of controlling the amount of electric charge to test anode characteristics. Unlike cathode materials, anode materials are charged up to voltages similar to those of lithium metals. As such, lithium intercalation and elution of lithium metals may occur simultaneously. Given the relatively flat potential profile, it is inappropriate to use the voltage cutoff control method, since a small change in potential leads to a significant change in electric charge. To resolve this problem, the required amount of electricity is charged followed by voltage control during discharging.

4.1.5
Constant Voltage (Potentiostatic) Method

Constant voltage [1–3] is an easier method than constant current for oxidative and reductive reactions of a cell to reach electrochemical equilibrium.

4.1.5.1 Constant Voltage Charging

Depending on the applied current, ion concentration gradients exist between the surface and the inner side of materials involved in intercalation. Unlike the electrode surface, charging is not fully complete within electrodes. If charging proceeds up to the rated charge capacity using only the constant current method, electrode materials may be damaged when the surface voltage exceeds the rated voltage. To prevent this problem, a constant current is applied for charging within the range of rated voltage followed by constant voltage charging. This method maximizes the energy storage capacity of batteries.

4.1.5.2 Potential Stepping Test

Potential stepping is a method of increasing or reducing the voltage of electrochemical cells by specific steps based on constant voltage control. The end-of-charge

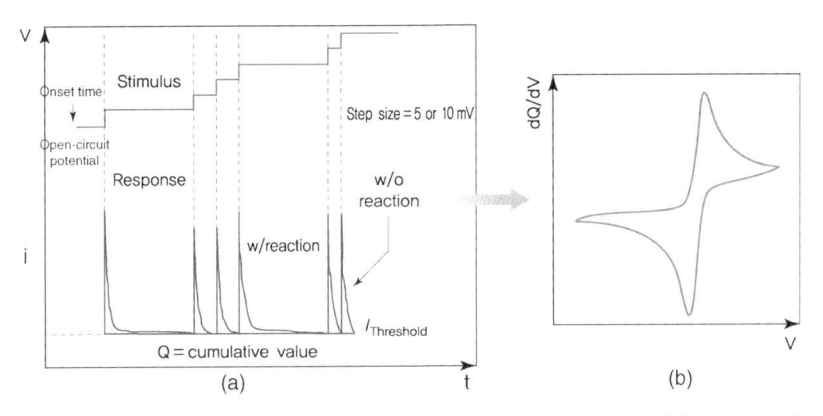

Figure 4.8 Current–time and differentiated capacity–voltage plots obtained from potential stepping.

current and times are set for each step. Using this method, we can obtain the open-circuit voltage and potential transient signal, which can then be used to derive differential capacity curves and diffusion rates.

Figure 4.8 shows the differential capacity curves obtained from the potential stepping test. The flow of current with increasing potential is shown in Figure 4.8a, while the plot of differentiated charge against potential can be seen in Figure 4.8b. Table 4.1 compares the characteristics of cyclic voltammetry, constant current charging/discharging, and a potential stepping test.

Table 4.1 Comparison of characteristics between cyclic voltammetry, constant current charging/discharging, and a potential stepping test.

	Cyclic voltammetry	**Constant current charge/discharge**	**Potential stepping test**
Data	I–V	dQ/dV–V	dQ/dV–V
Measurements	Measuring current as voltage is changed at a constant rate	Measuring capacity as a function of time under constant current charging or discharging	Derivative capacity when voltage is changed as a step function
Peak property	Peak can be shifted according to a scan rate due to ohmic polarization	Constant peak shift due to constant current and ohmic polarization	—
Analysis clarity	Well separable	Complex	Well separable
Effect of overpotential	Dependent	Constant	—

4.1.6
GITT and PITT

4.1.6.1 GITT

GITT (galvanostatic intermittent titration technique) is a type of constant current method that measures changes in open-circuit voltage induced by cutting off the current supply after applying constant current at each step during the charging and discharging process [3, 4]. When a constant current is applied to electrode materials, lithium is intercalated into or deintercalated from particles, resulting in concentration gradients at the surface and inner side of the electrodes. By measuring changes in the voltage over time, we can calculate the rate of change in concentration. From this, we are able to determine the diffusion coefficient of lithium.

Figure 4.9 shows the changes in current and voltage over time during a GITT experiment. The sharp increase or decrease in voltage when current is applied or cut off can be interpreted as a drop in iR. Changes in voltage with time are related to the rate of lithium diffusion. The diffusion coefficient of lithium can be obtained based on GITT using the following equation:

$$D^{GITT} = \frac{4}{\pi\tau}\left(\frac{m_B V_M}{M_B S}\right)^2 \left(\frac{\Delta E_s}{\Delta E_t}\right)^2$$

Here, τ is the time during which constant current is applied, m_B is the mass of the electrode material, V_M is the molar volume of the electrode material, M_B is the molar

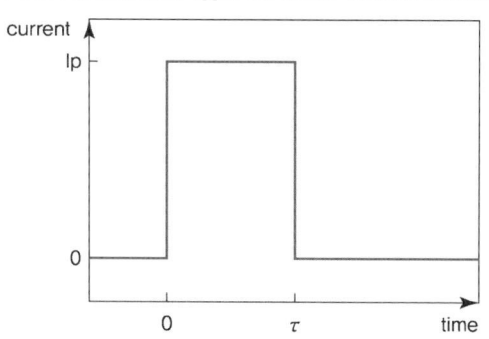

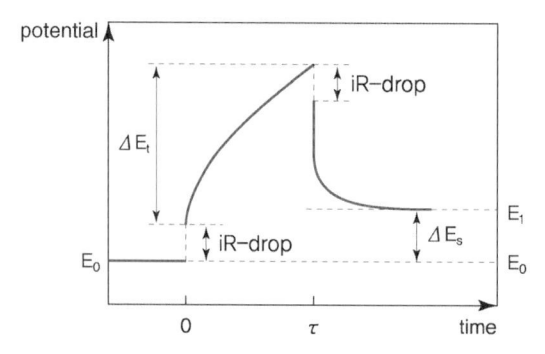

Figure 4.9 Changes in current and voltage during a GITT experiment [5]. Reprinted from [5] Copyright 2005, with permission from Elsevier.

mass of the electrode material, S is the electrode–electrolyte interfacial area, ΔE_s is the change in voltage at each step, ΔE_t is the change in overall voltage under constant current conditions, and $m_B V_M / M_B$ is the volume of the electrode material.

4.1.6.2 PITT

In PITT (potentiostatic intermittent titration technique), constant voltage is applied at each potential step to measure changes in current, which are then used to calculate the diffusion rate [3, 6–8]. From this, the ion concentration at the surface can be obtained. Testing at each step ends when the current drops below a set value, and a new potential step is applied to measure current changes in the next step. Figure 4.10 shows the change in current flow with varying potential of the LiMn$_2$O$_4$ cathode material in a lithium secondary battery.

PITT results can be interpreted by studying the linear behavior in current–time graphs. If active materials are of globular form, the transition time (t_T) can be used to calculate the diffusion coefficient as given in the following equation [6]:

$$D^{\text{PITT}} = \left(\frac{(I\sqrt{t})\max \sqrt{\pi}\, r_1}{\Delta Q} \right)^2$$

Here, I is the current, t is the time of measurement, r_1 is the radius of active material, τ is the diffusion time, and $\Delta Q = \int_{t=0}^{\infty} I(t)\,dt$ is the amount of electricity at each step.

4.1.7
AC Impedance Analysis

4.1.7.1 Principle

AC impedance analysis [9–12] is an electrochemical method used to examine current response under AC voltage to obtain values for resistance, capacitance, and inductance. As given in Eq. (4.1), AC voltage changes periodically over time. From

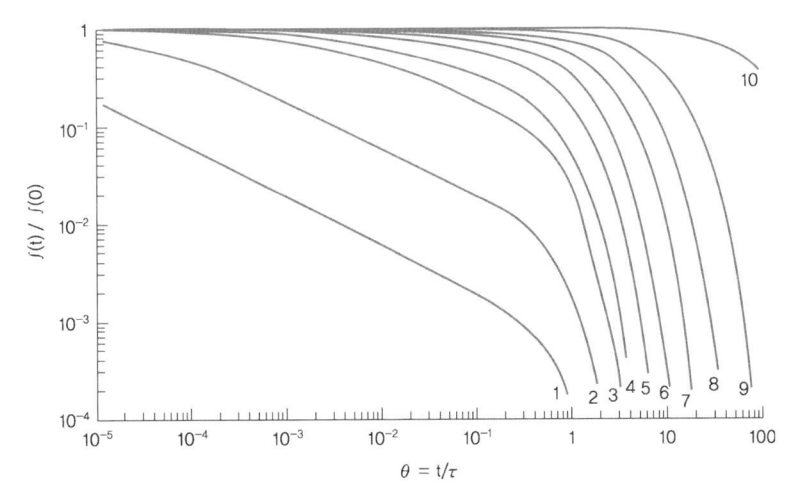

Figure 4.10 Change in current over time during a PITT experiment [7]. Reprinted from [7] Copyright 2004, with permission from Elsevier.

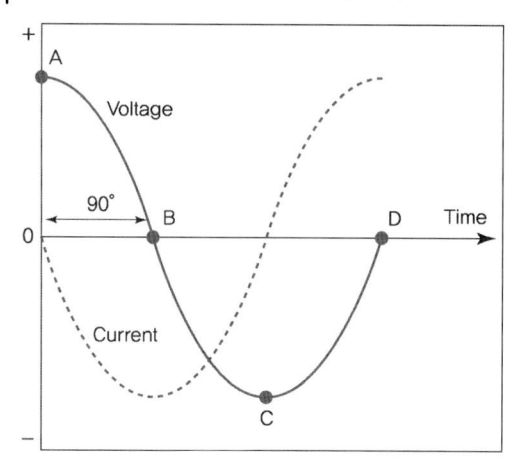

Figure 4.11 Phase difference between AC voltage and current.

Figure 4.11, we can see that a phase difference exists between the voltage and the current.

$$V(t) = V_m \cdot \sin(\omega t) \tag{4.1}$$

$$\omega = 2\pi v \quad (v : \text{frequency}) \tag{4.2}$$

Here, V_m is the maximum voltage and ω is the angular frequency. The current response (I) has a phase difference of θ with respect to AC voltage and can be obtained from Eq. (4.3).

$$I(t) = I_m \cdot \sin(\omega t - \theta) \tag{4.3}$$

The amplitude of AC voltage and current can be expressed as complex exponential functions, as shown in Eqs. (4.4) and (4.5).

$$V(t) = V_m \cdot \exp(j\omega t) \tag{4.4}$$

$$\begin{aligned} I(t) &= I_m \cdot \exp[j(\omega t - \theta)] \\ j &= \sqrt{-1} = \exp(j\pi/2) \end{aligned} \tag{4.5}$$

Impedance (Z) is defined in Eq. (4.6) and its magnitude is given by Eq. (4.7).

$$Z(\omega) = \frac{V(t)}{I(t)} \tag{4.6}$$

$$|Z(\omega)| = \frac{V_m}{I_m} \tag{4.7}$$

Based on the definition in Eq. (4.6), impedance can be broken down into the real part (Z') and the imaginary part (Z''). The real part is the resistance, and the imaginary part is the reactance, which consists of capacitance and inductance.

4.1 Electrochemical Analysis

$$Z = a + jt$$
$$= Z' + jZ'' \quad (Z': \text{real part of } Z, Z'': \text{imaginary part of } Z) \quad (4.8)$$

Using the phase difference θ, the above real and imaginary parts can be expressed as Eqs. (4.9) and (4.10). The phase difference (θ) is given by Eq. (4.11). The magnitude of impedance is expressed as Eq. (4.12).

$$Z' = |Z|\cos(\theta) \quad (4.9)$$

$$Z'' = |Z|\sin(\theta) \quad (4.10)$$

$$\theta = \tan^{-1}(Z''/Z') \quad (4.11)$$

$$|Z| = \sqrt{(Z'^2/Z''^2)} \quad (4.12)$$

The above rectangular coordinates can be converted to polar coordinates based on Euler's formula [$\exp(j\theta) = \cos(\theta) + j\sin(\theta)$]. This relationship is shown in Figure 4.12.

$$Z(\omega) = |Z|\exp(j\theta) \quad (4.13)$$

Figure 4.12 is a vector diagram in the complex plane and is known as a Nyquist plot or Cole-Cole plot. Figure 4.13 shows the changes in Z′ and Z// with frequency.

4.1.7.2 Equivalent Circuit Model

The flow of current through substances involves resistance and capacitance. When only resistance exists, $\theta = 0$, and impedance is given by the real part ($Z(\omega) = Z'(\omega)$). As shown in Eq. (4.14), it does not depend on frequency (Figures 4.14 and 4.15.

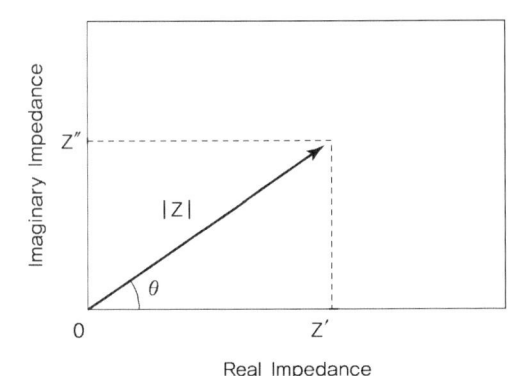

$$Z(t) = V(t)/I(t) = R \quad (4.14)$$

Figure 4.12 Complex plane plot of impedance.

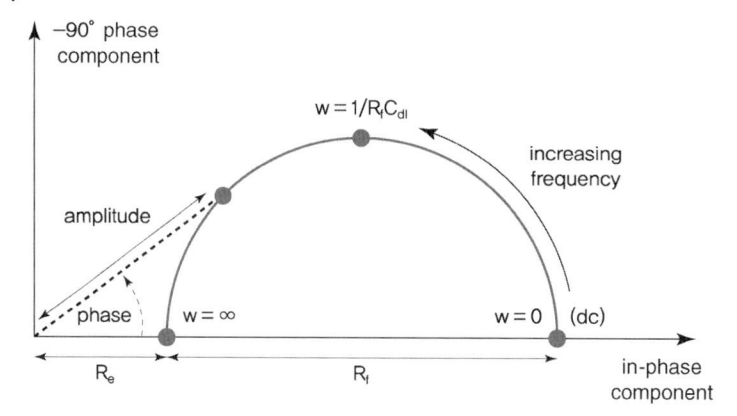

Figure 4.13 Change in impedance with frequency.

If there is only capacitance, the electrostatic capacity (Q) is given by Eq. (4.15). Using Eq. (4.1), this can also be expressed as Eq. (4.16).

$$Q = CV \qquad (4.15)$$

$$Q = C \cdot V_m \cdot \sin(\omega t) \qquad (4.16)$$

Since current is the change in electric charge over time, it can be expressed as Eq. (4.17), where I_m is the maximum current.

$$\begin{aligned} I(t) = dQ/dt &= C \cdot V_m \cdot \omega \cdot \cos(\omega t) \\ &= I_m \cdot \cos(\omega t) \end{aligned} \qquad (4.17)$$

From the above equations, we can see that the voltage and current follow sine and cosine functions, respectively. This means that the phase difference between the two is $\pi/2$ radian.

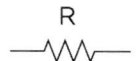

Figure 4.14 Resistance component.

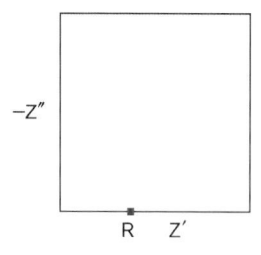

Figure 4.15 Nyquist plot of a resistance component.

C

—| |—

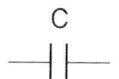

Figure 4.16 Capacitance (C) component.

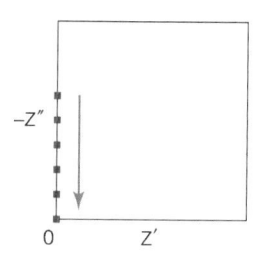

Figure 4.17 Nyquist plot of a capacitance component.

Capacitance (Figures 4.16 and 4.17) is denoted as X_C and defined by Eq. (4.18).

$$X_C = Z = \frac{V_m}{I_m} = \frac{1}{(\omega C)} = \frac{1}{(2\pi v C)} \tag{4.18}$$

Inductance is caused by the magnetic field generated from passing current through coils (Figures 4.18 and 4.19). It is denoted as L and measured in Henry units. Voltage and current are related to inductance, as shown in Eqs. (4.19) and (4.20). The inductance X_L is given by Eq. (4.21).

$$V(t) = L(dI(t)/dt) \tag{4.19}$$

$$I(t) = (1/L)\int V(t)dt = (V_m/\omega L)\sin(\omega t - \pi/2) \tag{4.20}$$

$$X_L = Z = V_m/I_m = \omega L \tag{4.21}$$

As shown in Eq. (4.20), current lags voltage by $\pi/2$ radian.

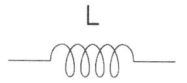

Figure 4.18 Inductance component.

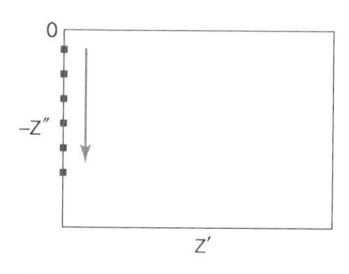

Figure 4.19 Nyquist plot of an inductance component.

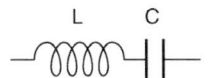

Figure 4.20 Equivalent circuit of an inductance–capacitance component.

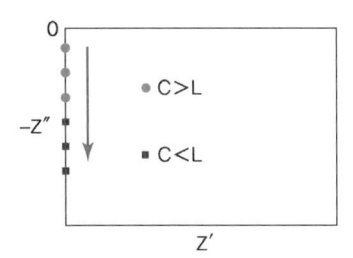

Figure 4.21 Nyquist plot of an inductance–capacitance component.

If the above capacitance (X_C) and inductance (X_L) are connected in series, the equivalent circuit will be as shown in Figure 4.20. The Nyquist plot is given in Figure 4.21.

In a system consisting of resistance, capacitance, and inductance, the total reactance component X is given by Eq. (4.22), with X_C and X_L belonging to the imaginary part (Figures 4.22 and 4.23). In other words, the imaginary part denoted by Z'' corresponds to the capacitive and the inductive reactance.

$$X = Z = R + jX_L - jX_C = R + j(X_L - X_C)$$
$$= R + j\left(\omega L - \frac{1}{\omega C}\right) \quad (4.22)$$

If the reactance is negligible, Eq. (4.22) can be simplified to Eq. (4.23).

Figure 4.22 Resistance–reactance–capacitance (RLC) component.

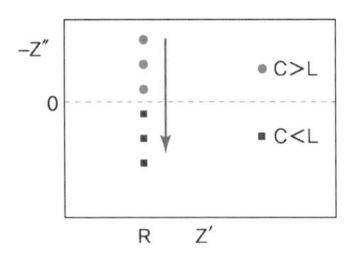

Figure 4.23 Nyquist plot of a resistance–reactance–capacitance (RLC) component.

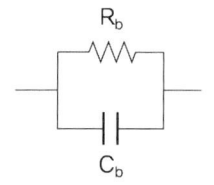

Figure 4.24 Equivalent circuit of parallel resistance–capacitance (R_b–C_b).

$$Z = R - \frac{j}{\omega C} \tag{4.23}$$

The impedance for a parallel connection of resistance and capacitance (Figure 4.24) can be expressed as Eqs. (4.24) and (4.25).

$$Y = Z^{-1} = G + j\omega C = Y' + jY''$$
$$(Y = 1/Z,\ G = 1/R,\ Y' = G = 1/R,\ Y'' = \omega C) \tag{4.24}$$

$$Z = \frac{1}{Y} = \frac{R}{RY} = \frac{R}{1 + j\omega RC} \tag{4.25}$$

Multiplying Eq. (4.25) with the complex conjugate of $(1 - j\omega RC)$ gives Eq. (4.26). The real and imaginary parts of impedance are shown in Eqs. (4.27) and (4.28).

$$Z = R \frac{1}{1 + j\omega RC} = R \frac{1 - j\omega RC}{1 + \omega^2 R^2 C^2} \tag{4.26}$$

$$Z' = \frac{R}{1 + \omega^2 R^2 C^2} \tag{4.27}$$

$$Z'' = \frac{-\omega R^2 C}{1 + \omega^2 R^2 C^2} \tag{4.28}$$

In Eq. (4.28), Z''_{max} occurs when $\omega_{max} RC = 1$ and takes on the maximum value of $R/2$. The corresponding Nyquist plot is shown in Figure 4.25.

Figure 4.26 shows the equivalent circuit of parallelly connected resistance (R_b) and capacitance (C_b) with another serial connection of resistance (R_s). From the Nyquist plot in Figure 4.27, we can see that the capacitance semicircle shifts by R_s. Figure 4.28

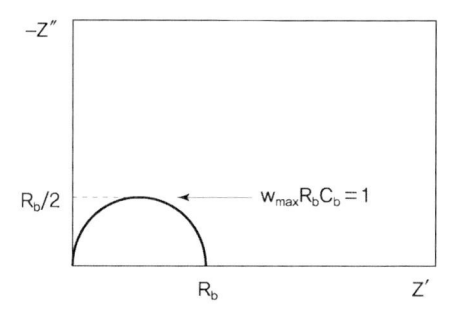

Figure 4.25 Nyquist plot of parallel resistance–capacitance (RC).

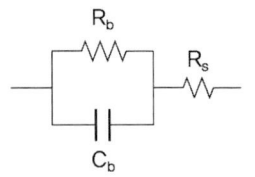

Figure 4.26 Equivalent circuit of series (parallel resistance–capacitance (R_b–C_b)) resistance (R_s).

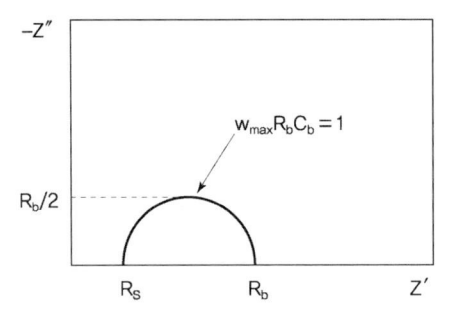

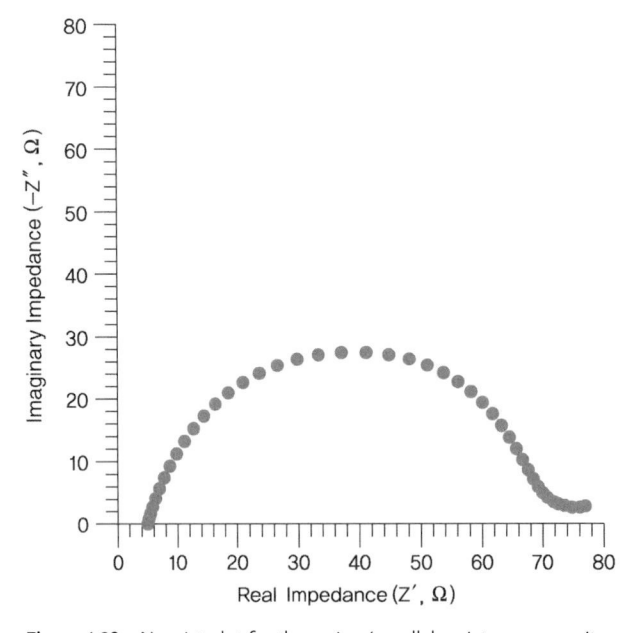

Figure 4.27 Nyquist plot for the series (parallel resistance–capacitance (R_b–C_b)) resistance (R_s) equivalent circuit.

is the Nyquist plot of a Li/PAN-SPE/Li cell for a parallel (series R_1–C)–R_2 equivalent circuit.

The equivalent circuit shown in Figure 4.29 is capacitance (C_e) connected in series to a parallel of resistance (R_b) and capacitance (C_b). The Nyquist plot

Figure 4.28 Nyquist plot for the series (parallel resistance–capacitance (R_b–C_b)) resistance (R_s) equivalent circuit.

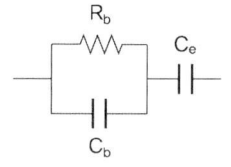

Figure 4.29 Equivalent circuit of capacitance + (resistance–capacitance).

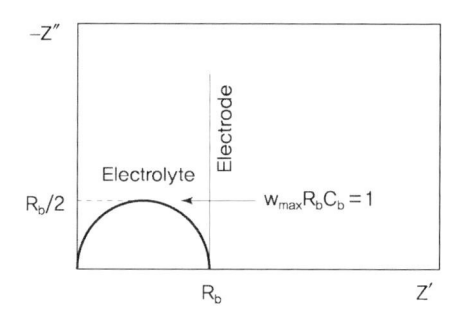

Figure 4.30 Nyquist plot for the capacitance + (resistance–capacitance) equivalent circuit.

in Figure 4.30 shows a semicircle corresponding to capacitance in the high-frequency range and the imaginary part for the series connected capacitance in the low-frequency range.

When C_b decreases in the equivalent circuit of Figure 4.29, the semicircle in Figure 4.30 becomes distorted. For neglectable values of C_b, the circuit exhibits characteristics similar to a R_b–C_e equivalent circuit.

4.1.7.3 Applications in Electrode Characteristic Analysis

A single porous electrode of a lithium secondary battery is shown in Figure 4.31, and its equivalent circuit is shown in Figure 4.32. Figure 4.33 shows the corresponding Nyquist plot, consisting of a high-frequency range, a middle frequency range (Warburg impedance) for diffusion control, and a low-frequency range for charge saturation.

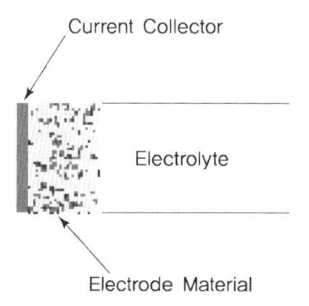

Figure 4.31 An electrode of a lithium secondary battery.

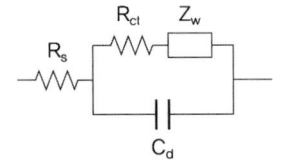

Figure 4.32 Equivalent circuit for an electrode of a lithium secondary battery.

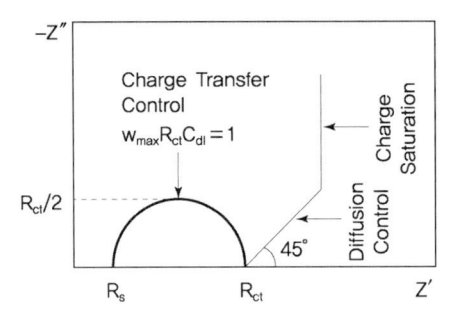

Figure 4.33 Nyquist plot for the equivalent circuit of a lithium secondary battery electrode.

Figures 4.34 and 4.35 show the impedance spectra for a carbon/electrolyte/lithium cell and a $LiCoO_2$/electrolyte/lithium cell, respectively. They differ from theoretical spectra as explained above since impedance characteristics are presented by both the measured (working) electrode and the lithium counter electrode. Given the current flow between the working electrode and the counter electrode, it

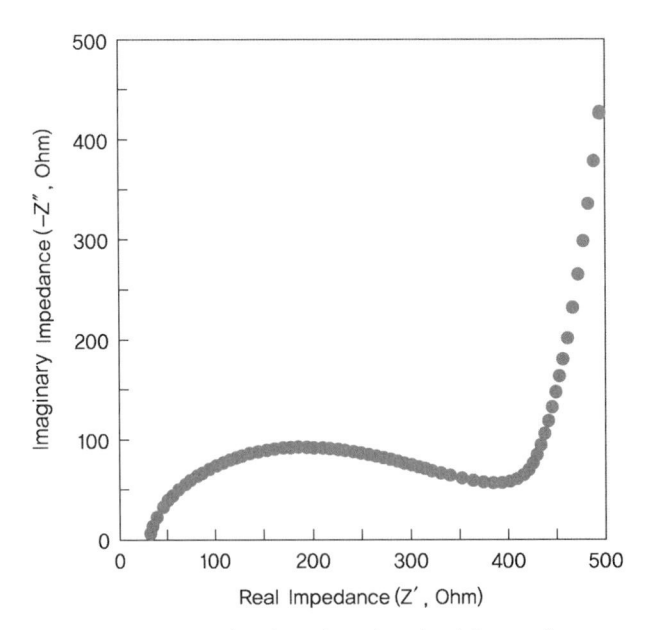

Figure 4.34 Nyquist plot of a carbon/electrolyte/lithium cell.

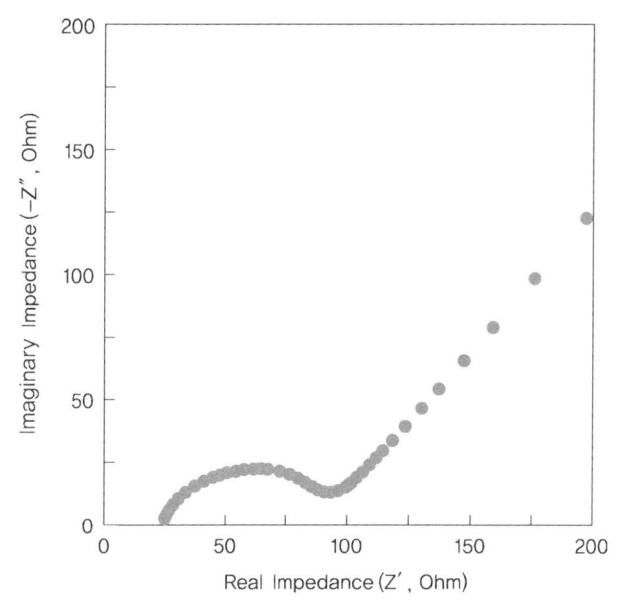

Figure 4.35 Nyquist plot of a LiCoO$_2$/electrolyte/lithium cell.

is impossible to distinguish among the results directly. When measuring AC impedance, the results are the same even if the working electrode and the counter electrode are switched.

4.1.7.4 Applications in Al/LiCoO$_2$/Electrolyte/Carbon/Cu Battery Analysis (1)

The above cell is shown in Figure 4.36 and its equivalent circuit in Figure 4.37. In this system, an electric double layer is formed at the electrode–electrolyte interface. Assuming the presence of charge transfer reactions, an impedance analysis can be applied as follows.

Impedance, as defined in Eq. (4.29), can be expressed in terms of resistance, capacitance, inductance, and Warburg impedance.

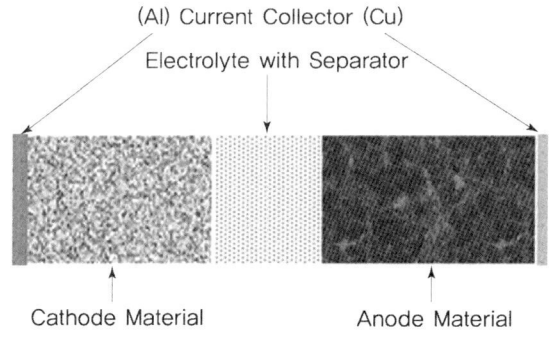

Figure 4.36 Cell composition.

4 Electrochemical and Material Property Analysis

Figure 4.37 Equivalent circuit of a LiCoO$_2$/carbon cell.

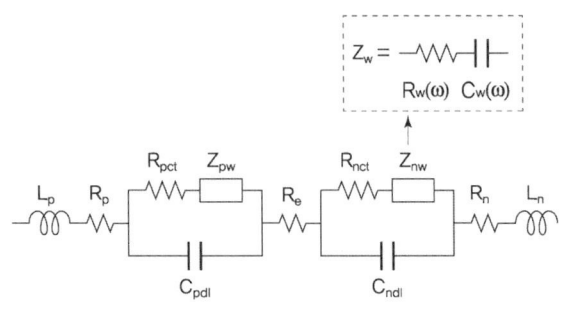

C: Double Layer Capacitance, Zf: Charge Transfer Rxn
R: Resistance Caused by Electrolyte, Current Collector, etc.
L: Inductance Cased by Cell Configuration
n, p, e: Anode, Cathode, Electrolyte

$$Z = R_s + \left(\sum_i^{n,p} \frac{1}{j\omega C_{dl(i)} + (1/Z_{\omega(i)})} \right) + j\omega L_s$$
$$R_s = R_p + R_e + R_n$$
$$L_s = L_p + L_n \qquad (4.29)$$
$$Z_{pw} = R_{pw} + C_{pw}$$
$$Z_{nw} = R_{nw} + C_{nw}$$

Here, Z_w is the Warburg impedance, which is created by diffusion within the cell. Z_{pw} at the cathode is a function of frequency if interfacial reactions are dominated by charge transfer and one-dimensional diffusion. As shown in Eqs. (4.30), (4.31) and (4.32), it is expressed as a series connection of the resistance term R_f and capacitance C_f.

$$Z_{pw} = \frac{S_p}{\sqrt{\omega}} - j \frac{S_p}{\sqrt{\omega}} \qquad (4.30)$$

$$R_f(\omega) = R_p + \frac{S_p}{\sqrt{\omega}} \qquad (4.31)$$

$$C_f(\omega) = \frac{1}{S_p \sqrt{\omega}} \qquad (4.32)$$

Here, S_p is a function related to the diffusion coefficient.

$$S_p = \frac{\beta}{\sqrt{2} n F A \sqrt{D}} \qquad (4.33)$$

Here, A is the area, n is the electron number, F is the Faraday constant, D is the diffusion coefficient, and β is $\partial E/\partial C$.

An impedance analysis of the anode is similar to that of the cathode and the impedance is defined by Eq. (4.34). The Cole-Cole plot in Figure 4.38 is obtained by placing Eq. (4.34) on a complex plane.

1. High-Frequency Range
2. Middle-Frequency Range
3. Low-Frequency Range

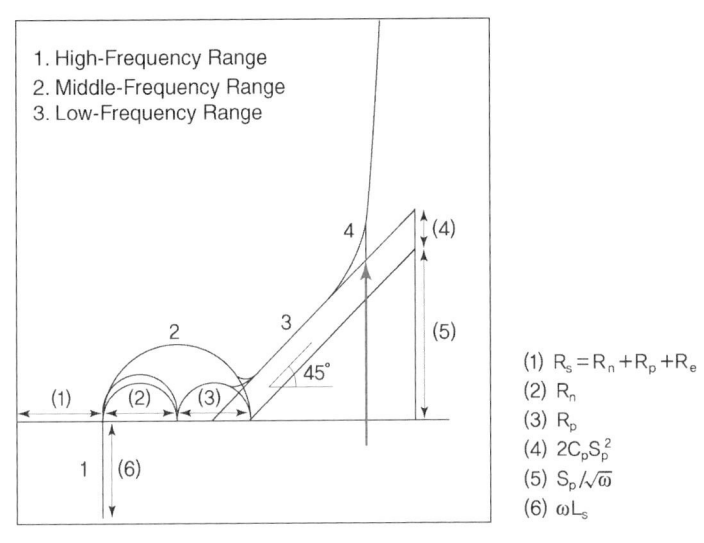

(1) $R_s = R_n + R_p + R_e$
(2) R_n
(3) R_p
(4) $2C_p S_p^2$
(5) $S_p/\sqrt{\omega}$
(6) ωL_s

Figure 4.38 Cole–Cole plot of a lithium secondary battery.

$$Z = R_s + \cfrac{1}{j\omega C_p + \cfrac{1}{R_p + \cfrac{S_p}{\sqrt{\omega}} - \cfrac{jS_p}{\sqrt{\omega}}}} \\ + \cfrac{1}{j\omega C_n + \cfrac{1}{R_n + \cfrac{S_n}{\sqrt{\omega}} - \cfrac{jS_n}{\sqrt{\omega}}}} + j\omega L_s \quad (4.34)$$

High-Frequency Region This impedance range is related to the movement of electrons at a frequency where the Warburg term and the RC parallel circuit are neglected. Ionic movement is not possible and the impedance is given by Eq. (4.35).

$$Z \approx R_s + j\omega L_s \quad (4.35)$$

Middle-Frequency Region At a frequency where the Warburg and inductance terms can be neglected, the impedance is given by Eq. (4.36).

$$Z = R_s + \cfrac{1}{j\omega C_p + \frac{1}{R_p}} + \cfrac{1}{j\omega C_n + \frac{1}{R_n}} \quad (4.36)$$

The above equation corresponds to the equivalent circuit of 2 RC parallel circuits connected in series to resistance (R_s), and the term containing ω can be removed if $C_p R_p (\equiv \tau_p) = C_n R_n (\equiv \tau_n)$ (τ: $1/\omega_{max}$, corresponds to relaxation time). To construct a Nyquist plot, this can also be expressed as Eq. (4.37). The Nyquist plot shows a semicircle with a radius of $(R_p + R_n)/2$.

4 Electrochemical and Material Property Analysis

$$\left(R - R_s - \frac{\theta}{2}\right)^2 + X^2 = \left(\frac{\theta}{2}\right)^2 \quad (\theta = R_p + R_n) \tag{4.37}$$

Electrodes do not have the same relaxation time due to the use of different materials for the anode and the cathode. For $\tau_p \gg \tau_n$, the arc for the smaller τ is found in the high-frequency range and the two arcs are separated. Interfacial information of the electrode and electrolyte is highly affected by the size of τ_p and τ_n. Considering incomplete semicircles and distributed relaxation times, nonideal components can be accounted for as shown in Eq. (4.38).

$$Z = R_s + \sum_i^{n,p} \frac{R_i}{1 + j(\omega \tau_i)^{1-h_i}} \quad (h = 2\alpha/\pi) \tag{4.38}$$

Figure 4.39, Table 4.2, and Figure 4.40 correspond to the two parallel RC equivalent circuits, model conditions, and the Nyquist plot. If the resistance and capacitance are the same, the two RC circuits behave as a single RC circuit, resulting in the Nyquist plot (a) shown in Figure 4.40. If one side has a small capacitance, the Nyquist plot takes the form shown in Figure 4.40(d). In many cases, batteries display an intermediate curve between Figure 4.40(a) and Figure 4.40(d).

Low-Frequency Region Dominated by Warburg Impedance In this region, all other terms can be omitted except $\omega^{-1/2}$.

$$Z = R - jX$$
$$R = R_s + \theta + S_p \omega^{-1/2} \quad (\theta = R_p + R_n) \tag{4.39}$$
$$X = S_p \omega^{-1/2} + 2S_p^2 C_p$$

When ω term is removed, the plot has a slope of 45° on a complex plane in the low-frequency range. This region is dominated by ionic diffusion.

$$X = R - R_s - \theta + 2S_p^2 C_p \tag{4.40}$$

Very Low-Frequency Region The impedance trajectory in the complex plane tends to rise rapidly in the low-frequency range and can be expressed as follows. As shown in Eq. (4.33), β represents the concentration change ($= \partial E / \partial C$) of the electrode potential.

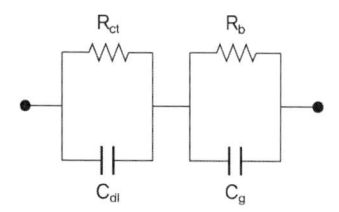

Figure 4.39 Series connection of two RC equivalent circuits.

Table 4.2 Model conditions of two RC equivalent circuits.

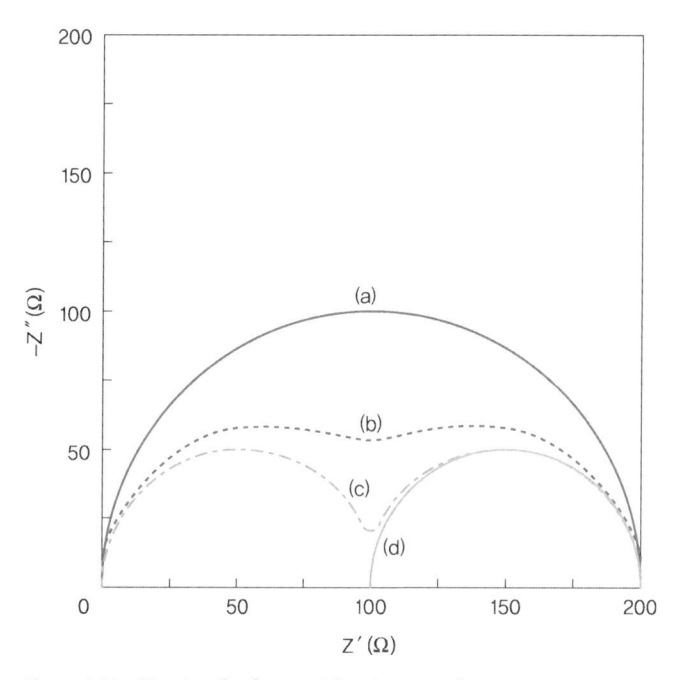

A	B	C	D
$R_b = 100\ \Omega$	$R_b = 100\ \Omega$	$R_b = 100\ \Omega$	$R_b = 100\ \Omega$
$C_g = 10^{-6}\ F$	$C_g = 10^{-7}\ F$	$C_g = 10^{-8}\ F$	$C_g = 10^{-11}\ F$
$C_{dl} = 10^{-6}\ F$	$C_{dl} = 10^{-6}\ F$	$C_{dl} = 10^{-6}\ F$	$C_{dl} = 10^{-6}\ F$
$R_b = 100\ \Omega$	$R_b = 100\ \Omega$	$R_b = 100\ \Omega$	$R_b = 100\ \Omega$

Figure 4.40 Nyquist plot for two RC series equivalent circuits.

$$X = |Z_f|\sin\beta$$
$$= \frac{1}{\omega C_L} \quad (\beta : \text{constant}) \tag{4.41}$$
$$R = |Z_f|\cos\beta$$
$$= R_L$$

4.1.7.5 Applications in Al/LiCoO₂/Electrolyte/MCMB/Cu Cell Analysis (2)

Figure 4.41 shows the Nyquist plot for a fuel cell consisting of a $LiCoO_2$ electrode ($LiCoO_2$:conductive agent (superpure black):PVDF = 94 : 4 : 4 wt%), MCMB electrode (MCMB-20-28:PVDF = 92 : 8 wt%), and electrolyte. The figure shows different results for liquid and polymer electrolytes.

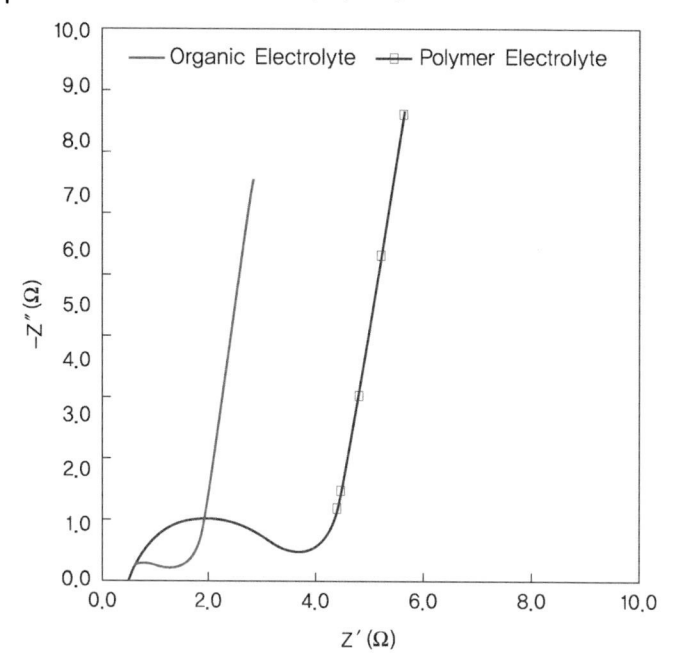

Figure 4.41 Nyquist plot of a Al/LiCoO$_2$/electrolyte/MCMB/Cu lithium secondary battery.

The middle-frequency region shows an incomplete semicircle and the combined characteristics of electrodes having different values for resistance and capacitance. Polymer electrolytes exhibit higher resistance than liquid electrolytes.

4.1.7.6 Relative Permittivity

Relative permittivity (ε_r) is an important characteristic of solvents and vacuum permittivity (ε_0) has a defined value of 8.854×10^{-12} F/m. As shown in Eq. (4.42), ε_0 is a constant that determines the force acting between q_1 and q_2 separated by a distance r.

$$f_{vac} = \frac{q_1 \cdot q_2}{4\pi\varepsilon_0 r^2} \tag{4.42}$$

In liquid dielectrics, the force acting between two particles is lower than that under vacuum due to interactions with the surrounding solute and solvent molecules. The relative permittivity defined in Eq. (4.43) gives a value larger than 1.

$$f = \frac{f_{vac}}{\varepsilon_r} \qquad \varepsilon_r = \frac{f_{vac}}{f} \tag{4.43}$$

Based on the above definition, $\varepsilon_r = 1$ under vacuum and $\varepsilon_r > 1$ in liquids. ε_r is found to be greater than 15–20 in polar solvents but smaller in nonpolar solvents.

For permittivity measurements, an electrochemical cell is set up as shown in Figure 4.42. The cell includes two electrodes (A: contact area with dielectric substance) and a dielectric substance is placed between the two electrodes (separated by L).

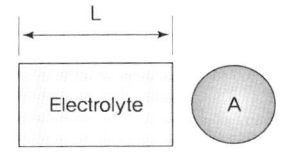

Figure 4.42 An electrochemical cell for permittivity measurements (A is electrode area).

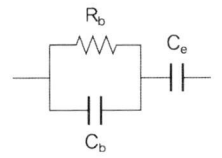

Figure 4.43 Equivalent circuit of an electrochemical cell for permittivity measurements.

The equivalent circuit is shown in Figure 4.43, and the total impedance is given by Eq. (4.44). Figure 4.44 shows the corresponding Nyquist plot, where Z'' is maximum when Z' is $R_b/2$. At this point, R_b and C_b can be obtained since $w_{max} R_b C_b = 1$.

$$Z_{total} = R_b \cdot \frac{1}{1 + (\omega R_b C_b)^2} - j\left(\frac{\omega R_b^2 C_b}{1 + (\omega R_b C_b)^2} - \frac{1}{\omega C_e}\right) \quad (4.44)$$

Here, ω is $2\pi f$, R_b is the bulk resistance, C_b is the bulk capacitance, and C_e is the interfacial capacitance.

Relative permittivity is the ratio of capacitance (C_0 and C_b) measured with a dielectric to that measured with vacuum (Eq. (4.45)). With C_b defined in Eq. (4.46), the ε_r can be obtained from an AC impedance analysis.

$$\varepsilon_r = \frac{C_b}{C_0} \quad (4.45)$$

$$C_b = \frac{\varepsilon_r \cdot \varepsilon_0 \cdot A}{l} \quad (4.46)$$

$$\varepsilon_r = \frac{C}{C_0} = \frac{C_b \cdot l}{\varepsilon_0 \cdot A} \quad (\varepsilon_0 = 8.854 \times 10^{-14} \, F/cm) \quad (4.47)$$

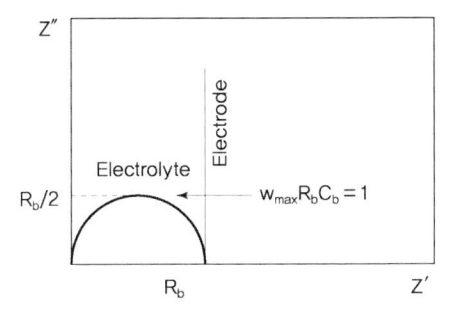

Figure 4.44 Nyquist plot of an electrochemical cell for permittivity measurements.

4.1.7.7 Ionic Conductivity

As shown in Eq. (4.48), ionic conductivity can be derived using R_b introduced above.

$$\sigma = \frac{l}{R_b \cdot A} = \sum n_i z_i u_i \quad (\text{S/cm}) \tag{4.48}$$

Here, l is the length of the sample in the direction of ionic movement, A is the area of the sample allowing ions to pass through, n_i is the number of moles per unit volume of component i, z_i is the quantity of electric charge of component i, and μ_i is the mobility of component i. Since the ionic conductivity depends on concentration, we need to look at the molar ionic conductivity, which is the ionic conductivity divided by concentration. In Eq. (4.49), the value of 1000 converts molar concentration to dm units. Ionic mobility (u_i) is inversely proportional to concentration and highest at infinite dilution. The molar conductivity at this point (Λ_0) and that of a general solvent (Λ) can be expressed in terms of Λ_0 and concentration, with activity (α) given by (Λ/Λ_0).

$$\Lambda = \frac{1000\sigma}{M} \quad (\text{Scm}^2/\text{mol}) \tag{4.49}$$

$$\Lambda = n_+ \Lambda_+ + n_- \Lambda_- \tag{4.50}$$

$$\Lambda_0 = n_+ \Lambda_{0+} + n_- \Lambda_{0-} \tag{4.51}$$

$$\Lambda = \Lambda_0 - (A\Lambda_0 + B)C^{1/2} \tag{4.52}$$

$$\alpha \cong \frac{\Lambda}{\Lambda_0} \tag{4.53}$$

Here, n_+ is the number of moles of cations, n_- is the number of moles of anions, Λ_+ is the molar conductivity of cations, Λ_- is the molar conductivity of anions, Λ_{0+} is the limiting molar conductivity of cations, Λ_{0-} is the limiting molar conductivity of anions, and A and B are constants. Equation (4.52) is the Onsager equation derived by considering relaxation and electrophoretic effects and shows the dependency of molar ionic conductivity on concentration. The transference number (t_i) is the contribution of ions to conductivity and a fraction of charge carried by ions. In lithium secondary batteries, the ideal transference number is 1, and charge transfer should not be induced by cations or anions. Since the concentration of ions is uniform in 1:1 electrolytes, such as NaCl, HCl, and KOH, t_+ is defined as follows:

$$t_i = \frac{\Lambda_i C_i}{\sum \Lambda_i C_i} \tag{4.54}$$

$$t_+ = \frac{\Lambda_+}{\Lambda_+ + \Lambda_-} \quad (1:1 \text{ electrolytes}) \tag{4.55}$$

$$t_+ = 1 - t_- \quad (1:1 \text{ electrolytes}) \tag{4.56}$$

4.1.7.8 Diffusion Coefficient

The diffusion coefficient, measured in the unit of m^2/s, represents the degree of diffusion of a substance in a solid, liquid, or gas. The diffusion coefficient of ions or molecules in aqueous solutions is $\sim 10^{-9}$ m^2/s and 10^{-10} m^2/s in nonaqueous solutions. These values can be derived from the Nernst–Einstein equation (Eq. (4.57)) and Stokes–Einstein equation (Eq. (4.58)). Electrochemical methods such as GITT, PITT, and AC impedance analysis can be used to obtain the diffusion coefficient for electrode materials.

$$D_i = \frac{\Lambda_i RT}{(n_i F)^2}, \quad (\Lambda_i\text{: molar ionic conductivity, } F\text{: faraday constant}) \quad (4.57)$$

$$D_i = \frac{kT}{6\pi r_i \eta}, \quad (r_i\text{: effective radius of solvated ion, } \eta\text{: viscosity}) \quad (4.58)$$

Also, Eq. (4.59) can be used to obtain the diffusion constant after obtaining the frequency (f_T) from Figure 4.43.

$$\tilde{D}_{Li+} = \frac{\pi f_T r^2}{1.94} \quad (4.59)$$

4.1.8 EQCM Analysis

An EQCM (electrochemical quartz crystal microbalance) analysis involves *in situ* monitoring of changes in the mass of electrodes during electrochemical reactions. This method is based on the principle that accumulated weight changes proportionally with resonance frequency [13, 14]. Figure 4.45 shows an EQCM device.

By obtaining the equivalent weight of adsorbed material using EQCM results, we are provided with an estimate of the species of material. Redox reactions at electrodes lead to elution, electrolyte decomposition, and formation of surface layers. Owing to mass changes at the electrode surface during these reactions, EQCM is an appropriate method of analysis.

Piezoelectric quartz crystals used in EQCM induce elastic lattice strain and shear strain by changing the dipole moment when subjected to mechanical compression or tension. The perturbation potential produces vibrations that are parallel to the surface. Transverse acoustic waves generated from these vibrations are able to propagate across the thickness of the crystal. The wavelength (λ) is given by Eq. (4.60).

$$\lambda = 2t_q \Rightarrow \lambda/2 = t_q \quad (4.60)$$

When electrochemical reactions create a new surface layer on the crystal, the thickness of this layer is t_r. The wavelength is shown in Eq. (4.61) and changes are shown in Figure 4.46.

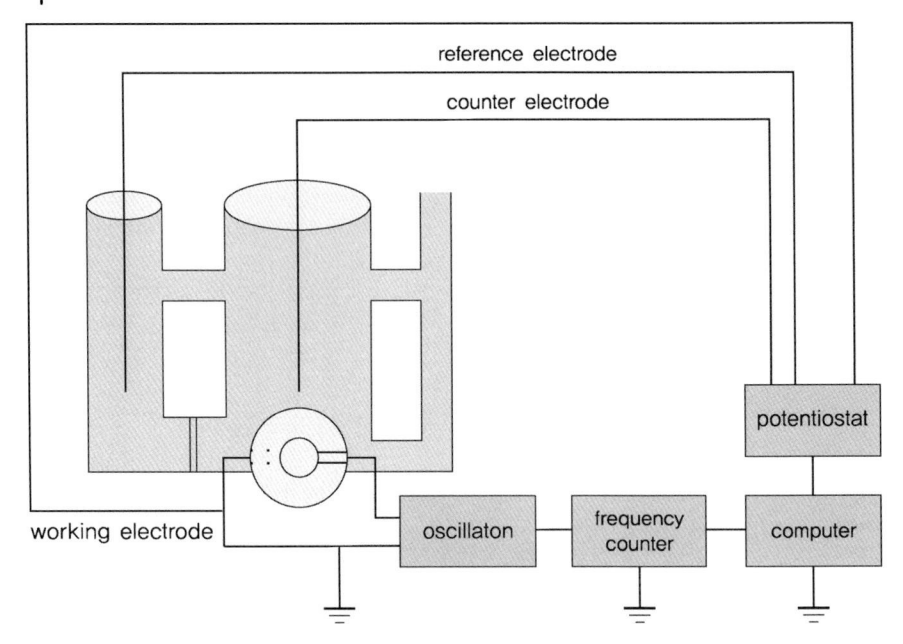

Figure 4.45 An EQCM device. Adapted with permission from [14] Copyright 1992 American Chemical Society.

$$\lambda/2 = t_q + t_r \tag{4.61}$$

The oscillation frequency is dependent on the wavelength and directly related to mass changes at the surface, thus leading to the Sauerbrey equation in Eq. (4.62).

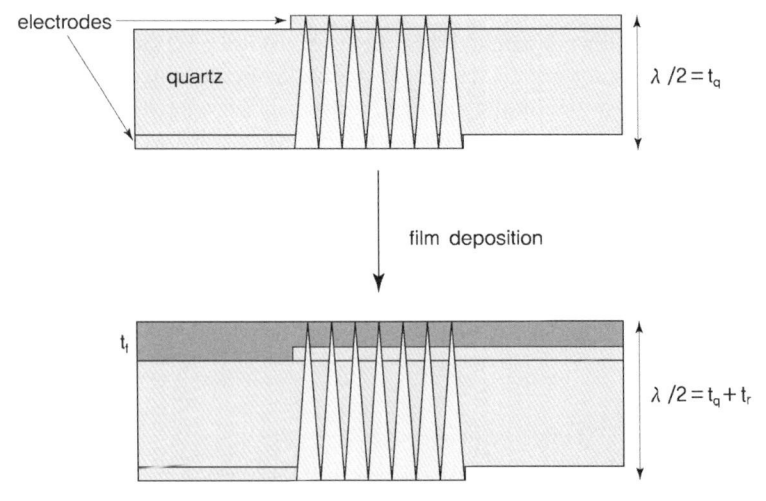

Figure 4.46 Comparison of transverse acoustic waves in quartz crystals with and without film deposition [13]. Adapted with permission from [13] Copyright 1992 American Chemical Society.

$$f_0 = v_{tr}/2t_q = (\mu_q^{1/2}/\rho_q^{1/2})/2t_q$$

$$\Delta f/f_0 = -\Delta t/t_g = -2f_0\Delta t/v_{tr} \tag{4.62}$$

$$\Delta f = 2f_0^2 \Delta m/A(\mu_q\rho_q)^{1/2}$$

Here, f is the oscillation frequency of the crystal, f_0 is the resonance frequency, Δm is the mass change, A is the piezoelectrically active area, ρ_q is the density of the quartz, μ_q is the shear modulus, and Δt can be expressed as the areal density $\Delta m/A$ since $\Delta t = \Delta m/\rho_q A$ [12, 13]. The above equations can be simplified to Eq. (4.63).

$$-\Delta f = C \cdot \Delta m \tag{4.63}$$

Here, C is the quartz crystal constant [15]. Mass per equivalent (mpe) is the mass per mole of electrons transferred and given by Eq. (4.64).

$$\text{mpe} = \frac{\Delta f \cdot F}{C \cdot Q} \tag{4.64}$$

By applying Eqs. (4.63) and (4.64), we can obtain the instantaneous mpe as shown in Eq. (4.65).

$$\begin{aligned}\text{Instantaneous mpe } (W'/z) &= -F\left(\frac{\Delta m}{\Delta Q}\right) \\ &= -\left(\frac{F}{i}\right)\left(\frac{\Delta m}{\Delta E}\right)\left(\frac{\Delta E}{\Delta t}\right)\end{aligned} \tag{4.65}$$

Here, W' is the instantaneous molecular weight, z is the number of atoms, F is the Faraday constant, Δm is the mass change, ΔQ is the amount of charge transferred to the electrode, i is the current flowing in a given voltage range, $\Delta M/\Delta E$ is the ratio of mass change to voltage, and $\Delta E/\Delta t$ is the scan rate of cyclic voltammetry [16].

If charge transfer from electrochemical reactions is caused by an increase in electrode mass, the value of mpe becomes the mass of materials adsorbed on the electrode surface. By deriving mpe as a function of E or t, we can determine the products and nature of electrochemical reactions at each stage.

This method of analysis has been widely used in the study of electrode–electrolyte interactions in lithium batteries. For instance, studies have covered mass change during the formation of surface layers from reactions between a lithium anode and an electrolyte [15].

The corrosion of aluminum current collectors has also been analyzed. Corrosion reactions result in the elution of electrode materials to the electrolyte or adsorption of products on the electrode surface. EQCM can be used to observe such mass changes [16].

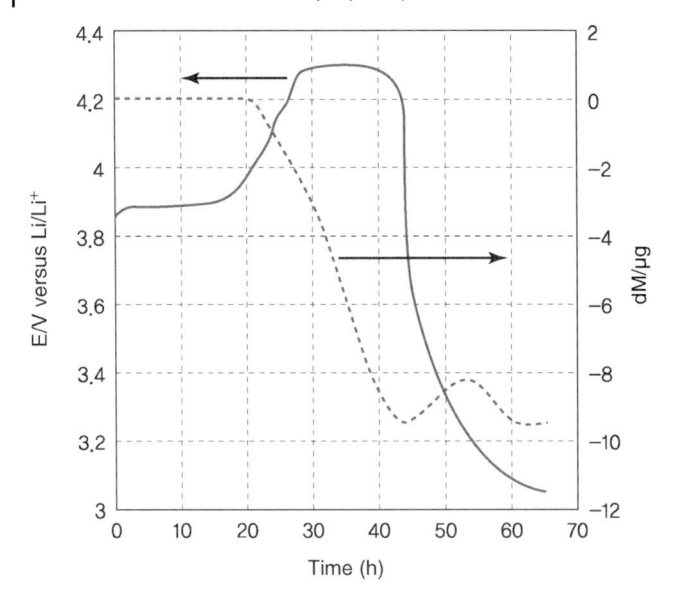

Figure 4.47 OCV and mpe changes of a LiMn$_2$O$_4$ film in 50 °C 1 M LiPF$_6$/PC/EC (1 : 1) electrolyte with storage time [17]. Reprinted from [17] Copyright 2001, with permission from Elsevier.

$$2H_2O + 2PF_6^- \rightarrow 2POF_3 + 4H^+ + 6F^- \quad (1)$$

$$2LiMn_2O_4 + 4H^+ \rightarrow 3\lambda - MnO_2 + Mn^{2+} + 2H_2O \quad (2)$$

As shown in Figure 4.47, EQCM was used to measure changes in mpe over time of the LiMn$_2$O$_4$ film. We can see that part of Mn was eluted during the cycle. By simultaneously measuring the open-circuit voltage, the different types of chemical reactions can be identified. The elution of MnO$_2$ results in a mass decrease [17].

References

1 Beak, W.K. and Park, S.M. (2003) *Electrochemistry*, CheongMonGak.
2 Wang, J. (2006) *Analytical Electrochemistry*, 3rd edn, Wiley-VCH Verlag GmbH.
3 Pyun, S.I. (2001) *Introduction to Material Electrochemistry*, SigmaPress.
4 Jung, K.N. and Pyun, S.I. (2007) *Electrochim. Acta*, **52**, 5453.
5 Deiss, E. (2005) *Electrochim. Acta*, **50**, 2927.
6 Kim, S.W. and Pyun, S.I. (2002) *J. Electroanal. Chem.*, **528**, 114.
7 Vorotyntsev, M.A., Levi, M.D., and Aurbach, D. (2004) *J. Electroanal. Chem.*, **572**, 299.
8 Deiss, E. (2002) *Electrochim. Acta*, **47**, 4027.
9 Orazem, M.E. and Trobollet, B. (2008) *Electrochemical Impedance Spectroscopy*, John Wiley & Sons, Inc.
10 Scully, J.R., Silverman, D.C., and Kendig, M.W. (1993) *Electrochemical Impedance: Analysis and Interpretation*, ASTM International.
11 Barsoukov, E. and Macdonald, J.R. (2005) *Impedance Spectroscopy*, 2nd edn, Wiley–Interscience.
12 White, R.E., Bockris, J.O'M., and Conway, B.E. (1999) *Modern Aspects of Electrochemistry*, Kluwer Academic/Plenum Publishers, p. 32, Chapter 2.

13 Buttry, D.A. and Ward, M.D. (1992) *Chem. Rev.*, **92**, 1355.
14 Bard, A.J. (1991) *Electroanalytical Chemistry*, Marcel Dekker, New York.
15 Aurbach, D. and Mashkovich, M. (1998) *J. Electrochem. Soc.*, **145**, 2629.
16 Song, S.W., Richardson, T.J., Zhuang, G.V., Devine, T.M., and Evans, J.W. (2004) *Electrochim. Acta*, **49**, 1483.
17 Uchida, I., Mohamedi, M., Dokko, K., Nishizawa, M., Itoh, T., and Umeda, M. (2001) *J. Power Sources*, **97–98**, 518.

4.2
Material Property Analysis

4.2.1
X-ray Diffraction Analysis

4.2.1.1 Principle of X-ray Diffraction Analysis

X-ray diffraction analysis is a method used to determine the phase and crystal structure of a solid sample by observing the scattered pattern of an X-ray beam from a lattice plane consisting of a uniform arrangement of atoms.

Within the solid crystal sample, lattice spaces are distanced some Å apart. An X-ray beam is an electromagnetic wave with a wavelength of 1 Å, which is similar to the size of atoms. When this X-ray beam is incident on the solid crystal, X-rays scattered from each atom interfere with each other to produce a diffraction pattern. Let d (A) be the spacing between the planes in the atomic lattice and θ be the angle between the incident ray and the scattering planes. Then, $2d \sin \theta$ corresponds to the light path difference with scattered waves, and diffraction occurs when this is equal to the wavelength (λ) multiplied by an integer n. This condition, known as Bragg's law, is expressed as follows:

$$n\lambda = 2d \sin \theta \qquad (4.66)$$

Bragg's law can be used to derive basal spacing (d), which is the spacing between lattice planes at 2θ from diffraction peaks (Figure 4.48) [1–3]. By analyzing the position of all peaks in the diffraction pattern, we can predict the distribution of lattice planes in the crystal and learn more about the crystal structure.

In general, the following information can be obtained from a XRD analysis:

1) Crystalline phase identification: The crystal structure can be examined by determining the space group and unit cell. The crystalline phase of the sample can be detected using peak positions and peak intensity. A useful reference is the powder diffraction pattern database assembled by the Joint Committee on Powder Diffraction Standards (JCPDS) and International Center for Diffraction Data (ICDD). The JCPDS card contains information on the crystalline phase of known substances, space group, unit cell, and peaks of diffraction patterns. Quantitative determination of impurities is also possible. For an actual sample, small amounts of impurities may overlap with the pattern of a single-phase substance. Since the XRD detection limit of impurities is 2 wt%, it is difficult to

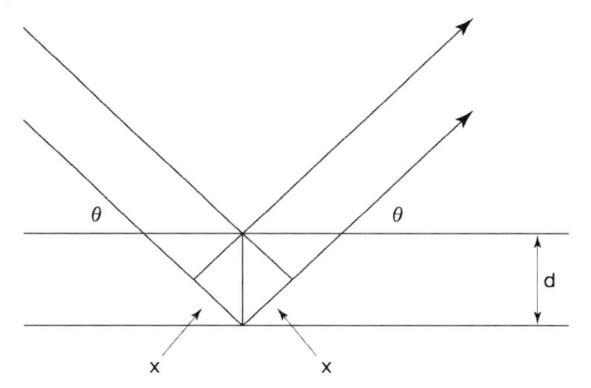

Figure 4.48 Bragg's law of diffraction.

detect lower quantities using a XRD analysis. For nanostructured powder or film samples, the peak size of oriented lattice planes tends to be larger than other peaks.

2) Crystallinity: While substances with high crystallinity exhibit sharp peaks (narrow half-width) in the diffraction pattern, amorphous liquids or glass shows widely distributed peaks. The diffraction pattern of polymers displays semicrystalline characteristics since part of the crystal exists within the sample. Thus, crystallinity can be derived with powder XRD measurements of peak shape and intensity.

3) Crystallite size: Particle size can be obtained from the Scherrer equation, which is based on the line broadening effect of X-ray peaks. Diffracted rays are clearly observed for highly crystalline primary particles of a powder sample with particle size averaging 0.2–20 µm. When the particle size of primary particles falls below 0.2 µm to tens of nanometers, the width of diffracted rays becomes larger. Noncrystalline diffraction patterns are exhibited if the particle size drops further to 20 Å. The Scherrer equation uses the above phenomenon to calculate average particle size, as given by Eq. (4.67).

$$t = k\lambda / B \cos \theta \tag{4.67}$$

Here, t is the particle size, k is the peak shape function (usually 0.9), λ is the X-ray wavelength, B is the full width of the peak at half-maximum (given in radians), and θ is the incident angle.

In addition, we can examine random strain and nonuniform distortion of particles. Local distortion of the crystal changes the distance between lattice planes, thus increasing the width of diffracted rays. Since this effect is more significant with larger angles of diffraction, the nonuniformity of crystals can be determined by studying the angle dependence of diffraction widths.

As shown in Figure 4.49, an XRD device consists of an X-ray source, goniometer, filter, sample, and a detector that gathers diffracted X-rays. The X-ray target produces X-rays when a high voltage is applied to the target filament, which is usually made of Cu.

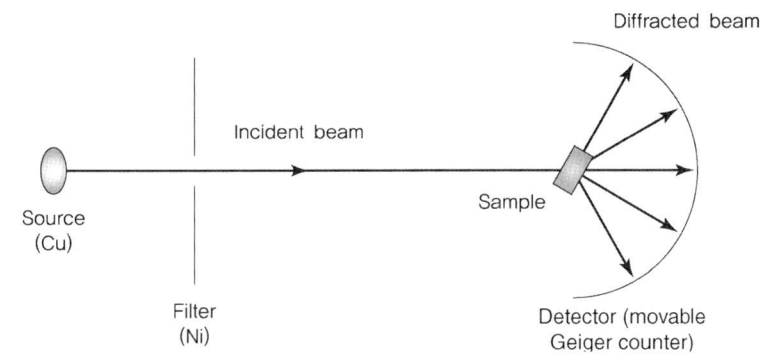

Figure 4.49 XRD device.

When electrons moving at high speed collide with atoms, electrons from inner shells near the nucleus jump out and create empty sites, and these sites are filled with electrons from outer shells. Electrons moving from a higher to a lower orbit generate an electromagnetic wave, called an X-ray, corresponding to the energy difference between orbits. X-rays produced from the filling of empty sites in shell K by electrons in shell L are known as K_α, while those originating from shell M are K_β lines. Cu-K_α lines are commonly used as the X-ray diffraction source and K_β lines can be filtered. To remove K_β lines, atoms lower in atomic number by 1–2 are used as filters. For instance, Ni-filtered Cu-K_α is often used as a light source in XRD since the Ni film strongly absorbs K_β lines of Cu.

4.2.1.2 Rietveld Refinement

Rietveld refinement is a method used in the determination of crystal structure by comparing diffraction patterns of a powder sample to that obtained from calculations and minimizing the difference between the two patterns. Owing to the symmetry of crystal space groups in a solid sample, peak positions are located at 2θ. The structure of the solid sample can be determined using peak intensity, shape, width, and position. In the past, the analysis of crystal structure was limited to single-crystal samples. However, detailed studies on the local structure of powder samples are now possible with the introduction of Rietveld refinement. While Rietveld refinement is less accurate than single-crystal X-ray diffraction, it is especially useful for substances that are not single crystal or single phase [3].

Rietveld refinement uses a least squares approach to match the observed intensity I_{obs} with the calculated intensity I_{cal} for the Bragg angle 2θ of a powder diffraction pattern based on various factors, including the lattice constant of a unit cell, position (x, y, z) of atoms and site occupancy, thermal parameters, baseline, and peak shape (Table 4.3). As can be seen in Figure 4.50, the measured diffraction pattern overlaps with the calculated diffraction pattern and the difference is represented by the bottommost line [4]. The vertical lines below the diffraction patterns are peak positions corresponding to *hkl* Miller indices. Refinement parameters for the diffraction data are scale factor, zero point, background, lattice constants (a, b, c)

Table 4.3 Information on LiFePO$_4$ crystal structure obtained from Rietveld refinement.

Atom	Site	g	x	y	z	B (Å^2)
Li(1)	4a	1	0	0	0	1
Fe(1)	4c	1	0.28 223(12)	1/4	0.9748(4)	0.6
P(1)	4c	1	0.0955(2)	1/4	0.4177(5)	0.6
O(1)	4c	1	0.0948(6)	1/4	0.7440(12)	1
O(2)	4c	1	0.4565(7)	1/4	0.2074(11)	1
O(3)	8d	1	0.1661(5)	0.0472(7)	0.2835(7)	1

Space group: Pnma, $a = 10.3234(8)$ Å, $b = 6.0047(3)$ Å, $c = 4.6927(3)$ Å, $R_{wp} = 8.85$.

of unit cells, atomic position (site, x, y, z), thermal parameter (B), peak shape, peak half-width, and degree of filling. Here, the degree of filling is the probability of each atom being located at the corresponding site and 1 is equivalent to 100%. The degree of filling changes when the same atomic position is filled by a second atom and the crystal structure becomes distorted.

Peak shapes can be expressed by several functions such as Gaussian, Lorentzian, Pearson, and pseudo-Voigt. Among these, the pseudo-Voigt function is the most widely used. Refinement is performed by minimizing R_{wp} (weighted pattern R factor), which is the difference between the measured and calculated profiles. R_{wp} is defined by Eq. (4.68).

$$R_{wp} = [W_i(Y_i(\text{obs}) - (1/c)Y_i(\text{cal})^2)]/[W_i(Y_i(\text{obs}))^2]^{1/2} \tag{4.68}$$

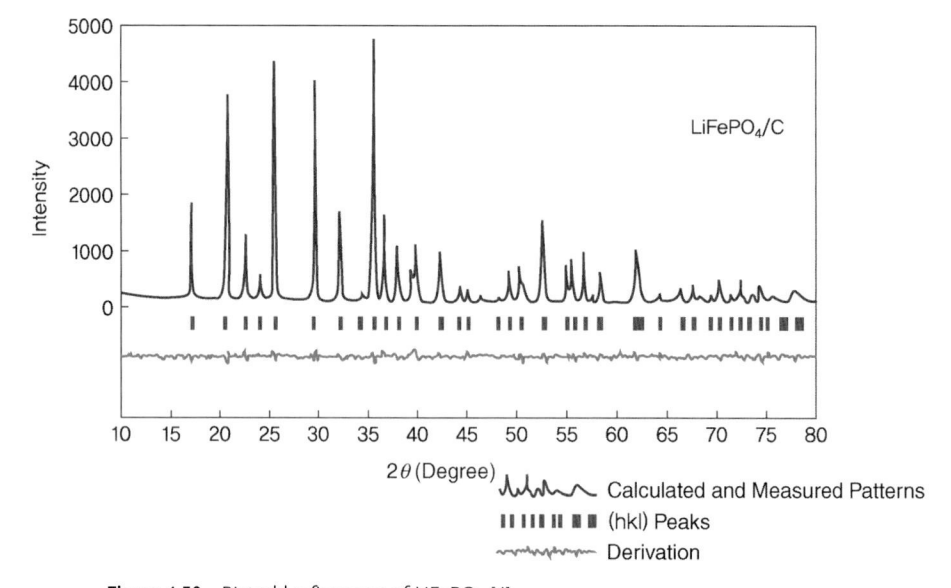

Figure 4.50 Rietveld refinement of LiFePO$_4$ [4].

Here, Y_i is the intensity at the *i*th step of the scanned pattern, c is the scale factor, and W_i is the weight factor. The results of Rietveld refinement are considered reliable when R_{wp} obtained using a least squares approach is found to be less than 10.

As shown in Table 4.3, Rietveld refinement reveals important information on crystal structure. The position of each atom within the crystal is determined by x, y, z coordinates, and the degree of filling g is also obtained. The thermal parameter B increases with thermal activity of an atom and takes on larger values for smaller atomic numbers. The R values below the table have been minimized after refinement. A smaller R indicates a smaller difference between measured and calculated diffraction patterns.

4.2.1.3 In Situ XRD

The intercalation and deintercalation of lithium ions cause changes in the crystal structure of electrode materials for lithium secondary batteries. These changes can be traced from changes in peak location or intensity in the XRD pattern. To make observations with *ex situ* XRD, the cell should be washed and dried at the end of electrochemical reactions. In this process, the cell may be transformed into a thermodynamically stable state, making it difficult to carry out an accurate analysis of structural changes that have occurred. On the other hand, *in situ* XRD allows real-time monitoring of changes in crystal structure.

To conduct an *in situ* experiment, we need to prepare an electrochemical cell for an *in situ* analysis, as shown in Figure 4.51. Even though XRD is used for bulk sample analysis [5], it should be noted that the potential observed during the experiment is caused by reactions at the particle surface. The time taken for surface reactions to reach an equilibrium state within particles must be minimized. In addition, there should be a uniform distribution of current density across particles, minimal internal resistance of the battery, and a constant supply of electrolyte throughout the cycle [6]. A transparent window is required for XRD beams to pass through. In order to keep the liquid electrolyte completely sealed, Be (beryllium) windows are used owing to their high mechanical strength and insulating properties.

In situ cells have been prepared by directly painting electrode materials on Be windows using the doctor blade technique [7]. Since Be oxidation occurs at potentials higher than 4.2 V [7], it is difficult to analyze anode materials that are easily oxidized. This problem has been resolved through the use of Bellcore cells, in which an air gap is introduced between electrodes to prevent Be from coming into direct contact with the electrolyte [5, 8].

XRD patterns for CoO_2 and NiO_2 were obtained after complete deintercalation of $LiCoO_2$ and $LiNiO_2$ using *in situ* cells. Changes in crystal structure with varying amounts of deintercalation can be identified and analyzed from the XRD patterns shown in Figures 4.52 and 4.53 [5, 9].

As a different example, alloy-type Sn-based cathode materials were deposited on a glass plate by sputtering. Figure 4.54 shows the results of an *in situ* XRD analysis [10]. XRD patterns with respect to time and voltage during charging and discharging are presented at the bottom and corresponding results are displayed above.

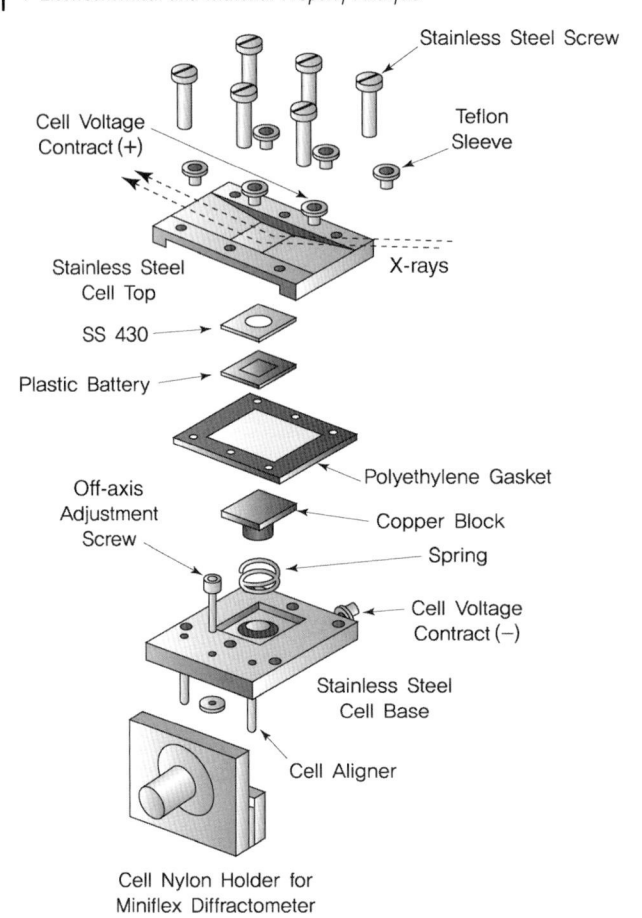

Figure 4.51 Schema of an *in situ* XRD cell [5]. Reproduced by permission of ECS – The Electrochemical Society.

As shown in Figure 4.55, another method is to use a cell that allows X-ray beams to pass through it. Despite requiring a synchrotron X-ray source, cell production is relatively easy and the X-ray penetration depth can be adjusted by varying the illumination angle [11, 12]. XRD patterns may exhibit peaks for copper or aluminum current collectors, but this can be resolved with smaller current collector widths. Such problems have not occurred in commercialized current collectors that are 10–25 μm thick.

Figure 4.56 shows *in situ* synchrotron diffraction results of a PLIon™ cell. When oxidation takes place with lithium deintercalation from the $LiMn_2O_4$ spinel anode, a new intermediate phase is formed at 3.3 and 3.95 V. The intermediate phase ceases to exist due to reduction by lithium intercalation. By analyzing these reactions, the intermediate crystal structure was identified as a double hexagon [13]. In Figure 4.56, (a–f) show pattern changes with charging and discharging of $LiMn_2O_4$.

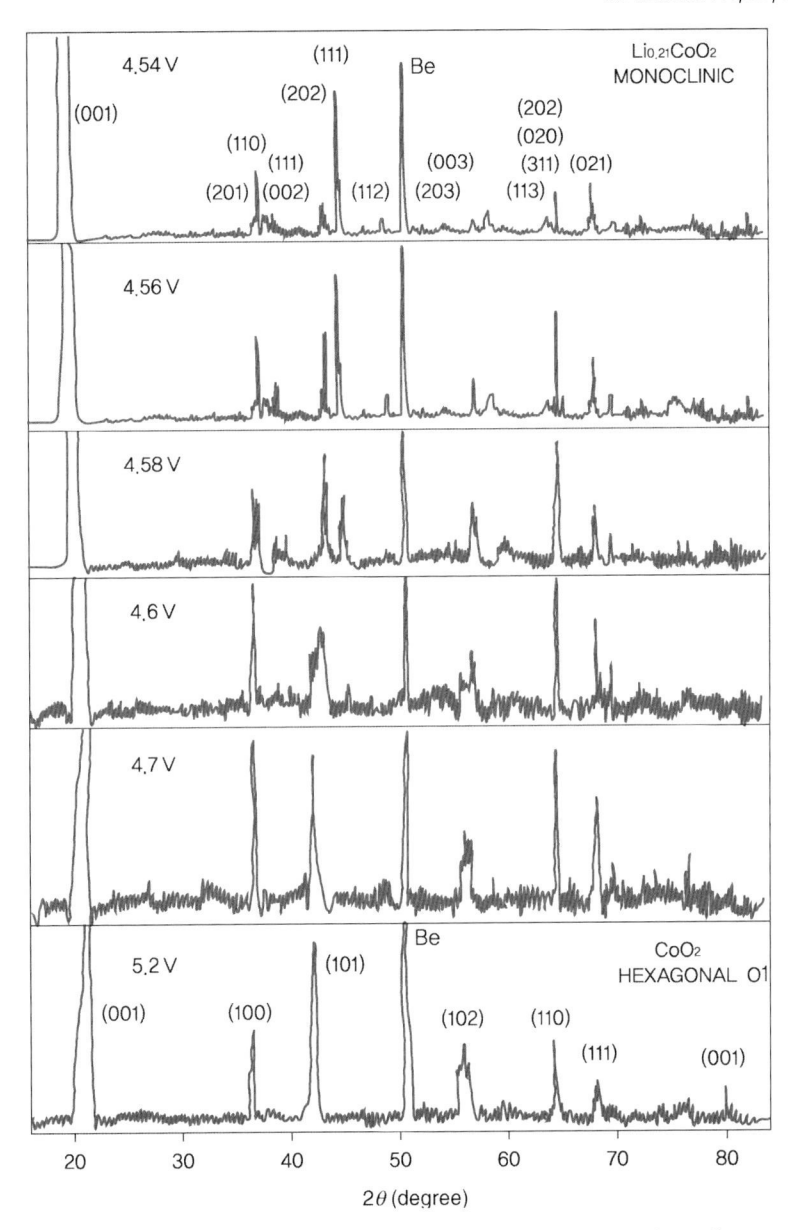

Figure 4.52 *In situ* XRD patterns measured while varying lithium intercalation from $LiCoO_2$ [5].

4.2.2
FTIR and Raman Spectroscopy

Depending on energy and frequency, electromagnetic radiation can be classified into various types as shown in Figure 4.57. Infrared and visible light correspond to the

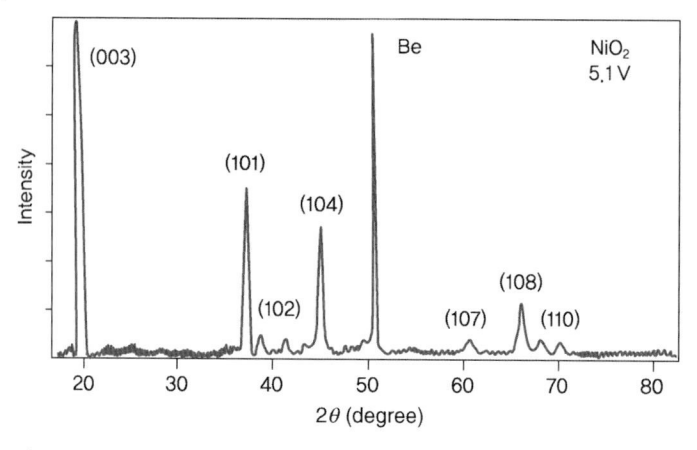

Figure 4.53 *In situ* XRD patterns of NiO$_2$ measured after lithium deintercalation from LiNiO$_2$ [15].

vibrational energy of molecules [14]. FTIR and Raman spectroscopy are common measurement techniques for electromagnetic radiation.

4.2.2.1 FTIR Spectroscopy

Compounds with covalent bonds absorb electromagnetic waves in the infrared range of the electromagnetic spectrum. The covalent bonds are then subjected to stretching or bending by infrared rays. Infrared absorption takes place when the frequency of

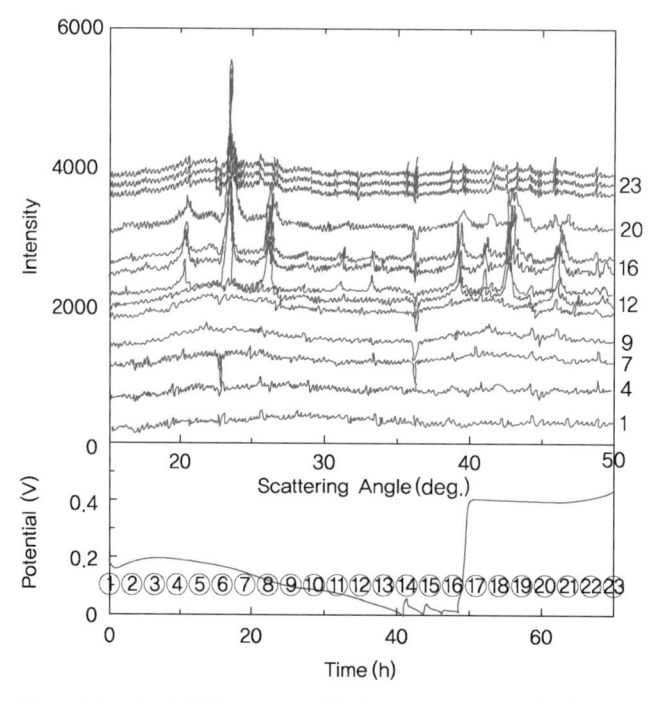

Figure 4.54 *In situ* XRD patterns of Sn-based anode material [10].

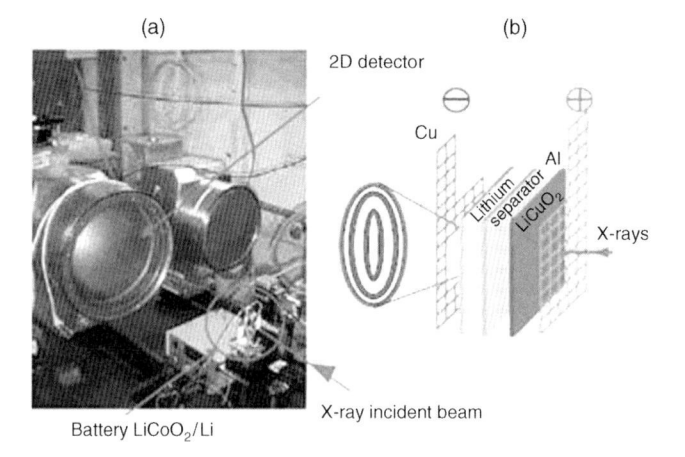

Figure 4.55 A thin plastic PLIon *in situ* cell allowing X-ray transmission [9]. Reprinted from [9] Copyright 2002, with permission from Elsevier.

molecular vibrations is the same as infrared frequency [15, 16]. The absorbed infrared energy induces transitions between levels according to the process described below (Figure 4.58).

When a molecule in the ground state ($n = 0$) becomes excited ($n = 1$), a fundamental transition occurs and the amplitude of vibrations increases. The wavelength of the fundamental transition is given as follows:

$$v(0 \to 1) = v_0/c[1-2X_a] \tag{4.69}$$

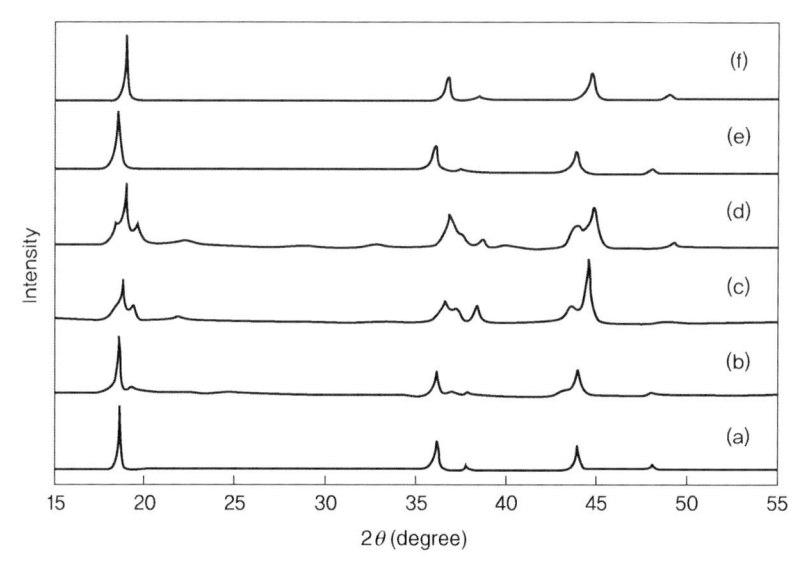

Figure 4.56 *In situ* XRD patterns measured with varying amounts of lithium deintercalation/intercalation from $LiMn_2O_4$. (a) $LiMn_2O_4$, (b) $Li_{0.8}Mn_2O_4$, (c) $Li_{0.5}Mn_2O_4$, (d) $Li_{0.1}Mn_2O_4$, (e) $LiMn_2O_4$ (4.8V after charging, 3.0V discharged), (f) $Li_{0.8}Mn_2O_4$ (initial state: $Li_{1.05}Mn_{1.95}O_4$) [13].

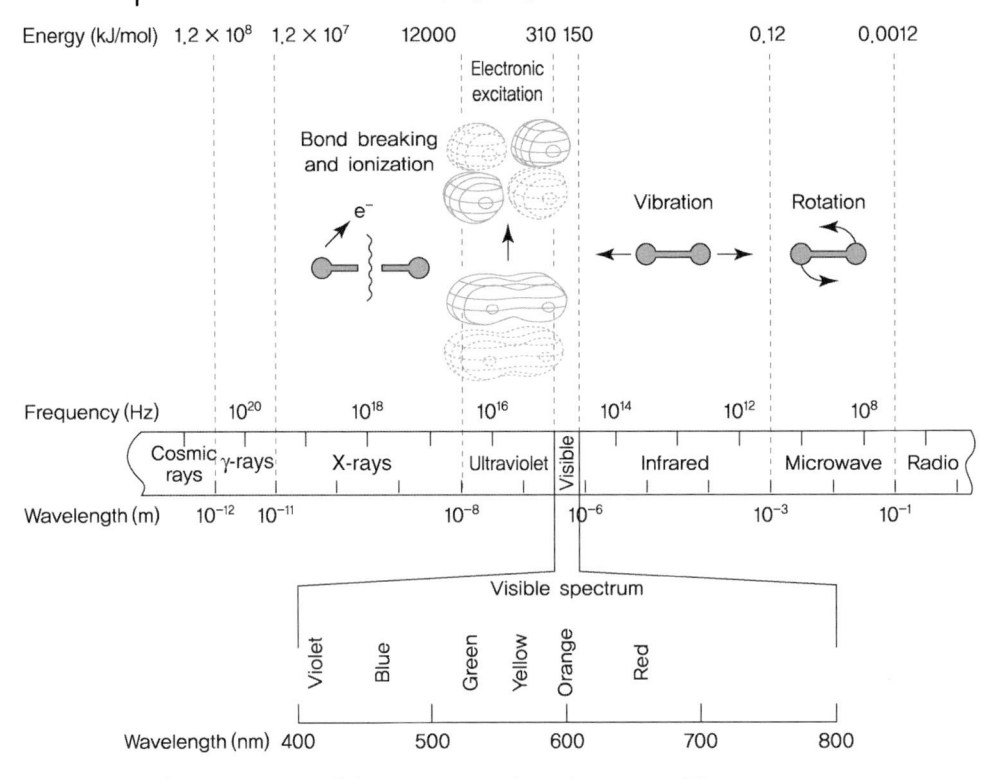

Figure 4.57 Types of electromagnetic radiation by energy and frequency.

$\nu(0 \rightarrow 1)$ is the transition from the ground state to the first excited state, X_a is the fraction that undergoes transition, ν_0 is the wavelength in the ground state, and c is the speed of light. Overtone transitions to higher excited states ($n = 2, 3, 4$) are not clearly represented in the spectrum due to weak absorption. The vibrational transition probability is proportional to the square of the transition dipole moment.

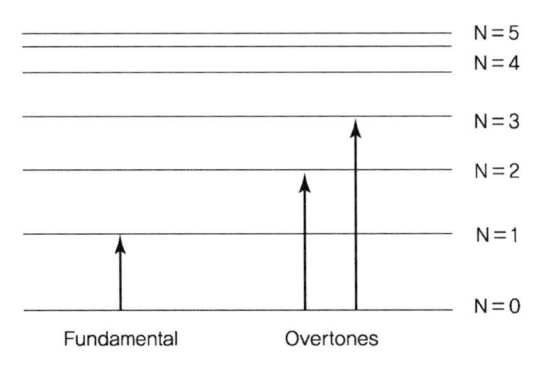

Figure 4.58 Transition of molecular vibrational energy with infrared absorption.

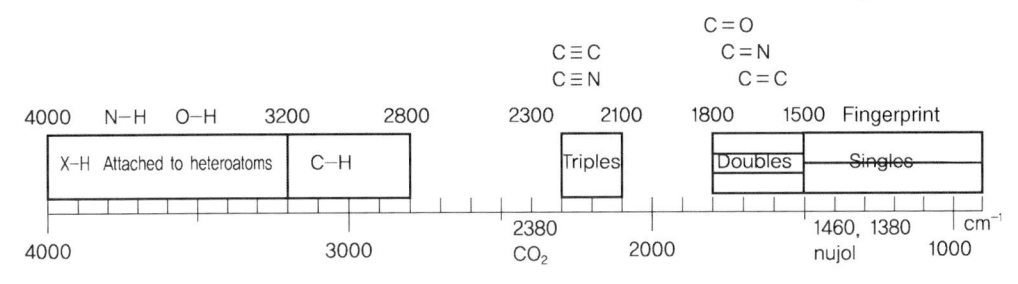

Figure 4.59 Observation of functional groups using infrared spectroscopy.

$$M_{vv'} = \psi(v)\mu\psi(v')d\tau \tag{4.70}$$

Here, $\Psi(v)$ and $\Psi(v')$ are the initial and final vibrational wave functions, and μ is the dipole moment. IR absorption takes place only with changes to the vibrational mode or dipole moment. The intensity of an absorption band (or peak) is defined as follows:

$$I_{IR} = (\delta\mu/\delta q)^2 \tag{4.71}$$

Here, μ is the dipole moment and q is the normal coordinate.

The infrared absorption spectrum is also called the fingerprint region, as chemical bonds of various organic functional groups can be observed in the mid-infrared range of 4000–400 cm^{-1} (Figure 4.59). Changes in the movement of absorption band wave number or intensity imply changes in the chemical structure of compounds or the surrounding chemical environment.

The infrared spectrum can be obtained by various methods including transmission, reflection, diffuse reflectance, and internal reflection [17].

1) Transmission mode can be applied only to transparent boards or samples, and incident light is not absorbed. Pellets prepared by mixing the powder sample and KBr are pressed into the form of a transparent disk. Absorbed infrared radiation can be expressed in terms of transmittance and absorbance.
2) Reflection mode is also known as reflection absorption infrared spectroscopy (RAIRS) or infrared reflection absorption spectroscopy (IRAS). It is performed at near-grazing incidence for metal single-crystal samples.
3) In diffuse reflectance, incident light is scattered from the rough surface of the powder, and the scattered light is then collected to obtain an absorption spectrum. This method is useful for samples with low transmittance.
4) In internal reflection, continuous reflection occurs when light enters internal reflection elements (IRE), such as diamond, germanium, and ZnSe. When a sample is coated onto an IRE, light is absorbed by the sample due to its low reflection index in the infrared region. The reflectivity of the IRE decreases, resulting in attenuated total reflection (ATR). Since infrared is absorbed mainly at the sample surface for each internal reflection, a surface analysis can be performed.

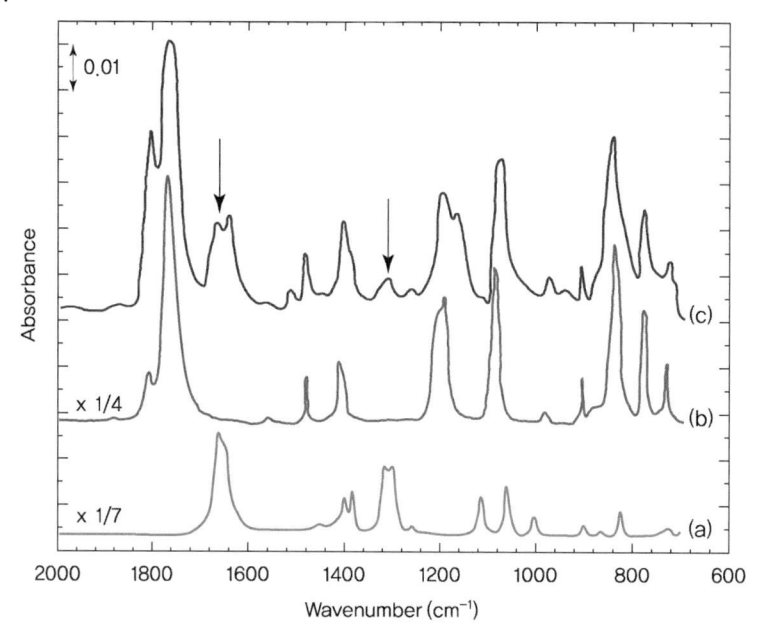

Figure 4.60 FTIR spectrum of (a) synthesized lithium ethylene carbonate, (b) EC solvate, and (c) Ni electrode surface after cycling at 0.5–2.5 V [18]. Adapted with permission from [18] Copyright 2005 American Chemical Society.

A FTIR analysis is conducted on lithium secondary batteries to obtain the following:

1) Compositional analysis and local structure of liquid and polymer electrolytes
2) Local structure of inorganic electrode materials
3) *Ex situ* and *in situ* analysis to determine SEI layer composition at the electrode surface

Figure 4.60 is an example of *ex situ* internal reflection spectroscopy for a FTIR analysis and shows the results of a compositional analysis on SEI layers formed at the electrode surface from reduction by a 1 M $LiPF_6$/EC:EMC (3:7) electrolyte. By comparing the spectrum of the reference sample and that of the electrode surface, SEI layers were found to contain lithium ethylene dicarbonate $(CH_2OCO_2Li)_2$ [18]. The existence of lithium alkyl carbonate $(LiOCO_2R)$ in SEI layers and whether it is transformed to anhydride $(ROCOR')$ or carboxylate $(R–CO_2-)$ are strongly debated issues. However, in the infrared spectrum, the lithium ethylene dicarbonate peak is clearly observed, with the calculated frequency matching the measured frequency. The corresponding molecular structure is shown in Figure 4.61.

Electrolyte decomposition occurs with redox reactions at the electrode surface and results in the formation of SEI layers. *In situ* FTIR is the most effective method to obtain information about these layers. Figure 4.62 shows the spectrum derived from subtractive normalized interfacial Fourier transform infrared spectroscopy (SNIFTIRS) on an *in situ* FTIR cell. IR spectra obtained from PC decomposition at different

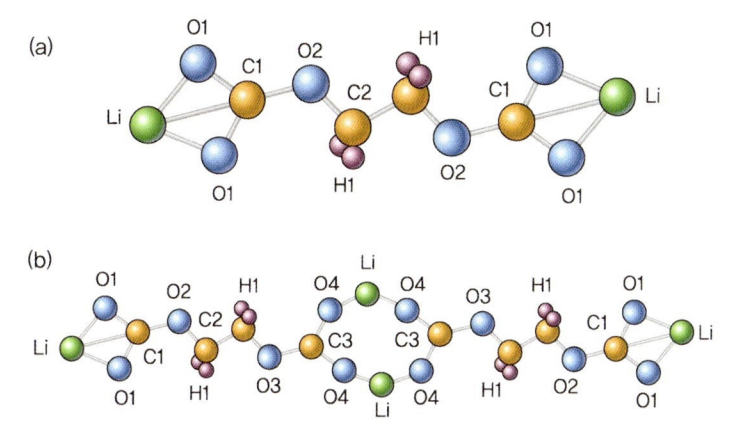

Figure 4.61 Molecular structure of (a) lithium ethylene carbonate and (b) a dimer (chemically different forms of C and O atoms are numbered 1–4) [18]. Adapted with permission from [18] Copyright 2005 American Chemical Society.

potentials were analyzed to obtain more information on the mechanism underlying SEI layer formation [19].

In Figure 4.62b, the peak of 1731 cm^{-1} observed at 4.2 V corresponds to the carbonyl group, while 1413 and 1221 cm^{-1} peaks correspond to C–O stretches in carboxyl groups. From this, we can see that the oxidation of PC to carboxylic groups begins at 4.2 V.

4.2.2.2 Raman Spectroscopy

In Raman spectroscopy, light intensity is weaker than that of infrared spectroscopy but single-wavelength lasers can be easily focused on specific areas of the sample, and there is less interference from water or carbon dioxide. Raman scattering is the inelastic scattering of light having lower energy than the incident light. This scattering occurs because part of the energy is used for molecular vibrations within the sample. The incident photon induces an electric dipole that forms a new energy level by interacting with molecular vibrations or vibrational energy levels. Stokes or anti-Stokes is observed since the electric dipole eventually resonates in the frequency of the incident light and either absorbs or loses the vibrational or rotational energy [20, 21].

As shown in Figure 4.63, electrons are excited into a virtual state during mutual interactions between photons and the molecule. Light is released as electrons jump from the virtual energy level to the ground state. This phenomenon is called Rayleigh scattering when the incident and emitted photons have the same energy, Stokes scattering if energy is lost and photons have higher vibrational energy levels, and anti-Stokes scattering if energy is gained and photons have lower vibrational energy levels. Let the electric field be $E = E_0 \cos 2\pi n_0 t$ (n_0: the frequency(Hz) of the incident electromagnetic wave) and the dipole moment be $M = \alpha E$ (α: polarizability, $\alpha = \alpha_0 + \frac{\partial \alpha}{\partial Q} dQ = Q_0 \cos(2\pi n_m t)$). The dipole moment of the molecule in the electric field can then be expressed as Eq. (4.72), which consists of Rayleigh scattering and Raman scattering.

$$M = \alpha_0 E_0 \cos(2\pi n_m t) + \left(\frac{\partial \alpha}{\partial Q} \frac{Q_0 E_0}{2}\right) \{\cos[2\pi(n_0 - n_m)t] + \cos[2\pi(n_0 + n_m)t]\}$$

(4.72)

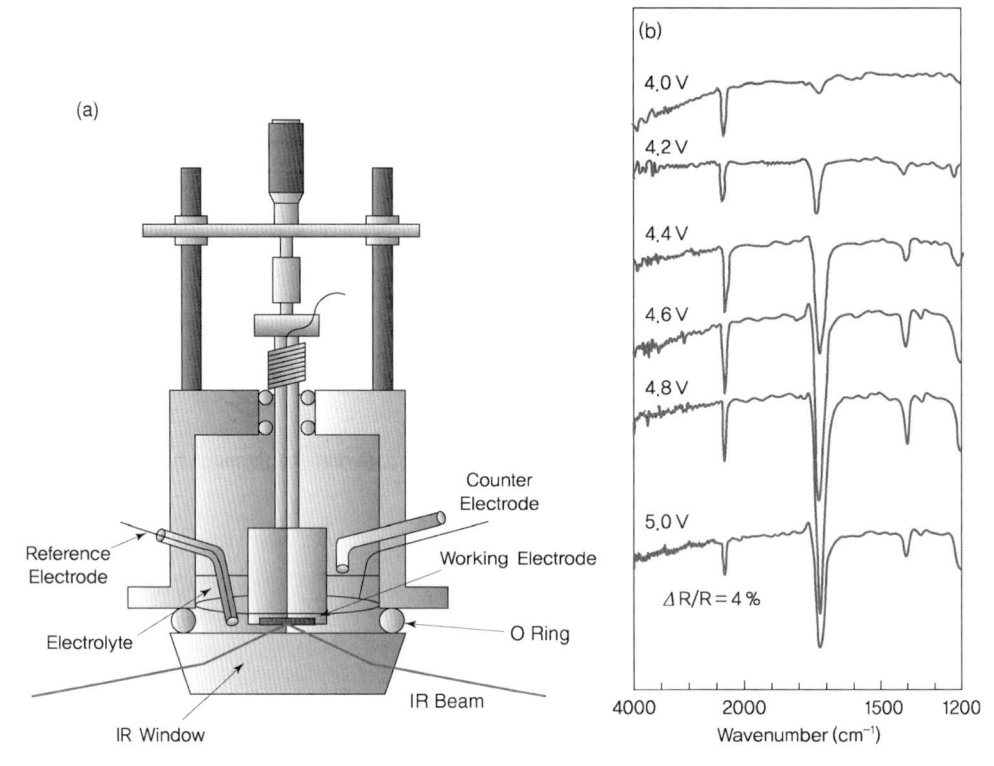

Figure 4.62 Changes in *in situ* IR spectrum by PC decomposition in (a) an *in situ* IR cell. Reproduced by permission of ECS – The Electrochemical Society. (b) A lithium cell consisting of 1 M LiClO$_4$/PC electrolyte and Ni electrodes [19]. Reproduced by permission of ECS – The Electrochemical Society.

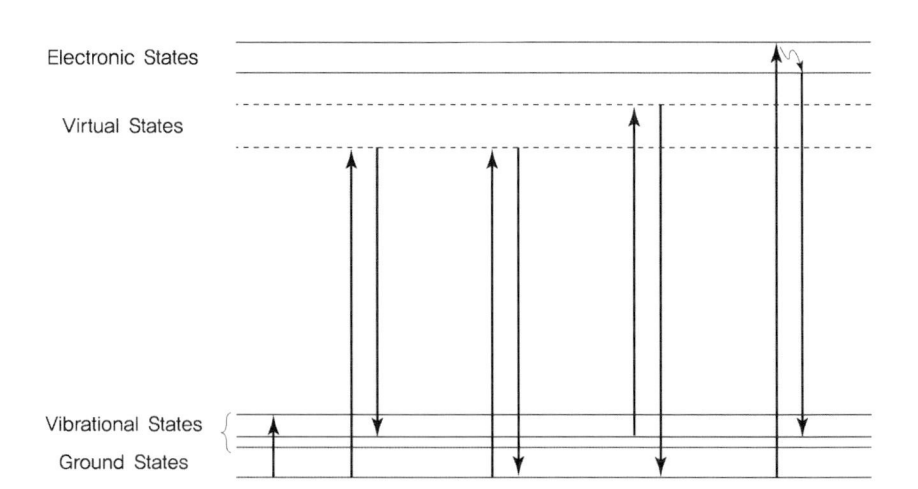

Figure 4.63 Vibrational energy transition by energy absorption in Raman spectroscopy.

With Rayleigh scattering as the reference, longer wavelengths (low frequency) are called Stokes lines and shorter wavelengths are anti-Stokes lines. Since most compounds exist in the ground state at room temperature, interactions take place with light corresponding to Stokes lines. Raman shift is the difference between Rayleigh lines and Stokes lines. The Raman spectrum is represented by the Raman shift frequency and the intensity of Raman lines.

Raman spectroscopy is used to observe molecular vibrations corresponding to the near-infrared, mid-infrared, and far-infrared regions and provides information on sample structure and composition. This method complements infrared spectroscopy, and vice versa. While infrared spectroscopy measures vibrational energy through changes in the transition dipole moment, Raman spectroscopy adheres to a different selection rule with measurements based on polarizability changes.

When a laser beam illuminates a sample such that molecular absorption is maximized, a strong Raman spectrum is obtained with molecular vibrations of chromophores caused by resonance Raman scattering. The strong resonance Raman spectrum facilitates analysis of samples existing in small amounts and low concentrations.

Micro-Raman spectrometers are equipped with an optical microscope, excitation laser, monochromer, and high sensitivity CCD (charge-coupled detector). By illuminating a laser beam on a small section of the sample, we can acquire a high-resolution spectrum in a short time and make detailed observations.

Another form of Raman spectroscopy is hyperspectral imaging, which collects information from thousands of Raman spectra. The heterogeneity of the sample is easily visible to the human eye since different compositions in specific areas are represented in unique colors. Figure 4.64 shows a hyperspectral image with

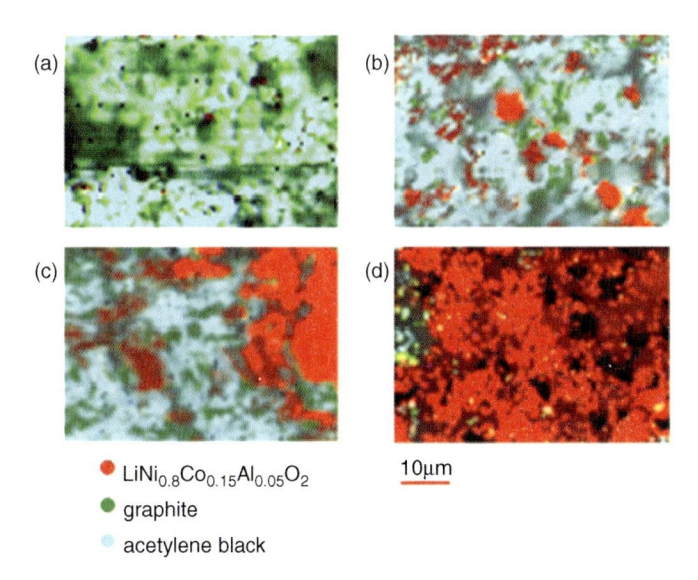

Figure 4.64 Raman microscope image of a cathode consisting of $LiNi_{0.80}Co_{0.15}Al_{0.05}O_2$–graphite–acetylene black. (a) Before charging/discharging. (b) Cathode with an output loss of 10%, (c) 34%, and (d) 52% [22]. Reproduced by permission of ECS – The Electrochemical Society.

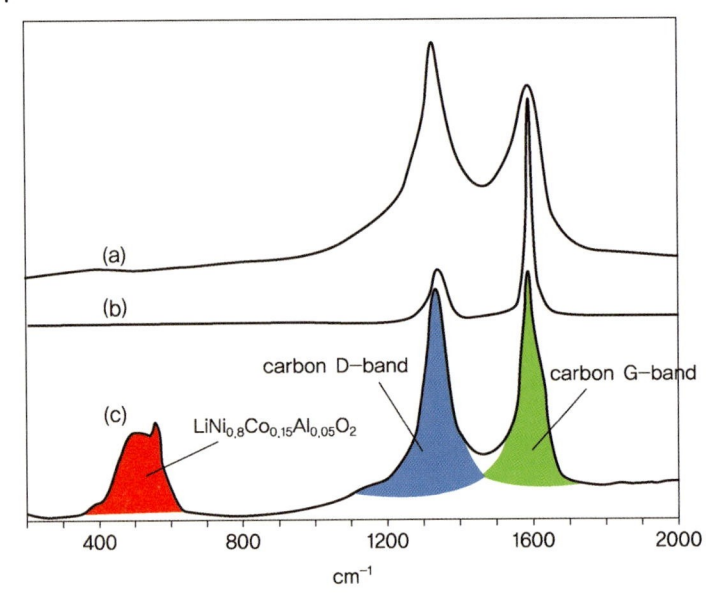

Figure 4.65 Raman spectrum of a cathode consisting of LiNi$_{0.80}$Co$_{0.15}$Al$_{0.05}$O$_2$–graphite-type carbon–acetylene black. Surface mainly composed of (a) acetylene black, (b) graphite, and (c) LiNi$_{0.80}$Co$_{0.15}$Al$_{0.05}$O$_2$ [22].

different colors used for LiNi$_{0.80}$Co$_{0.15}$Al$_{0.05}$O$_2$, graphite, and acetylene black. From the figure, it is easy to detect changes in composition as electrodes deteriorate over time [22]. Graphite and acetylene black exist in the bulk, and the cell coated with LiNi$_{0.80}$Co$_{0.15}$Al$_{0.05}$O$_2$ showed the largest decrease in capacity. This shows that cycle characteristics are affected by connectivity between particles, as well as uniform mixing with carbon.

Confocal microscopy uses point illumination along the z-axis of a sample and allows measurements in the sample depth direction while requiring only a small sample area. In addition to a side and depth resolution of 250 nm, this method provides a high spatial resolution of up to several micrometers. Based on automated XYZ stages for confocal performance, the spectra of various depth profiles can be obtained. We can also analyze the local structure of the sample, compositional distribution, and different phases.

For lithium secondary batteries, the structure of lithium metal oxides and carbons at the cathode can be analyzed. Recently, confocal microscopy has been used to examine structural changes at electrodes with varying potential before or after *ex situ* cycling. The cathode shown in Figure 4.65 is composed of LiNi$_{0.80}$Co$_{0.15}$Al$_{0.05}$O$_2$ with acetylene black and graphite carbons. With the Raman spectrum acquired from various parts, we can observe a nonuniform distribution depending on positions within the cathode.

Similar to FTIR, *in situ* Raman spectroscopy can be used as shown in Figure 4.66 to make effective observations of reactions at the electrode surface [23]. Using the *in situ*

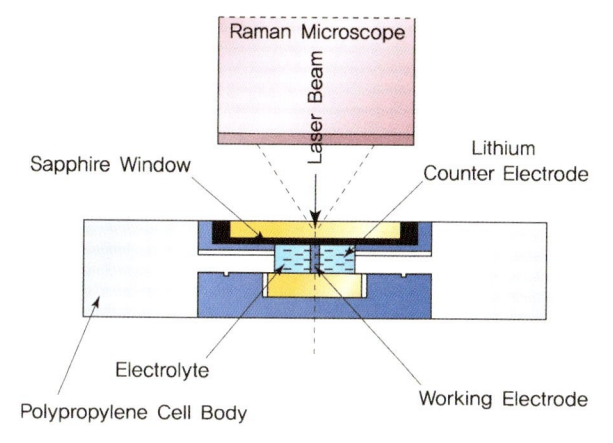

Figure 4.66 An *in situ* Raman cell [23].

Raman cell, a Raman spectrum can be obtained by varying the potential of the $LiCoO_2$ electrode. As shown in Figure 4.67, higher potentials lead to weaker intensities of the A_{1g} mode in $LiCoO_2$. Since the A_{1g} mode corresponds to vibrations along the *c*-axis [24], the results indicate changes in the lattice plane orientation of $Li_{1-x}CoO_2$ particles.

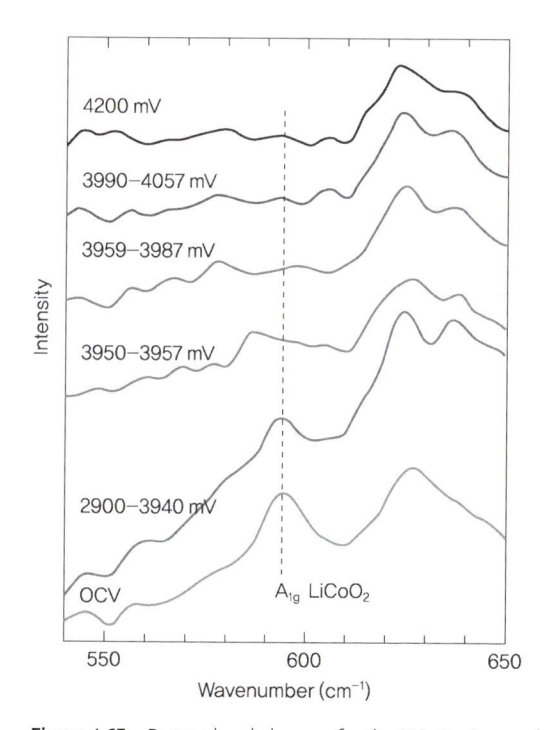

Figure 4.67 Raman band changes for the $LiCoO_2$ A_{1g} mode with varying potential [23].

4.2.3
Solid-State Nuclear Magnetic Resonance Spectroscopy

Nuclear magnetic resonance (NMR) observes the splitting of spin energy levels in overlapping nuclei within a magnetic field, as well as energy transitions resulting from resonant absorption by nuclear spins at the radio frequency range.

The additional splitting of a spectral line in the presence of an external magnetic field was observed by Zeeman in the late nineteenth century. In 1924, Pauli proposed the concept of a magnetic moment induced by the spin of a charged particle with angular moment. Zaviosky detected electron paramagnetic resonance using $CrCl_3$ in 1944, and the first NMR signals were observed in 1946 by Block and Purcell. Theoretical spin characteristics proposed by Pauli were later proved to be quantum properties under Dirac's quantization.

The splitting of energy levels from reactions between the magnetic field and the magnetic moment of the nucleus is a Zeeman interaction proportional to the intensity of the external magnetic field. The final NMR energy level is determined by smaller mutual interactions between magnetic moments that provide detailed information on the molecular structure.

The resonance frequency detected by NMR spectroscopy is the accumulation of various interactions between magnetic moments. It reacts with great sensitivity to nuclei bonds and chemical structure and presents important information on surrounding molecular structures. Molecular dynamics can be derived by analyzing changes in peak shapes or relaxation after resonance transition.

The magnetic moment μ of a nucleus is given by the following equation:

$$\mu = \gamma P = \gamma \hbar [I(I+1)]^{1/2} \tag{4.73}$$

Here, the angular momentum of the nucleus is proportional to the magnitude of P, and the proportionality constant γ is the gyromagnetic ratio, which takes on different values depending on the nucleus. I is the nuclear spin quantum number determined by internal composition. Elements with an odd-numbered atomic number and even-numbered atomic weight such as ^2D ($I=1$) and ^6Li ($I=1$) have integer values 1, 2, 3, while those with odd-numbered atomic weight such as ^1H ($I=1/2$), ^{13}C ($I=1/2$), ^7Li ($I=3/2$) have half-integer values of 1/2, 3/2, and so on. The z-axis components can have values of $-I, -I+1, \ldots, I-1, I$, and quantization occurs in $(2I+1)$ directions.

$$U = -\mu \cdot B = -\mu_z B = -\gamma P_z B = -\gamma \hbar m_I B \tag{4.74}$$

Assuming the z-axis to be the direction of the magnetic field, the Zeeman interaction can be expressed as follows. Here, m_I represents the z-axis component of the nuclear spin quantum number [25, 26]. The Zeeman interaction is the interaction of the magnetic moment of the nucleus with an applied magnetic field, and it varies according to the type of nucleus. Magnetic moment can be arranged in ascending order.

$$H_{Total} = H_{Zeeman} + H_{Quadrupolar} + H_{Fermi\ contact} + H_{Dipolar} + H_{CSA} + H_{J\text{-}coupling} \tag{4.75}$$

Interactions between nuclear spins are complex and provide extensive information on the internal structure of molecules. A basic understanding of such interactions is necessary to accurately interpret results of NMR spectroscopy.

H_{Zeeman} is as large as up to several hundred MHz due to the above-mentioned interactions. While its interactions are not caused by a magnetic moment, $H_{Quadrupolar}$ must be considered when interpreting results from a solid-state NMR analysis as it contributes to the NMR resonance frequency. When the nuclear spin quantum number is greater than 1, the nucleus takes on an asymmetric structure. The charged nonspherical nucleus is affected by the magnetic field and quadrupolar interaction in the lattice of the solid sample, along with the orientation of the nucleus, and is found to be in the range of several MHz. Such interactions must be considered when analyzing the sample using ^{6}Li ($I = 1$) and ^{7}Li ($I = 3/2$) NMR. $H_{Fermi\ contact}$ is the interaction between the spin of an electron existing inside a nucleus and that of the nucleus and can be found in paramagnetic substances. $H_{Dipolar}$ is the direct dipolar interaction between a nuclear spin and neighboring nuclear spins. It is usually as large as tens of KHz and has directionality. H_{CSA} is the interaction including chemical shift that is commonly used in NMR spectroscopy. It represents the tendency of the electron cloud to shield the nucleus from external magnetic fields. This value has directionality and changes according to electron preference and structure. $H_{J-coupling}$ is the direct and the indirect interaction between two nuclear spins through an electron pair. Unlike dipolar interaction, J-coupling has no directionality and is as small as hundreds of Hz. It is considered insignificant in solid-state NMR experiments but provides valuable information on connectivity with neighboring atoms.

In a liquid sample with active molecular activity, spin interactions having directionality are canceled out and detailed information on molecular orientation is lost. However, a high-resolution analysis is possible. In the solid state, interactions that change with molecular orientation are accumulated and wide peaks larger than hundreds of kHz are observed. These peaks contain extensive information but are too complex for analysis.

When the angle between the direction of orientation and the magnetic field is θ, H_{CSA} and $H_{Dipolar}$ interactions are proportional to $(1-3\cos^2\theta)$. In the liquid state, this is averaged and removed due to the active Brownian motion of molecules, thus allowing high-resolution results. In the powder form, individual powder grains are uniformly distributed in all directions, and nuclear spin interactions also follow a wide distribution. From an analysis of the relationship between θ and the magnitude of interactions, $(1-3\cos^2\theta)$ was eliminated at the magic angle of 54.74°. This was confirmed by setting the angle at 54.74° followed by magic angle spinning (MAS). Even in solid samples, high-resolution results can be derived by canceling out interactions of nuclear spins having directionality. High-resolution solid-state NMR spectroscopy presents several advantages, but loses detailed information on interactions between nuclear spins. Given the various methods of selective detection and analysis of nuclear spin interactions, it is important to select an appropriate technique depending on the sample and required information.

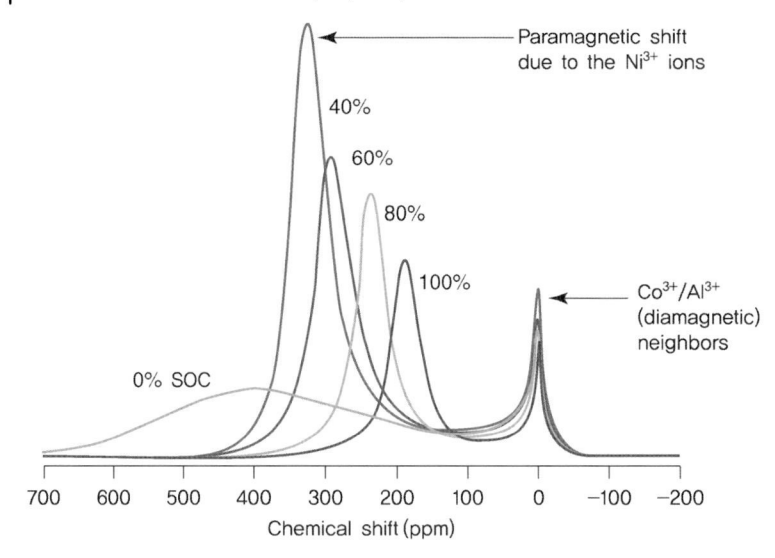

Figure 4.68 Changes in the ^{7}Li MAS NMR spectrum with charge state of LiNi$_{0.8}$Co$_{0.15}$Al$_{0.05}$O$_2$.

NMR for lithium batteries involves the analysis of structure and composition of various components, such as electrolytes, binders, and electrode materials. Structural changes in electrode composition can also be observed during charging and discharging. Figure 4.68 shows changes in the NMR spectrum with varying lithium deintercalation during the charging of LiNi$_{0.8}$Co$_{0.15}$Al$_{0.05}$O$_2$ [27].

Fermi contact and dipolar interactions are caused by the paramagnetic Ni^{3+}, and a wide distribution is seen with shifts in peak positions. As lithium ion content decreases during charging, Ni is transformed from its oxidized state of Ni^{3+} to the diamagnetic Ni^{4+}. The interactions are thus weakened, resulting in smaller peak widths and diamagnetic shifts of peak positions. As shown in Figure 4.68, lithium ions are deintercalated throughout the charging process and peak sizes become smaller. Figure 4.69 shows changes in the NMR spectrum for LiCuO$_2$ as the diffusion of lithium ions becomes easier with increasing temperature [28].

Lithium ions engage freely in diffusion within the lattice structure of LiCuO$_2$. The diffusion becomes active at higher temperatures, and peak widths are reduced when nuclear spin interactions are averaged and canceled out. From this, we can derive the activation energy and diffusion coefficient.

NMR spectroscopy is not limited to the example of anode materials presented above. It has been successfully implemented in studying the state of lithium ions in cathode materials, electrolyte composition, and the diffusion of ions.

4.2.4
X-ray Photoelectron Spectroscopy (XPS)

Special techniques are needed to carry out surface analysis of a sample. The thickness of the surface layer that can be observed differs according to the method of analysis.

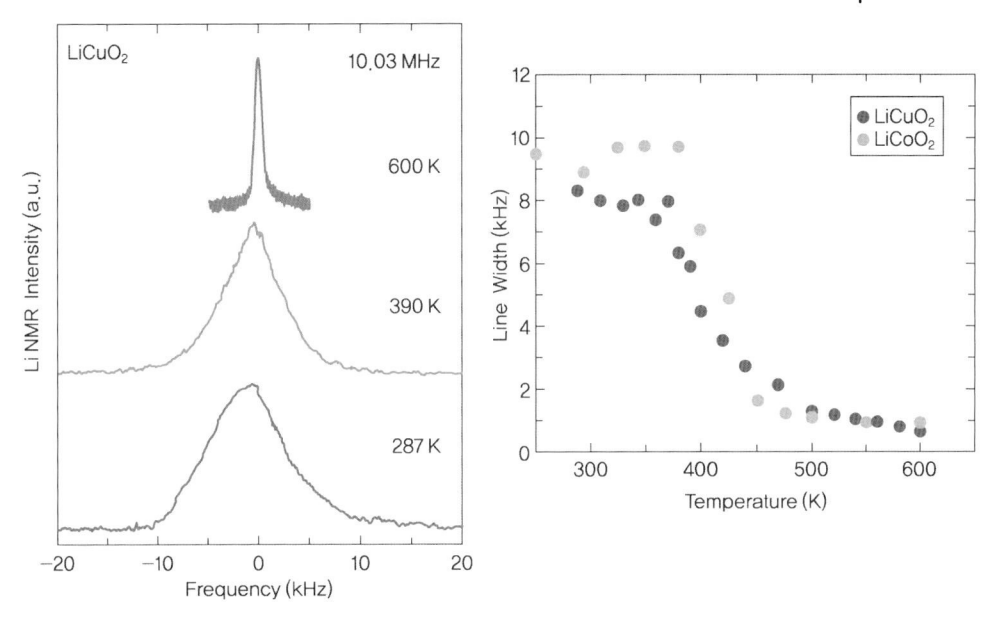

Figure 4.69 Changes in the ^{7}Li static NMR spectrum and peak width with varying temperature in $LiCuO_2$ [28]. Reprinted from [28] Copyright 2005, with permission from Elsevier.

X-ray photoelectron spectroscopy (XPS) is particularly useful for thin surface layers less than 100 Å, and this is also called electron spectroscopy for chemical analysis (ESCA). Figure 4.70 illustrates the principle of XPS, in which photoelectrons are emitted from a surface element when an X-ray having uniform energy ($h\nu$) is illuminated on the sample.

Given the uniform energy of the X-ray, the electron binding energy (E_B) of the emitted electron can be obtained by measuring the kinetic energy of the photoelectron. This binding energy is a unique property from which element types can be derived [29, 30]. Quantitative analysis is possible with binding energy measurements, and the bonding state of atoms can be determined from changes in binding energy.

An XPS system consists of a sample processing chamber, measurement chamber, and signal processing chamber. Functions of the sample processing chamber include exhaust gas processing, Ar ion etching at the surface, temperature control, and Au deposition. The measurement chamber focuses a weak X-ray on the sample transferred from the sample processing chamber. The released photoelectrons are moved to the energy analyzer. To prevent scattering of emitted electrons, the measurement chamber maintains a pressure of 10^{-9} Torr. The signal processing chamber measures the strength of photoelectrons and produces a photoelectron spectrum relating to binding energy.

Information obtained from XPS can be summarized as follows.

1) Element analysis (qualitative analysis): The X-ray energy (E_x) illuminated on the sample has the following relationship with electron binding energy (E_B) and

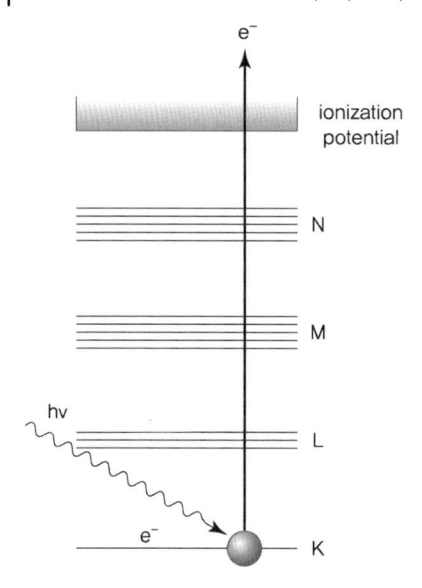

Figure 4.70 XPS system showing emission of core-level electrons by X-ray absorption.

photoelectron kinetic energy (E_k). Since E_x and E_k are obtained from measurements, E_B can be calculated to provide information on element type, electrode composition, and elements of impurities existing in small amounts.

$$E_B = E_x - E_k \tag{4.76}$$

2) Quantitative analysis: Several oxidation numbers can exist for a specific element and quantitative information can be derived by the peak resolution for that element. During this process, the binding energy of compounds containing the specific element and peak shapes (proportion of Gaussian versus Lorentzian) must be considered. After peak splitting, the relative amount of oxidation numbers can be calculated based on the area percentage of peaks. Figure 4.71 shows the Mn 2p spectrum peak resolution for $LiMn_2O_4$ with varying temperature. Binding energies of various manganese compounds are presented in Table 4.4. The areas of Mn^{3+} and Mn^{4+} peaks correspond to their respective concentrations. The average oxidation number of M can be determined from the ratio of Mn^{3+}/Mn^{4+} [31]. The nonlinear background is eliminated by the Shirley method [32].

3) Depth profiling: Ar ion etching allows analysis of elements and concentration change from the surface to the core. From this, we can identify differences in the element distribution between the surface and deeper layers. Figure 4.72 is an example of depth profiling performed to determine the thickness and composition of SEI layers on the graphite surface [33]. The change in concentration of elements is shown with Ar sputtering time (etching from the surface to the interior).

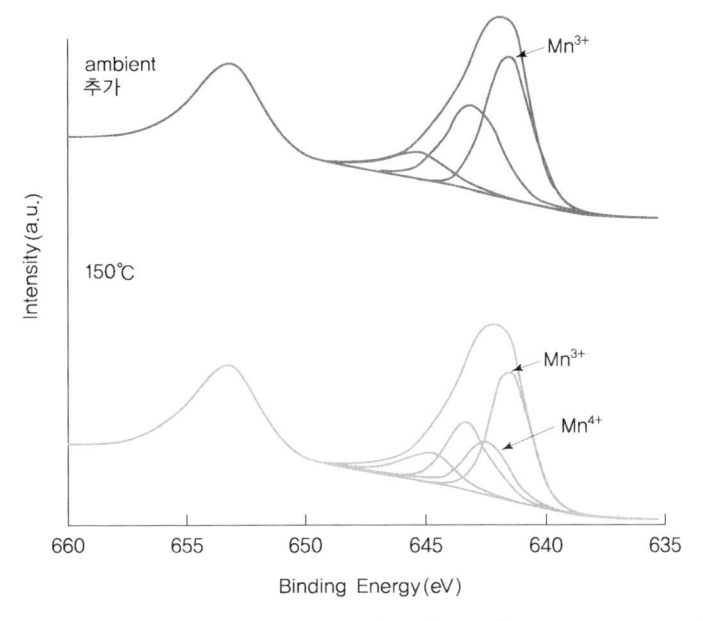

Figure 4.71 Mn 2p$_{3/2}$ XPS spectrum peak resolution of manganese compounds [30].

4.2.5
X-ray Absorption Spectroscopy (XAS)

X-ray absorption spectroscopy (XAS) is a method of determining the absorption coefficient based on the photon energy absorbed by a specific element of a substance. The absorption coefficient (μ) can be obtained from the emergent and incident radiation intensities by the following equation:

$$\mu = \ln(I_0/I) \tag{4.77}$$

Here, I_0 is the intensity of incident radiation and I is the intensity of transmitted radiation. When high-energy photons are absorbed by an atom, the atom is ionized as core–shell electrons are released. The resulting photoelectrons travel in the form of spherical waves shown in Figure 4.73. This wave is scattered by neighboring atoms and undergoes constructive or destructive interference.

Table 4.4 Mn 2p$_{3/2}$ binding energy (eV) values of various manganese compounds.

Mn	MnO	Mn$_2$O$_3$	MnO$_2$
—	—	641.5	642.4
639.2	641.0	641.9	642.5
—	641.7	641.8	642.4
638.2	640.9	641.8	642.5
—	641.0	641.7	642.2

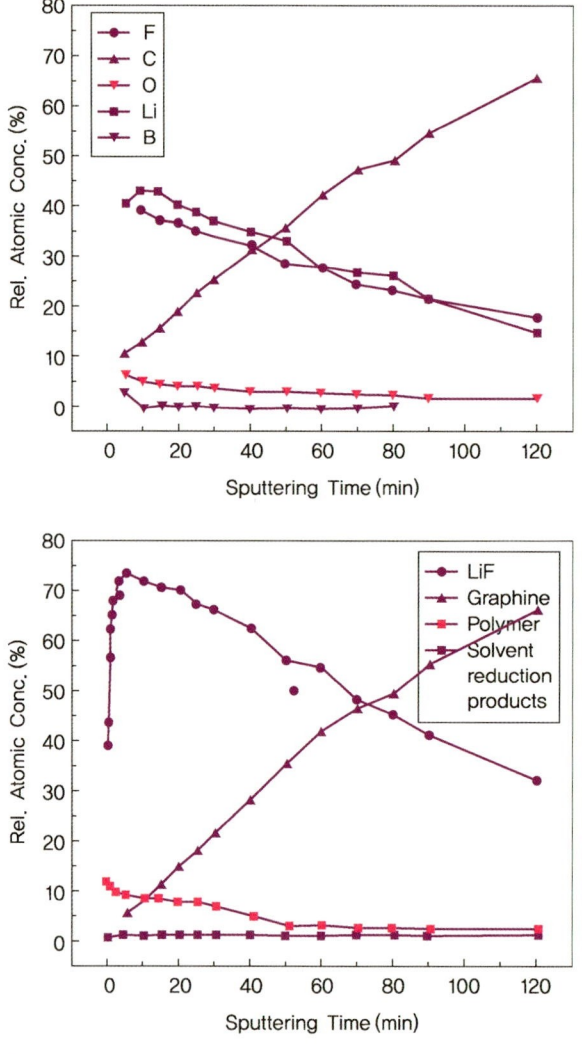

Figure 4.72 Depth profile of graphite after two cycles in 1 M LiPF$_6$/EC:DMC. Given as (a) elements and (b) solvent reduction products, polymers, and LiF on the graphite surface with Ar sputtering time [33]. Reprinted from [33] Copyright 2003, with permission from Elsevier.

The absorption coefficient of emitted photoelectrons changes, as shown in Figure 4.74, depending on the energy of X-ray incident radiation. This XAS spectrum can be classified into X-ray absorption near-edge structure (XANES) and extended X-ray absorption fine structure (EXAFS) regions.

The absorption coefficient increases sharply in the range known as the edge. The XANES region is the low-energy range from the pre-edge to about 40 eV. The remaining higher energy range is known as EXAFS. Since electron transitions in XANES can be described by orbital functions from the core to the Fermi level, we can

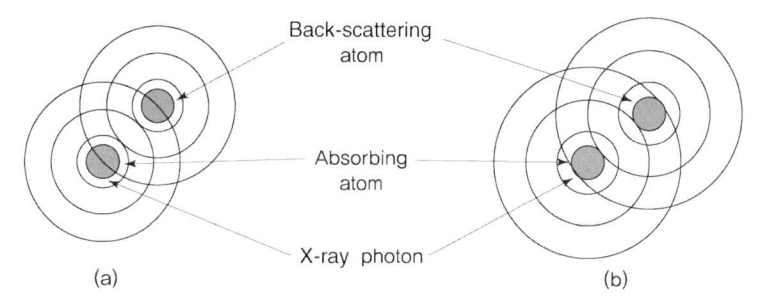

Figure 4.73 (a) Constructive interference and (b) destructive interference between photons and neighboring atoms.

obtain information on electrons and the dimensional structure of absorbed atoms. For example, the 1s → np transition corresponds to the K-edge. At the absorption edge greater than 1200 eV, the EXAFS absorption coefficient changes in the form of a sine curve depending on the photon energy. This provides detailed information on coordination number, distance between atoms, and degree of disorder [34–36]. Previously, such information was available only for single crystals. With recent developments in XAS technology, we can now determine the crystal structure of powder samples regardless of crystallinity or form. Furthermore, the crystal structure can be thoroughly examined by analyzing the spectra of all elements constituting a substance.

4.2.5.1 X-ray Absorption Near-Edge Structure (XANES)

The fine structure of XANES contributes to the transition of photoelectrons emitted from the core to partially filled orbital functions near the Fermi level. Overlapping scattering is caused by long mean free paths even though the low energy of released

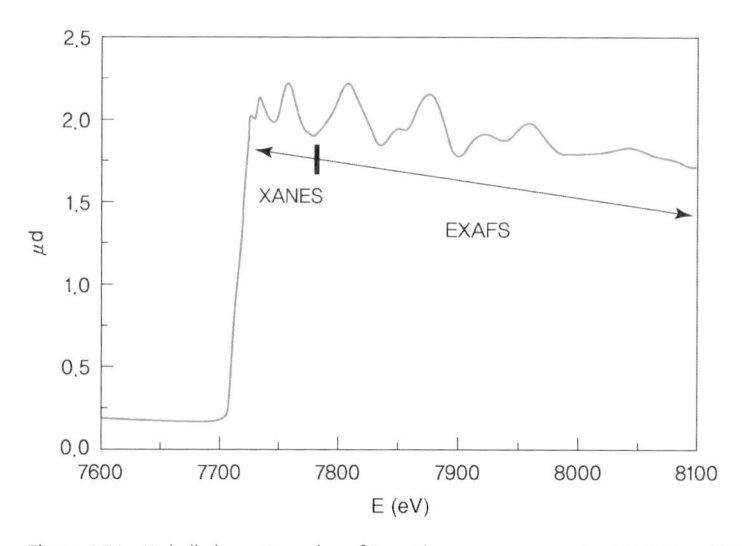

Figure 4.74 K-shell absorption edge of Co with arrows representing XANES and EXAFS regions.

photoelectrons is insufficient to counteract the influence of atoms. The edge shape provides information on the symmetry of ligands around an excited atom. In addition, the change in oxidation state of atoms can be seen through edge shifts, which arise from interactions between the nucleus and the hole.

The physical mechanism of XANES is explained by Fermi's golden rule. The transition probability (μ) from the initial to the final state due to a perturbation is given as follows [34].

$$\mu = (4\pi^2 \omega e^2/c) N_a /\langle \psi_i/z/\psi_f \rangle / \varrho(E_f) \qquad (4.78)$$

Here, μ is proportional to the final state density $\varrho(E_f)$ and z is an X-ray photon. According to the dipole assumption, transitions are subject to $\Delta l = \pm 1$ and $\Delta j = \pm 0$ or 1 [37]. At the K-edge of most period 1 transition metals, a transition occurs from 1s to 4p. If inversion symmetry exists, a small pre-edge peak is seen when there is a weak quadrupole ($l = \pm 2$) transition. A 1s → 3d transition takes place when a p orbital is combined with T_d, C_{2v}, C_{4v}, and D_{2d} symmetries, or for a quadrupole transition with O_h symmetry. This is generally observed in transition metals having an asymmetric orientation and empty d orbitals. These findings explain the crystal field splitting effect on the XANES spectrum. When covalent bonds are formed from the combination of p and d orbitals, they are reflected in the XANES spectrum as pre-edge peaks. The local structure obtained from XANES is under the influence of bonding angle, change in oxidation number, and geometric orientation. The example in Figure 4.75 shows the P (phosphorus) K-edge *ex situ* XANES of LiFePO$_4$ and quantitative analysis results for x in two-phase reactions of (xLiFePO$_4$ + (1 − x) FePO$_4$) [38].

4.2.5.2 Extended X-ray Absorption Fine Structure (EXAFS)

EXAFS contributes to backscattering of electrons in the excited state due to surrounding atoms. As shown in Figure 4.73, EXAFS signal vibrations arise from interference between outgoing waves from the core and those backscattered from surrounding atoms. The wave frequency and width, respectively, correspond to the distance between atoms and the number of neighboring atoms. The EXAFS spectrum with background removed from the absorption coefficient is given by the following equation. The difference between μ and μ_0 affects the local structure of absorbed atoms.

$$\chi(k) = \mu(E) - \mu_0(E)/\mu_0(E) \qquad (4.79)$$

Here, μ is the absorption coefficient in the presence of surrounding atoms, while μ_0 is the background absorption coefficient when there are no surrounding atoms.

To obtain EXAFS, the pre-edge background is first subtracted from the XAS spectrum. The cubic spline function is then used to separate low-frequency oscillations in the background. Figure 4.76 shows the EXAFS separation process.

Using Eq. (4.80), energy is transformed into a wave vector to obtain structural information based on EXAFS.

$$k = \sqrt{2m(E-E_0)/(2\pi/h)^2} (\approx \sqrt{0.263(E-E_0)}, \text{eV units}) \qquad (4.80)$$

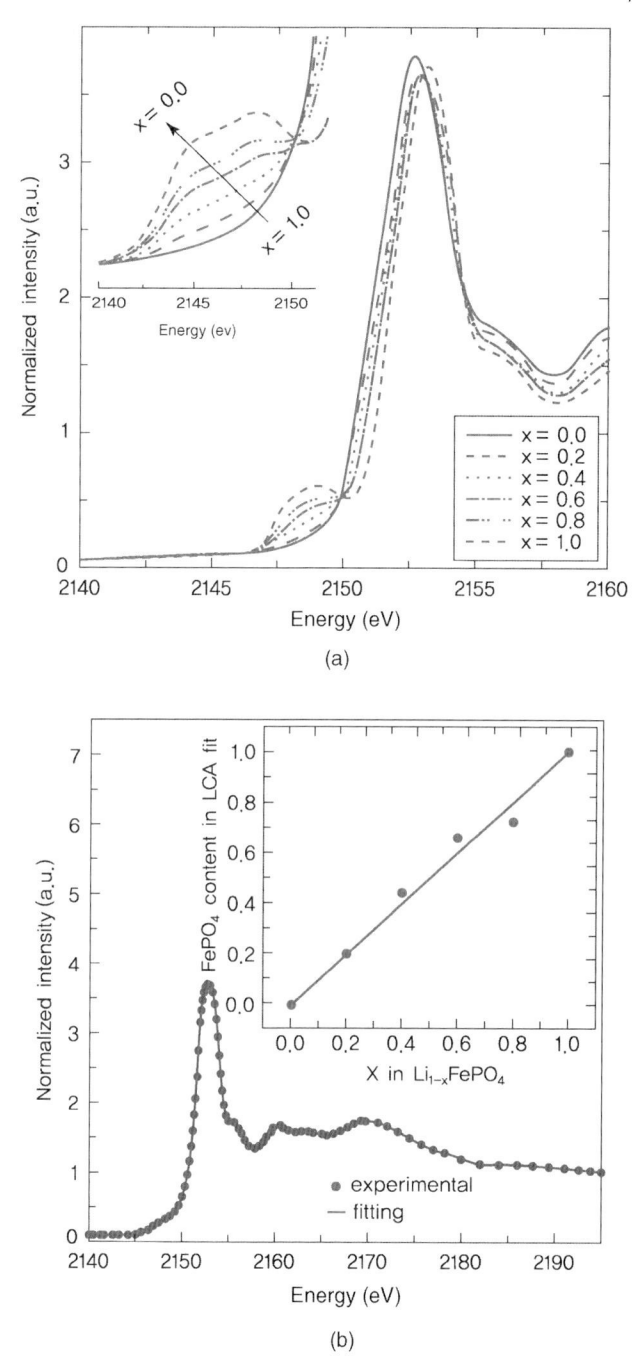

Figure 4.75 (a) P (phosphorus) K-edge XANES of the $Li_{1-x}FePO_4$ electrode with varying lithium amount x and (b) determination of x from P K-edge XANES spectrum fitting [38].

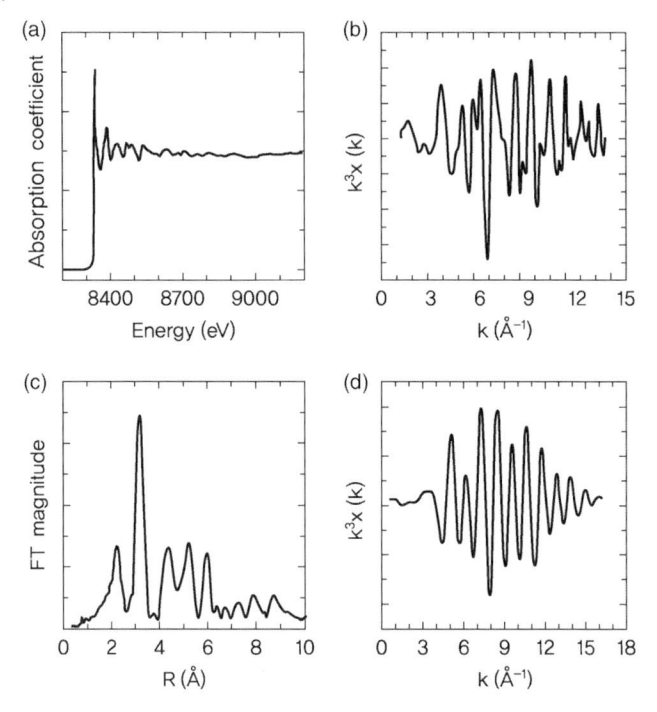

Figure 4.76 EXAFS analysis: (a) spectrum obtained from measurements, (b) k^3-weighted EXAFS oscillation, (c) spectrum obtained from Fourier transformation, and (d) spectrum obtained from inverse Fourier transformation.

Here, m is the electron mass and E_0 is the threshold energy of excited atoms in the core–shell.

The resulting EXAFS spectrum appears with backscattering from shells of different configurations. The following equation shows the relationship between EXAFS and structural factors:

$$\chi(k) = -S_0^2 \sum_i N_i F_i(k) \exp\{-2\sigma_i^2 k^2\} \exp\{-2R_i/\lambda(k)\} \\ \sin\{2kR_i + \varphi_i(k)\}/(kR_i^2) \tag{4.81}$$

Here, i is the ith shell, S_0^2 is the amplitude reduction factor indicating multi-electron excitations (shake up/off) of absorbed atoms, N_i is the average coordination number, $F_i(k)$ is the backscattering amplitude of an atom in the ith shell, σ_i^2 is the Debye–Waller factor static, R_i is the average distance between an absorbed atom and surrounding atoms in the ith shell, $\lambda(k)$ is the mean free path showing coherence loss due to multielectron excitations, and $\varphi_i(k)$ is the sum of phase transitions during scattering [34–36, 39]. In the above equation, the term $\exp\{-2R_i/\lambda(k)\}$ represents the nonelastic loss caused by the interaction medium. On the other hand, $\exp\{-2\sigma_i^2 k^2\}$ arises from thermal vibrations and the degree of static disorder. Ultimately, an EXAFS analysis involves the derivation of R_i, N_i, and σ_i^2. Other factors such as ϕ_i, and F_l can be obtained from using standard compounds or through theoretical calculations. Parameters that are closely related can be divided into two groups:

$$\{F(k), \sigma, \lambda, N, S\} \text{ and } \{\varphi(k), E_0, R\}$$

The spectrum obtained from a Fourier transform (FT) shows peaks of R space, where R is the distance between absorbed atoms. This value is smaller than the configuration distance due to the phase potential $w_i(k)$. Since the EXAFS spectrum is the sum of shell x in different configurations, FT in EXAFS is also the sum of FTs from each shell x. As such, by performing an inverse FT on a peak of interest in the R space, a single shell can be isolated from other shells. The obtained data follow a sine EXAFS function and can be used to derive R_i, N_i, σ_i^2, ΔE_0. Such coupling is decoupled with k^1 and k^3 fitting in k and R spaces. This is because σ_i^2 and ΔE_0 do not affect k^3 and k^1, respectively. Also, R_i and ΔE_0 tend to have opposite effects on imaginary parts. Thus, these structural factors must be considered when interpreting EXAFS data.

In situ XAS is used to track changes in the local structure of atoms during charging and discharging of lithium secondary batteries (Figure 4.77). The structure of an *in situ* cell is as follows [40]. On the basis of *in situ* EXAFS measurements, we can identify local structural changes of Ni and Co atoms during lithium deintercalation from anode materials containing various elements such as $LiNi_{0.85}Co_{0.15}O_2$. Figure 4.78 shows the Ni EXAFS FT spectrum where the 1.5 A peak arises from the first shell of Ni–O bonding, and the peak intensity increases with the amount of lithium deintercalation. This is because the conversion of Ni^{3+} to Ni^{4+} is accompanied by a less significant Jahn–Teller effect and greater symmetry in the Ni configuration. The Ni–Ni bond represented by the 2.6 A peak shifts to the lower R regions, which indicates shorter Ni–Ni bond lengths. In the same manner, Co–O and Co–Co peaks are observed in the FT spectrum of CO. Unlike Ni, Co–O peak intensity and Co–Co

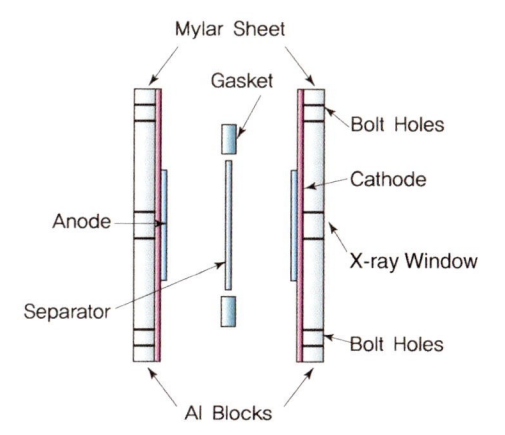

Figure 4.77 An *in situ* XAS measurement cell [40]. Reprinted from [40] Copyright 2001, with permission from Elsevier.

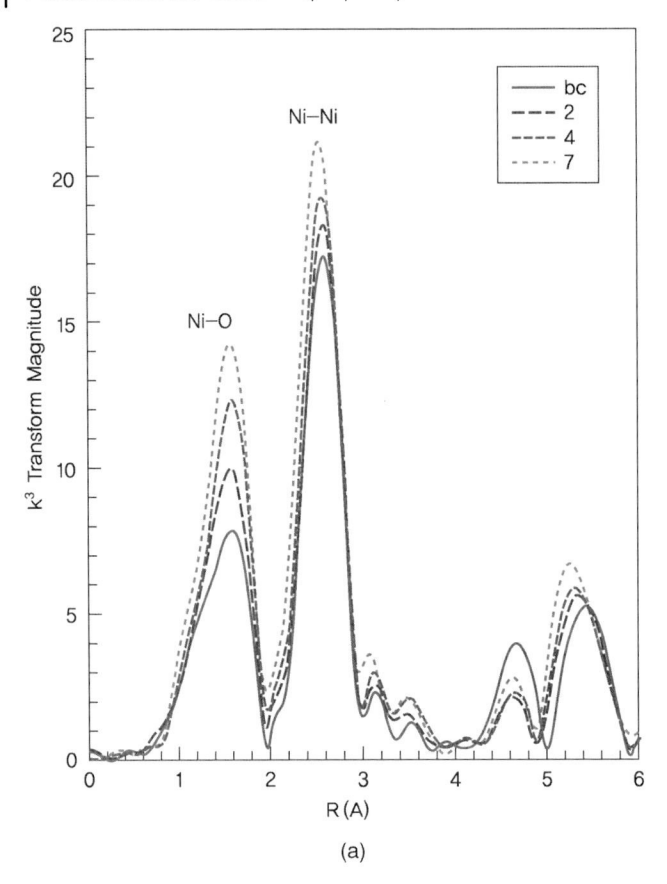

Figure 4.78 (a) Ni EXAFS spectrum (k^3 weighted) after Fourier transform. Reprinted with permission from American Chemical Society Copyright 2001 and (b) Co EXAFS spectrum (k^3 weighted) after Fourier transform of Co–Ni oxide measured during charging [40]. Reprinted from [40] Copyright 2001, with permission from Elsevier.

peak locations do not undergo major changes. The Co–Co peak intensity decreases slightly due to structural disorder in the CoO_2 layer resulting from lithium deintercalation. By analyzing this example, we can see that Ni is more easily oxidized than Co, leading to shorter bond lengths and local structural changes.

4.2.6
Transmission Electron Microscopy (TEM)

Transmission electron microscopy (TEM) is a microscopy technique whereby a beam of 120–200 kV is passed through a specimen and allowed to interact, thus obtaining information on grain size, crystallinity, microstructure, and crystal structure (Figure 4.79). Electrons generated from a tungsten filament are negatively charged and attracted to the cathode, with the speed determined by the potential difference between electrodes. The electrons are accelerated and concentrated on

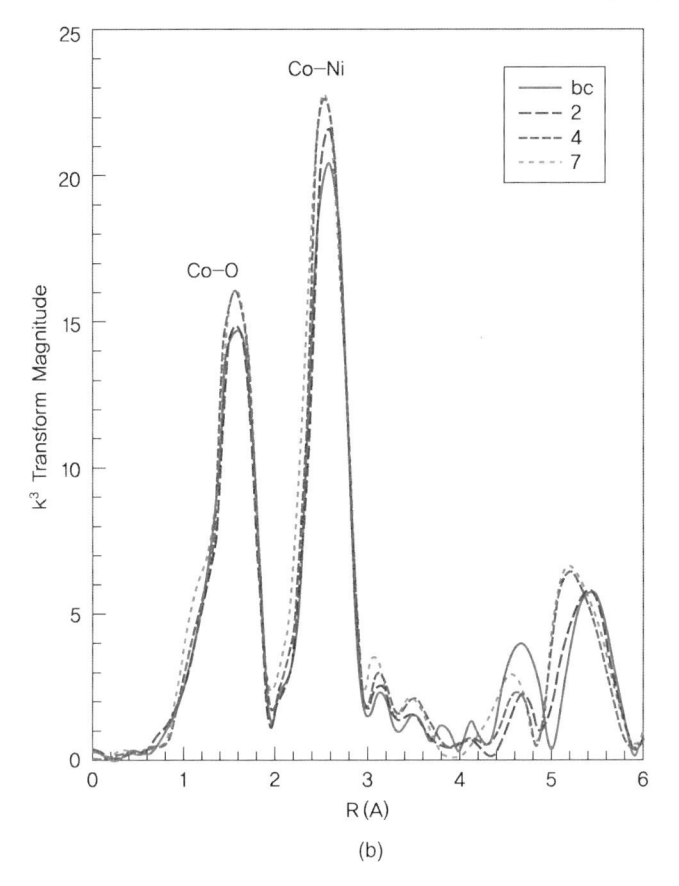

Figure 4.78 Continued

the specimen by electromagnetic lenses. The wavelength of an accelerated electron beam moving at v can be expressed as $\lambda = h/mv$. A TEM is composed of an electron gun, a condenser system to collect electron beams, an imaging system, a projection system, and an observation and recording system. The condenser system controls the intensity and angle of electron beams incident on the specimen. The objective lens produces an image that is gradually magnified and focused onto a fluorescent screen.

Crystalline materials interact with electron beams through diffraction and not absorption. The diffraction strength depends on the orientation of lattice planes in the interacting material. Specific diffraction patterns can be derived from TEMs equipped with goniometers that allow movement or two-axis tilting of sample holders. In addition, diffraction patterns in a specific direction are possible with apertures immediately below the specimen. By placing the aperture such that only unscattered electrons pass through, we can obtain information on the crystal structure of the specimen by observing electron density in high contrast images.

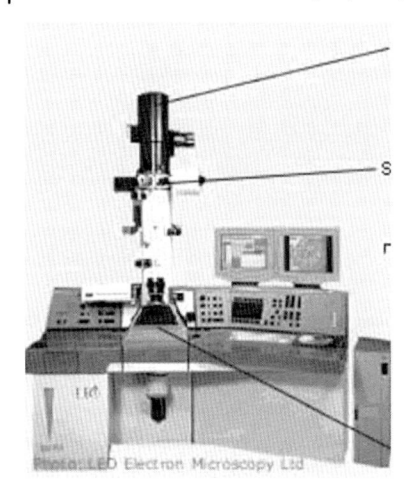

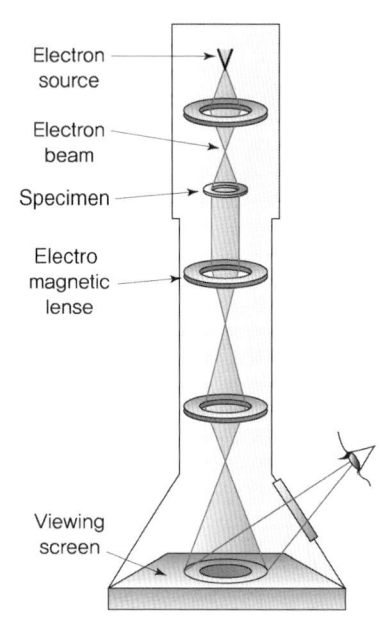

Figure 4.79 Structure of TEM.

TEM consists of an imaging mode and a diffraction mode. In the imaging mode, micrograins of the specimen are magnified and an image is formed by focusing the projector lens onto the image side of the objective lens. The diffraction mode focuses the projector lens on the rear focal plane such that diffraction patterns are projected onto a fluorescent screen. The imaging mode can be classified into high-resolution TEM and conventional TEM (CTEM). A high-resolution image refers to the diffraction image formed from differences in the electron phase after passing through a thin specimen. Known as phase contrast, this method obtains interference patterns by

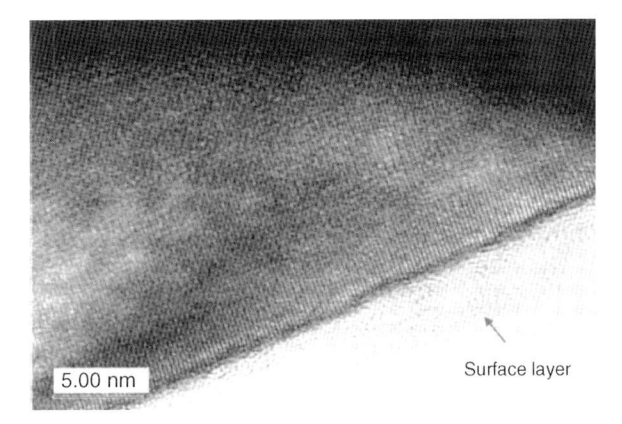

Figure 4.80 SEI layer image formed on the graphite surface [43]. Reproduced by permission of ECS – The Electrochemical Society.

passing two or more beams through an aperture, and contrast arises from the phase difference of electron beams. The CTEM mode does not produce lattice patterns and is used to analyze defects (e.g., stacking faults, grain boundaries) within the crystal. From TEM, we can obtain bright field (BF) images, dark field (DF) images, and selected area electron diffraction (SAED) patterns. BF is particularly useful in observing highly ordered crystal lattices with dislocations and other defects that cause changes in electron density. By moving the aperture to the position of deflected electrons or tilting the electron beam for deflected electrons to pass through, DF produces images based on deflected electrons alone. One advantage of this method is that images of deflected electrons can be generated for specific crystal lattice planes.

The crystalline specimen undergoes strong Bragg scattering in a specific direction and forms diffraction patterns on the rear focal plane of the objective lens. An SAED aperture is used to observe diffraction patterns in a limited area. By restricting the observation region ($\sim$0.2 μm), diffraction patterns seen on the fluorescent plane correspond to the selected area only. When a condensing lens is used in place of an aperture, the probe size can be as small as 50 nm and a large amount of electron beams are utilized simultaneously. Compared to SAED, electrons having high convergence are able to produce diffraction patterns for a much smaller region. The highly convergent beam allows simultaneous injection of electrons from various angles, thus yielding CBED patterns instead of dot patterns [41, 42].

TEM is useful for lithium batteries in determining the structure and composition of new electrode materials. This technique allows the analysis of microstructures undetected by XRD. Figure 4.80 shows the TEM image of SEI layers formed on the graphite surface [43]. Figure 4.81 is a TEM image of a 3 nm SnO_2 cathode material particle [44].

From the above lattice fringes produced through high-resolution magnification, we can see that SnO_2 is mixed with tetragonal Sn metals that measure several nanometers in size. In the upper right of Figure 4.81, patterns of Sn metal can be clearly observed in the diffraction ring patterns of t-Sn and SnO_2.

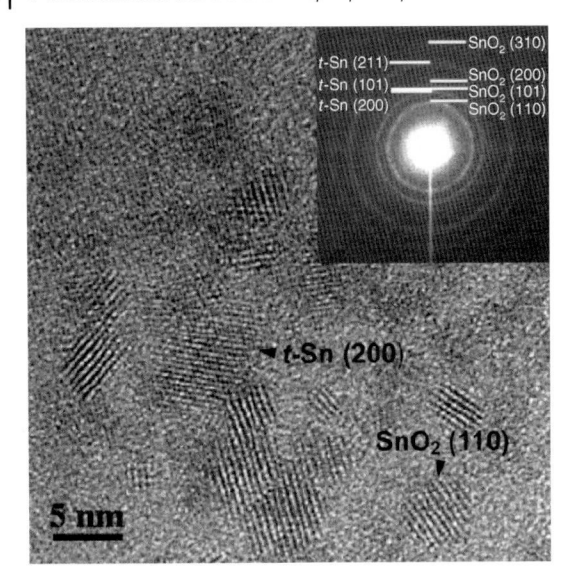

Figure 4.81 Lattice fringe TEM image and electron diffraction patterns of SnO_2 cathode material particle. Adapted with permission from American Chemical Society Copyright 2005.

The following example presents the results of TEM SAED patterns used to examine interactions between CoO and lithium in metal oxide cathode materials (Figure 4.82) [45]. TEM was used to examine changes in crystal structure and new components, leading to the following new conversion reaction mechanism:

$$\frac{\begin{array}{c} CoO + 2Li^+ + 2e^- \rightleftarrows Li_2O + Co \\ 2Li \rightleftarrows 2Li^+ + 2e^- \end{array}}{CoO + 2Li \rightleftarrows Li_2O + Co} \quad \begin{array}{c} 1 \\ \\ 2 \end{array} \quad (4.82)$$

4.2.7
Scanning Electron Microscopy (SEM)

A scanning electron microscope (SEM) is an electron accelerator that uses electromagnetic lenses to produce an image by focusing accelerated electron beams (Figure 4.83). An electron gun is fitted with a tungsten filament acting as an illumination source and electron beams are accelerated to 60–100 keV. When a high voltage is applied, the temperature of the filament rises to 2700 K and electrons are emitted from the tip of the filament. These field-emitted electrons are concentrated into a condenser lens and focused onto the sample by an objective lens, and then passed through a deflection coil of the scanning system. When incident beams collide with the specimen, secondary electrons are emitted. These secondary electrons are collected by a scintillation counter and the resulting signals are amplified for a CRT display.

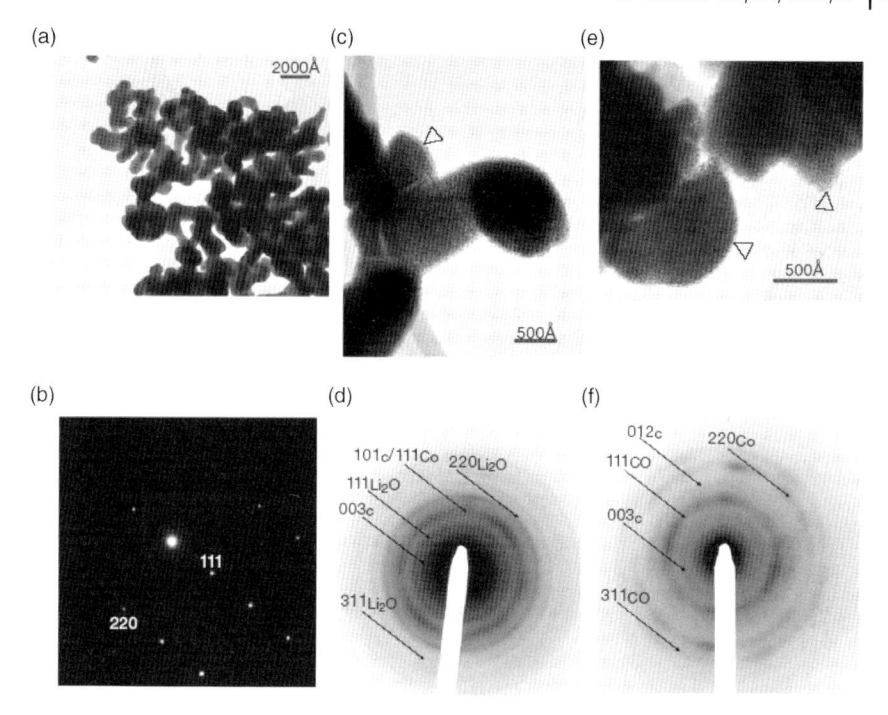

Figure 4.82 CoO cathode material SAED patterns for (a, b) a specimen before cycling, (c, d) a specimen extracted from a charged cell, and (e, f) a specimen extracted from a discharged cell [45].

The lens system and the sample are positioned in a vertical cylindrical chamber under vacuum conditions. This is to prevent scattering of electrons due to collisions with molecules in the air while moving from the filament to the sample. Since the mean free path of electrons is 125 cm at a pressure of 10^{-4} Torr, the minimum

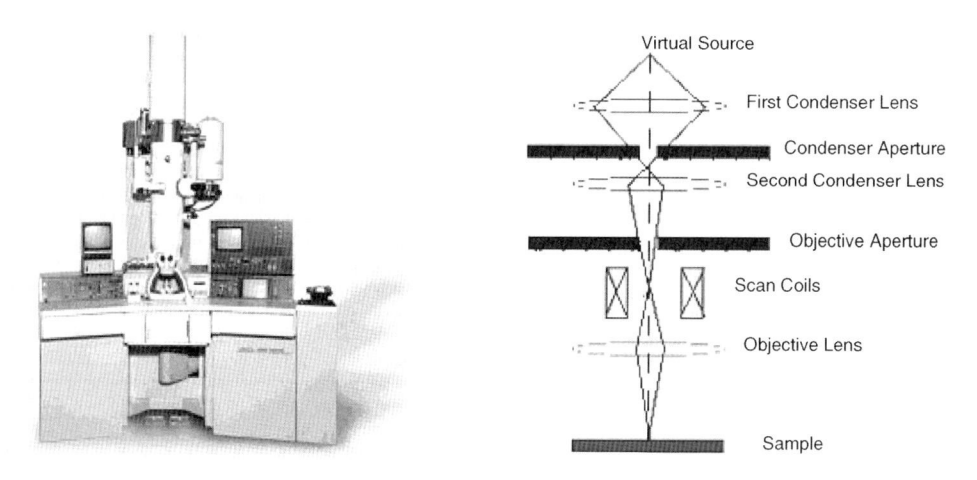

Figure 4.83 Structure of SEM equipment.

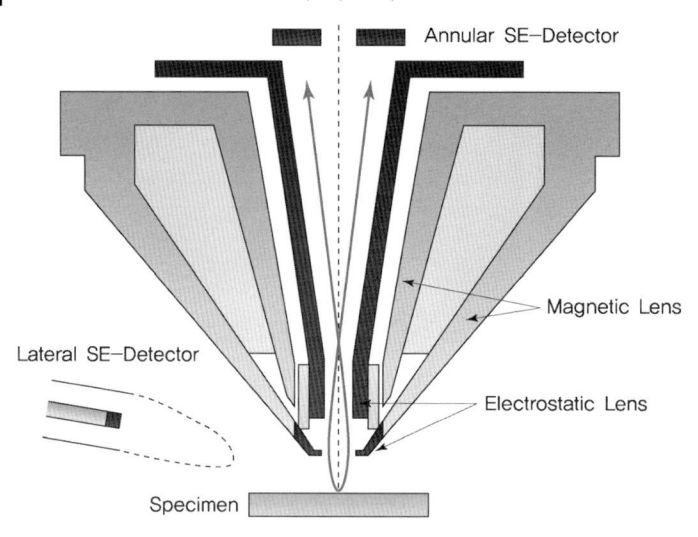

298 | *4 Electrochemical and Material Property Analysis*

Figure 4.84 FE-SEM equipment structure.

pressure should be 10^{-7} Torr. Unlike optical and transmission electron microscopes, condenser and objective lenses in SEMs do not image the specimen. Instead, the beam is focused to a spot and an image is generated by scanning the surroundings [46, 47]. SEM utilizes secondary electrons produced nearest to the surface of the sample and amplified signals are displayed on a CRT. The brightness of a spot in the CRT display is proportional to the number of secondary electrons generated from interactions between the electron beam and the sample.

The sample is fixed in the aluminum metal cylindrical holder and grounded to prevent charging during collisions with incident electron beams. Nonconductive samples may be discharged through the holder by sealing the surroundings with carbon tape, or using the sputtering method to coat the surface with a thin 20–30 nm layer of carbon, gold, or Pd metal.

In field-emission SEM (FE-SEM), secondary electrons are produced when the sample surface is scanned with primary incident electron beams, and clear images can be obtained with interception of secondary electrons by a weak retarding field (Figure 4.84). Field emission electron guns have higher electron energy than other electron guns. The narrow probes allow an image resolution less than 1.5 nm, which is 3–6 times better than existing SEMs. The generated image corresponds to regions near the surface because electrons with low kinetic energy do not travel far through the sample. Since acceleration voltages as low as several keV is able to produce high-quality images, even uncoated and nonconductive samples can achieve resolution of a few nm [48].

SEM images are similar to photos taken with a camera. This technique provides direct images that facilitate the analysis of electrode materials. Figure 4.85 shows a cross-sectional image of a typical lithium secondary battery [49].

The next example (Figure 4.86) shows an anode material in the form of a spherical core–shell, with $LiNi_{0.8}Co_{0.1}Mn_{0.1}O_2$ as the core and $LiNi_{0.5}Mn_{0.5}O_2$ as the shell [50].

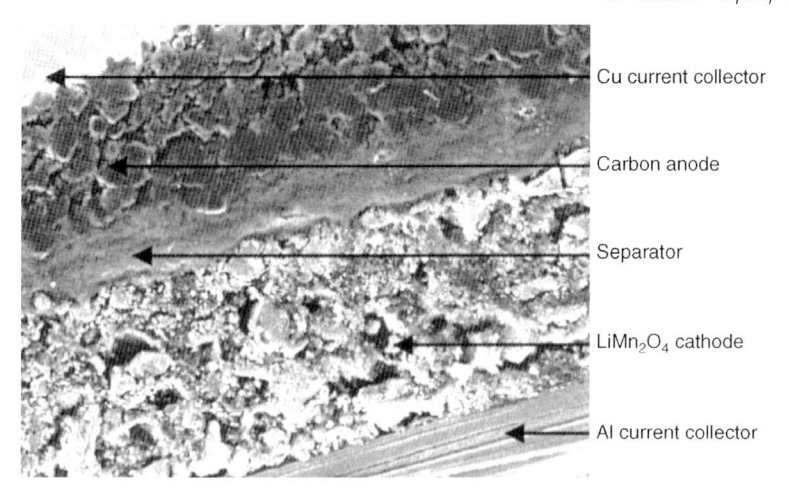

Figure 4.85 Cross-sectional SEM image of a lithium battery [49]. Reprinted from [49] Copyright 2000, with permission from Elsevier.

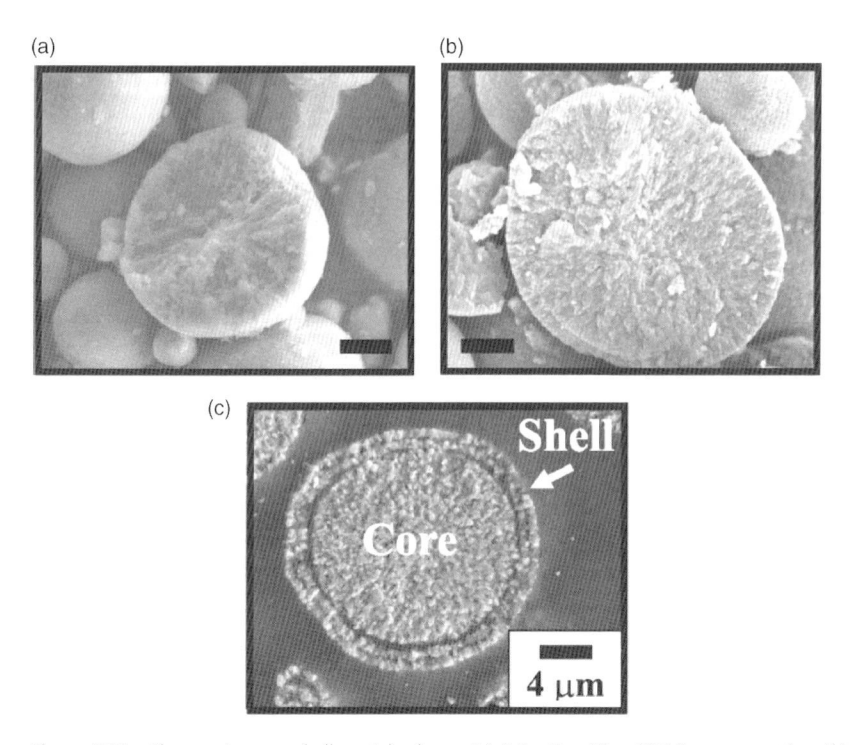

Figure 4.86 Changes in core–shell particle shape: (a) [$Ni_{0.5}Co_{0.1}Mn_{0.5}$](OH)$_2$ cross section, (b) [($Ni_{0.5}Co_{0.1}Mn_{0.5}$)$_{0.8}$($Ni_{0.5}Mn_{0.5}$)$_{0.2}$](OH)$_2$ core–shell, and (c) Li[($Ni_{0.5}Co_{0.1}Mn_{0.5}$)$_{0.8}$($Ni_{0.5}Mn_{0.5}$)$_{0.2}$]O$_2$ core–shell. The scale bar is 4 μm long [50]. Reproduced by permission of ECS – The Electrochemical Society.

From the cross-sectional image of the synthesized sample, we can see that the core is surrounded by a shell with a thickness of 1–1.5 μm.

In addition to making observations of the sample surface, SEM can be equipped with energy dispersive spectroscopy (EDS) to conduct quantitative and qualitative analyses of elements existing within the sample, as well as image observations using backscattered electrons.

EDS is a method of bulk sample analysis as it is able to detect particles in μm units with a 5–10% margin of error. Tilting the sample holder leads to diffraction reactions between electron beams and atoms within the sample. This allows identification of crystal structure, grain boundary, and directionality of crystals. Advantages of SEM over TEM include easier sample preparation and analysis of diverse materials.

4.2.8
Atomic Force Microscopy (AFM)

As shown in Figure 4.87, the atomic force microscopy (AFM) consists of a cantilever and a sharp probe made of Si or Si_3N_4 with nanometer precision. When the probe is brought close to the sample surface, repulsive forces between the tip and electrons on the surface result in a deflection of the cantilever. Photodiodes are then used to measure the intensity of the laser reflected from the cantilever.

While scanning the surface under constant force, the vertical movement of the cantilever is recorded and used to generate an image of the surface [51, 52]. If the cantilever spring constant k_N is known, the repulsive force F_N is as follows.

$$F_N = k_N \times \Delta Z \tag{4.83}$$

The vertical movement of the probe can be expressed in forces, and van der Waals interactions can be measured by applying force from 10^{-13} to 10^{-8} N. Types of forces include mechanical contact force, van der Waals forces, capillary forces, chemical bonding, and magnetic forces. These forces are affected by electrode characteristics, tip-to-sample distance, surface impurities, and probe shape. Since current does not have to flow between the probe and the surface, AFM allows analysis of all materials including insulators and conductors. It is particularly useful in analyzing insulating materials that cannot be imaged using scanning tunnel microscopy (STM). AFM operates in various modes such as contact mode, friction mode, tapping mode, and noncontact mode.

1) Contact mode maintains physical contact between the probe and the surface and is used to obtain clear images despite the risk of scratching the sample.
2) Friction mode measures the cantilever deflection by applying a lateral force to establish contact between the probe and the surface.
3) In tapping mode, a piezoelectric element drives the cantilever to oscillate at its resonance frequency. The amplitude of oscillations or phase changes can be measured with fewer scratches compared to the contact mode.
4) In noncontact mode, measurements are made without any physical contact and in the absence of repulsive forces. As such, it is more appropriate to use probes

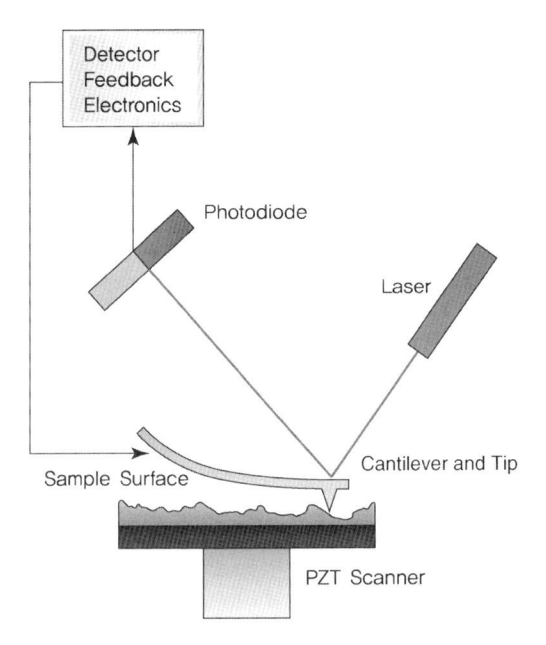

Figure 4.87 AFM equipment structure.

with conductive or magnetic characteristics. *In situ* AFM is used to make direct observations of the electrode surface during electrochemical reactions. To perform *in situ* experiments, the AFM is placed on a mat inside a glove box. The structure of an *in situ* cell is shown in Figure 4.88 [53].

4.2.9
Thermal Analysis

Thermal analysis measures changes in thermal characteristics of materials with varying temperature. Some common methods of thermal analysis are thermogravimetric analysis (TGA), differential thermal analysis (DTA), and differential scanning

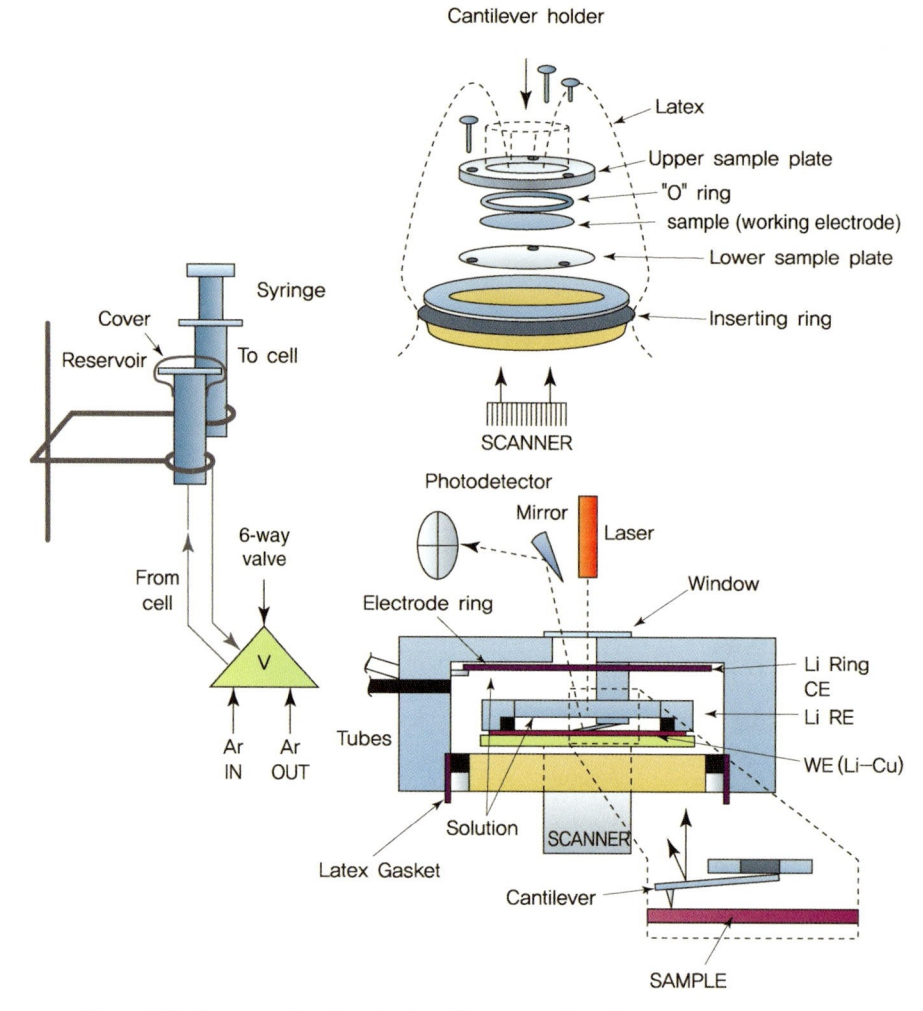

Figure 4.88 Structure of an AFM *in situ* cell [53]. Reproduced by permission of ECS – The Electrochemical Society.

calorimetry (DSC). Figure 4.89 shows the results of *in situ* CV and AFM on the surface of SEI layers formed at the graphite cathode due to electrolyte reduction and decomposition [54].

TGA observes changes in weight arising from a thermal analysis or volatilization when the temperature rises from room temperature to the target temperature. The point at which weight loss occurs can be identified through derivative weight loss curves. TGA is used to determine degradation temperatures of polymers, absorbed moisture content of materials, the level of inorganic and organic components in materials, and thermal decomposition temperatures of organic substances or solvents [55, 56].

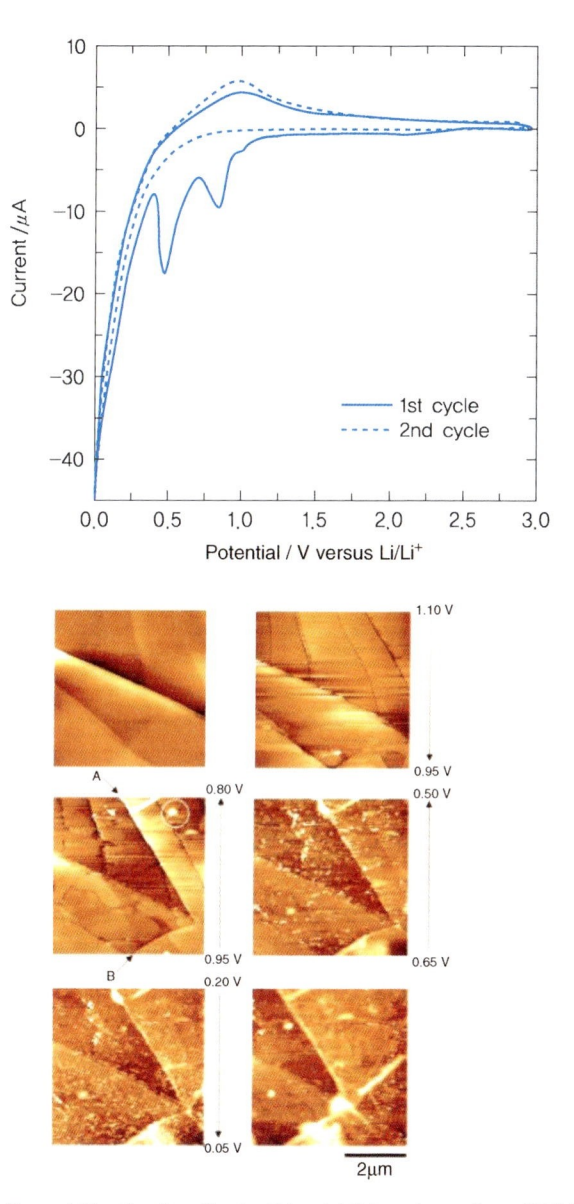

Figure 4.89 Results of *in situ* CV and AFM on the surface of SEI layers formed at the graphite cathode due to electrolyte reduction and decomposition [54]. Reproduced by permission of ECS – The Electrochemical Society.

As shown in Figure 4.90, TGA equipment consists of a pan loaded with the sample and a high precision balance. The sample is placed in a small electric furnace equipped with a thermocouple for accurate temperature measurements. To prevent oxidation or other undesirable reactions, specific gases (N_2, Ar, etc.) are purged into the atmosphere to ensure inert conditions.

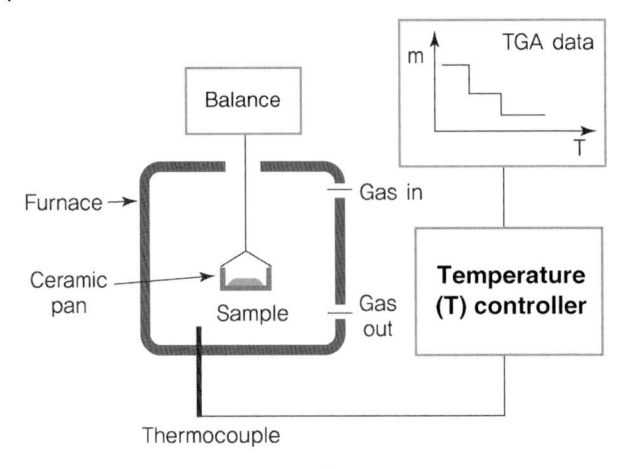

Figure 4.90 Schematic diagram of TGA equipment structure.

DTA measures thermal capacity and enthalpy change while measuring temperature difference between the sample and the (nonreactive) reference over time. From this, we can analyze thermodynamic characteristics and reaction kinetics of reactions with changes in temperature of the sample.

DSC measures the amount of absorbed or released heat relative to a reference sample when a sample undergoes physical changes (e.g., phase transition) [55, 56]. The heat capacity of the reference is known. Exothermic or endothermic reactions occur when the sample releases or absorbs more heat, with respective peaks appearing above or below the baseline in the DSC thermogram. As shown in Figure 4.91, the DSC consists of a sample pan and a reference pan. The sample pan is made of highly conductive aluminum metal, while the reference is an empty aluminum pan. The sample is usually 0.1–100 mg, and tests can be conducted under inert conditions. DSC is classified into heat flux DSC and power compensated DSC.

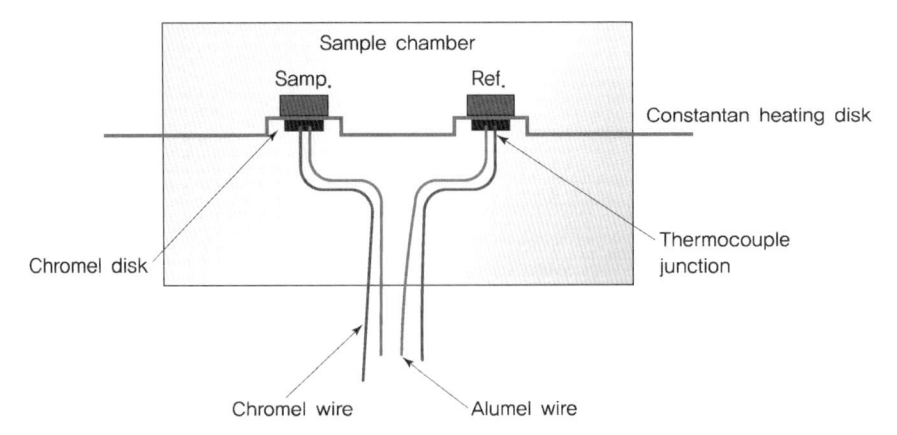

Figure 4.91 Heat flux DSC equipment structure.

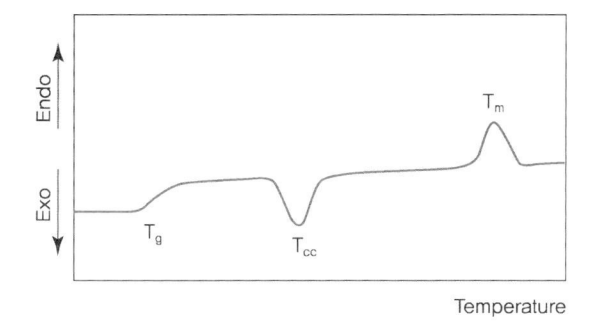

Figure 4.92 DSC thermogram of PET [56].

The heat flux type is also known as calorimetric DTA, where the sample and the reference are placed in the same electric furnace. Heat is passed to the sample and reference through silver (Ag) or alloy disks. These disks are able to detect temperature.

Unlike the heat flux type, power compensated DSC uses a separate electric furnace for the sample and the reference. This method operates by eliminating any difference in temperature between the sample and the reference. When the temperature of the sample drops during endothermic reactions, power is supplied to maintain a constant temperature. The amount of power supplied can be seen in the DSC thermogram.

As an example, Figure 4.92 shows the DSC analysis of polyethylene terephthalate (PET) [56]. The peaks corresponding to phase transitions in the DSC thermogram are differentiated to obtain area and enthalpy can be calculated by Eq. (4.84).

$$A = \Delta H \times k \tag{4.84}$$

Here, A is the area of the endothermic or exothermic peak, ΔH is the transition enthalpy, and k is the calorimetric constant. The value of k differs depending on the device and samples with a known transition enthalpy are used.

Figure 4.93 shows thermal characteristics of an electrolyte in a lithium battery obtained by DSC analysis. We can see that the polymer salt $LiPF_6$ melts at 200 °C and breaks down at 300 °C according to Eq. (4.85) [57].

$$LiPF_6(s) \Rightarrow LiF(s) + PF_5(g) \tag{4.85}$$

Irreversible decomposition occurs when $LiPF_6$ is added to the EC–DEC solvent, with endothermic reactions occurring at 230–250 °C and exothermic reactions at 250–320 °C. As can be seen in Figure 4.93, the resulting PF_5 from $LiPF_6$ decomposition reacts with a small amount of water to produce HF. The peak near 230 °C corresponds to EC decomposition arising from reactions between EC and HF. In addition, we can observe changes in the thermal characteristics of the electrolyte with varying concentrations of $LiPF_6$.

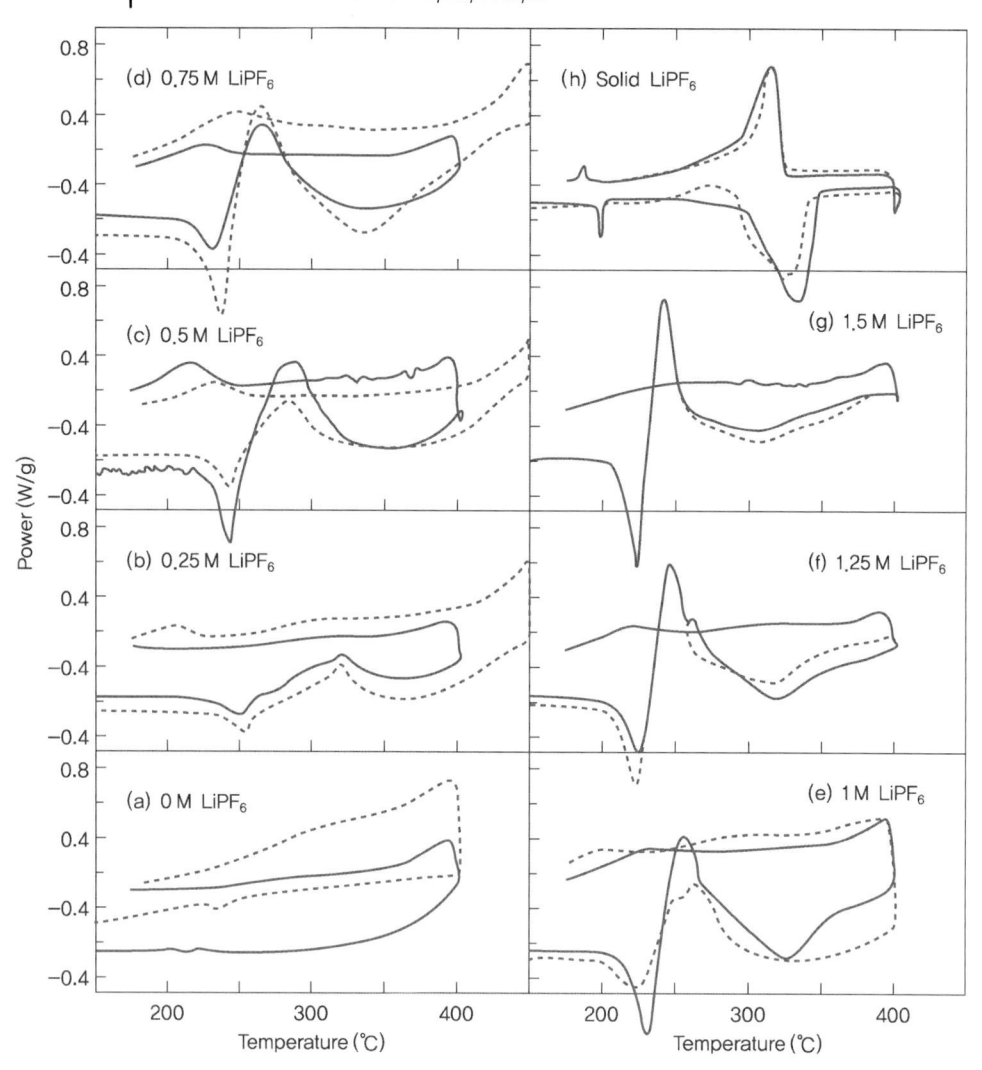

Figure 4.93 Changes in the DSC thermogram of EC/DEC with varying concentration of LiPF$_6$: comparisons for two different samples (solid line, dotted line) with temperature increasing at 5 °C/min [57].

4.2.10
Gas Chromatography-Mass spectrometry (GC–MS)

Chromatography is employed to perform quantitative and qualitative analysis of compounds in a liquid or gaseous state by comparison with standard reference materials. For lithium batteries, this method is used to analyze the stability of electrolytes at high temperatures and decomposition products (gas). When a mixture passes through the chromatograph column, molecules travel at different speeds and remain in the column for different amounts of time. This allows the molecules to be separated from each other. In gas chromatography (GC), the sample is a gas or liquid

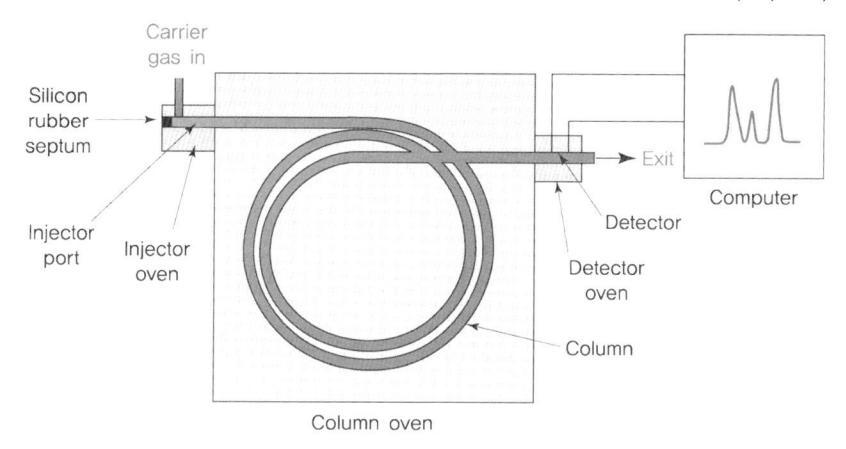

Figure 4.94 GC equipment structure.

that can be vaporized. As shown in Figure 4.94, the sample is vaporized when transported through a rubber septum into a heated port and analyzed in a thin long column containing the stationary phase [14].

Commonly used carrier gases include H_2, N_2, and He. Flux is adjusted using pressure and temperature. The open tubular column is made of silica (SiO_2) that acts as a liquid or gas stationary phase in column walls. The solid stationary phase can be activated carbon or liquefied forms (Table 4.5) that are selected depending on the polarity of the sample [14].

When the temperature of the column oven rises, the vapor pressure of the sample is increased and remains in the column for a short time. Molecules can then be separated depending on the boiling point and polarity of the sample. Common detectors are flame ionization detectors and thermal conductivity detectors. In flame ionization, carbon (C) produces CH radicals and eventually CHO^+ ions. We can then obtain the quantity of electric charge from the following equation:

$$CH + O \rightarrow CHO^+ + e^- \tag{4.86}$$

The thermal conductivity detector consists of a tungsten–rhenium filament. By introducing a gas sample, the thermal conductivity changes along with the electrical resistance and potential of the filament. GC results showing changes in the potential with thermal conductivity are given in Figure 4.95 [14].

GC uses mass spectrometry (MS) as a detector to obtain fragmentation patterns of substances in the gaseous phase. The structure of molecules can be directly studied through a qualitative analysis. For a mass analysis, molecules are ionized and sorted according to their mass-to-charge ratio (m/z). The most widely used MS is the transmission quadrupole mass spectrometer, which is connected to the GC, as shown in the left-hand side of Figure 4.96. Compounds eluted from the GC column pass through the connector to the ionization chamber. The molecules are ionized at energies below 15 keV before entering the quadrupole mass separator [14]. Ionization methods can be divided into electron ionization and chemical ionization.

Table 4.5 Types of stationary phase in capillary GC.

Structure		Polarity	Temperature range (°C)
(Diphenyl)$_x$(dimethyl)$_{1-x}$ polysiloxane	$x = 0$	Nopolar	−60 to 360
	$x = 0.05$	Nopolar	−60 to 360
	$x = 0.35$	Intermediate polarity	0–30
	$x = 0.65$	Intermediate polarity	50–370
(Cyanopropylphenyl)$_{0.14}$ (dimethyl)$_{0.86}$ polysilocane		Intermediate polarity	−20 to 280 °C
Carbowax(polyethylene glycol)		Strongly polar	40–250 °C
(Biscyanopropyl)0.9 (cyanopropylphenyl)0.1 polysiloxane		Strongly polar	0–275 °C

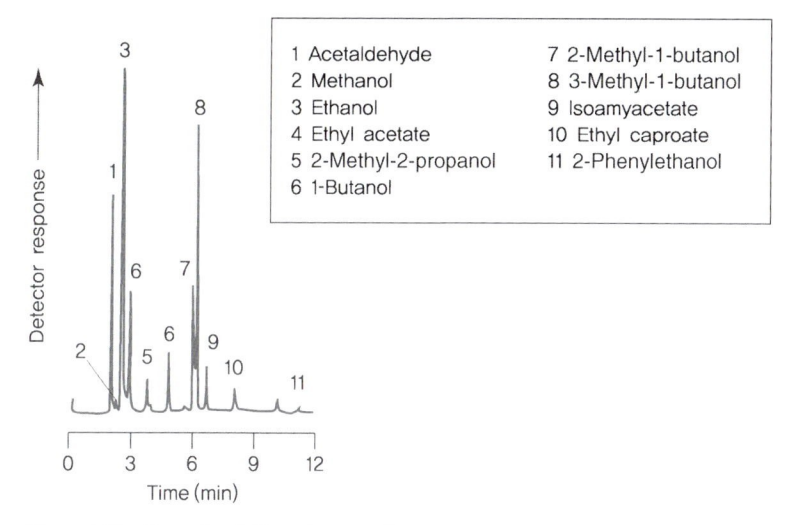

1 Acetaldehyde	7 2-Methyl-1-butanol
2 Methanol	8 3-Methyl-1-butanol
3 Ethanol	9 Isoamyacetate
4 Ethyl acetate	10 Ethyl caproate
5 2-Methyl-2-propanol	11 2-Phenylethanol
6 1-Butanol	

Figure 4.95 Example of GC analysis results.

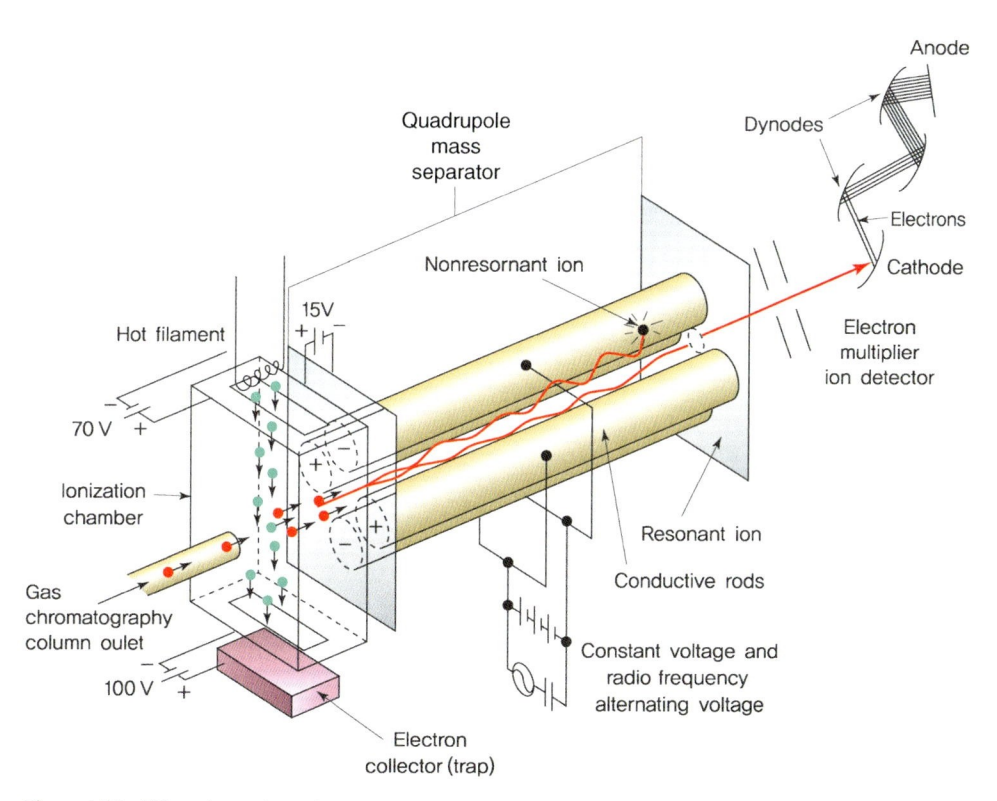

Figure 4.96 MS equipment structure.

In electron ionization, electrons emitted from a heated filament are accelerated to 70 kV and ionized according to Eq. (4.87).

$$M + e^-(70\,\text{eV}) \rightarrow M^+ + e^-(55\,\text{eV}) + e^- \qquad (4.87)$$

Here, M^+ is the resulting ion that undergoes further fragmentation.

In chemical ionization, gas of 1 mbar is ionized by electrons with energies of 100–200 eV. For methane, CH_4 becomes CH_5^+. As shown in Eq. (4.88), CH_5^+ reacts with the atoms of the molecule being analyzed to produce protonated molecules. The resulting spectrum and m/z values are different from those obtained from electron ionization.

$$CH_5^+ + M \rightarrow CH_4 + MH^+ \qquad (4.88)$$

To analyze the mass spectra, different fragmentation patterns should be compared against m/z values and peak locations. Figure 4.97 is an example illustrating how to derive fragmentation patterns [14].

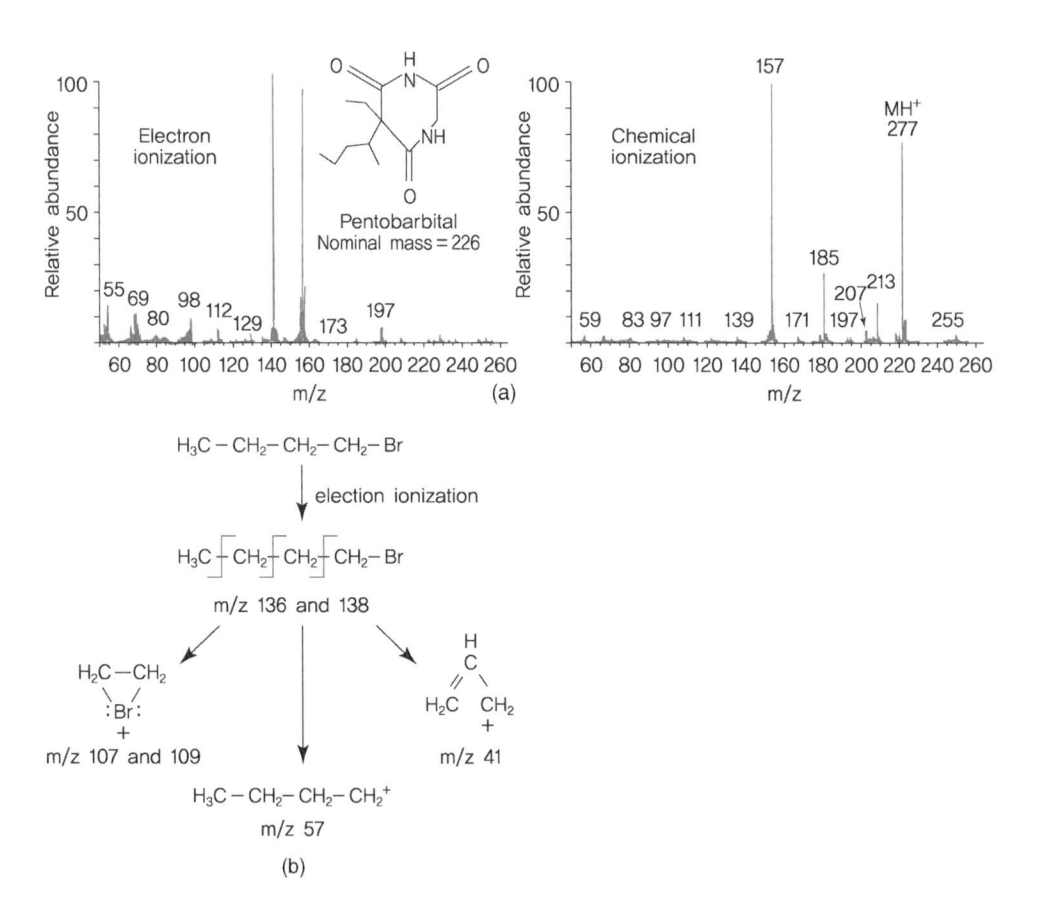

Figure 4.97 (a) MS spectrum and (b) example of fragmentation pattern analysis.

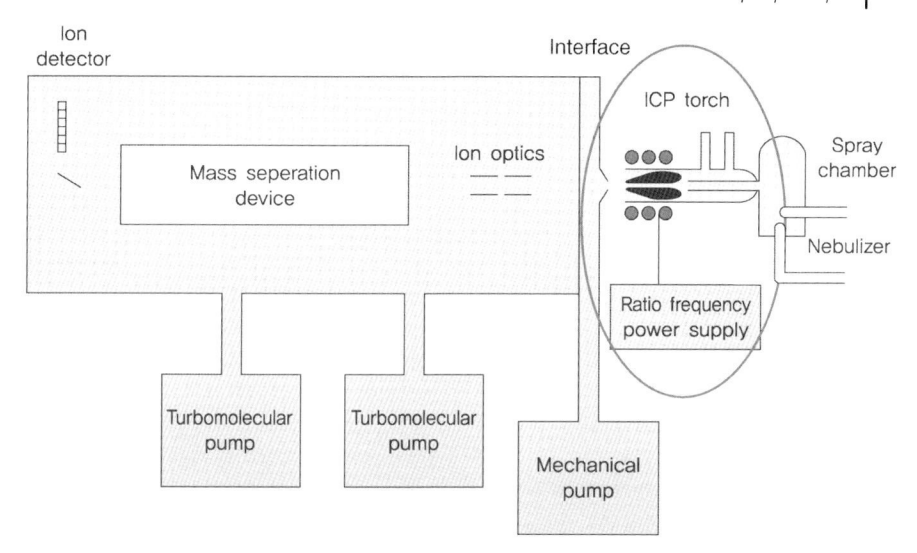

Figure 4.98 ICP equipment structure.

4.2.11
Inductively Coupled Plasma Mass Spectroscopy (ICP-MS)

Inductively coupled plasma mass spectrometry uses argon gas to perform a qualitative analysis of cations released from plasmas at 10 000 K. Inductively coupled plasma (ICP) is a trace elemental analysis method that transfers ions to a mass spectrometer, where they are expressed as m/z signals. Figure 4.98 shows the structure of ICP.

ICP produces ions from high-temperature plasmas with a radio frequency electric current. Plasma is formed from electric current via quartz tubes placed in an induction coil. The frequency is usually 27.12–40.68 MHz and the operating power is 800–1500 W. When ions generated from the plasma enter the mass spectrometer, they are separated according to their mass-to-charge ratio [14]. We can analyze samples simultaneously for all elements with atomic weights ranging from 7 (Li) to 250 (U). Solvents should be diluted to maintain detection limits in the ppm to ppb range. Solid samples are prepared by melting in a strong acid such as HNO_3. For an accurate elemental analysis, a standard solvent of known concentration is used to obtain a calibration curve, and that of the sample is derived from ICP-MS. Another calibration method is to observe changes in peak intensity after adding a standard reference material of known concentration that is not contained in the sample.

4.2.12
Brunauer–Emmett–Teller (BET) Surface Analysis

Since most electrode active materials for lithium ion batteries are in powder form, and lithium intercalation/deintercalation during charging and discharging occurs at the surface of these particles, it is important to examine the surface structure and surface area of these materials.

Gaseous molecules undergo condensation on the surface of a solid when there are changes in pressure or temperature. This phenomenon is caused by van der Waals forces and is known as adsorption. The vaporization of gaseous molecules from the solid is called desorption. Adsorption can be physical or chemical. Physical adsorption arises from van der Waals forces, whereas chemical adsorption is due to electron transfer or electron sharing between the solid and the gas. The surface area of the solid can be calculated by measuring the quantity of adsorbed gas. The amount of adsorbed gas differs according to temperature, pressure, the type of gas, and the type of solid. When a constant temperature is maintained, the adsorbed quantity depends on the pressure, which can be varied to obtain more information on porous materials.

BET is an acronym derived from the first initials of the family names Brunauer, Emmett, and Teller. According to the BET theory, the adsorbed gas quantity on a solid at constant temperature is a function of partial pressure of the gas. Before applying BET to evaluate the surface characteristics of materials, we need a basic understanding of Langmuir adsorption. The BET theory is based on the Langmuir adsorption equation and an expansion of the Langmuir theory. The Langmuir theory makes the following assumptions [58]:

1) Molecules are chemically adsorbed on the surface as a monolayer.
2) All sites are equivalent and an active site can be occupied only by one particle.
3) Active sites are independent and unaffected by the probability of one site being occupied.

The equilibrium constant K in the adsorption of gas A on solid B, A (g) + B (surface) $\leftrightarrow$ AB (surface), is given by Eq. (4.89), where k_a is the rate constant and k_b is the desorption constant.

$$K = \frac{k_a}{k_d} \tag{4.89}$$

In an equilibrium state, the coverage of the surface remains unchanged and the rate of change in $K_a[A]$ from adsorption is the same as the rate of change in $K_b[B]$. This principle is explained by the Langmuir isotherm, which is stated in Eq. (4.90) [58, 59].

$$\theta = \frac{N}{S} = \frac{KP}{1+KP} \tag{4.90}$$

Here, N is the number of adsorbed molecules, S is the number of sites, and P is the gas pressure. The Langmuir adsorption model describes an ideal situation, while the BET theory expands this model for actual systems. According to BET, molecular adsorption does not occur on a monolayer basis, but rather on a multilayer basis. The BET theory assumes the following [59, 60]:

1) Gas molecules (adsorbate) physically adsorb on a solid (adsorbent) in layers infinitely.

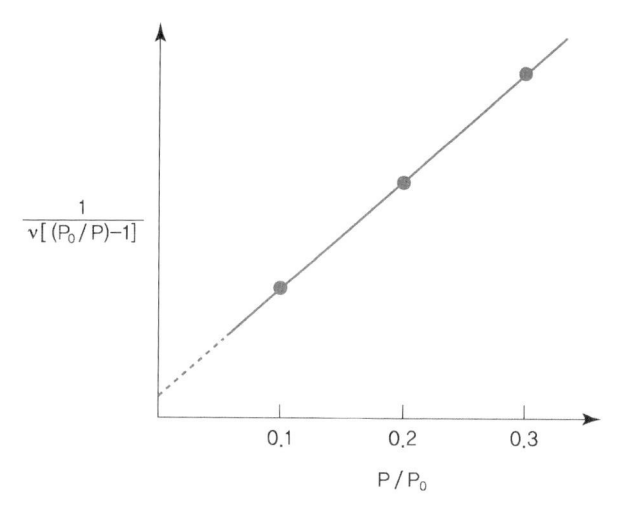

Figure 4.99 BET plot.

2) There is no interaction between each adsorption layer.
3) The Langmuir theory can be applied to each layer.

The BET equation is given by Eq. (4.91).

$$\frac{1}{v[(P_0/P)-1]} = \frac{c-1}{v_m c}\left(\frac{P}{P_0}\right) + \frac{1}{v_m c} \tag{4.91}$$

Here, P is the pressure, P_0 is the saturation pressure of adsorbates at the temperature of adsorption, v is the adsorbed gas quantity (volume), v_m is the monolayer adsorbed gas quantity, and c is the BET constant expressed by Eq. (4.92).

$$c = \exp\left(\frac{E_1 + E_L}{RT}\right) \tag{4.92}$$

Here, E_1 is the heat of adsorption for the first layer and E_L is that for the second and higher layers and corresponds to the heat of liquefaction. This adsorption isotherm can be plotted as a straight line with P/P_0 on the x-axis and $1/v[(P_0/P) - 1]$ on the y-axis. This plot is known as a BET plot (Figure 4.99).

The above linear relationship is maintained in the range of $0.05 < P/P_0 < 0.35$. The slope of this plot and the y-intercept are used to calculate the monolayer adsorbed gas quantity and BET constant c.

The BET method is used for the calculation of surface area of solids based on physical adsorption of gas molecules. The total surface area S_{total} and specific surface area S are obtained by Eqs. (4.93) and (4.94).

$$S_{total} = \frac{(v_m N s)}{M} \tag{4.93}$$

$$S = \frac{S_{total}}{a} \tag{4.94}$$

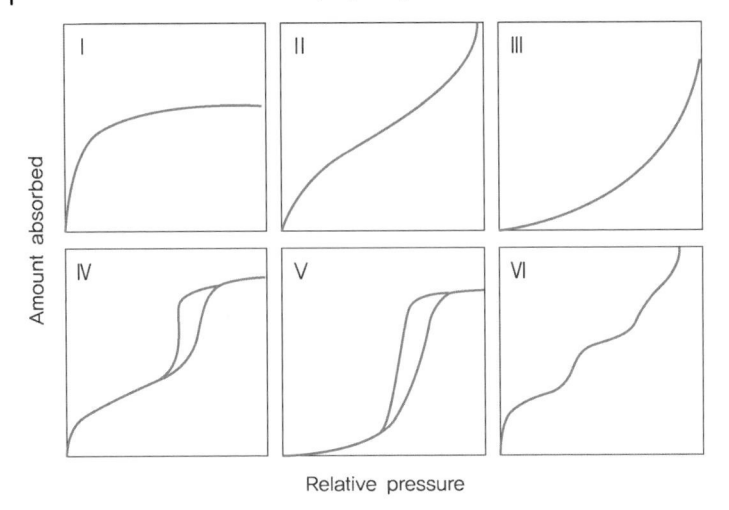

Figure 4.100 IUPAC classification of adsorption isotherms.

Here, N is Avogadro's number, s is the adsorption cross section, M is the molar volume of the adsorbate gas, and a is the mass of the adsorbent. BET measurements are usually taken during the adsorption of nitrogen gas at liquid nitrogen temperature.

Depending on pore size, porous materials are categorized as follows:

1) Macropores: pore size of 50–1000 nm
2) Mesopores: pore size of 2–50 nm
3) Micropores: pore size of 0.2–2 nm

These three categories are referred to as nanopores (pore size of 0.2–1000 nm).

As shown in Figure 4.100, BET isotherms can be classified into the following six patterns:

Type I: Adsorption is limited to a few molecular layers. Physical adsorption can be seen in microporous materials.

Type II: This type is obtained in case of a nonporous or macroporous adsorbent. The knee of the isotherm indicates the point at which monolayer coverage is complete and multilayer adsorption begins.

Type III: The adsorbate–adsorbent interactions are relatively weak and play an important role. This type is rarely observed.

Type IV: This type is typical for mesoporous materials. The most characteristic feature is the hysteresis loop caused by pore condensation. When pores are almost completely filled, a flat plateau is seen for high P/P_0. As shown in Figure 4.101, the hysteresis curve of type IV can be attributed to multilayer adsorption and pore condensation.

Type V: The initial part of this isotherm is similar to type III. Pore condensation and hysteresis are observed.

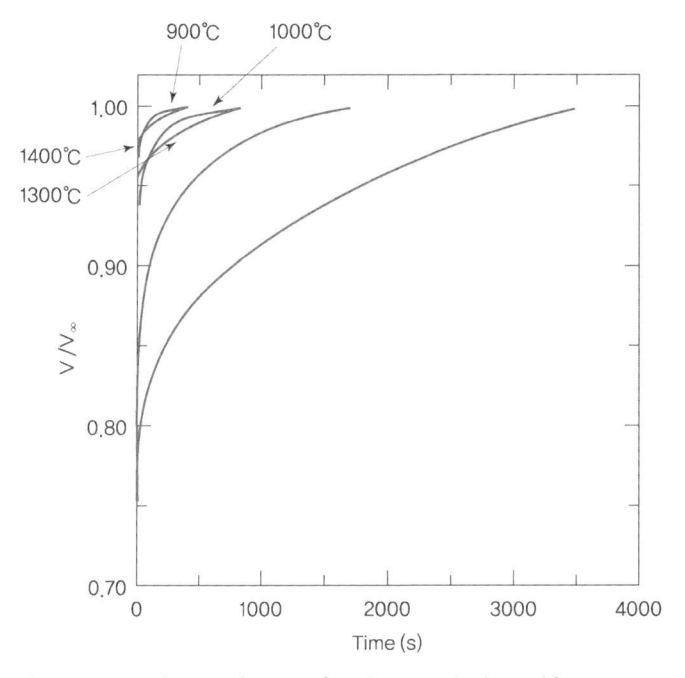

Figure 4.101 Adsorption kinetics of a carbon sample obtained from sintering at 900–1400 °C [62].

Type VI: This is a special case in which spherically symmetrical, nonpolar adsorptives are adsorbed as multilayers on a uniform, nonporous surface, resulting in the stepwise curve shown in the figure. The shape of the steps depends on the homogeneity of the gas, the solid, and the temperature [59, 61].

The BET method is widely used to measure the surface area and pore size of various carbon-based cathode and anode materials. As shown in Figure 4.101, when hard carbon at the cathode is heated to high temperatures, the micropores are partially closed and a small hysteresis is observed. The closing of micropores results in a slower gas adsorption rate [62].

References

1 Cullity, B.D. (1978) *Elements of X-ray Diffraction*, Addison-Wesley Publishing Company, Inc.
2 Bish, D.L. and Post, J.E. (1989) *Modern Powder Diffraction: Reviews in Mineralogy*, vol. **20**, Mineralogical Society of America.
3 Rokakuho, U. (1991) *X-ray Diffraction Analysis*, Bando Publishing.
4 Yonemura, M., Yamada, A., Takei, Y., Sonoyama, N., and Kanno, R. (2004) *J. Electrochem. Soc.*, **151**, A1352.
5 Amatucci, G.G., Tarascon, J.M., and Klein, L.C. (1996) *J. Electrochem. Soc.*, **143**, 1114.
6 Novak, P., Panitz, J.C., Joho, F., Lanz, M., Imhof, R., and Coluccia, M. (2000) *J. Power Sources*, **90**, 52.

7 Dahn, J.R. and Haering, R.R. (1981) *Solid State Commun.*, **40**, 245.
8 Richard, M.N., Koetschau, I., and Dahn, J.R. (1997) *J. Eelectrochem. Soc.*, **144**, 554.
9 Morcrette, M., Chabre, Y., Vaughan, G., Amatucci, G., Leriche, J.B., Patoux, S., Masquelier, C., and Tarascon, J.M. (2002) *Electrochim. Acta*, **47**, 3137.
10 Hatchard, T.D. and Dahn, J.R. (2004) *J. Electrochem. Soc.*, **151**, A838.
11 Tarascon, J.M., Gozdz, A.S., Schmutz, C., Shokoohi, F., and Warren, P.C. (1996) *Solid State Ionics*, **86–88**, 49.
12 Morcrette, M., Chabre, Y., Vaughan, G., Amatucci, G., Leriche, J.B., Patoux, S., Masquelier, C., and Tarascon, J.M. (2002) *Electrochim. Acta*, **47**, 3137.
13 Palacin, M.R., Chabre, Y., Dupont, L., Hervieu, M., Strobel, P., Rousse, G., Masquelier, C., Anne, M., Amatucci, G.G., and Tarascon, J.M. (2000) *J. Electrochem. Soc.*, **147**, 845.
14 Harris, D.C. (2003) *Quantitative Chemical Analysis*, 6th edn, W. H. Freeman and Company, New York.
15 Lin-Vien, D., Colthup, N.B., Fately, W.G., and Graselli, J.G. (1991) *The Handbook of Infrared and Raman Characteristic Frequencies of Organic Molecules*, Academic Press, San Diego.
16 Aldrich (2003) *Handbook of Fine Chemicals and Laboratory Equipment*, Aldrich Chemical Company, Milwaukee, p. 1494.
17 Kolasinsky, K.W. (2002) *Surface Science*, John Wiley & Sons, Ltd.
18 Zhuang, G.V., Xu, K., Yang, H., Jow, T.R., and Ross, P.N., Jr. (2005) *J. Phys. Chem. B*, **109**, 17567.
19 Kanamura, K., Toriyama, S., Shiraishi, S., and Takehara, Z. (1995) *J. Electrochem. Soc.*, **142**, 1383.
20 Wartewig, S. (2003) *IR and Raman Spectroscopy: Fundamental Processing*, Wiley-VCH Verlag GmbH.
21 Schrader, B. (1995) *Infrared and Raman Spectroscopy: Methods and Applications*, Wiley-VCH Verlag GmbH, New York.
22 Kostecki, R. and Mclarnon, F. (2004) *Electrochem. Solid State Lett.*, **7**, A380.
23 Panitz, J.C., Joho, F., and Novak, P. (2000) *Appl. Spectrosc.*, **53**, 1188.
24 Inaba, M., Iriyama, Y., Ogumi, Z., Todzuka, Y., and Tasaka, A. (1997) *J. Raman Spectrosc.*, **28**, 613.
25 Nelson, J.H. (2003) *Nuclear Magnetic Resonance Spectroscopy*, Prentice Hall, Upper Saddle River, NJ.
26 Fyfe, C.A. (1983) *Solid State NMR for Chemists*, CFC Press, Guelph.
27 Kerlau, M., Reimer, J.A., and Cairns, E.J. (2005) *Electrochem. Commun.*, **7**, 1249.
28 Nakamura, K., Moriga, T., Sumi, A., Kashu, Y., Michihiro, Y., Nakabayashi, I., and Kanashiro, T. (2005) *Solid State Ionics*, **176**, 837.
29 Ertl, G. and Kuppers, J. (1985) *Low Energy Electrons and Surface Chemistry*, Wiley-VCH Verlag GmbH, Weinheim.
30 Muilenberg, G.E. (1978) *Handbook of X-ray Photoelectron Spectroscopy*, Perkin-Elmer, Minesota.
31 Regan, E., Groutso, T., Metson, J.B., Steiner, R., Ammundsen, B., Hassell, D., and Pickering, P. (1999) *Surf. Interface Anal.*, **27**, 1064.
32 Shirley, D.A. (1972) *Phys. Rev. B*, **5**, 4709.
33 Andersson, A.M., Henningson, A., Siegbahn, H., Jansson, U., and Edstrom, K. (2003) *J. Power Sources*, **119**, 522.
34 Teo, B.K. (1986) *EXAFS: Basic Principles and Data Analysis*, Springer, Berlin.
35 Sayers, D.E. and Bunker, B.A. (1988) in *X-ray Absorption: Principles, Applications, Techniques of EXAFS, SEXAFS and XANES* (eds D.C. Koningsberger and R. Prins), Wiley–Interscience, New York.
36 Lytle, F.W. (1989) in *Applications of Synchrotron Radiation* (eds H. Winick et al..), Gordon and Breach Science, New York.
37 Mosset, A. and Galy, J. (1989) in *Applications of Synchrotron Radiation* (eds H. Winick et al..), Gordon and Breach Science, New York.
38 Yoon, W.S., Chung, K.Y., McBreen, J., Zaghib, K., and Yang, X. (2006) *Electrochem. Solid State Lett.*, **9**, A415.
39 Sayers, D.E., Stern, E.A., and Lytle, F.W. (1971) *Phys. Rev. Lett.*, **27**, 1204.
40 Balasubramanian, M., Sun, X., Yang, X.Q., and McBreen, J. (2001) *J. Power Sources*, **92**, 1.

41 Williams, D.B. (1996) *Carter, Transmission Electron Microscopy*, Plenum Press, New York.
42 Fultz, B. and Howe, J.M. (2001) *Transmission Electron Microscopy and Diffractometry of Materials*, Springer.
43 Striebel, K.A., Shim, J., Cairns, E.J., Kostecki, R., Lee, Y.J., Reimer, J., Richardson, T.J., Ross, P.N., Song, X., and Zhuang, G.V. (2004) *J. Electrochem. Soc.*, **151**, A857.
44 Kim, C., Noh, M., Choi, M., Cho, J., and Park, B. (2005) *Chem. Mater.*, **17**, 3297.
45 Poizot, P., Laruelle, S., Grugeon, S., Dupont, L., and Tarascon, J.M. (2000) *Nature*, **407**, 496.
46 Goldstein, J.I., Newbury, D.E., Echlin, P., Joy, C., Romig, A.D., Lyman, C.E., Fiori, C., and Lifshin, E. (1992) *Scanning Electron Microscopy and X-ray Microanalysis*, Plenum Press.
47 Reed, S.J.B. (1996) *Electron Microprobe Analysis and Scanning Electron Microscopy in Geology*, Cambridge University Press, Cambridge.
48 Jaksch, H.(Oct. 1996) Materials world.
49 Orsini, F., Dupont, L., Beaudoin, B., Grugeon, S., and Tarascon, J.M. (2000) *Int. J. Inorg. Mater.*, **2**, 701.
50 Sun, Y.K., Myung, S.T., Kim, M.H., and Kim, J.H. (2006) *Electrochem. Solid State Lett.*, **9**, A171.
51 Binning, G., Quate, C.F., and Gerber, C. (1986) *Phys. Rev. Lett.*, **56**, 930.
52 Cohen, S.H., Bray, M.T., and Lightbody, M.L. (1994) *Atomic Force Microscopy/Scanning Tunneling Microscopy*, Plenum Press, New York.
53 Aurbach, D. and Cohen, Y. (1996) *J. Electrochem. Soc.*, **143**, 3525.
54 Jeong, S.K., Inaba, M., Abe, T., and Ogumi, Z. (2001) *J. Electrochem. Soc.*, **148**, A989.
55 Haines, P.J. (2002) *Principles of Thermal Analysis and Calorimetry*, Royal Society of Chemistry.
56 Ford, J.L. and Mann, T.E. (2012) Adv. DrugDeliv. Rev., **64**, 422.
57 MacNeila, D.D. and Dahn, J.R. (2003) *J. Electrochem. Soc.*, **150**, A21.
58 Langmuir, I. (1928) *J. Am. Chem. Soc.*, **40**, 1368.
59 Lowell, S., Shields, J.E., Thomas, M.A., and Thomas, M. (2004) *Characterization of Porous Solids and Powders: Surface Area, Pore Size and Density*, Kluwer Academic Publishers.
60 Brunauer, S., Emmett, P.H., and Teller, E. (1938) *J. Am. Chem. Soc.*, **60**, 309.
61 Sing, S.W., Everett, D.H., Haul., R.A.W., Moscou, L., Pierotti, R.A., Rouquerol, J., and Siemieniewska, T. (1985) *Pure Appl. Chem.*, **57**, 603.
62 Buiel, E., George, A.E., and Dahn, J.R. (1998) *J. Electrochem. Soc.*, **145**, 2252.

5
Battery Design and Manufacturing

Depending on the requirements of application fields, batteries must be designed and manufactured to exhibit adequate capacity, power, and safety features. In the design of batteries, the electrochemical potential and capacity of cathodes and anodes should be first considered. Other factors to be secured are the current collection method, electrode kinetics, and basic battery safety [1–5]. The manufacturing process should take into account both design and process distribution. Some important variables include the uniformity of electrode composition and physical properties such as coating width and length. Individual manufacturing steps, which can be divided into assembly and activation processes, are then carried out for battery production. The initial activation process is determined by design conditions and has a powerful influence on battery characteristics [4, 5].

5.1
Battery Design

Batteries can be grouped according to the type of electrode/electrolyte configuration, electrolyte, and packaging. The two types of electrode/electrolyte configurations are the winding type and the stack type. Li-ion batteries use liquid electrolytes while Li polymer batteries require polymer electrolytes. Packaging is classified into the can type (cylindrical or rectangular) and pouch type.

While battery design may differ slightly depending on shape and function, its basic principles are as follows:

1) The total electric capacity of the cathode is identical to that of the anode during charging and discharging.
2) Battery voltage is proportional to its state of charge.
3) The electric capacity of battery depends on designed voltage range, and therefore the battery cannot be charged over the maximum cutoff voltage or discharged below the minimum cutoff voltage.

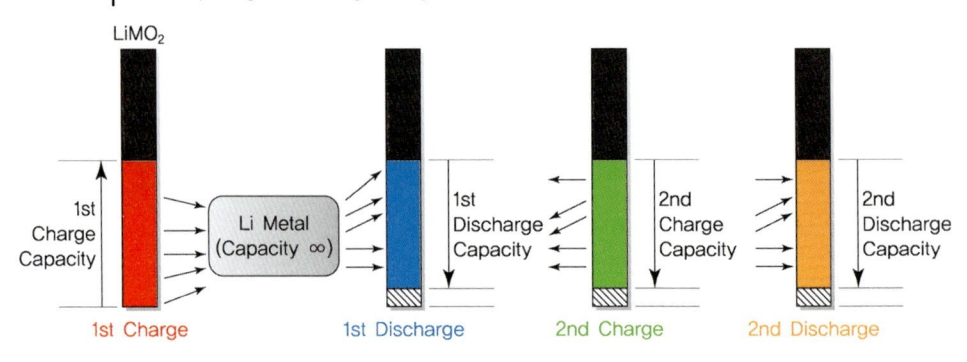

Figure 5.1 Capacity of a half cell using cathode material (lithium metal oxide).

5.1.1
Battery Capacity

Battery capacity can be determined from open-circuit voltages of half cells consisting of lithium metal and a cathode or an anode. Figure 5.1 shows the half cell capacity of a lithium metal oxide used as a cathode material.

The crystal structure of lithium metal oxides changes with lithium deintercalation when the cell is first charged, but does not return to the initial structure even if lithium is intercalated due to the initial irreversible capacity at the cathode. This value is associated with variables, such as the metal element, ratio of lithium to the metal element, and particle size. In general, the initial irreversible capacity of $LiCoO_2$ is 3–5 mAh/g and that of $LiNiO_2$ is 20–30 mAh/g. Coulombic efficiency is close to 100% after two charging/discharging cycles. The initial irreversible capacity of carbon-based anodes is attributed to SEI formation caused by electrolyte reduction at the anode surface. This is dependent on the crystallinity, structure, specific surface area, and particle size of the anode materials. Commercialized graphite anodes have an irreversible capacity value of 20-30 mAh/g. The coulombic efficiency of anodes is also near 100% after two charging/discharging cycles. Figure 5.2 illustrates the concept of half cell capacity for an anode material.

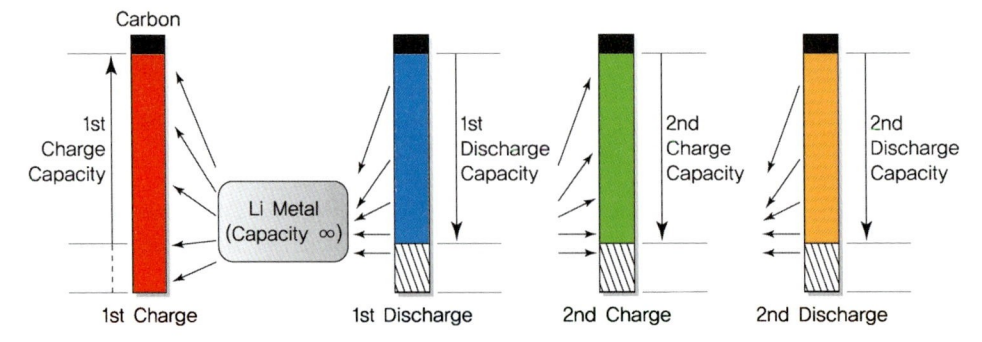

Figure 5.2 Capacity of a half cell using anode material.

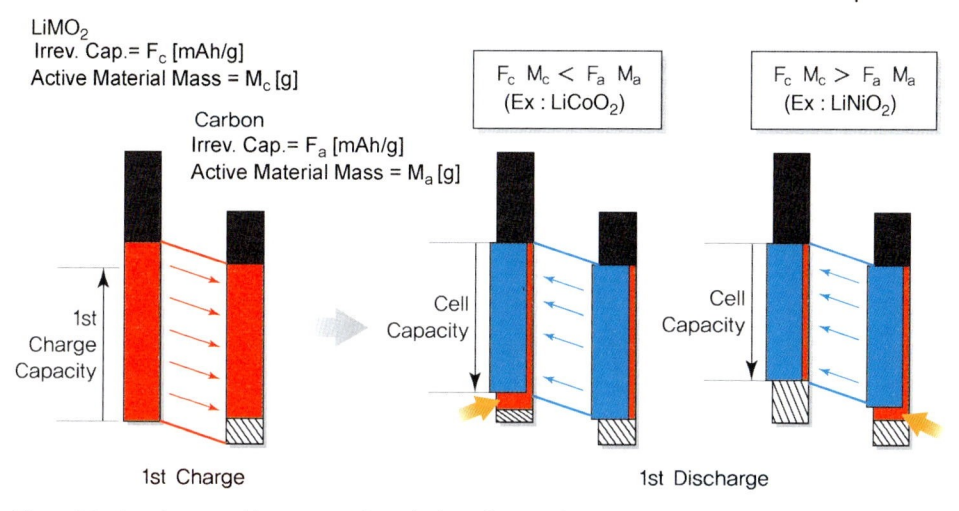

Figure 5.3 Initial irreversible capacity of a cathode and an anode.

Considering the initial irreversible capacity of a full cell consisting of a cathode and an anode, battery capacity can be expressed as shown in Figure 5.3. For this full cell, part of the lithium supplied from the cathode during initial charging is consumed in the initial irreversible reaction of SEI layer formation at the anode. To examine discharging in later stages, we can compare the relatively small initial irreversible capacity of $LiCoO_2$ to the relatively large initial irreversible capacity of $LiNiO_2$. The battery capacity of $LiCoO_2$ is the initial charged amount minus the irreversible capacity of the cathode, while that of $LiNiO_2$ is the initial charged amount minus the irreversible capacity of the anode. These are called the anode-limited design and cathode-limited design, respectively. They are attributed to the difference in the amount of lithium supplied to the half cell with lithium metal as the reference electrode and a full cell with a counter electrode. This is because lithium is limitless in the half cell but restricted by cathode materials in the full cell. Thus, the design of battery capacity is limited by the initial irreversible characteristics of electrode materials.

5.1.2
Electrode Potential and Battery Voltage Design

As shown in Figure 5.4, battery voltage is represented as the difference in potential between the cathode and the anode. This voltage is designed based on the open-circuit voltages of electrodes and reflects various conditions such as depth of discharge and temperature.

Depending on the situation, potential behavior of the cathode and anode may differ even if the same battery voltage is shown. Charge balance is affected not only by electrode potential, but also by the ratio of the cathode and anode within the battery. This implies that changing the ratio of the cathode and anode can lead to a different

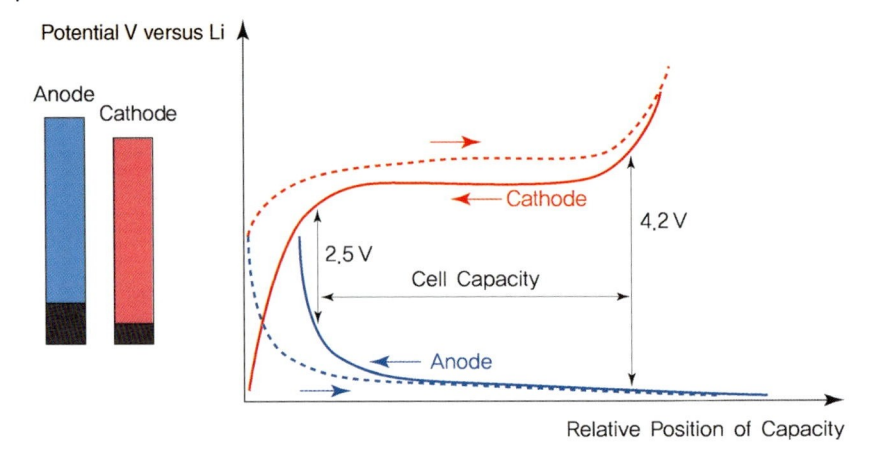

Figure 5.4 Relationship between battery voltage and electrode potential.

cutoff voltage. The concept of charge balance is presented in Figures 5.5 and 5.6. Figure 5.5 shows how to adjust the potential balance of a battery by increasing the initial irreversible capacity of the cathode.

The potential balance can also be adjusted by increasing the initial irreversible capacity of the anode, as shown in Figure 5.6. This adjustment can be understood in the context of adding surplus lithium to the cathode or anode to offset the irreversible capacity at the anode or cathode. Adjustments to the charge balance must be carefully executed as it is closely related to capacity, voltage, and safety characteristics. The following section describes factors to be considered in the design of charge balance.

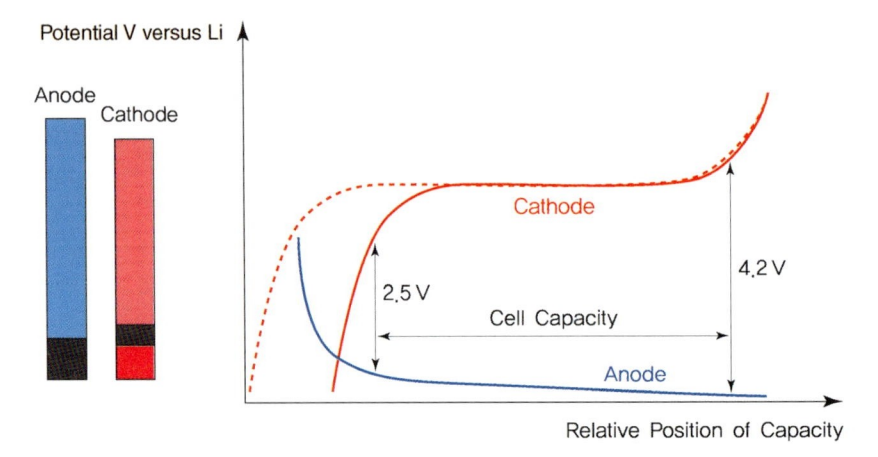

Figure 5.5 Adjustment of potential balance by increasing initial irreversible capacity of the cathode.

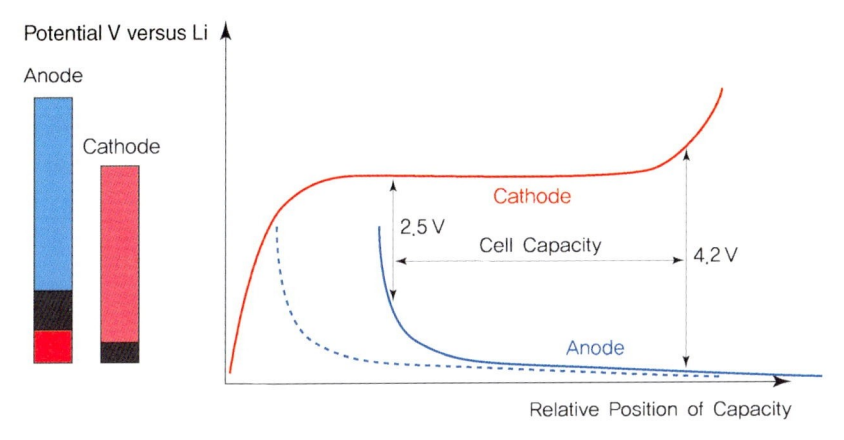

Figure 5.6 Adjustment of potential balance by increasing initial irreversible capacity of the anode.

5.1.3
Design of Cathode/Anode Capacity Ratio

The most important criterion in the design of battery capacity is that the anode must have a larger reversible capacity than the cathode. Although there may be some advantages such as greater battery capacity if the anode capacity is smaller than that of the cathode, safety issues may arise due to lithium deposition at the anode during charging. As can be seen from the cathode to anode capacity ratio in Figure 5.7, the battery is limited to smaller capacity even if an electrode with equivalent or larger capacity is used. If the anode to cathode capacity ratio, so-called N/P ratio (negative electrode capacity/positive electrode capacity) is set to 1 by allowing the electrodes to have the same initial irreversible capacity, the battery will have a capacity of 80 mAh. On the other hand, if the cathode uses an electrode with larger capacity and the ratio becomes 1.5, the battery capacity is reduced to 70 mAh. Of course, such extreme results are rare if the N/P ratio is adjusted appropriately.

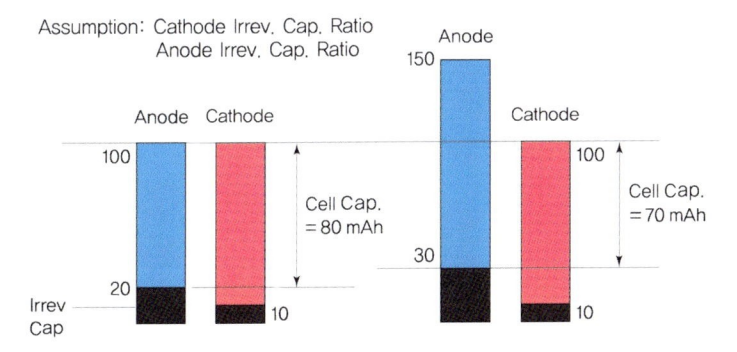

Figure 5.7 Relationship between battery capacity, initial irreversible capacity, and N/P ratio.

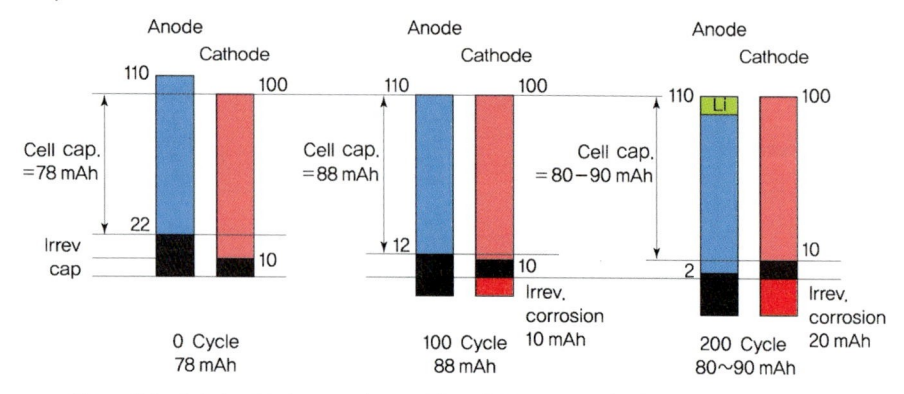

Figure 5.8 Relationship between battery life and N/P ratio: cathode degradation.

In this section, we examine the effect of N/P ratio on battery cycle life, which is determined by relatively complex variables. Capacity degradation may result from reactions between main components, including the cathode, anode, electrolyte, and separator. Since the deterioration of the electrolyte and separator is related to the cathode and anode, the following examples will be limited to cathode and anode degradation. Assuming a constant N/P ratio of 1.1 when the initial irreversible capacity of the anode is larger than that of the cathode, the effects of electrode degradation on battery cycle life and safety are presented in Figures 5.8 and 5.9. If we suppose that irreversible reactions at the cathode cause a 10 mAh drop every 100 cycles, the results will be as shown in Figure 5.8. As mentioned previously, the battery capacity is 78 mAh for an N/P ratio of 1.1. The actual capacity becomes 88 mAh even with cathode degradation after 100 cycles. At this point, the N/P ratio falls below 1 and the reversible capacity of the anode is 100% used. After 200 cycles, lithium is deposited at the anode and contributes to a capacity of 88 mAh. While capacity increases as cycling progresses, battery safety is seriously threatened.

Irreversible reactions at the anode resulting in a 10 mAh drop per 100 cycles are shown in Figure 5.9. For an N/P ratio of 1.1 and taking into account 10 and 22 mAh as

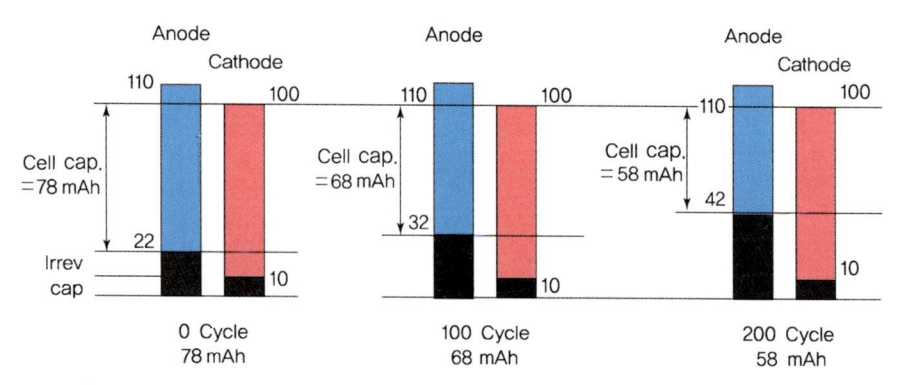

Figure 5.9 Relationship between battery life and N/P ratio: anode degradation.

the initial irreversible capacity of the cathode and anode, respectively, the initial battery capacity becomes 78 mAh. After 100 cycles, degradation of the anode leads to a smaller capacity of 68 mAh. At this point, the N/P ratio is greater than the initial value of 1.1. The capacity drops further to 58 mAh after 200 cycles. Although battery safety is not threatened, the design should be adjusted to prevent excessive loss of battery cycle life. Considering the above variables of battery design, we can see that a cautious approach is needed in setting design conditions.

5.1.4
Practical Aspects of Battery Design

In this section, the specifics of battery design are described based on actual examples. The physical and electrochemical design factors of cylindrical batteries are given in Tables 5.1 and 5.2. Similar tables can be obtained for noncylindrical batteries. Along with electrochemical factors, physical factors must be considered for efficient spatial arrangement of components including electrodes having uniform volume and weight. These physical and electrochemical design factors must be covered before looking at battery manufacturing and battery evaluation, as well as advantages and disadvantages of materials. While some terms may differ according to manufacturer, common terms will be used as much as possible.

First, we examine physical design factors outlined in Table 5.1. A standardized cylindrical battery has a diameter of 18 mm and consists of a cathode, an anode, and a

Table 5.1 Physical design factors of cylindrical batteries (example).

	Design parameters	Values (mm)
Jelly roll	Inner (jelly roll) diameter	17.6
	Mandrel diameter	3
Cathode	Total thickness	0.128
	Outer thickness	0.057
	Inner thickness	0.055
	Current collector thickness	0.015
	Width	58.0
Anode	Total thickness	0.113
	Outer Thickness	0.052
	Inner thickness	0.052
	Current collector thickness	0.008
	Width	59.0
	Separator thickness	0.020
	Cathode tab thickness	0.100
	Anode tab thickness	0.100
Coating Length	Outer cathode	813.7
	Inner cathode	783.5
	Inner anode	816.9
	Outer anode	785.4

separator, which are wound in a jelly roll configuration. This diameter should be designed to allow winding and can insertion. The thickness of electrodes is set considering the rate capability, and the cathode (~130 μm) is generally thicker than the anode (~110 μm). This is because the lithium diffusion coefficient of $LiCoO_2$ is larger than that of graphite. These thicknesses are generally accepted even though slight differences may exist depending on particle size and its distribution. Since the width of each electrode is related to safety, the capacity of the anode must always be larger than that of the cathode. As such, the width of the anode (~59 mm) is slightly larger than that of the cathode (~58 mm). The length of a standardized cylindrical lithium secondary battery is 65 mm, and the width of electrodes should be designed less than 65 mm for the insertion of other components such as the upper cap and safety pin. The separator is wrapped around the cathode and anode, thus preventing short-circuiting between electrodes. Each electrode has different values of inner and outer coating length. Since this is achieved by physical winding, the charge balance changes according to the location of the cathode and anode (inner or outer). The uncoated current collector is located at the outermost cathode (or anode) and sealed with finishing tape.

From the electrochemical design factors listed in Table 5.2, we can see that cathode and anode materials are present in different proportions. This is attributed to the difference in electrical conductivity between the cathode material ($LiCoO_2$) and the anode material (graphite). To overcome this difference, an appropriate amount of carbon is mixed with the cathode material. During this process, high mixing uniformity must be secured. The charge capacity corresponds to lithium deintercalation from the

Table 5.2 Electrochemical design factors of cylindrical batteries (example).

	Parameters	Values	Units
Cathode	Active material composition	96	%
	Charging capacity	175	mAh/g
	Initial reversible efficiency	95	%
	Initial irreversible capacity	8.75	mAh/g
	Usable capacity	160	mAh/g
	Thickness	0.128	mm
	Loading level	43	mg/cm^2
	Loading (or electrode) density	3.75	g/cc
	Total cathode loading weight	18.9	g
Anode	Active material composition	97	%
	Total charging capacity	350	mAh/g
	Thickness	0.113	mm
	Loading (or electrode) density	1.75	g/cc
	Loading level	22	mg/cm^2
	Total anode loading weight	9.9	g
Cell	Current density	3.2	mA/cm^2
	Nominal capacity	2950	mAh
	N/P ratio	1.1	dimensionless

LiCoO$_2$ cathode material. Theoretically, this is 0.5 mol but differs according to the charge voltage, compositional design, and structural stability of materials. The design should account for safety issues that may arise during overcharging. If NCA (Ni–Co–Al), NMC (Ni–Mn–Co), or a combination of both is used instead of LiCoO$_2$, the characteristics of each material must be considered.

Initial coulombic efficiency is a basic factor applied to both the cathode and the anode during the design of potential or capacity. Sections 5.1.1 and 5.1.2 discuss the initial irreversible capacity. Electrode density should be planned considering the physical composition, packing density, and porosity. If the electrode density is too high, it may affect battery safety and electrolyte injection process. Loading level, which is the amount of active material per unit area of an electrode, should reflect the diffusion coefficient of lithium ions, electrical conductivity of particles, and electric paths to the current collector. This factor also affects rate capability of the battery. Priority is given to features of the anode opposing the cathode. The capacity of the battery is determined by considering these physical arrangement and electrochemical design factors.

5.2
Battery Manufacturing Process

The manufacturing of lithium ion batteries differs according to battery types and such differences are listed in Table 5.3. Since the manufacturing processes for rectangular and cylindrical batteries are similar, this section focuses on the

Table 5.3 Comparison of manufacturing process by battery type.

	Li-ion batteries	Li-ion polymer batteries		
		Winding		Stack
Process	Cylindrical/rectangular	Physical gels	Chemical gels	Physical or chemical
Electrode	Mixing	Mixing	Mixing	Mixing
	Coating/pressing	Coating/pressing	Coating/pressing	Coating/pressing
		Separator coating		Separator coating
Assembly	Winding	Winding	Winding	Punching
	Jelly roll insertion	Jelly roll insertion	Jelly roll insertion	Stacking
	Welding	Welding	Welding	Welding
	Electrolyte injection	Electrolyte injection	Electrolyte injection	Electrolyte injection
	Welding	Pouch sealing	Pouch sealing	Pouch sealing
Posttreatment	None	Pressing	Thermal crosslinking	Pressing or thermal crosslinking
Formation	Aging	Aging	Aging	Aging
	CHG/DIS	CHG/DIS	CHG/DIS Degasing/resealing	CHG/DIS Degasing/resealing

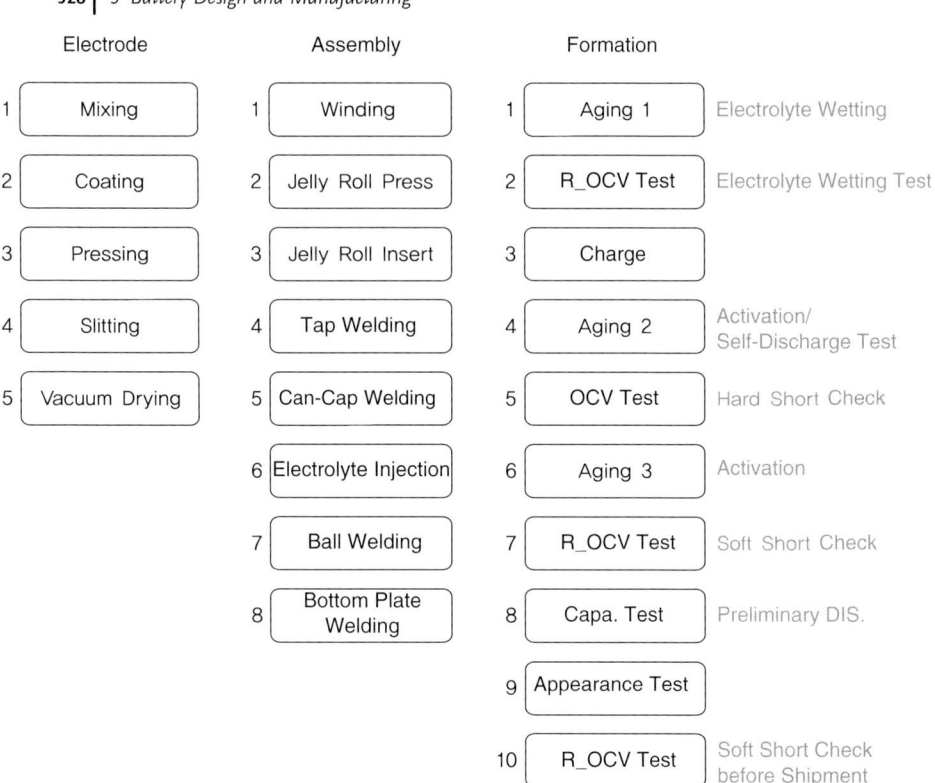

Electrode		Assembly		Formation	
1	Mixing	1	Winding	1	Aging 1 — Electrolyte Wetting
2	Coating	2	Jelly Roll Press	2	R_OCV Test — Electrolyte Wetting Test
3	Pressing	3	Jelly Roll Insert	3	Charge
4	Slitting	4	Tap Welding	4	Aging 2 — Activation/Self-Discharge Test
5	Vacuum Drying	5	Can-Cap Welding	5	OCV Test — Hard Short Check
		6	Electrolyte Injection	6	Aging 3 — Activation
		7	Ball Welding	7	R_OCV Test — Soft Short Check
		8	Bottom Plate Welding	8	Capa. Test — Preliminary DIS.
				9	Appearance Test
				10	R_OCV Test — Soft Short Check before Shipment

Figure 5.10 Flow chart of battery manufacturing process (for rectangular batteries).

manufacturing of rectangular-shaped batteries. Polymer batteries can be divided into chemical and physical gels depending on the type of polymer electrolyte. It should be noted that some polymer batteries are based on a stack structure and not of the winding type.

The manufacturing process can be largely divided into the electrode process, assembly process, and chemical process. Before battery manufacturing, the specifics of each process must be determined. When overall planning is complete, manufacturing proceeds first with the electrodes, followed by assembly of various components, final inspections of charging and discharging, and sorting for shipment [4, 5]. Figure 5.10 shows the manufacturing process of rectangular-type lithium ion batteries. Each process is described in detail in the following sections.

5.2.1
Electrode Manufacturing Process

5.2.1.1 Preparation of Electrode Slurry
In this process, a binder solution is produced by dissolving a polyvinylidene fluoride binder in N-methyl pyrrolidone (NMP) solvent. Active materials and conductive

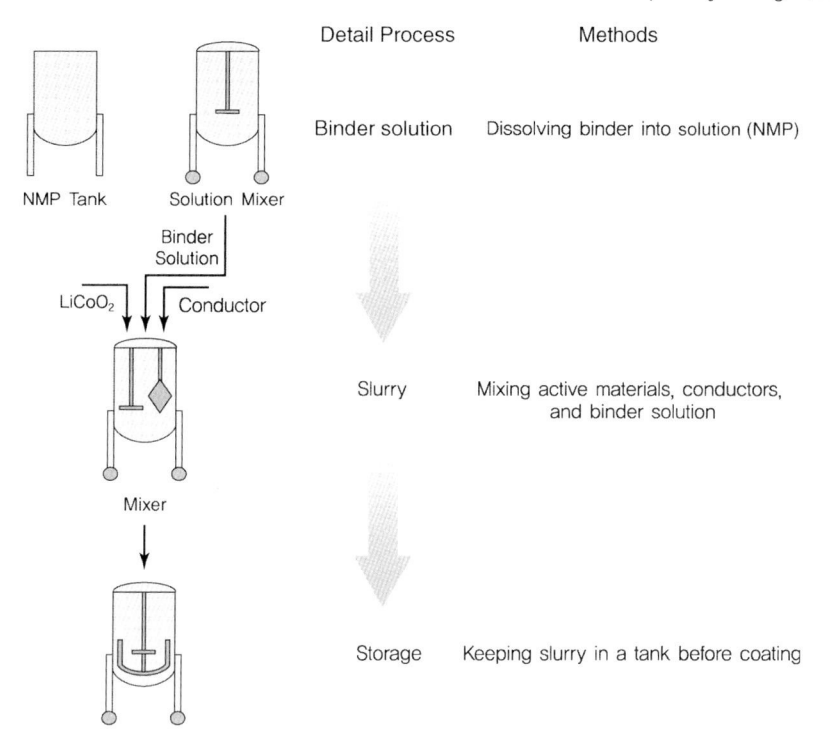

Figure 5.11 Preparation of active material slurry.

agents are then mixed/dispersed to prepare a homogeneous slurry. This process can be divided into binder solution preparation, binder solution transfer, slurry preparation, and slurry storage and transfer. Figure 5.11 shows the preparation of an active material slurry by stage. First, the binder and solvent are mixed to form a binder solution. The solution is then transferred to a buffer tank connected to slurry mixer. Finally, the electrode slurry is prepared by mixing active materials, electric conductor, and binder solution with optimized procedures.

During the slurry storage process, the prepared slurry is pumped to a storage mixer and stirred to prevent hardening and aggregation. If coating is involved, the slurry is transferred automatically from the storage mixer to the head tank, from which the slurry is moved to a coating head.

5.2.1.2 Electrode Coating

During the electrode coating process, the slurry is passed through the coater head and the metal current collector is coated in a given pattern and thickness followed by drying. This process consists of unwinding, coating, drying, density measurement, and winding. Coating is applied to the front and back of electrodes. This coating process is the same for both the cathode and the anode. Aluminum and copper films are used as the cathode current collector and anode current collector, respectively. The unwinding process prepares the metal current collector or one side-coated electrode

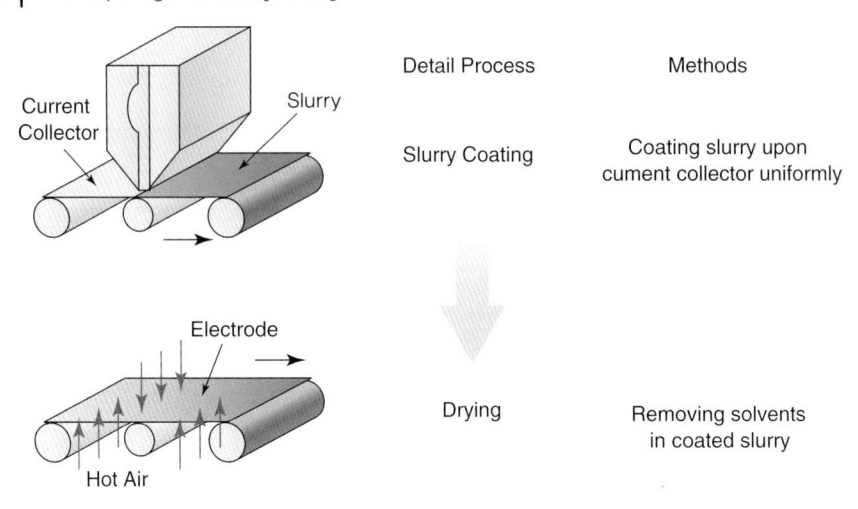

Figure 5.12 Coating of current collector with electrode slurry.

for coating. After the slurry prepared from the mixing process passes through the coater head, the metal current collector is coated in a given pattern and constant thickness. The drying process removes solvents and moisture from within the slurry coated on the metal current collector. The density (thickness) measurement process examines the amount of slurry coated on electrodes. In the winding process, electrodes are wound in roll form while under an appropriate tension. Meanwhile, the back coating process repeats the above steps on the back of a front coated electrode. Figure 5.12 shows the coating of a metal current collector with slurry using a die coater.

5.2.1.3 Roll Pressing Process

The roll pressing process increases electrode density after the coating process, improves adhesion between the current collector and active materials, and flattens electrodes between two heated or cold rolls. This process is comprised of unwinding, trimming, preheating, cleaning, and winding. In the unwinding process, electrodes in jumbo roll are released for the roll pressing while maintaining tension. The trimming process trims away the uncoated edge of the electrodes to remove creases arising from thickness differences between coated and uncoated areas. Preheating warms up the electrodes before insertion into the roll so as to press electrodes well. The cleaning process removes impurities from the electrode surface with a cleaning nonwoven before winding of electrodes. The winding process winds electrodes into the roll while a constant tension is applied. Figure 5.13 shows a roll pressing machine used to press coated electrodes into metal rolls.

5.2.1.4 Slitting Process

In the slitting process, electrodes are cut to uniform widths and prepared for winding in the assembly process. Slitting process can be broken down into unwinding, cleaning, and winding. The unwinding process maintains an appropriate tension while feeding

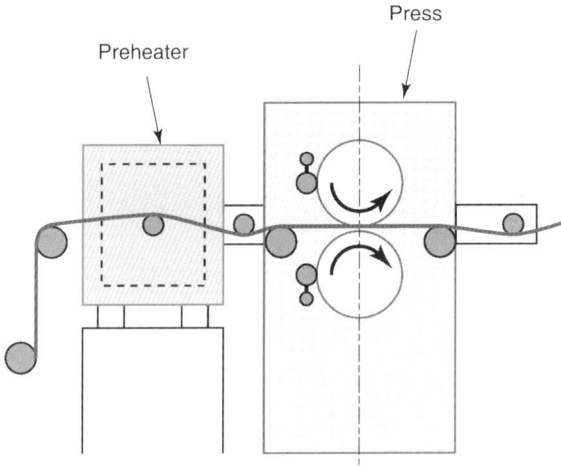

Figure 5.13 A roller and the roll pressing process.

electrodes into the cutter. The cleaning process removes impurities from the electrode surface with a cleaning nonwoven prior to winding of the electrodes. After cleaning, the winding process winds electrodes into the roll while maintaining tension. Figure 5.14 shows a slitter used in the cutting of pressed electrodes to specified widths.

5.2.1.5 Vacuum Drying Process
In the vacuum drying process, the electrode wound on a reel is dried for a given period in a vacuum chamber. Moisture and excess stress from the rolling process are removed through thermal treatment.

5.2.2
Assembly Process

5.2.2.1 Winding Process
The winding process produces a jelly roll by attaching a tab to the electrode and placing the separator between cathode and anode followed by cylindrical winding. At

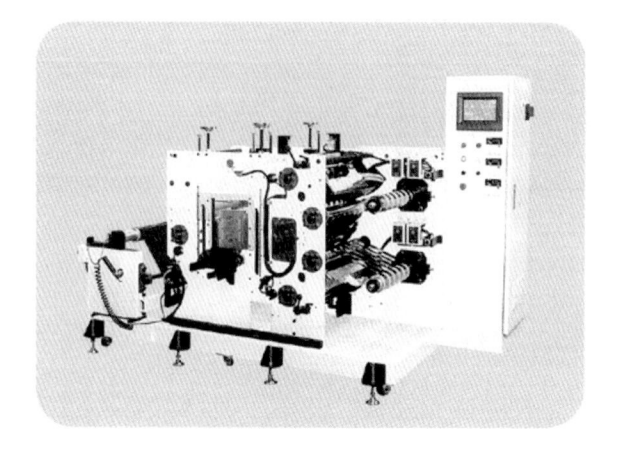

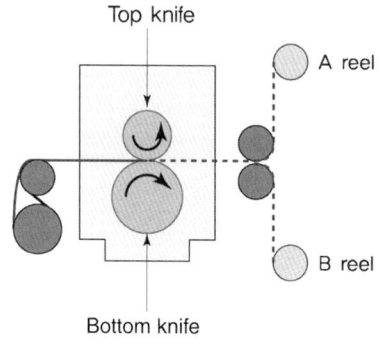

Figure 5.14 A slitter and the slitting process.

this point, the jelly roll is completely wound and ready for insertion into the can. This process is comprised of tab ultrasonic welding, center reforming, and internal short-circuit inspection. The tab ultrasonic welding process uses ultrasonic welding to attach an aluminum tab to the cathode and a nickel tab to the anode. The center reforming process removes creases at the center of the jelly roll and creates space to insert the welding tip. The internal short-circuit inspection process measures the resistance of the jelly roll to prevent defects by ensuring a resistance higher than tens of MΩ. For rectangular batteries, the jelly roll may be pressed and inserted into the can. Figure 5.15 shows a winder used for jelly roll winding and the winding process.

5.2.2.2 Jelly Roll Insertion/Cathode Tab Welding/Beading Process

The jelly roll insertion process inserts the jelly roll to a certain depth in the can and measures the extent of insertion using X-ray inspection. The anode tab welding process bends the anode tab for welding to the floor of the can. In the assembly of cylindrical batteries, the beading process creates a bent groove for the gasket to reach the top of the can containing the jelly roll.

Jelly Roll

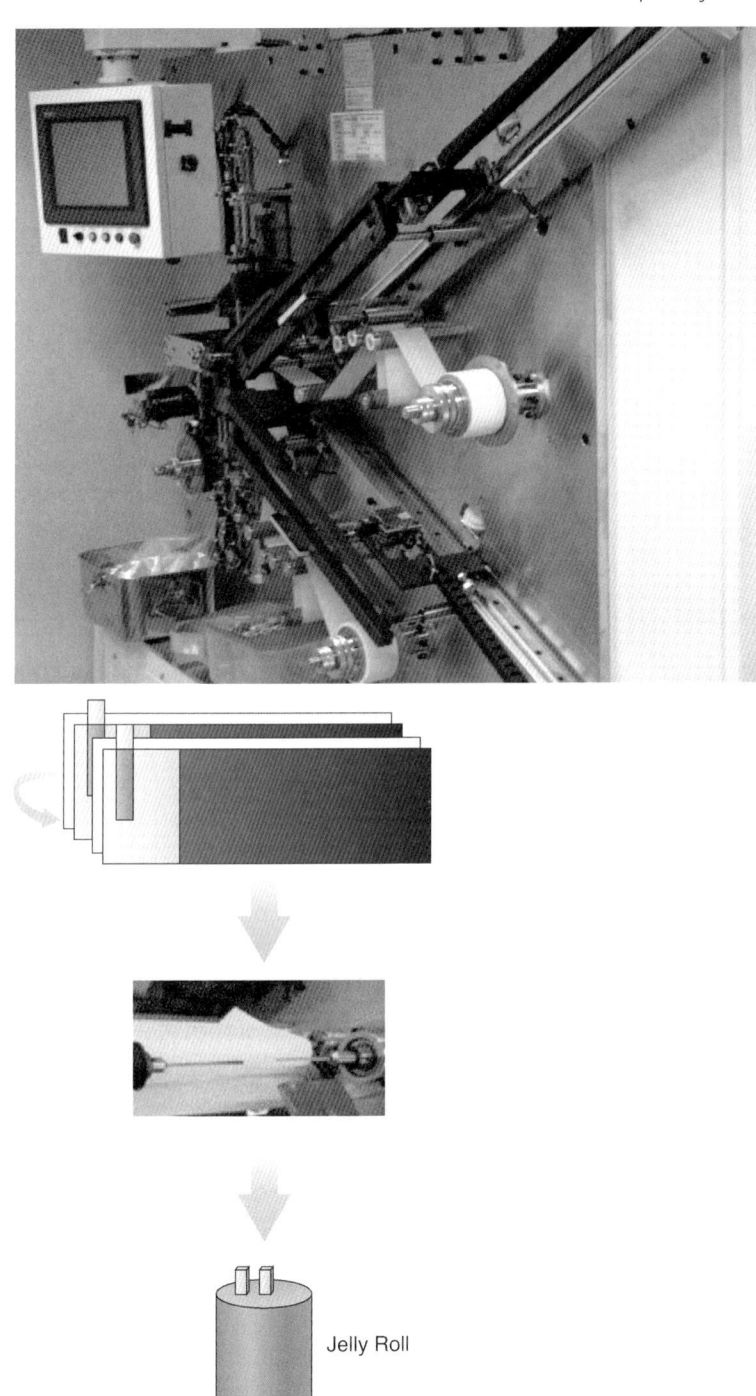

Figure 5.15 A jelly roll winder and the winding process.

5.2.2.3 Electrolyte Injection Process

During electrolyte injection, the internal pressure after injecting the electrolyte should be kept lower than atmospheric pressure and the electrolyte is allowed to be impregnated into the jelly roll by introducing air into the can. After electrolyte injection, electrolyte surrounding the cathode tab and beading area is wiped with a cleaning nonwoven. The gasket is inserted to complete the process.

5.2.2.4 Cathode Tab Welding/Crimping/X-ray Inspection/Washing Process

The cathode tab welding process includes welding of the cathode tab and the bottom of the current break assembly. Current break is enabled by ultrasonic welding of the central area of the safety vent assembly. The crimping process applies pressure to the top part of the battery containing the current break assembly, positive temperature coefficient (PTC), and cap-up. After sealing the current break assembly, PTC, and cap-up, the crimped battery is pressed to maintain a constant height. X-ray inspection examines the internal electrode arrangement after assembly and checks for any defects. The washing process uses water to remove electrolyte and other impurities from the battery surface. The battery is then dried to eliminate moisture. Finally, the battery is imprinted with the manufacturing factory, line number, and date before entering the formation process.

5.2.3 Formation Process

5.2.3.1 Purpose of the Formation Process

The formation process is divided into activation, removal of out-of-spec batteries, and capacity selection. First, in the activation process, the assembled battery is stabilized through charging, aging, and discharging. SEI layers are formed on the anode surface at 3.3 V during initial charging. This process facilitates the movement of lithium ions at the electrolyte–electrode interface and suppresses electrolyte decomposition. The extent of activation affects various characteristics, such as battery performance, life, and safety. Faulty batteries can be detected by measuring the open-circuit voltage during aging and discharge capacity after aging. The main cause of faulty batteries is internal short circuits, which lead to drop in the open-circuit voltage and discharge capacity. Such batteries must be discarded to avoid performance or safety issues. Capacity selection is important when lithium ion batteries are connected in series or parallel and used as a pack. If batteries within the pack do not have the same characteristics (capacity, voltage), performance, battery life, and safety may be negatively affected. In general, the batteries in the same group should have similar capacities within 3%.

5.2.3.2 Procedures and Functions

The main procedures and functions of the formation process are as follows:

1) **External inspection**: Batteries are examined for any leakage, impurities, or distortion.

2) **Aging 1**: Constant temperature and humidity are maintained for electrolyte intrusion into jelly roll. Strict maintenance is required to prevent corrosion and other side effects that may arise over time.
3) **OCV 1**: This process checks for electrolyte wetting, occurrence of side effects, and defects in assembly components.
4) **Charging**: The batteries are fully charged to remove faulty ones from previous steps.
5) **Standing**: This process stabilizes charged batteries for OCV 2 and usually takes up to several hours.
6) **OCV 2**: This process detects faulty batteries.
7) **Aging 2**: Completely charged batteries are kept under constant temperature and humidity for a week.
8) **OCV 3**: Similar to OCV 2, this process detects faulty batteries.
9) **Aging 3**: Batteries are further stabilized by maintaining constant temperature and humidity for 28 days.
10) **Discharge/charging for shipment**: Batteries are completely discharged to determine capacity requirements and then charged for shipment.
11) **Capacity selection**: Batteries are classified according to discharge capacity. In general, 3% capacity corresponds to 1 level.

References

1 Lim, D.J. (2000) Battery Technology Symposium in Korea, The Korean Society of Industrial & Engineering Chemistry.

2 Schalkwijk, W.A. and Scrosati, B. (2002) *Advances in Lithium-Ion Batteries*, Kluwer Academic, New York.

3 Hong, Y.S. (2006) Advanced Secondary Battery Technologies, The Korea Industrial Technology Association.

4 Kim, S.S. (2008) Industrial Trends on Lithium Secondary Batteries for Mobile IT. Electronic Times, Seoul.

5 Kim, M.H. (2008) Symposium on Chemical Materials for Energy Conversion, Korea Research Institute of Chemical Technology, Daejeon.

6
Battery Performance Evaluation

6.1
Charge and Discharge Curves of Cells

This section begins with a definition of terms related to battery performance, followed by a discussion of their significance.

6.1.1
Significance of Charge and Discharge Curves

Battery characteristics are exhibited through charge and discharge curves. The charge and discharge curves shown in Figure 6.1 contain extensive information. In general, the charge and discharge curves (Figure 6.1a) of a lithium secondary battery represent the difference in capacity and potential between curves of the anode and cathode derived from a half cell (Figure 6.1b) with lithium metal as its reference electrode. In Figure 6.1a, the vertical axis indicates the change in voltage while the horizontal axis gives the capacity. Higher values in the vertical axis and horizontal axis lead to improved output and operating time, respectively.

The voltage of lithium secondary batteries is 2.5 times higher than that of existing NiCd or NiMH batteries. NiCd or NiMH batteries containing aqueous electrolytes have a voltage of 1.5 V, whereas lithium secondary batteries with a $LiCoO_2$ cathode and graphite anode have voltages higher than 3.6 V. This is due to the large potential difference between the transition metal oxide cathode, which uses lithium ions as a charging and discharging medium, and the graphite anode having a low voltage of 0.05 V versus lithium. Since mobile phones and other communication devices involving voice and video transmission require a minimum voltage of 3.0 V, lithium secondary batteries are the most appropriate choice for large data transmission.

As shown in Figure 6.2, charge and discharge curves take on various forms depending on the cathode material, and these are indicative of the amount of electrical energy that can be delivered by the battery. The shapes of charge/discharge curves vary according to the crystal structure of oxides during lithium deintercalation or intercalation, bonding between atoms, and the energy state determined by the

Principles and Applications of Lithium Secondary Batteries, First Edition. Jung-Ki Park.
© 2012 Wiley-VCH Verlag GmbH & Co. KGaA. Published 2012 by Wiley-VCH Verlag GmbH & Co. KGaA.

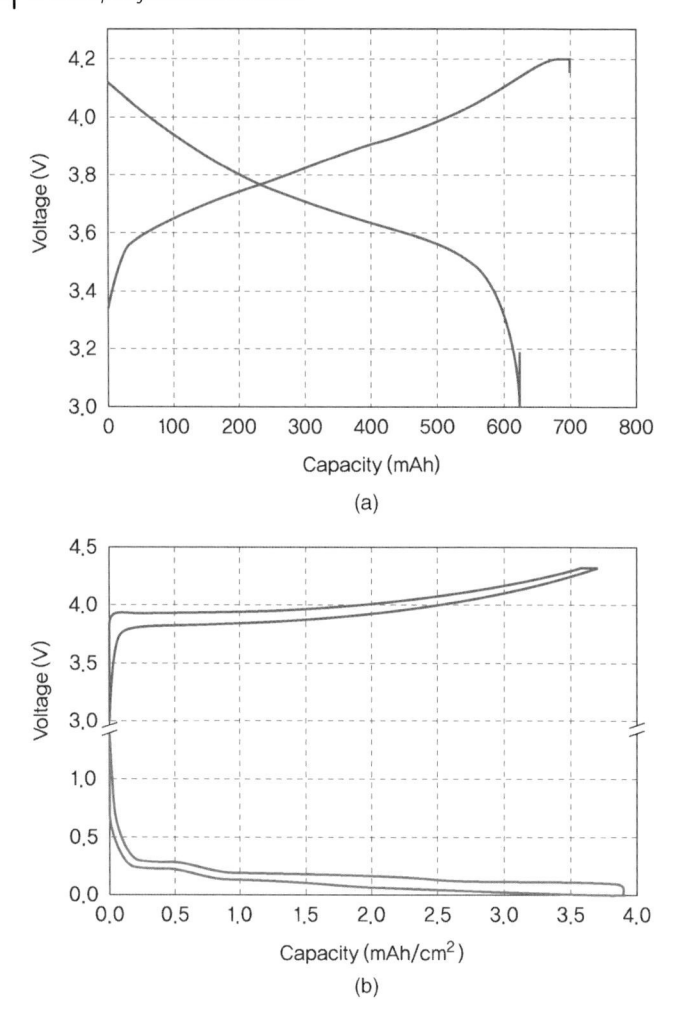

Figure 6.1 Charge and discharge curves of a lithium secondary battery: (a) full cell and (b) half cell.

bonding of electrons in the d orbital of the transition metal and p orbital of oxygen. This has already been covered under the crystal structure of cathode materials in Section 3.1. The actual voltage of batteries differs from the theoretical value due to polarization. iR drop is representative of such polarization effects.

iR drop depends on battery design. Figure 6.3 relates voltage to electrode thickness. Batteries with thick electrodes designed for high capacity experience an imbalance of potential within electrodes. Electrical resistance occurs when lithium passes through the space between the separator and the current collector. Since resistance increases with thickness, this must be considered in the design of high-capacity batteries. Owing to the significant iR drop at each electrode, it is important to maintain a balance between the cathode and the anode.

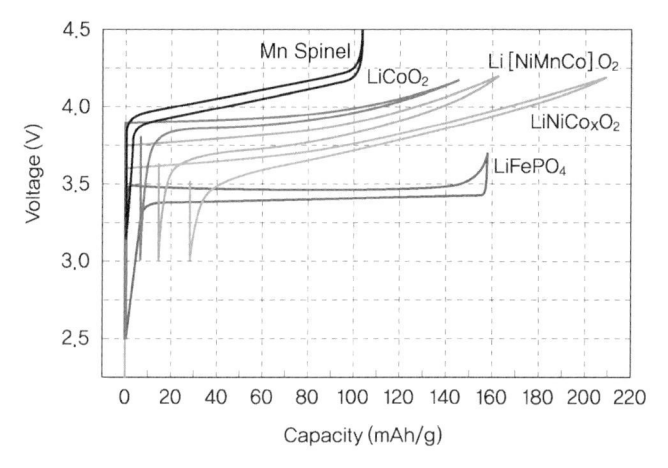

Figure 6.2 Charge and discharge curves for different cathode materials.

6.1.2
Adjustment of Charge/Discharge Curves

Charge/discharge curves of batteries are determined by the reaction voltage of electrodes with lithium. In addition, they are closely related to the electrical devices being used. A minimum voltage of 3.0 V is required to reduce noise and maintain

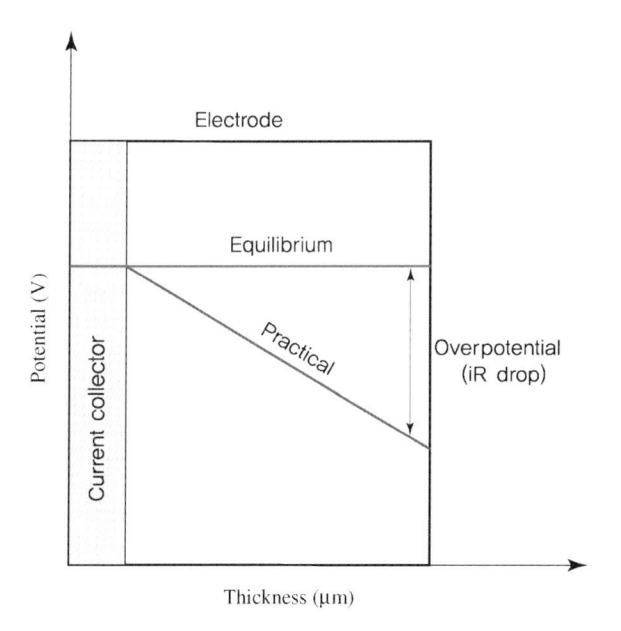

Figure 6.3 iR drop within a thick electrode.

high audio quality for audiovisual transmission devices. Requirements of the final electrical product should be carefully examined to select appropriate battery materials.

As noted previously, the voltage of cathode materials is determined by the bonding force between atoms, especially the bonding force between d orbital electrons of transition metals and p orbital electrons of oxygen, as well as the crystal structure of transition metal oxides. The anode voltage changes according to the crystallinity of carbon layers in graphite. To understand the concept of charge/discharge curves, we should be aware that they are attributed to the difference between such curves at each electrode. A battery of 0 V indicates that there is a difference of 0 V between the cathode and the anode, but this does not provide absolute voltage for each electrode. The absolute voltage of each electrode can be measured by using the three-electrode system with lithium as the reference electrode. This method is used to analyze changes in the absolute voltage of the cathode and anode during charging and discharging.

Most commercialized lithium secondary batteries contain $LiCoO_2$ as the cathode and graphite as the anode. This is because the $LiCoO_2$–graphite system has high voltage, good stability, and excellent performance at high temperatures. However, the recent price increase of cobalt (Co) and thermal instability of $LiCoO_2$ have motivated active studies on the Ni–Mn–Co three-component system. These materials have a slightly lower voltage compared to $LiCoO_2$, but they are highly stable and affordable. The amounts of Ni, Co, and Mn can be adjusted to maximize the advantages of capacity, safety, and cost. An increase in the amount of Ni leads to higher capacity and lower voltage, while higher Mn content allows greater safety. Co facilitates the movement of electrons and enhances battery performance. The choice of materials is related to final product requirements. For example, $LiNi_{1-x}M_xO_2$ ($0 < x < 0.2$) exhibits extremely high capacity but has a voltage lower than $LiCoO_2$ by 0.2 V. This is negligible for a single battery, but becomes apparent when several batteries are connected in series (battery pack for laptops) and significantly affect the device. As such, requirements of the electrical device must be first considered. On the basis of this understanding, we can then examine charge/discharge curves and approximate the voltage that can be produced.

6.1.3
Overcharging and Charge/Discharge Curves

A charger is responsible for battery charging and contains a three-level protection system. If this three-level protection system does not function properly, the battery may exceed its normal charging range and become overcharged. This is known as the overcharged state. Overcharging is usually caused by problems with the charger or other components. The peak voltage in the normal range is 4.3 V but does not eliminate the possibility of overcharging.

What happens when a battery is overcharged? First, the temperature rises as heat is produced due to electrical resistance within the battery. More lithium is produced

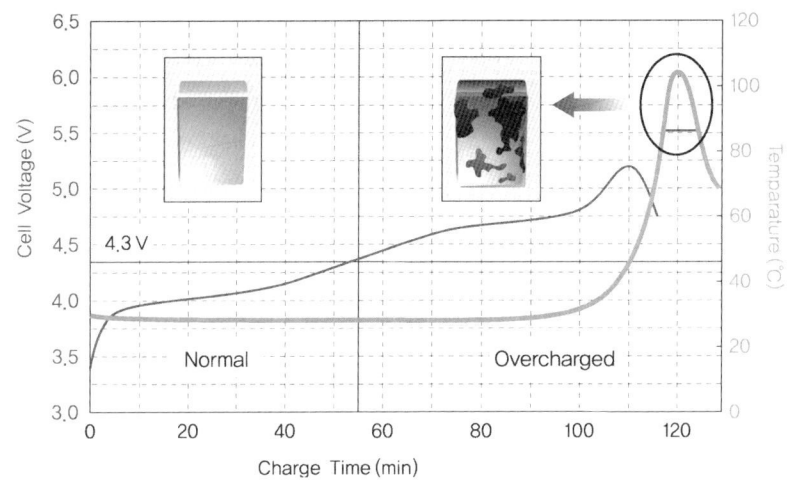

Figure 6.4 Overcharging behavior of a battery.

from the cathode than in the normal state and this is deposited as lithium metal on the anode. Structural collapse at the cathode during overcharging produces thermal energy and oxygen. Figure 6.4 shows the effects of overcharging through changes in the curve.

As shown in Figure 6.4, overcharging occurs when a constant current (CC) is applied and the battery exceeds the specified voltage. The temperature rises rapidly at this point. The separator starts to melt due to heat produced by the electrodes and results in an internal short circuit. The battery becomes dangerous and may even explode. The example in Figure 6.4 contains no protection mechanism and clearly demonstrates the dangers of overcharging. To prevent fires or explosions, the electrolyte viscosity can be increased to restrict ion movement under overcharged conditions, or insulating materials can be produced on the surface of electrodes from the decomposition of electrolyte additives. Even if the extent of overcharging is not extreme, the battery is considered to be in an abnormal state. If the battery does not perform as designed, the cathode voltage increases continuously and lithium is deposited at the anode, resulting in a voltage close to 0 versus lithium. At this point, the excessively high voltage at the cathode causes the electrolyte to decompose and release combustible gases. With the rapid drop in thermal stability of cathode materials, the structural collapse is accompanied by a large amount of heat release.

Since overcharging arises not from the battery itself but from charger defects, the use of original chargers is recommended. Even though batteries are equipped with overcharging protection, the risks cannot be ignored. Recently, there has been active research on functional separators, electrolyte additives, and safer cathode materials even upon overcharging. To resolve possible safety issues that may emerge in applications requiring large-sized batteries, more research should be conducted in this area.

6.2
Cycle Life of Batteries

6.2.1
Significance of Cycle Life

Cycle life is an indicator of how long the battery can be used and is expressed in cycle numbers. It shows the number of times that the battery can be charged and discharged. One cycle refers to charging and completely discharging. The concept of speed (or rate) should be considered in the evaluation of battery cycle life. With the increasing functions of mobile devices and laptops, superior battery performance is required to meet the demands for high output and energy. It is difficult to rely on existing battery designs to fulfill the requirements of high power and life characteristics for hybrid electric vehicles and power tools. From the change in capacity presented in Figure 6.5, we can see that higher c-rate leads to a greater decrease in capacity. If the current density is very low when discharged, the battery voltage approaches the equilibrium voltage while discharge capacity becomes close to the theoretical capacity. However, as discharge current rises, the overcharging voltage increases and battery capacity is reduced.

6.2.2
Factors Affecting Battery Cycle Life

Factors affecting battery cycle life can be divided into intrinsic properties of materials and design properties. Intrinsic properties are related to core materials (cathode, anode, separator, and electrolyte), while design factors include the design balance of the cathode and anode.

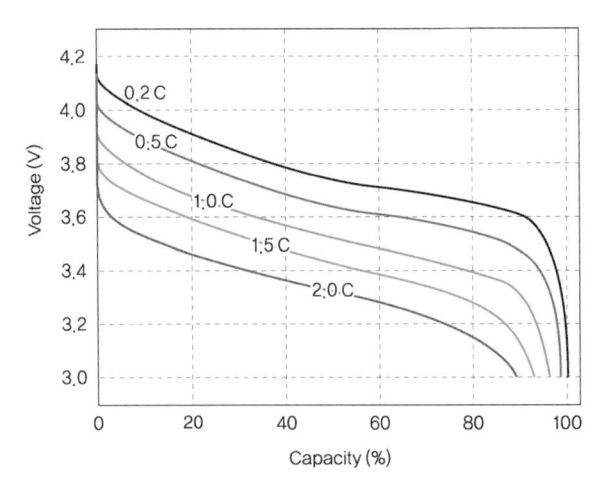

Figure 6.5 Discharge performance of a battery as a function of c-rate.

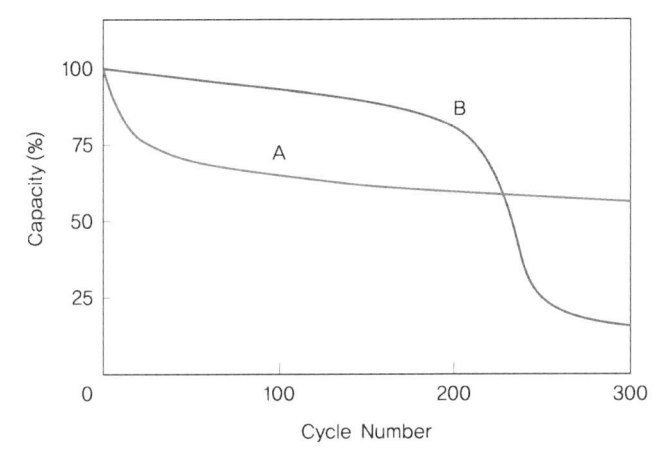

Figure 6.6 Typical cycle life problems in a battery.

Recovery is impossible if the battery cycle life is decreased by the deterioration of core components. As the temperature of the battery rises, the deterioration process is accelerated and cycle life is rapidly worsened. In this case, the fundamental properties of materials must be improved. On the other hand, where design factors are the cause of cycle life deterioration, the problem usually arises from a thermodynamic or electrochemical imbalance between the cathode and the anode. For instance, selecting a cathode and an anode of high performance does not necessarily result in a battery with excellent cycle life, and vice versa. Battery design is considered challenging because various factors have to be taken into account and a substantial amount of experience is required. Figure 6.6 shows the two forms of cycle life deterioration.

Curve A in Figure 6.6 represents a good design balance of electrodes after lithium deposition at the anode. Superior cycle life is shown when lithium is deliberately deposited in the initial stage. In this case, cycle life deterioration can be overcome by adjusting the design balance of the cathode and anode. On the other hand, curve B shows excellent cycle life initially but rapidly deteriorates after a certain point. Since this involves a complex process, an analysis of the cell performance must be carefully carried out. The capacity recovery should be checked when the battery is cycled at a low rate (C/10). If the capacity improves, the cause of deterioration can be traced to inadequate design. Otherwise, it is likely due to degradation of core materials within the battery. For a curve B in which capacity improves with low-rate charging/discharging, the cell resistance increases as the cycle progresses. This is because of the formation of resistive layer at the electrode interface or the resistance increase of electrolyte. Electric resistance can be neglected as electrons flow through an external circuit at relatively high speeds. The disruption of lithium ions within the battery is caused by resistance at the interface of electrode active materials, a clogging in separator pores, or depletion of the electrolyte. Interfacial resistance is usually attributed to decomposition products of the electrolyte. Electrolyte decomposition reactions produce gases or increase electrolyte viscosity, thereby increasing battery resistance.

To avoid this phenomenon, recent attempts to prevent electrolyte decomposition include surface modification of active materials or the use of electrolyte additives to form stable SEI layers during charging. Oxidative reactions on the cathode occur at a relatively high voltage, whereas reductive reactions for the anode take place near the lithium potential. Electrolyte decomposition proceeds with oxidative reactions at the cathode material surface due to the high voltage, and large amounts of gases are released at this point. The gases released from reactions between the cathode material and the electrolyte are affected by impurities (LiOH, Li_2CO_3) and structural defects existing on the cathode surface. In the latter case, the battery capacity gradually decreases as lithium ions are consumed, with structural collapse resulting from changes in the crystal lattice or reactions with the electrolyte.

The anode is subject to large volume expansion and contraction during charging and discharging. As such, SEI layers formed on the surface of anode materials must demonstrate flexibility toward volumetric change. If SEI layers are degraded due to anode expansion during charging, new layers are formed from electrolyte decomposition. The electrolyte is then gradually depleted and lithium ions are consumed, thus leading to a drop in capacity as the cycle progresses. Other problems pertaining to both the cathode and the anode are attributed to electrode thickness. As shown in Figure 6.3, thick electrodes may result in an imbalance of potential within electrodes. Decomposition reactions can be accelerated at the regions with higher resistance, which causes large overpotential. For the anode, the voltage may drop until the point of lithium deposition, and lithium ions existing in the electrolyte are easily deposited. Both cases have a negative impact on battery cycle life.

As mentioned previously, the electrolyte is continuously consumed with the repeated process of contraction and expansion at the anode, leading to SEI deterioration followed by reformation. To enhance cycle life, SEI layers resulting from reactions with the electrolyte and anode surface must possess superior qualities. SEI layers are usually influenced by electrolyte additives, and the use of various additives is being explored.

6.3
Battery Capacity

6.3.1
Introduction

Battery capacity refers to the charge amount stored by the battery and is expressed in mAh or Ah units. In other words, it is a measure of the charge amount that can be extracted from the battery under a constant current. This is related to the amount of cathode material within the battery, as lithium ions are derived from cathode materials. Capacity per unit volume is used for batteries in mobile phones or laptops, while capacity per unit mass is preferred for applications unconstrained by volume.

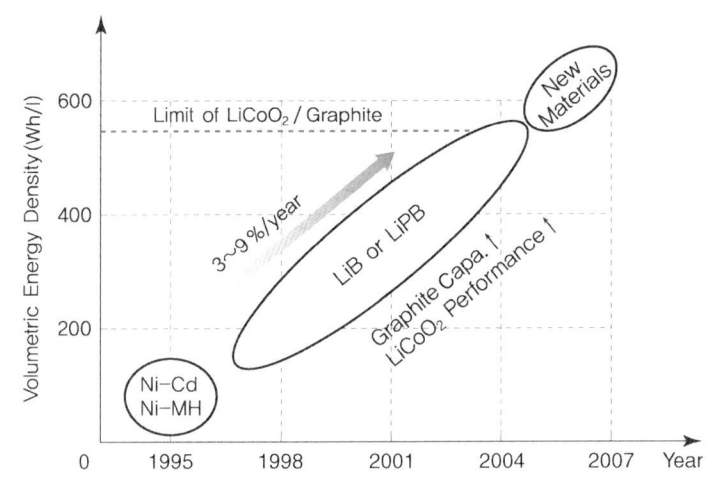

Figure 6.7 Developments in energy density of lithium secondary batteries.

Figure 6.7 shows the change in capacity of lithium secondary batteries throughout history.

6.3.2
Battery Capacity

Most mobile devices including cellular phones and laptops require high capacity per unit volume since high capacity is expected for small batteries. The design should allow not only large capacity per unit volume, but also high performance and safety. Battery capacity is affected by the following factors. First, active materials should have high capacity per unit mass or volume for high battery capacity. Regardless of the adequacy of the design, high capacity cannot be achieved if cathode materials are accompanied by small capacity. Second, active materials must have high tap density. A higher tap density allows improved packing density of electrodes. To prepare electrodes, active materials are mixed with a binder or conductive agent and coated on the current collector followed by roll pressing. If packing is not done properly, the electrode will take up more volume than necessary and high capacity cannot be obtained. Third, active materials must have small specific surface areas. If the specific surface area is large, there will be a high liquid content when active materials are coated on electrodes. Furthermore, electrical flow will be disrupted by the large surface between particles, and a large amount of binder and conducting agent is needed for adhesion. Higher amounts of conductive agent and binder are required to lower resistance and improve adhesion. This leads to a smaller amount of active materials, and high capacity cannot be attained in limited space. Fourth, it is necessary to reduce the thickness of the separator or current collector and minimize the space taken up by the electrolyte. However, this should be carefully executed as these components are closely related to battery safety. For instance, the current collector cannot be too thin as it should possess sufficient mechanical strength to withstand impact from coating. Since the current collector provides a path for the movement of electrons on the cathode or the anode, excessive thinning may cause increased resistance and overheating. The separator is directly related to battery safety, and hence an appropriate thickness should be maintained to avoid sharp declines in safety. In the design of batteries, active materials must be properly integrated while minimizing any loss in function or performance of other components.

6.3.3
Measurement of Battery Capacity

In general, battery capacity is obtained through a charge/discharge measuring device. Owing to the various types of batteries used in cellular phones, it is difficult to achieve standardization in capacity measurement. A standard cylindrical cell is used in laptops and represented as 18 650, which is explained in Figure 6.8.

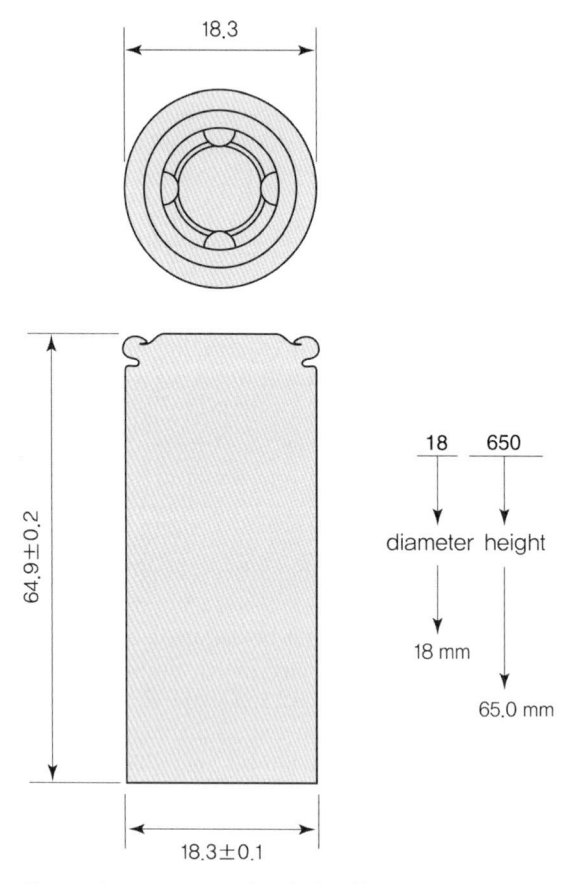

Figure 6.8 Dimensions of a cylindrical battery.

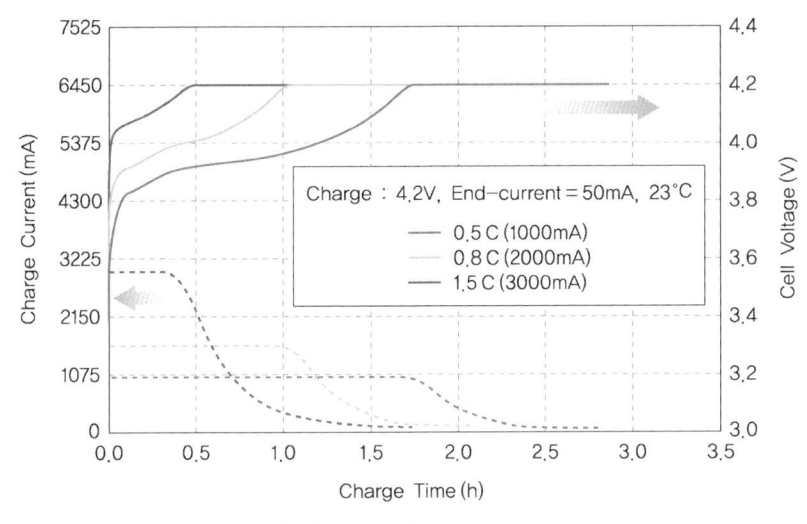

Figure 6.9 Voltage/current behavior during charging under various c-rates.

Battery capacity is generally expressed by the nominal capacity claimed by the manufacturer. Capacity differs according to c-rate conditions and C/5 is set for the nominal capacity. High c-rate capability is required to maintain high output in laptops. Since the operational rate is higher than C/2, cycle life should be evaluated at a faster rate.

In the evaluation of battery capacity, CC/CV (constant current/constant voltage) mode is used during charging while CC is appropriate for discharging. This difference is attributed to cell characteristics. If charging is carried out under CC, a great amount of heat is produced within the cell, and resistance of the battery leads to an overpotential at each electrode. Even though the battery reaches the predetermined cutoff voltage during charging, each electrode does not attain an equilibrium potential due to kinetics. As such, an equilibrium voltage can be formed at the end of charging by decreasing current under a constant cutoff voltage. Figure 6.9 shows the typical method of charging. Under high c-rate conditions during charging, the CC range is relatively short compared to CV. A longer CV range indicates that more time is needed to reach an equilibrium voltage due to larger overpotential resulting from the high current. As shown in Figure 6.9, the amount of current in the CV range gradually decreases over time. When current is reduced to a certain point, an equilibrium voltage is reached and charging becomes complete as there is almost no current flow even as charging continues. Discharging is a simpler process that involves only CC.

6.4
Discharge Characteristics by Discharge Rate

Rate capability refers to the maintenance of capacity with increases in charging/discharging rates. The charge/discharge rate is expressed as x C, and 1.0 C means that the nominal capacity of a battery is used up in 1 h. A battery discharged at 2.0 C can be used in 30 min, 4.0 C in 15 min, and C/2 in 2 h.

With diversified functions of the latest electrical devices, there is an increasing demand for high rate capability. Higher performance is thus expected of batteries as the main energy source. Existing cellular phones require a charge/discharge rate of C/2, but this will be increased to 1.0 C in the near future. Owing to the larger current in laptops compared to cellular phones, such batteries must be equipped with high capacity and rate capability. In particular, a motor should be operated to save works on a running computer or for initialization after starting up. High rate capability is necessary to obtain high power of the motor. Usually, a discharge rate of 1.0 C is needed for 4C/5 charging. Hybrid electric vehicles require output c-rate of 40 C, while power tools need at least 10 C. We can see that rate capability varies according to the type of application. Rate capability of batteries is also closely related to battery design. They are affected by various factors, such as the electrolyte, separator, type of active material, particle size, and so on. Among them, the electrode thickness is the main factor to affect rate capability. Rate capability is greatly improved for thin electrodes because of less electrical and ionic resistances within electrodes. Another influential factor is particles of the active material. However, thinning of the electrodes results in a smaller amount of active materials (taken up by the current collector, separator, and electrolyte) and thus decreases the battery capacity. In accordance with the application of batteries, these factors must be appropriately taken into consideration. The core technological challenge of batteries is to increase rate capability without any reduction in battery capacity, and this area of study will be a high priority in future research.

When a battery is used for a long time, its cycle life and rate capability are significantly reduced. This is because the internal resistance of the battery continuously increases over time. As mentioned earlier, this increase in resistance is caused by various factors, such as the formation of products from interfacial reactions, deterioration of active materials and the electrolyte, and higher contact resistance of electrodes.

$$I = V/R \tag{6.1}$$

Here, I is the current, V is the voltage, and R is the resistance. As shown in Eq. (6.1), an increase in R leads to a drop in current. To enhance rate capability, the resistance within the battery should be kept to a minimum.

According to Eq. (6.2), the battery resistance can be divided into two components based on whether it arises from electrons or lithium ions.

$$R_{tot} = R_e + R_{Li} \tag{6.2}$$

Here, R_e is the resistance to electrons and R_{Li} is the resistance to lithium ions. Resistance due to electrons increases at high currents and contact resistance becomes the main factor. This increased resistance is highly significant for batteries used in HEVs, and hence it is necessary to minimize resistance of electrical connections while maintaining a proper current collector thickness. On the other hand, resistance from lithium ions is more apparent in cellular phones and laptops that are equipped with compact batteries requiring lower rate capability.

6.5 Temperature Characteristics

Temperature characteristics of batteries are an indicator of the reliability of battery performance. The deterioration of performance can be evaluated through changes in the surrounding temperature. Temperature characteristics can be classified into performance reliability in the normal temperature range and battery safety outside the normal temperature range. At the normal working temperature, electrical resistance produces heat when the battery is operated. The battery is examined for performance deterioration after the temperature is set to balance heat dissipation and heat generation. Outside the normal temperature range, the battery is checked for safety issues that may arise due to mishandling or its surroundings, and the extent of recovery is compared with the initial value when returned to normal temperatures. Temperature performance is becoming increasingly important with the rise in operating temperature of electrical devices. This section describes the different types of temperature characteristics and effects on battery performance.

6.5.1 Low-Temperature Characteristics

One example of low-temperature characteristics is whether the release of energy is sufficient to make a phone call during a cold winter day or in a freezer. The relevant characteristics can be grouped into discharge characteristics and cycle life characteristics. Low-temperature discharge is a measure of the battery capacity and can be obtained from one discharge cycle. General temperature conditions are from -10 to $-20\,°C$ for small batteries and $-30\,°C$ for batteries in HEVs. Low-temperature cycle life characteristics examine cycle life in the range of $0–10\,°C$ and battery performance at low temperatures for long hours. With future applications in artificial intelligence robots and industries, batteries will be exposed to more extreme conditions. While low-temperature discharge is considered an important factor in the design of batteries for mobile devices, high power consumption devices that are directly exposed to the external environment must also focus on charging properties at low temperatures. Batteries acting as the main power source in specialized environments including aircraft and military vehicles must exhibit excellent performance with varying temperature, humidity, and pressure.

Factors affecting temperature characteristics are similar to those involved in cycle life. One difference is in maintaining the mobility of substances well at low temperatures. For example, smaller particle size of active materials will impart a battery enhanced low-temperature performance due to the increased passage of

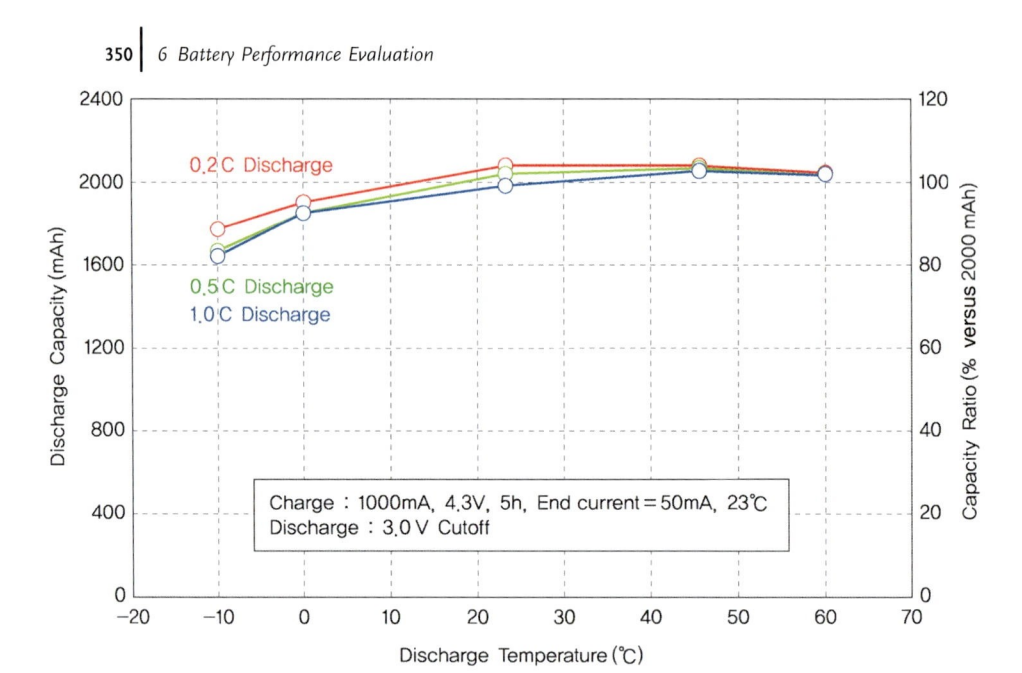

Figure 6.10 Discharge performance of a battery with varying temperatures.

lithium ions, which compensates for less lithium ion mobility at low temperature. The typical discharge performance of a battery at low temperature is presented in Figure 6.10.

6.5.2
High-Temperature Characteristics

High-temperature characteristics refer to the maintenance of initial performance when a battery is operated at a temperature higher than its normal operating range. At high temperatures, this can be divided into long-term storage and cycle life characteristics. The former evaluates the performance of a battery kept at high temperatures, while the latter involves a rise in temperature from the battery itself or its surroundings during summer or in other high-temperature environments. In both cases, the battery should be able to withstand thermal stress.

At high temperatures, electrochemical deterioration proceeds rapidly due to increased reactivity of components such as the electrolyte, active materials, and separator. Low-temperature and high-temperature characteristics tend to oppose each other. To enhance low-temperature performance, smaller active materials or electrolytes with low viscosity are used. This increases the reactivity between active materials and electrolytes at high temperatures, thus causing a rapid decline in cycle life. In other words, high-temperature performance can be improved by impeding low-temperature characteristics. This can be achieved through smaller specific surface area of active materials, electrolytes resistive to high temperatures, and the use of various electrolyte additives to form stable SEI layers at the anode.

6.6
Energy and Power Density (Gravimetric/Volumetric)

6.6.1
Energy Density

Energy density is the amount (Wh/kg, Wh/l) of energy (charge) stored per unit mass or volume. Energy density per unit volume is useful in small batteries and other systems constrained by volume. For power storage applications with no volumetric constraints, energy density per unit mass is more appropriate. This is because power storage batteries occupy fixed positions and are used in a wide area.

6.6.2
Power Density

Power density is the amount of energy released per unit time. Cellular phones and laptops require at most 2.0 C, but this figure should be much higher in HEVs. Since high power density involves the release of high power in a given amount of time, it is necessary to have high voltage and high current. As shown in Eq. (6.3), electrical resistance must be minimized to increase power output.

$$P = I^2 R \tag{6.3}$$

Here, P is the power, I is the current, and R is the resistance. Among the various factors affecting resistance, electrode thickness is the most significant at high rates. A different approach should be adopted when designing batteries for new fields (HEV, PHEV, EV, etc.). In addition to selecting materials that minimize resistance within electrodes, other factors to be considered are energy and power output. If electrodes are made to be thinner so as to achieve better output, manufacturing cost will increase for current collectors, separators, and electrolytes. The battery becomes closer to a supercapacitor, as the amount of energy that can be stored is greatly reduced. The complex manufacturing process for thin electrodes also affects productivity.

6.7
Applications

Lithium secondary batteries are used in various types of mobile devices, including communication equipment, computers, entertainment devices, power tools, toys, games, lighting, and medical devices. Batteries in mobile applications exhibit high power density, which is the most important condition for mobile environments. These batteries are suitable for devices requiring high energy density and high power density and can be expanded to HEVs and other transportation fields.

6.7.1
Mobile Device Applications

The mobile device market grew rapidly with the commercialization of lithium secondary batteries in the early 1990s. Lithium secondary batteries were the most adequate energy source to meet the requirements for small size and high capacity. Changes in mobile consumer behavior have ushered in a new paradigm of ubiquitous digital convergence, indicating a transition from passive consumption to two-way communication. With the demand for high energy density arising from content diversification and proliferation of software, we have entered the age of lithium secondary batteries. In addition to high capacity, design flexibility is another requirement. Lithium polymer batteries are an example of lithium secondary batteries that satisfy these conditions (size, form, flexibility). Since greater stability is indispensable for higher energy densities, battery markets are beginning to implement safety regulations.

6.7.2
Transportation

The limited supply of fossil fuels has led to an increase in oil prices and sparked interest in environmental protection. The development of high-power lithium secondary batteries was stimulated by commitment to reduce CO_2 emissions in the Kyoto Protocol and demand for eco-friendly vehicles through CARB (California Air Resource Board). The Toyota Prius was the first commercial HEV. This vehicle uses NiMH batteries for power sources and presents a solution to the problem of high oil prices. However, they cannot compete against lithium secondary batteries in terms of small size and high power density. Large batteries for vehicles are one of the most challenging applications of lithium secondary batteries, as they demand both high energy density and high power density. In addition to performance, high safety requirements must be met. This is because larger batteries are more likely to experience safety issues due to increased amounts of electrical energy. Cycle life should also be enhanced to match the standard for vehicles, which is far longer than that of mobile devices. The term used for vehicles is calendar life. In battery operated vehicles, circuits and other components play important roles. In particular, it is important to have a battery management system (BMS) for proper management of battery packs containing hundreds of batteries connected in series or parallel.

The objective of HEVs is to improve fuel efficiency by combining a combustion engine and a secondary battery system. These vehicles are repeatedly charged and discharged on the road and thus do not need to be separately charged. As such, high power density and safety are top priorities. Since EVs operate based on electricity alone, lithium secondary batteries require high energy density, high power density, and excellent safety. For future applications in this field, lithium secondary batteries must be further developed beginning with the selection of appropriate materials.

6.7.3
Others

While less dominant than mobile and transportation applications, the following fields show some promise in relation to lithium secondary batteries. Examples are UPS batteries, emergency batteries, alternative energy storage batteries, and military/aircraft batteries. These batteries require not only high energy density and long cycle life, but also high-temperature characteristics to ensure performance during exposure to extreme surroundings.

Index

a

absorption coefficient 286
AC impedance analysis 239, 255, 257
– Al/LiCoO$_2$/electrolyte/carbon/Cu battery analysis 249–253
– Al/LiCoO$_2$/electrolyte/MCMB/Cu cell analysis 253, 254
– diffusion coefficient 257
– electrochemical method 239
– electrode characteristic analysis, applications 247–249
– equivalent circuit model 241–247
– ionic conductivity 256
– phase difference 240
– principle 239–241
– relative permittivity 254, 255
active material slurry
– electrode production 99
– preparation of 329
adsorption isotherms, IUPAC classification 314
adsorption kinetics, of carbon sample 315
AFM. *See* atomic force microscopy (AFM)
AFM equipment structure 301
AFM *in situ* cell, structure 302
Al current collector foil, C–H groups, comparison 224
alloys 120
– change in potential
– – of Sn–Li and Si–Li with varying lithium composition 122, 125
– cracking in metal alloys 123
– discharge capacity 121
– Li, Li–Al, and Li–Si alloys 120
– lithium alloying of loosely arranged micrometal particles 124
– metal/alloy–carbon composites 128

– minimizing volume expansion 123
– multiphase lithium alloys 125
– Sn–Co–C alloying 128, 129
– Sn–Li equilibrium phase 122
aluminum corrosion 226, 227
aluminum current collectors, corrosion 259
aluminum–electrolyte interfacial reactions 226
aluminum metal cylindrical holder 298
aluminum metal, thermodynamically unstable 225
ammonium chloride (NH$_4$Cl) 2
amorphous carbon 100
– carbon raw materials, and carbonization 117
– electrochemical reactions
– – of low-crystalline carbon 102–108
– – of noncrystalline carbon 108–110
– gaseous carbonization 117
– liquid-phase carbonization 117, 118
– reactions involving electrolytes 110–114
– solid-phase carbonization 118
– structure of 100–102
– – structural model 101
– thermochemical characteristics 114–117
animal electricity 1
anode capacity 323
– design of 323
– potential balance, adjustment of 323
anode degradation 324
anode-electrolyte interfacial reactions
– additives effect 212–214
– interfacial reactions
– – at graphite (carbon) 209–211
– – noncarbonaceous anode and electrolytes 214–216
– of lithium metal 203–209

– SEI layer thickness 211, 212
anode materials 89
– amorphous carbon 100–118
– carbon materials 92–118
– characteristics of 91
– conditions 90, 91
– development history of 89
– graphite 92–100
– half cell capacity 320
– initial irreversible capacity of 321
– lithium metal 91, 92
– nitride anode materials 135–137
– noncarbon materials 118–120
– – Li, Li–Al, and Li–Si alloys 120–123
– – metal/alloy-carbon composites 128, 129
– – metal oxides 130–135
– – metal thin-film electrodes 130
– overview of 90, 91
anode voltage 340
anodic reactions, in organic electrolyte solutions 227
anti-Stokes lines 277
Armand reaction 19
atomic force microscopy (AFM) 300, 301
ATR. *See* attenuated total reflection (ATR)
attenuated total reflection (ATR) 273
audio quality, voltage 340
audiovisual transmission devices 340
Avogadro's number 314

b

Baghdad battery 2
basal-to-edge ratio 210
batteries 9
– capacity 15, 16, 320, 321, 344–347
– – anode 323
– – determined from open-circuit voltages of 320
– – evaluation of 347
– – of $LiCoO_2$ 321
– – measurement of 346, 347
– characteristics 15
– – cycle life 17
– – discharge curves 17–19
– – energy density 16
– – power 16, 17
– components 9
– design 319, 320
– – basic principles 319
– – cathode/anode capacity ratio 323–325
– – electrode potential/battery voltage design 321–323

– – practical aspects of 325–327
– discharge performance of 350
– disruption of lithium ions 343
– resistance 344
– types 2
battery cycle life 17
– factors affecting 344
– problems 343
– significance of 342
battery manufacturing process 327
– assembly process 331–334
– – cathode tab welding/crimping/x-ray inspection/washing process 334
– – electrolyte injection process 334
– – jelly roll insertion/cathode tab welding/beading process 332, 333
– – winding process 331, 332
– electrode coating 329, 330
– electrode manufacturing process 328
– electrode slurry, preparation of 328, 329
– flow chart of 328
– formation process
– – procedures and functions 334, 335
– – purpose of 334
– roll pressing process 330
– slitting process 330, 331
– vacuum drying process 331
battery performance 337
– charge/discharge curves 341
– – adjustment of 339, 340
– – significance of 337–339
– charge/discharge rate 347–349
– discharge performance of 342
– electrical energy 337
– energy density 351
– lithium secondary batteries 351
– mobile applications 351, 352
– overcharging 341
– power density 351
– temperature characteristics
– – high 350
– – low 349, 350
– transportation applications 352
– typical cycle life problems 343
battery safety 325
– basics of 65–68
– and cathode materials 68, 69
– reactions involving 69
– thermal reactions 67
battery voltage 321. *See also* voltage
– electrode potential, relationship 322
BET. *See* Brunauer-Emmett-Teller (BET)
binders 181
– functions 181, 182

Index

- PVdF binders 185–187
- requirements of 182–185
- SBR/CMC binders 187–189

Born–Oppenheimer approximation 78
Bragg angle 265
Bragg scattering 295
Bragg's law 263
- basal spacing 263
- of diffraction 263, 264

Brunauer-Emmett-Teller (BET) 312, 315
- isotherms 314
- plot 313
- surface analysis 311–315

c

cadmium 3
California Air Resource Board (CARB) 352
capacity 15, 16
- degradation 324
- effect of temperature on 19
- metal elements, form alloys with lithium 119

CARB. *See* California Air Resource Board (CARB)

carbon anode
- lithium ion batteries, design of 55
- lithium secondary battery, redox reactions of 23
- SEI characteristics of 213

carbonate solvents
- electrochemical decomposition of 195
- lithium secondary batteries 148
- LUMO energy levels 204
- oxidative decomposition reactions of 196
- reduction reactions of 196

carbon–electrolyte interface 210
carbon/electrolyte/lithium cell 248
carbonization reactions 117
- carbon raw materials and 117
- gaseous 117
- liquid-phase 117, 118
- solid-phase 118

carbon (SWCNT)/lithium battery
- cross-sectional sem image of 299
- electrolyte, thermal characteristics of 305

cathode capacity, potential balance 322
cathode degradation 324
cathode–electrolyte interfacial reactions 216
- interfacial reactions
- – phosphate cathode materials 223–225
- oxide cathode materials
- – interfacial reactions 218–223

- – native surface layers of 217, 218
- – SEI layers of 218

cathode materials 21, 23
- battery characteristics of 28
- charge and discharge curves 339
- demand characteristics of 26, 27
- development history 21
- discharge potential curves 24–26
- half cell capacity 320
- initial irreversible capacity of 321
- layered structure compounds 27–30
- – $LiCoO_2$ 30–34
- – $LiMO_2$(M=Mn, Fe) 37–40
- – $LiNi_{1-x}Co_xO_2$ 34–37
- – lithium-rich phases 44–46
- – Ni–Co–Mn three-component system 40, 41
- – Ni–Mn System 41–44
- – Li_xTiS_2, structure 22
- octahedral and tetrahedral sites 29
- olivine composites 52
- – $LiFePO_4$ 52–55
- – $LiMPO_4$ (M=Mn, Co, Ni) 55–57
- principle cathode materials 27
- redox reaction of 23, 24
- spinel composites 46
- – $LiMn_2O_4$ 46–51
- – $Li M_x Mn_{2-x} O_4$ (M=Transition Metal) 51, 52
- structure of densest oxygen layer 28
- structure of layered $LiMO_2$ 30
- TEM image 217
- thermal stability of 65, 69, 70
- – $LiCoO_2$ complex oxide 70
- – $LiFePO_4$ active material 74, 75
- – Ni-Co-Mn three-component oxide 71, 73
- – spinel $LiMn_2O_4$ 73, 74
- vanadium composites 57, 58

cathode physical properties 75–77
- first-principles calculation 77
- – prediction from 76
- potential difference 76
- redox couple 76

cathode tab welding process 334
cathode voltage 341
CCV. *See* closed-circuit voltage
cell composition 249
cell technology, development of 3
chalcogen compounds 22
charge balance 321, 322
charge-coupled detector 277
charge transfer 250
- from electrochemical reactions 259
- in electrode materials 231

– Faraday constant   259
– lithium secondary batteries   256
– rate-limiting process   14
– SEI layer   201
charging voltage   13
chemical ionization   310
chromatography   306
closed-circuit voltage (CCV) measures   231
CO
– gas   200
– K-shell absorption edge   287
C–O bond, β-decomposition   198
Co–Co peak intensity   292
Cole-Cole plot   241, 250
– lithium secondary battery   251
conducting agents   189
– dispersibility   190
– modification of   191
– types of   189, 190
– and wettability of electrodes   190
confocal microscopy   278
constant capacity cutoff control   236
constant current/constant voltage (CC/CV)   341, 347
constant voltage charging   236
conventional TEM (CTEM)   294
CoO cathode material, SAED patterns   297
core–shell particle shape   299
Coulomb energy   78, 79
Coulomb interactions   78
covalent bonds   270
– electromagnetic waves   270
– graphene layers   92
– infrared rays   270
– Li_2 107
CRT display   298
crystalline materials   293
crystalline phase identification   263, 264
crystallinity   264
crystallite size   264
CTEM. See conventional TEM (CTEM)
current   14, 16, 17, 19, 156, 178, 232, 348, 351. See also voltage
current break   334
current collectors   191
– aluminum corrosion   226–228
– aluminum surface, passive layers formation   228
– anode   192
– cathode   192
– lithium bis-perfluoroalkylsulfonylimide ($LiN(SO_2CF_3)_2$)   226
– native layer of aluminum   225, 226

– requirements   192
– role of   191
current density   15, 18, 57, 60, 78, 79, 148, 267, 342
current–voltage curves   232
CV. See cyclic voltammetry (CV)
cyclic carbonates
– free radical state, resonance structure   199
– oxidative decomposition reactions   199, 200
– vinylene carbonate   200
cyclic voltammetry (CV)   232
– characteristics of   237
– – comparison   237
– current–voltage   233, 234
– graphite anode   232
– redox reactions   232
– reduction peak   211
– in situ   303
cylindrical batteries
– dimensions of   346
– electrochemical design factors of   326
– physical design factors   325

d
dark field (DF) images   295
Debye–Waller factor   290
DEC. See diethylene carbonate
density functional theory (DFT)   79, 196
– calculations   197
– for EC   196
DFT. See density functional theory
diethylcarbonate   231
diethyl carbonate
– DSC thermogram of   306
– lithium ethylene dicarbonate   207
diethylene carbonate (DEC)   196
differential scanning calorimetry (DSC)   302
– of cathode, anode, and electrolyte   68
– for EC   207
– heat flux, equipment structure   304
– of Li_xCoO_2 71, 72
– Li_xFePO_4 75
– Li_xNiO_2 73, 74
– thermogram   304
– – of EC/DEC   306
– – polyethylene terephthalate   305
differential thermal analysis (DTA)   301
– calorimetric   305
– thermal capacity and enthalpy change   304
diffraction
– Bragg's law of   263, 264
– X-ray   265

diffusion coefficient 250
– of ions 257
– of lithium 238
diffusion, one-dimensional 250
dimethyl carbonate (DMC) 6, 196, 232
– electrochemical reduction potential of 196
– graphite Timrex KS 44, charge/discharge curves 111
– lithium metal/platinum electrodes, cell consists of 232
– organic solvents, physicochemical properties of 143
– oxidation potential 196
dione 200
discharging 10
– lithium metal alloys 121
– voltage 13
DMC. See dimethyl carbonate (DMC)
DSC. See differential scanning calorimetry (DSC)
DTA. See differential thermal analysis

e

EC. See ethylene carbonate (EC)
EDS. See energy dispersive spectroscopy (EDS)
electric current 1
electric dipole 275
electric potential 11, 12
– battery 16
– versus capacity 23
– cathode material 24, 25, 75
– circuit voltage 12
– current flow 16
– d-orbital electrons 25
– electrodes 12
– voltage 12
electric resistance 343
electric vehicles (EVs) 3
electrochemical analysis 231, 234
– constant current method
– – constant capacity cutoff control 236
– – cutoff voltage control 234–236
– constant voltage method
– – charging 236
– – potential stepping test 236, 237
– cyclic voltammetry 232–234
– linear sweep voltammetry 232
– open-circuit voltage 231, 232
electrochemical cells 9
electrochemical decomposition, nonaqueous electrolytes 195–200
electrochemical equilibrium 236
electrochemical oxidation 10

electrochemical quartz crystal microbalance (EQCM) analysis 257
– charge transfer, from electrochemical reactions 259
– corrosion reactions 259
– device 258
– electrochemical reactions 257
– film deposition 258
– $LiMn_2O_4$ film 260
– mass changes 259
– oscillation frequency 258
– piezoelectric quartz crystals 257
electrochemical reaction 11, 12
– discharge 11
– – curves 18
– electrode/cathode 11, 14
– of graphite 94
– kinetics 195
– $LiMn_2O_4$ 218
– of Li_2MnO_3 46
– lithium channel 55
– lithium ion batteries 142
– of low-crystalline carbon 102
– of noncrystalline carbon 108
– redox reactions 10
– separators 173
electrode coating process 329
electrode density 327
electrode/electrolyte configurations 319
electrode-electrolyte interfacial reactions 195, 200
– electrode materials and electrolytes 195
– Li^+ ions 200
– lithium metals 172
– polymer electrolytes 172
– SEI layer formation 201
electrode physical properties
– application programs 83
– battery voltage 80
– input files 83, 84
– lithium diffusion 80–82
– prediction, using first-principles calculation 79
– structural stability of electrode materials 80
electrode slurry
– coating of 330
– current collector, coating 330
– manufacturing based on nanosized particles 134
– preparation of 328
electrolyte–electrode interface 334
electrolytes 10, 209
– lithium secondary batteries 195

– oxidative decomposition reactions 196, 208, 222, 274
– oxidative reactions, at cathode material surface 344
– solvents, electrochemical stability 198
electromagnetic radiation 269
– measurement techniques for 270
– types 272
electromagnetic wave 275
electron binding energy 283
electron gun 296
electrostatic capacity 242
energy dispersive spectroscopy (EDS) 300
EQCM analysis. See electrochemical quartz crystal microbalance (EQCM) analysis
equivalent circuit
– of $LiCoO_2$/carbon cell 250
– resistance–capacitance 247
– of series resistance 246
ethylene carbonate (EC) 196, 231
– cyclic carbonates 199
– decomposition 305
– DEC solvent 305
– DFT calculations for 196
– DMC electrolyte 214
– DSC thermogram of 306
– electrochemical reduction potential of 196
– Li^+ reductive decomposition mechanism 198
– lithium ethylene dicarbonate 207
– oxidation potential 196
– thermal decomposition 199
EVs. See electric vehicles (EVs)
EXAFS. See extended X-ray absorption fine structure (EXAFS)
exchange–correlation energy 79
exothermic/endothermic reactions 304
extended X-ray absorption fine structure (EXAFS)
– absorption coefficient 287
– analysis of 290, 291
– backscattering of electrons 288
– structural factors 291

f

Fermi contact 282
Fermi level 286, 287
FE-SEM. See field-emission SEM (FE-SEM)
field-emission SEM (FE-SEM) 298
– equipment structure 298
– field emission electron guns 298
fingerprint region 273
first principles calculation 40, 57
– application programs 83
– prediction of cathode material from 77
– structural stability of electrode materials 80
– understanding of 77–79
Fourier transform (FT) 291
– Co–Ni oxide 292
– infrared spectroscopy 274
– Ni EXAFS spectrum 292
– spectrum 290, 291
Fourier transform infrared spectroscopy (FTIR) 270–275
– analysis of 204
– ex situ internal reflection spectroscopy 274
– of $LiMn_2O_4$ cathode surface 221
– lithium ethylene carbonate 274
– lithium methyl carbonate 205
fragmentation pattern analysis 310
FT. See Fourier transform (FT)
FTIR. See Fourier transform infrared spectroscopy (FTIR)
full cell 11

g

galvanostatic intermittent titration technique (GITT) 238, 239
– current and voltage changes 238
– electrochemical methods 257
– experiment 238
gas chromatography (GC) 306
– analysis of 309
– equipment structure 307
– mass spectrometry (GC–MS) 306–311
– stationary phase 308
gaseous molecules, vaporization 312
gasket 334
Gaussian vs. Lorentzian 284
GC. See gas chromatography (GC)
generalized gradient approximation (GGA) 79
GGA. See generalized gradient approximation (GGA)
GITT. See galvanostatic intermittent titration technique (GITT)
grain boundary 300
graphite 92. See anode materials
– anistropic behavior 92, 93
– charge/discharge curves 98
– depth profile 286
– design of graphite particles 94–99
– discharge curves of batteries with 120
– electrochemical reaction 94

Index

– – galvanostatic curve 95
– – staging effects during lithium intercalation 95
– – voltammetric curve 95
– impedance spectrum and equivalent circuit of 210
– Raman spectroscopy 103
– SEI layer formation in 124
– structure 92–94
– – in-plane structure 96
– – MPCF artificial graphite 98
– – particle shapes 96, 99
graphite anodes, exfoliation 217
graphite/lithium cell
– charge/discharge capacity and coulombic efficiency 235
– differential capacity curves of 236
– voltage controlled constant current charge–discharge curve 235
graphite surface, SEI layer image 295
green energy 7
gyromagnetic ratio 280

h

half cells 11
Hartree–Fock calculations 204
Hartree–Fock method 78
Hartree potential 79
HEVs. See hybrid electric vehicles (HEVs)
$H_{\text{Fermi contact}}$ 281
$H_{\text{J-coupling}}$ 281
Hohenberg–Kohn theorem 78
HOMO energy level 218
hybrid electric vehicles (HEVs) 3, 51, 53, 342, 348
– anode material 107
– application 3
– batteries 134, 348, 349, 351
– commercial 352
– objective of 352
H_{Zeeman} 281

i

ICP. See inductively coupled plasma (ICP)
imide-based lithium salts 227
impedance
– complex plane plot of 241
– definition of 249
– with frequency 242
– high-frequency region 251
– low-frequency region, by Warburg impedance 252
– middle-frequency region 251, 252
– very low-frequency region 252, 253
inductively coupled plasma (ICP) 311
– equipment structure 311
– mass spectrometry (ICP-MS) 311
information technology (IT) 1
infrared 269
– absorption 270, 273
– electromagnetic spectrum 270
– Fourier transform 274
– functional groups, observation of 273
– molecular vibrational energy, transition of 272
– Raman spectroscopy 275
– reflection absorption infrared spectroscopy (RAIRS) 273
– spectrum 273
infrared reflection absorption spectroscopy (IRAS) 273
interfacial reactions, at graphite (carbon) 209–211
internal reflection elements (IRE) 273
International Center for Diffraction Data (ICDD) 263
IRAS. See infrared reflection absorption spectroscopy (IRAS)
iR drop 338
– polarization 14, 19, 338
– within thick electrode 339

j

Jahn–Teller effect 291
jelly roll insertion process 331
jelly roll winder 332, 333
Joint Committee on Powder Diffraction Standards (JCPDS) 263

k

kinetic energy 79
Kohn–Sham equations 79

l

LDA. See local density approximation (LDA)
lead–acid batteries 2, 3
Leclanché (or manganese) cell 2
$LiBF_4$
– corrosion/passive layer formation 228
LiBOB decomposition 212
LIBs. See lithium ion batteries (LIBs)
$Li/(CF)_n$ batteries 21
Li_2CO_3, cathode surface 217

LiCoO$_2$ A$_{1g}$ mode, Raman band changes 279
LiCoO$_2$/electrolyte/lithium cell 248
LiCoO$_2$–graphite system 340
LiFePO$_4$
– electrochemical cycle characteristics 223
– FTIR analysis 224
– FTIR spectra 225
– rietveld refinement of 266
light intensity 275
Li-ion batteries 319
– manufacturing process 327
Li/LiNiO$_2$ cell
– capacity–potential curve of 223
Li/MnO$_2$ batteries 21
LiN(CF$_3$SO$_2$)$_2$/PC electrolyte
– aluminum corrosion mechanism 227
linear carbonate 144
– C=O, positive charge 197
– dimethyl carbonate, exception of 187
– oxidative decomposition reactions 199
– resonance structure of the free radical state 199
– solvents 169
linear sweep voltammetry (LSV) 148, 226, 232, 233
LiNi$_{0.8}$Co$_{0.15}$Al$_{0.05}$O$_2$
– ^{7}Li MAS NMR spectrum with charge state of 282
LiNi$_{0.80}$Co$_{0.15}$Al$_{0.05}$O$_2$–graphite–acetylene black
– Raman microscope image of 277
LiNi$_{0.80}$Co$_{0.15}$Al$_{0.05}$O$_2$–graphite-type carbon–acetylene black
– Raman microscope image of 278
LiNiO$_2$, lithium deintercalation 270
LIPBs. See lithium ion polymer batteries (LIPBs)
LiPF$_6$/EC/PC electrolytes 215
liquid electrolytes 142
– characteristics of 147–149
– components of 143–147
– development trends 161, 162
– electrolyte additives 153–157
– enhancement of thermal stability 157, 160, 161
– ionic liquids 149–153
– requirements of 142
Li/SO$_2$ batteries 21
Li/SOCl$_2$ batteries 21
^{7}Li static NMR spectrum, thermal effect 283
LiTFSI salts 228
lithium alkoxide 219

lithium alkyl bicarbonates 196
lithium alkyl carbonate 207, 208
lithium alloying 214
lithium anions, hydrolysis of 208
lithium batteries 196, 306
– cross-sectional SEM image of 299
– electrode-electrolyte interactions in 259
lithium bis-perfluoroalkylsulfonylimide 226
lithium cells 196. See also lithium batteries
– cycle life 200
– cycling of 218
– inert metal electrodes and electrolytes 196
– SEI layer formation 201
lithium deintercalation/intercalation
– in situ XRD patterns 271
lithium deposition 323
lithium, diffusion coefficient 238
lithium ethylene carbonate, molecular structure 275
lithium ethylene dicarbonate 207, 208
lithium intercalation–deintercalation cycle 21
lithium ion batteries (LIBs) 5
lithium ion mobility 350
lithium ion polymer batteries (LIPBs) 5
lithium iron oxide 223
lithium metal alloys 209
– discharge capacity 121
– solvent, reactions between 209
lithium metal oxides, crystal structure 320
lithium methyl carbonate 206
– FTIR spectrum 205
– thermodynamically stable compound 206
lithium polymer batteries 352
lithium salts
– acidic nature 218
– electrolytes, reductive decomposition 216
lithium secondary batteries 1, 3–7, 278, 337, 351, 353
– capacity 345
– changes in energy density and 4
– characteristics 6
– charge and discharge curves of 338
– charge transfer 256
– commercialization of 22
– discharge curves of 120
– DSC characteristics, of components 116
– electrode of 247

– electrolytes 5, 142, 195
– energy density of 345
– equivalent circuit for electrode 248
– future of 7
– key components 6
– lithium ion, diffusion coefficient 22
– movement of Li$^+$ 4
– nonaqueous electrolytes 195
– performance, deterioration of 195
– range of potential for reactions with lithium 121
– shapes 5
– using MoS$_2$ as cathode material 22
– voltage of 337
local density approximation (LDA) 79
LSV. *See* linear sweep voltammetry (LSV)
LUMO energy 146, 147, 204, 218

m

magic angle spinning (MAS) 281
magnetic moment 280
manganese compounds, binding energy 284, 285
manganese dioxide (MnO$_2$) cathode 2
manganese ions 22
MAS. *See* magic angle spinning (MAS)
mass spectrometry (MS)
– equipment structure 309
– gas chromatography (GC) 306–311
– inductively coupled plasma (ICP) 311
– spectrum 310
material property analysis 263
metal oxides 130–135
– anatase titanium dioxide 131
– change in potential 131
– charge/discharge curves 132
– Li$_{2.6}$Co$_{0.4}$N 135
– rutile TiO$_2$ 131, 132
– volume change in 130
metal thin-film electrodes 130
methyl carbonate anions 206
microcells 7
molar conductivity
– cations 256
– ionic 256
MS. *See* mass spectrometry (MS)

n

negative electrode capacity/positive electrode capacity (N/P) ratio 323, 324
– battery life, relationship 324
Nernst–Einstein equation 257
Nernst equations 14

NiCd batteries 2, 3, 337
Ni EXAFS spectrum 292, 293
NiMH batteries 3, 337
NiMH cells. *See* NiMH batteries
Ni–Mn–Co three-component system 340
Ni–Ni bond lengths 291
nitride anode materials 135–137
– Li$_{2.6}$Co$_{0.4}$N
– – charge/discharge curves 135
– – rate capability characteristics 136
N-methyl pyrrolidone (NMP) solvent 328
NMR spectroscopy. *See* nuclear magnetic resonance (NMR) spectroscopy
nonaqueous electrolytes, electrochemical decomposition 195–200
nonaqueous organic solvents
– one-electron transfer of 197
– reduction from one-electron transfer 197
noncarbon materials 118–120
– composites of lithium-reactive/nonreactive metals 125–128
– Li, Li–Al, and Li–Si alloys 120–129
– metal/alloy–carbon composites 128, 129
– metal thin-film electrodes 130
– micrometal particles in lithium reactions 123–125
– multiphase lithium alloys 125
nuclear magnetic resonance (NMR) spectroscopy 282
– resonance frequency 280, 281
– solid-state 280–282
nuclear spin quantum number 280
Nyquist plot 241
– Al/LiCoO$_2$/electrolyte/MCMB/Cu lithium secondary battery 254
– capacitance component 243
– for capacitance + (resistance–capacitance) equivalent circuit 247
– carbon/electrolyte/lithium cell 248
– for charged Li/LiMn$_2$O$_4$ cell 219
– electrochemical cell for permittivity measurements 254
– equivalent circuit 246
– for equivalent circuit of lithium secondary battery electrode 248
– for fuel cell 253
– inductance–capacitance component 244
– inductance component 243
– of LiCoO$_2$/electrolyte/lithium cell 249
– parallel resistance–capacitance (RC) 245
– RC series equivalent circuits 253

– resistance component 242
– resistance–reactance–capacitance (RLC) component 244

o

OCV. *See* open circuit voltage (OCV)
ohmic polarization. *See* iR drop
open circuit voltage (OCV) 13, 231
– battery capacity 320
– electrochemical cell 231
– electrode materials 231
– $LiMn_2O_4$ film, mpe changes of 260
– oxidation number of lithium ions 25
– single-walled carbon nanotubes (SWCNTs) 231, 232
– Sn–Li and Si–Li 122
overcharging 340
– battery capacity 342
– behavior of battery 341
– charging of battery 38
– oxidation number of Co 33
– redox additives 156
oxidative electrode 11
oxidative reactions 232

p

parallel resistance–capacitance
– equivalent circuit 245, 246
– Nyquist plot of 245, 246
PC. *See* propylene carbonate (PC)
permittivity measurements
– electrochemical cell 255
– – equivalent circuit 255
– – Nyquist plot 255
PET. *See* polyethylene terephthalate (PET)
PITT. *See* potentiostatic intermittent titration technique (PITT)
plug-in hybrid electric vehicles (PHEVs) 3
polarization 13, 14, 15
– effect of current density on 15
polyethylene terephthalate (PET)
– DSC analysis of 305
– DSC thermogram of 305
polymer electrolytes 162, 328
– characteristics of 171–173
– development trends 173
– preparation of 169–171
– types of 162–169
positive temperature coefficient (PTC) 334
potassium hydroxide (KOH) 2
potential stepping
– current–time and differentiated capacity–voltage plots 237

potentiostatic intermittent titration technique (PITT) 238, 239
– change in current 239
– diffusion coefficient 239
propylene carbonate (PC) 195
– decomposition
– – *in situ* IR cell 276
– electron transfer, oxidation reactions 199
– graphite interfacial reactions 212
– lithium bisoxalato borate (LiBOB) 212
– oxidation reactions from electron transfer 199
– reduction reactions of 195
– ring-opening reactions 196
– undergoes ring-opening oxidative reactions 220
PTC. *See* positive temperature coefficient (PTC)

r

radiation intensities 284
RAIRS. *See* reflection absorption infrared spectroscopy (RAIRS)
Raman cell, *in situ* 279
Raman scattering 275, 277
Raman spectroscopy 270, 275–279
– for crystalline graphite 101
– – carbon materials 103
– energy absorption 276
– hyperspectral imaging 277
– light intensity 275
– used to observe molecular vibrations 277
Randles circuit 218
Rayleigh–Ritz variational theorem 78
Rayleigh scattering 275
RC. *See* resistance-capacitance (RC)
redox reactions at electrodes 10
reductive electrode 11
reflection absorption infrared spectroscopy (RAIRS) 273
resistance–capacitance (RC) 245, 247
– equivalent circuits 252
– model conditions of 253
resistance–reactance–capacitance (RLC) component 244
roll pressing process 331

s

Sauerbrey equation 258
scanning electronmicroscope (SEM) 296–300
– cross-sectional 299
– dry process/wet process 181
– electron accelerator 296

– energy dispersive spectroscopy (EDS)   300
– equipment, structure of   297, 298
– field-emission   298
– MCMB-25-28 artificial graphite   97
– microporous film   180
– MPCF-3000 artificial graphite   99
scanning tunnel microscopy (STM)   300
Schrödinger equation   77, 78
SEI layer. *See* solid electrolyte interphase (SEI) layer
selected area electron diffraction (SAED)   295
SEM. *See* scanning electronmicroscope (SEM)
separators   173
– basic characteristics   174–176
– cycle performance   178
– development of materials   179, 180
– effects on battery assembly   176
– functions   173, 174
– manufacturing process   180, 181
– oxidative stability   176, 177
– prospects for   181
– thermal stability   178
single-walled carbon nanotubes (SWCNTs)   231
Si/Sb-based alloys   215
Si/Sn/Sb-based metals   214
slitting process   330, 332
slurry storage process   329
Sn-based anode material
– *in situ* XRD patterns of   270
SnO_2 cathode material particle
– electron diffraction patterns of   296
Sn–Sb–Cu–graphite alloy anode   215
solid electrolyte interphase (SEI) layer   200
– at anode   321
– anode–electrolyte interface, Randles circuit   218
– battery performance   217
– cathode–electrolyte interface, Randles circuit   218
– charge/discharge of lithium batteries   203
– collector–electrolyte interface   203
– disproportionation reactions   222
– at electrode surface   200–203
– electrolyte decomposition   210
– electrolyte/gold electrode, cyclic voltammetry of   211
– formation   201, 209, 210, 211, 303, 334
– – in metals   124
– FTIR spectrum of   215
– Li^+ ion transport   202
– of $LiMn_2O_4$ 219
solvent reduction, potential values   196

sulfonyl amide lithium salts   226
surface–graphite interface   211
surface modification, surface modification   58–60
– layered structure compounds   60, 61
– olivine compounds   64, 65
– spinel compound   61–64

t
TEM structure   294
– conventional TEM (CTEM)   294
tetrahydrofuran (THF)   185, 195, 196
TGA. *See* thermogravimetric analysis (TGA)
thermal analysis   301–306
thermal conductivity   307
thermal decomposition   302
thermal stability, cathode materials   65
thermodynamic equilibrium   231
thermogravimetric analysis (TGA)   301
– equipment structure, schematic diagram   304
THF. *See* tetrahydrofuran (THF)
THF/$LiClO_4$ electrolyte   196
transmission electron microscopy (TEM)   292–296
transverse acoustic waves   257
– in quartz crystals   258
tungsten filament   292, 296

v
vacuum drying process   331
VC. *See* vinylene carbonate (VC)
vinylene carbonate (VC)   196, 200
– oxidation potential   196
– ring-opening polymers   213
voltage   12. *See also* current
– anode   340
– audio quality   340
– cathode materials   340
– effect of current density on   18
– metal elements, form alloys with lithium   119
voltage/current behavior, during charging   347
voltaic pile   2

w
Warburg impedance   249, 250
wavefunction   78
welding tip   332
winding process   331
– winds electrodes   330, 331
wireless charging   7

x

XAS measurement cell
– *in situ* 291
XPS system 283, 284
X-ray absorption 284
X-ray absorption near-edge structure (XANES) 287, 288
– LiFePO$_4$ 288
– physical mechanism of 288
– P K-edge 289
X-ray absorption spectroscopy (XAS) 285–287
– extended x-ray absorption fine structure (EXAFS) 288–292
– X-ray absorption near-edge structure (XANES) 287, 288
X-ray beam 263
X-ray diffraction analysis
– principle of 263–265
– Rietveld refinement 265–267
– *in situ* 267–269
– – LiCoO$_2$ 269
X-ray incident radiation 286
X-ray photoelectron spectroscopy (XPS) 282–285
– peaks for carbon materials 114
X-ray transmission 271
XRD analysis 263
XRD device 265

z

Zaviosky detected electron paramagnetic resonance 280
Zeeman interaction 280
zinc anode 2
zinc chloride (ZnCl$_2$) 2